Springer Series in Materials Science

Volume 219

The Springer Series in Materials Science covers the complete spectrum of materials physics, including fundamental principles, physical properties, materials theory and design. Recognizing the increasing importance of materials science in future device technologies, the book titles in this series reflect the state-of-the-art in understanding and controlling the structure and properties of all important classes of materials.

More information about this series at http://www.springer.com/series/856

Dusan Losic · Abel Santos
Editors

Nanoporous Alumina

Fabrication, Structure, Properties and Applications

Editors
Dusan Losic
School of Chemical Engineering
The University of Adelaide
Adelaide, SA
Australia

Abel Santos
School of Chemical Engineering
The University of Adelaide
Adelaide, SA
Australia

ISSN 0933-033X ISSN 2196-2812 (electronic)
Springer Series in Materials Science
ISBN 978-3-319-20333-1 ISBN 978-3-319-20334-8 (eBook)
DOI 10.1007/978-3-319-20334-8

Library of Congress Control Number: 2015943361

Springer Cham Heidelberg New York Dordrecht London

Printed on acid-free paper

Springer International Publishing AG Switzerland is part of Springer Science+Business Media (www.springer.com)

Preface

Written by an outstanding group of experts in the field, this book presents an exciting and fresh compilation of the last advances and developments in one of the most popular nanostructured materials, nanoporous anodic alumina. The electrochemical and synthetic methods, as well as characterization techniques and applications will be discussed in detail throughout this book.

Nanoporous anodic alumina was discovered during the first decades of the twentieth century and widely used in industry for corrosion protection, car industry, and metal decoration purposes for almost a century. The advent of high resolution scanning electron microscopy and other surface characterization techniques revealed the unique structural properties of this nanomaterial, which after a short time became one of the most explored nanostructures across a broad range of disciplines and fields. As an example, more than 3000 journal papers on nanoporous anodic alumina were published in last 20 years. This intensive research work can be explained by its simplicity of fabrication, unique self-ordered nanoporous structure, and a plethora of outstanding properties, which have spread the use of nanoporous anodic alumina in a broad range of applications and disciplines, including optics and photonics, electronics, membrane science, fundamental research, materials science, engineering, medicine, and industry.

The aim of this book is to present the recent progress in nanoporous anodic alumina, with special focus on the understanding of its properties, preparation methods, structural engineering, and applications. The book covers selected topics with 11 Chapters that we believe will be the most significant contribution to this emerging and fast developing field.

Chapter 1 presents the mechanisms of nanoporous alumina formation and self-organized growth, with special focus on the different concepts and aspects involved in this unresolved process. A more detailed insight into the theoretical models describing this electrochemical process is presented in Chap. 2. Chapter 3 is devoted to the synthesis of nanoporous anodic alumina by electrochemical anodization of low purity aluminum substrates, which is a critical factor for spreading the use of this nanomaterial in industrial applications. Chapter 4 compiles an outstanding insight into the different electrochemical approaches used to tailor the internal pore structure of nanoporous anodic alumina. The applicability of this nanomaterial is highly dependent on its surface chemistry. In that respect, Chap. 5 presents the different soft and hard modifications of nanoporous anodic alumina aimed to improve its chemical and physical properties. The fundamental aspects of the optical properties of nanoporous anodic alumina are shown in Chap. 6 and Chap. 7 compiles a variety of examples of applicability of nanoporous anodic alumina as optical biosensing platform, which is recognized as one of the most promising applications for this nanomaterial. Chapter 8 is devoted to optofluidic applications using nanoporous anodic alumina. The applicability of this nanomaterial as a platform to develop electrochemical sensors, which is a very promising area to develop cost-competitive and simple devices for point-of-care biomedical and environmental analysis, is presented in detail in Chap. 9.

Another exciting topic on the application of nanoporous anodic alumina is presented in Chap. 10, where nanoporous alumina membranes for chromatography and molecular transporting are presented. Finally, Chap. 11 presents the most recent advances in the use of nanoporous anodic alumina for drug delivery and biomedical applications. This chapter shows new concepts and future perspectives towards advanced medical therapies including orthopedic and dental implants, heart/coronary/vasculature stents, immunoisolation, skin healing, tissue engineering, and cell culture.

As a result of the highly interdisciplinary nature of this book, it should be of profound and immediate interest for a broad audience, including undergraduate students, academics and industrial scientists and engineers across many disciplines, ranging from physics, chemistry, engineering, materials science, bioengineering, and medicine. We believe that this book will also be valuable to many entrepreneurial and business people, who are in the process of trying to better understand and valuate nanotechnology and new nanomaterials for future high-tech emerging applications and disrupting industries.

Adelaide, Australia

Dusan Losic
Abel Santos

Contents

Contributors

Moom Sinn Aw School of Chemical Engineering, The University of Adelaide, Adelaide, Australia

Manpreet Bariana School of Chemical Engineering, The University of Adelaide, Adelaide, Australia

Claudio L.A. Berli INTEC, UNL-CONICET, Santa Fe, Argentina

Chuan Cheng Department of Mechanical Engineering, The University of Hong Kong, Hong Kong, People's Republic of China

Alfredo de la Escosura-Muñiz ICN2—Nanobioelectronics & Biosensors Group, Institut Catala de Nanociencia i Nanotecnologia, Bellaterra (Barcelona), Spain

Marisol Espinoza-Castañeda ICN2—Nanobioelectronics & Biosensors Group, Institut Catala de Nanociencia i Nanotecnologia, Bellaterra (Barcelona), Spain

Josep Ferré-Borrull Departament d'Enginyeria Electrònica, Universitat Rovira i Virgili, Tarragona, Spain

Anisah Shafiqah Habiballah Chemistry Department, Faculty of Applied Sciences, Universiti Teknologi MARA, Arau, Perlis, Malaysia

Abdul Mutalib Md Jani Chemistry Department, Faculty of Applied Sciences, Universiti Teknologi MARA, Arau, Perlis, Malaysia

Tushar Kumeria School of Chemical Engineering, The University of Adelaide, Adelaide, Australia

Elżbieta Kurowska-Tabor Department of Physical Chemistry and Electro chemistry, Faculty of Chemistry, Jagiellonian University in Krakow, Krakow, Poland

Woo Lee Department of Physics, Hankuk University of Foreign Studies (HUFS), Yongin, Korea

Yi Li Department of Electronic Materials Science and Engineering, Institute of Materials Science and Engineering, South China University of Technology, Guangzhou, People's Republic of China

Zhiyuan Ling Department of Electronic Materials Science and Engineering, Institute of Materials Science and Engineering, South China University of Technology, Guangzhou, People's Republic of China

Dusan Losic School of Chemical Engineering, University of Adelaide, Adelaide, SA, Australia

Abdul Hadi Mahmud Chemistry Department, Faculty of Applied Sciences, Universiti Teknologi MARA, Arau, Perlis, Malaysia

Lluis F. Marsal Departament d'Enginyeria Electrònica, Universitat Rovira i Virgili, Tarragona, Spain

Arben Merkoçi ICN2—Nanobioelectronics & Biosensors Group, Institut Catala de Nanociencia i Nanotecnologia, Bellaterra (Barcelona), Spain; ICREA, Institucio Catalana de Recerca i Estudis Avançats, Barcelona, Spain

A.H.W. Ngan Department of Mechanical Engineering, The University of Hong Kong, Hong Kong, People's Republic of China

Josep Pallarès Departament d'Enginyeria Electrònica, Universitat Rovira i Virgili, Tarragona, Spain

Abel Santos School of Chemical Engineering, The University of Adelaide, Adelaide, Australia

Grzegorz D. Sulka Department of Physical Chemistry and Electrochemistry, Faculty of Chemistry, Jagiellonian University in Krakow, Krakow, Poland

Raúl Urteaga IFIS-Litoral (UNL-CONICET) Güemes 3450, Santa Fe, Argentina

Ewa Wierzbicka Department of Physical Chemistry and Electrochemistry, Faculty of Chemistry, Jagiellonian University in Krakow, Krakow, Poland

Elisabet Xifré-Pérez Departament d'Enginyeria Electrònica, Universitat Rovira i Virgili, Tarragona, Spain

Hanani Yazid Chemistry Department, Faculty of Applied Sciences, Universiti Teknologi MARA, Arau, Perlis, Malaysia

Leszek Zaraska Department of Physical Chemistry and Electrochemistry, Faculty of Chemistry, Jagiellonian University in Krakow, Krakow, Poland

Chapter 1
Mechanisms of Nanoporous Alumina Formation and Self-organized Growth

Zhiyuan Ling and Yi Li

Abstract The synthesis of nanoporous alumina based on anodization process is a field of active research. In this chapter, some fundamental scientific topics toward formation mechanisms of nanoporous alumina will be discussed in detail. (1) The intrinsic mechanisms of mild and hard anodization processes: the definition of mild anodization (MA) and hard anodization (HA); the critical condition between MA and HA processes. (2) The origins of self-ordering phenomenon: the structure models of nanoporous alumina; the influence of electrical field and internal stress. (3) The growth models of nanoporous alumina: the classical growth models of nanoporous alumina; the advantages and disadvantages of present models; the physical and chemical processes during aluminum anodization. (4) The intrinsic relationship between anodization conditions and anodization processes: the influence of anodization voltage and current density; the influence of electrolyte system; the influence of other anodization conditions. (5) Controllable fabrication of nanoporous alumina: the relationship between anodization conditions and structural parameters (e.g., interpore distance, pore size); fabrication of regular nanoporous alumina (i.e., with highly ordered pores); fabrication of nanoporous alumina with complex pore structure. In addition to nanoporous alumina, the findings and conclusions in this chapter may also be applied in other valve metal anodization processes (e.g., Ti, Zr, W anodization).

Z. Ling (✉) · Y. Li
Department of Electronic Materials Science and Engineering,
Institute of Materials Science and Engineering, South China University of Technology,
510640 Guangzhou, People's Republic of China
e-mail: imzyling@scut.edu.cn

D. Losic and A. Santos (eds.), *Nanoporous Alumina*,
Springer Series in Materials Science 219, DOI 10.1007/978-3-319-20334-8_1

1.1 Introduction

Anodic aluminum oxide (AAO) with ordered honeycomb-like pore arrangement is one of the most-studied membrane materials. In the past decades, great efforts have been paid, trying to make clear its intrinsic relationship between the morphological structure and the anodization conditions. In this chapter, the fundamental knowledge and recent progress on mechanisms of AAO formation and self-ordering growth are reviewed.

1.2 Types of Anodic Aluminum Oxides (AAO) Membranes

AAO membranes can possess two different morphologies: nonporous-type and porous-type (Fig. 1.1). Which one will be formed depends on the chemical nature of electrolytes [1, 2]. Generally, in neutral electrolytes (pH 5–7), such as borate, oxalate, and etc., nonporous AAO membranes are formed [3]. However, in acidic electrolytes, such as sulfuric, oxalic, phosphoric, and etc., porous AAO membranes are formed [2, 4, 5].

1.2.1 Nonporous AAO Membranes

Nonporous AAO is also called barrier-type AAO. In this case, anodic oxide layer is compact, amorphous, uniform in thickness up to several hundred nanometers, and completely insoluble in the formed electrolytes. During anodization growth process of nonporous AAO membranes under constant voltage, anodization current density (*J*) decreases exponentially with time (*t*) (Fig. 1.2a).

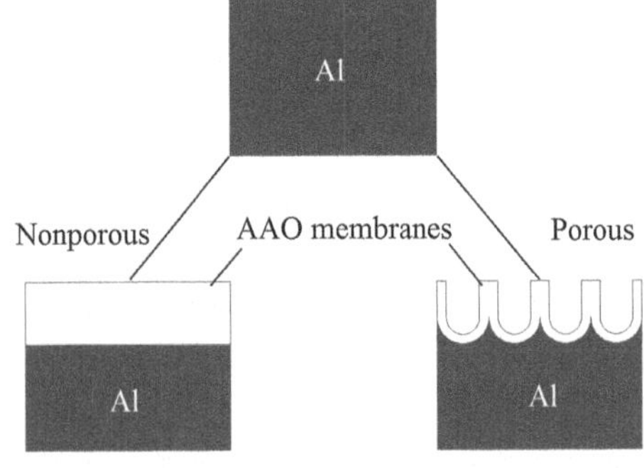

Fig. 1.1 Two types of AAO membranes formed in neutral and acidic electrolytes, respectively

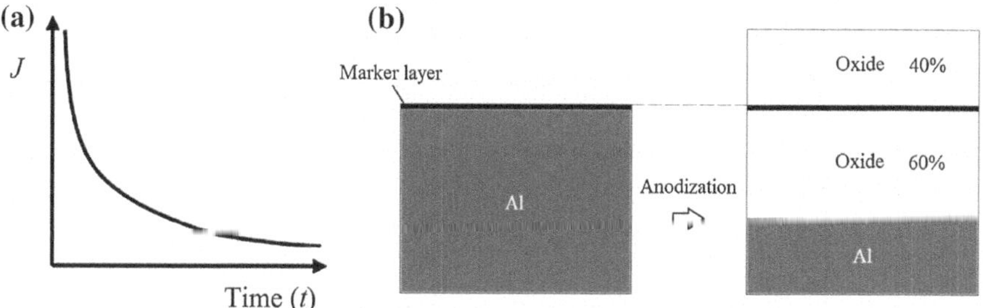

Fig. 1.2 Schematic diagrams of nonporous AAO membranes formed under constant voltage. **a** *J*-*t* transient. **b** Dimensional changes before and after anodization

The J consists of electronic current (J_e) and ionic current (J_i). At a high electric field strength (E) in the oxide, $J = J_e + J_i \approx J_i$. J_i can be described by the exponential law of Guntherschulze and Betz as [6, 7]

$$J_i = J_0 \exp(\beta E) \tag{1.1}$$

where J_0 and β are temperature-dependent material constants; E is a function of the anodization voltage (U), the potential drop at the metal/oxide ($U_{m/o}$) and oxide/electrolyte ($U_{o/e}$) interfaces, respectively [8].

$$Et_b = U - U_{m/o} - U_{o/e} \tag{1.2}$$

where t_b is the thickness of barrier layer. In general, $U_{m/o}$ and $U_{o/e}$ are much small compared to the value of U. Therefore, E can be expressed approximately as [9]

$$E \approx U/t_b \tag{1.3}$$

Obviously, (1.1) and (1.3) place an upper limit on the thickness of nonporous AAO membranes.

Ideally, the barrier oxide can be formed at 100 % current efficiency (η_J). If an immobile marker layer, such as ^{125}Xe, is implanted on the Al surface before anodization, one will find that the oxide grows simultaneously at both the interfaces of oxide/electrolyte and metal/oxide [10–12]. 40 % of the membrane thickness is formed above the marker layer by outwardly migrating Al^{3+} cations and 60 % below the marker layer by inwardly migrating O^{2-} anions (Fig. 1.2b) [13].

1.2.2 Porous AAO Membranes

To be different from the nonporous AAO membranes, the thickness of porous AAO membranes is independent on the anodization voltage (U) and can be up to several

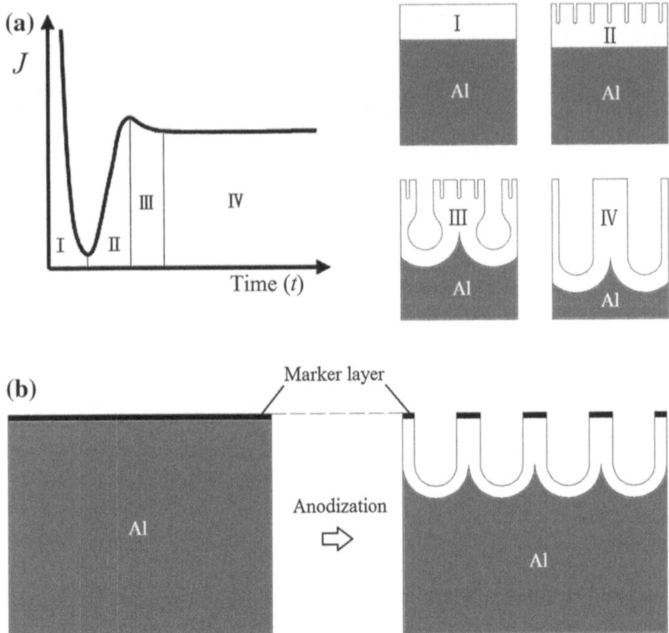

Fig. 1.3 Schematic diagrams of porous AAO membranes formed under constant voltage. **a** *J-t* transient and the kinetics of porous oxide growth on aluminum. **b** Dimensional changes before and after anodization

hundred micrometers. Under a constant *U*, the *J-t* curve can be divided into four stages (Fig. 1.3a).

At the moment after applying a constant *U*, the *J* will reach a high value quickly, which can be attributed to the occurrence of electrolytic process of water. Immediately, a thin compact oxide barrier layer begins to form on the aluminum surface which in contact with the electrolyte (Stage I). At this stage, the thickness of the oxide barrier layer increases rapidly, which means that the total resistance will increase correspondingly. Therefore, the *J* will decrease abruptly to reach the minimum value under the potentiostatic mode, when the barrier layer thickness increases to a certain value (Stage II) [9, 14, 15]. In this stage, Parkhutik et al. suggested that relatively fine-featured pathways can be formed in the initial oxide barrier layer prior to the formation of real pores [14]. The possible formation mechanism of the pathways have been investigated and reported by many groups. O'Sullivan et al. have reported that the *J* concentrates on defects of the initial oxide barrier layer, thus resulting in non-uniform barrier layer thickening. The pathways and real pores can grow at the thinner part of the barrier layer [16]. In addition, Thompson et al. suggested that the pathways may be originated from the local cracking of the initial oxide barrier layer because of the cumulative tensile stress [3, 17, 18]. Figure 1.4 shows the cross-section image of the porous AAO, corresponding to the initial pore formation process. It can be seen that some of the

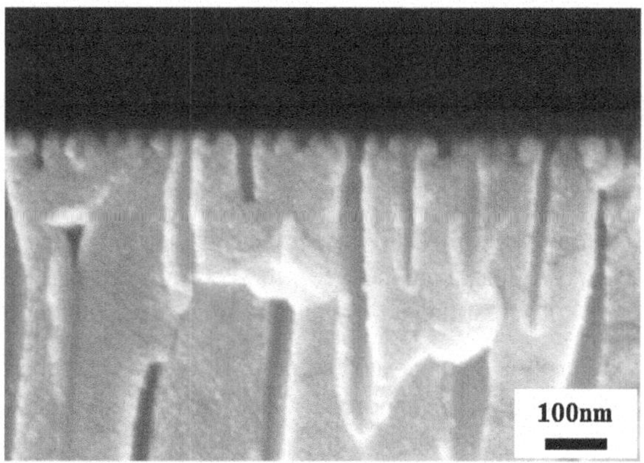

Fig. 1.4 The fracture surface image of the porous AAO, corresponding to the initial stages of the pore formation process (40 V, 0.3 M oxalic acid electrolyte, 5 °C)

pathways continue to grow, thus forming real pores, while other pathways stop their growing. According to these results, the J will gradually increase to a regional maximum value because of the decreasing total resistance, i.e., pathways and pore channels formation in the initial oxide barrier layer (Stage III). As the anodization conducts, the pore size will increase by merging with neighboring pores, resulting in a decreasing pore density. Therefore, the J will decrease from the regional maximum value to a stable value (Stage IV), due to the dynamic equilibrium of the forming and dissolving of AAO [9, 15, 19].

In this case, the anodic oxide is slightly soluble in the formed electrolyte. Ideally, η_J is just above 60 %. If an immobile marker layer is implanted on the Al surface, the marker layer will be located above the original metal surface after anodization (Fig. 1.3b). ^{18}O tracer studies showed that outwardly migrating Al^{3+} cations do not contribute to the oxide growth, but are all shed into the electrolyte by field-assisted ejection and dissolution processes [20, 21]. The thickness of porous AAO membrane is proportional to the integral of current density to time.

$$Q = \int J dt \tag{1.4}$$

where Q is the charge passed during time Δt, and J is the current density at time t.

1.3 Unit Cell Structure of Porous AAO Membranes

In 1995 and 1997, Masuda and co-workers proposed a two-step anodization process and a pre-texturing process for fabrication of porous AAO membranes, respectively [22, 23]. By forming dents on aluminum, highly ordered pore arrangement was obtained.

1.3.1 Unit Cell Structure

Porous AAO membranes are made up of numerous hexagonal unit cells (Fig. 1.5a). Each unit cell contains three different distinct parts:

(1) A hexagonal inner layer, also called the "skeleton", which is made up of the common internal walls between the unit cells.
(2) An outer layer, between the central pore and the inner layer.
(3) An interstitial rod inside the inner layer, at the triple cell junction.

Interpore distance (D_{int}), pore diameter (D_p), barrier layer thickness (t_b), pore wall thickness (t_w), pore density (ρ_p), and porosity (P) are important structural parameters (Fig. 1.5b). For porous AAO membranes with highly ordered pore arrangement, the relationship among these parameters can be expressed as follows:

$$D_{int} = D_p + 2t_w \tag{1.5}$$

$$\rho_P = \left(\frac{2}{\sqrt{3}D_{int}^2}\right) \times 10^{14}\,\mathrm{cm}^{-2} \tag{1.6}$$

$$P(\%) = \left(\frac{\pi}{2\sqrt{3}}\right)\left(\frac{D_p}{D_{int}}\right) \times 100 \tag{1.7}$$

These parameters are mainly dependent on the electrolyte type, anodization voltage (U), anodization current density (J), and temperature (T) [2, 16, 24, 25]. e.g., D_{int} is linearly dependent on the U with a proportionality constant ζ, $D_{int} = \zeta U$ at a given temperature. For porous AAO membranes formed by mild anodization (MA) and hard anodization conditions, $\zeta_{MA} \approx 2.5$ nm/V and $\zeta_{HA} \approx 2.0$ nm/V at 1 °C in 0.3 M oxalic acid electrolyte, respectively [25, 26].

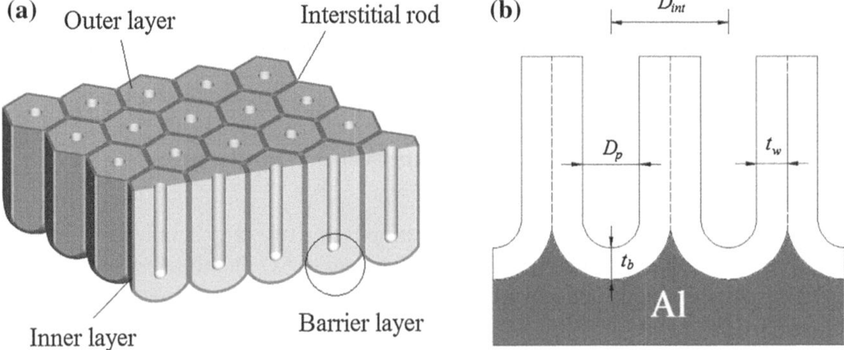

Fig. 1.5 Schematic diagrams of the AAO membrane (**a**) and cross-sectional structure (**b**)

1.3.2 Chemical Composition of Unit Cell

Thompson and Wood suggested that the chemical composition of the inner layer is relatively pure and dense alumina, whereas that of the outer layer is aluminum hydroxide with acid radical ion contamination [5]. The thickness ratio of inner to outer layers is decided by the electrolyte types, in the order: sulfuric acid < oxalic acid < phosphoric acid < chromic acid.

Recent microscopic chemical analyses of Le Coz and co-workers revealed that porous AAO membranes are mainly composed of aluminum hydroxide $Al(OH)_3$, oxy-hydroxide AlOOH and hydrated alumina Al_2O_3 (Fig. 1.6) [27]. Different parts of the unit cell have a heterogeneous chemical composition. For porous AAO membranes formed in phosphoric acid electrolyte, the chemical compositions are $Al_2O_3 \cdot x_1H_2O$ for the inner layers, $Al_2O_3 \cdot 0.24AlPO_4 \cdot x_3H_2O$ for the outer layer, and $Al_2O_3 \cdot 0.018AlPO_4 \cdot x_2H_2O$ for the interstitial rod. Under a strong electron-beam irradiation, the inner layer is more easily crystallized than the outer layer.

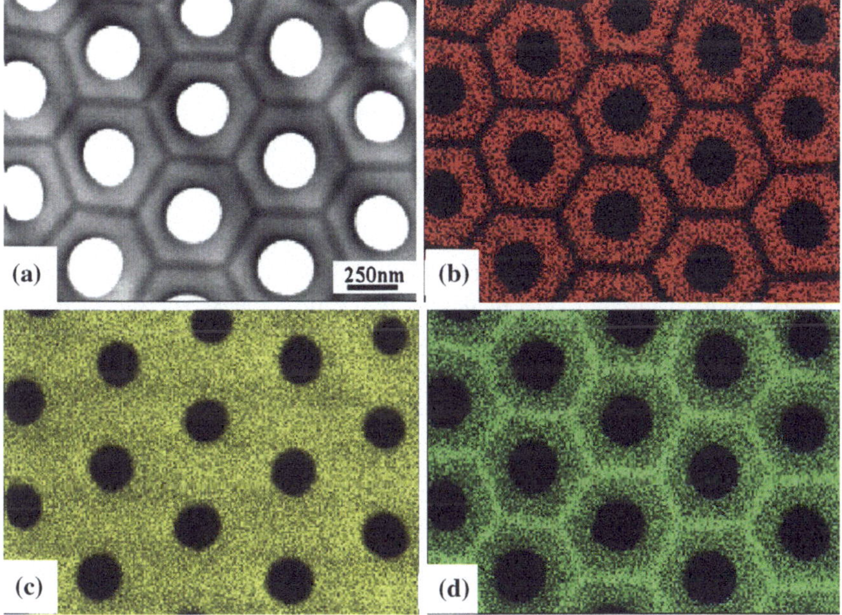

Fig. 1.6 TEM plan view (**a**) of H_3PO_4–AAO membrane and corresponding X-ray maps of the elements: **b** phosphorus, **c** oxygen and **d** aluminum. Reproduced from [27]

1.3.2.1 Outer Layer

One of the most-important features of porous AAO membranes is that the pore diameter (D_p) can be controlled precisely by chemical etching time (t_{etch}) and temperature [28]. As shown in Fig. 1.7, D_p is linearly dependent on the t_{etch} at a given temperature. The temperature is high, the etching rate is high.

The outer layer can be dissolved to form the $Al(OH)_3$ gel during the hydrothermal process. As the $Al(OH)_3$ is much less stable, the following chemical reaction can occur at relatively low temperature:

$$Al(OH)_3 \rightarrow AlOOH + H_2O \tag{1.8}$$

Therefore, by a simple hydrothermal treatment and subsequently selective chemical etching of inner layer, highly uniform aluminum oxy-hydroxide (AlOOH) nanofibers, nanorods and nanotubes can be synthesized (Fig. 1.8) [29–31]. The morphologies of the hydrothermal products are decided by the D_p, the outer layer thickness, and hydrothermal temperature, simultaneously.

1.3.2.2 Inner Layer

If the pores are widened and a proper hydrothermal treatment is conducted to improve the crystallinity of inner layer, the high etching contrast between the inner and outer layers will facilitate the extraction of inner layer [32].

In recent years, by taking the porous AAO membranes as the experimental targets, Chang and coworkers widened the pores to avoid the pore sealing, hydrothermally treated to enhance the inner layer crystallinity, and subsequently selectively etched in a tri-sodium citrate solution. The outer layers were completely removed from all the unit cells of porous AAO membranes. The obtained inner layer kept a hexagonal honeycomb-like structure with uniform wall thickness. To

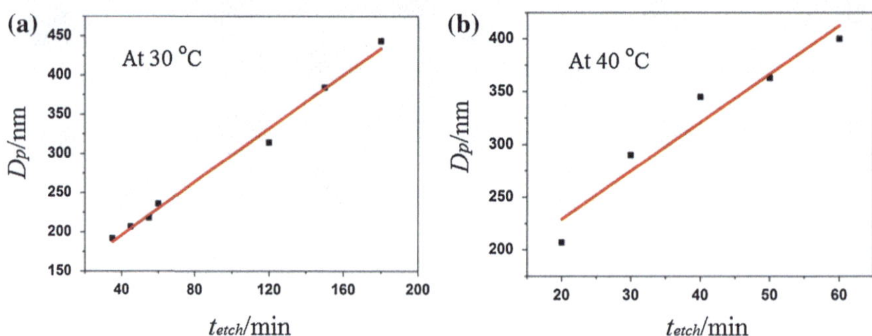

Fig. 1.7 Dependence of the pore diameter (D_p) of porous AAO membranes formed in 1 wt% phosphoric acid electrolyte with 0.01 M Alox addition at 205 V on the wet-chemical etching time (t_{etch}) in 5 wt% H_3PO_4 at 30 °C (**a**) and 40 °C (**b**)

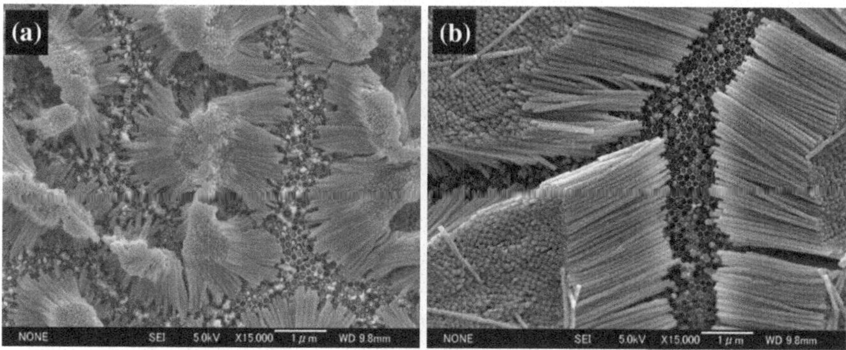

Fig. 1.8 Nanofibers obtained from 20 min etching of a hydrothermally sealed porous oxide film formed after: **a** anodizing at 50 V; **b** anodizing at 50 V, followed by 2 h pore-widening at 25 °C. The size of the nanofibers in **a** and **b** is about 55 and 75 nm, respectively. Reproduced from [29]

anneal the as-prepared samples at high temperature, deformation-free porous α-Al_2O_3 membranes were fabricated (Fig. 1.9).

1.3.2.3 Interstitial Rod

It has been known from the previous reports of Ono and et al. that the interstitial rod is a composition of voids and oxide [33, 34]. The formation of voids is due to high cation vacancy diffusivity below the barrier layer and tensile stress at the triple cell junction. During anodization at high voltage (U) or high anodization current density, cation vacancies are produced, condensed, and merged at the triple cell junction, resulting in a string of voids appeared inside the inner layer (Fig. 1.10).

In the work of Zhao and coworkers, it is found that the oxide in the interstitial rods is easily dissolved in aqueous $HCl/CuCl_2$ solution, while the oxide in barrier layer is not [35]. By completely dissolving the oxide in the interstitial rods, small pores can be observed clearly. If the barrier layer of the large pores are also removed in H_3PO_4 etching solution, it can be seen that each large pore is surrounded by smaller pores in a six-membered ring structure (Fig. 1.11).

1.4 Chemical Reactions During the Steady-State Growth of Porous AAO Membranes

During the steady-state growth of porous AAO membranes, All the structural parameters of unit cell including D_{int}, D_p, t_w, t_b, and the thickness ratio ($R_{i/o}$) of inner and outer layers keep constant. That is to say, the formation and dissolution rates of the inner and outer layer reach a dynamic equilibrium. At this moment, a

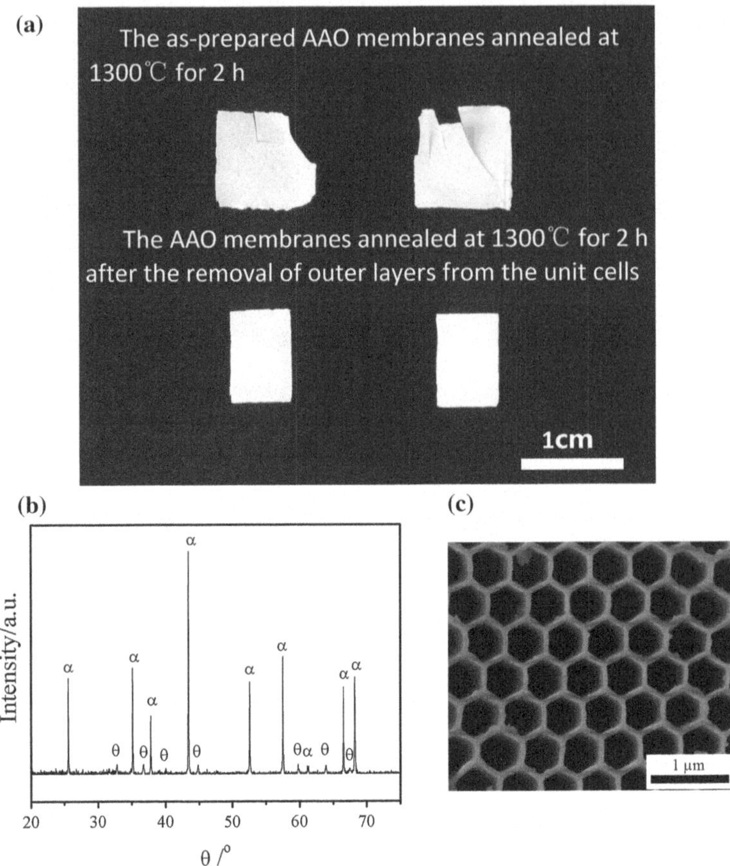

Fig. 1.9 A photograph of the porous α-Al_2O_3 membranes obtained by annealing the porous AAOs with and without outer layers (**a**). **b** and **c** XRD spectrum and SEM image of porous Al_2O_3 membranes obtained by annealing the porous AAOs after removal of outer pore walls at 1300 °C for 2 h in air. The *scale bars* are 1 μm

series of chemical reactions are possible occurring at the outer layer/electrolyte, inner layer/metal, and outer/inner layer interfaces [3, 10, 11, 13, 36].

(a) At the outer layer/electrolyte interface

Heterolytic dissociation of water molecules will occur, producing a large number of $O^{2-}_{(ox)}$, $OH^{-}_{(ox)}$ and $H^{+}_{(aq)}$ ions.

$$2H_2O_{(l)} \rightarrow O^{2-}_{(ox)} + OH^{-}_{(ox)} + 3H^{+}_{(aq)} \tag{1.9}$$

In an electric field, $O^{2-}_{(ox)}$ and $OH^{-}_{(ox)}$ anions leave the outer layer/electrolyte interface, migrating inwardly to the inner layer/metal and inner layer/outer layer

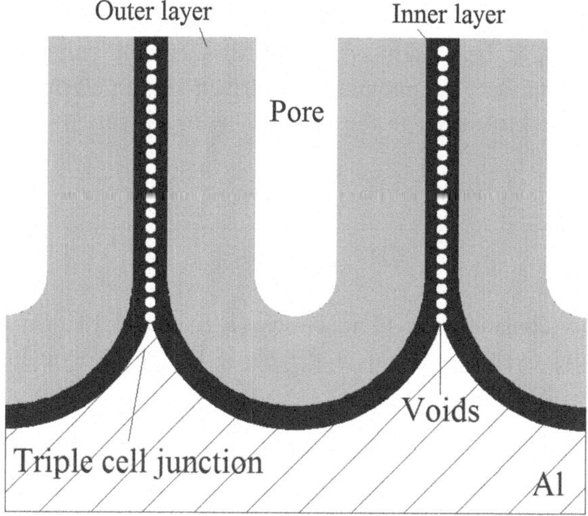

Fig. 1.10 Schematic diagram of the interstitial rod structure

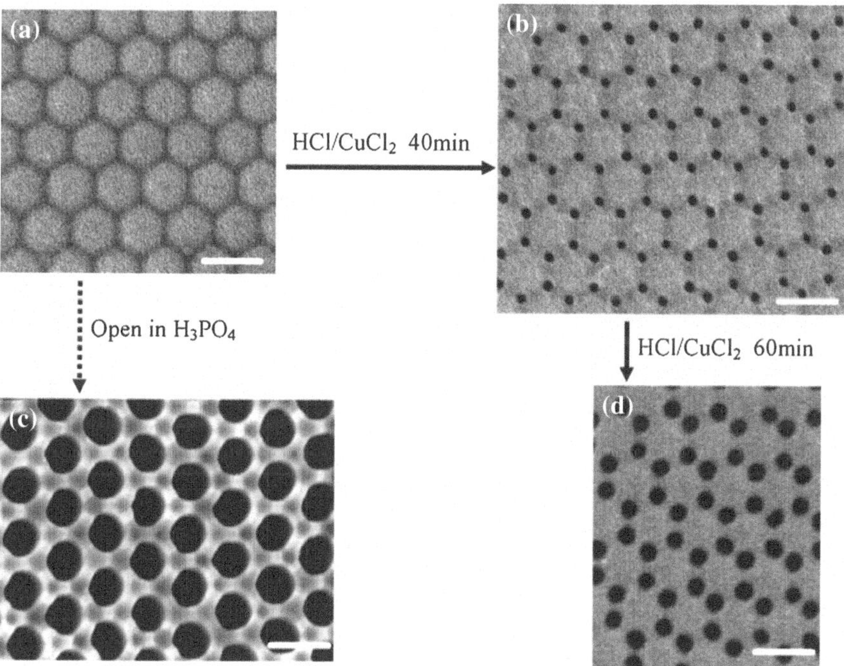

Fig. 1.11 Bottom view of the barrier layers of porous AAO membranes. **a** After removal of the Al substrate; **b** after an immersion in aqueous $HCl/CuCl_2$ solution; **c** after an immersion in 5 % H_3PO_4 solution; **d** extending the immersion time in aqueous $HCl/CuCl_2$ solution. The *scale bar* is 100 nm. Reproduced from [35]

interfaces, respectively. Simultaneously, part of $H^+_{(aq)}$ cations thins the outer layer thickness by Joule's heat induced oxide dissociation reaction at the outer layer/electrolyte interface; the remainder is driven by the electric field, gathering near the cathode and releasing in the form of hydrogen gas.

$$Al_2O_3 + 6H^+_{(aq)} \rightarrow 2Al^{3+}_{(aq)} + 3H_2O_{(l)} \tag{1.10}$$

$$2H^+ + 2e \rightarrow H_{2(g)} \tag{1.11}$$

On the whole consideration of reactions (1.9) and (1.10), Su and coworkers proposed that the overall reaction at the oxide and electrolyte interface can be expressed as [37, 38]

$$Al_2O_3 + nH_2O_{(l)} \rightarrow 2Al^{3+}_{(aq)} + (3 - n - x)O^{2-}_{(ox)} + xOH^-_{(ox)} + (2n - x)H^+_{(aq)} \tag{1.12}$$

where n not only denotes the amount of water dissociated per molar of Al_2O_3 that is dissolved at the same time, but also is the key factor in governing the porosity (*P*) of porous of AAO membranes.

$$P = \frac{3}{n + 3} \tag{1.13}$$

(b) At the inner layer/metal interface

As $O^{2-}_{(ox)}$ anions driven by electric field through barrier layer migrate to the inner layer/metal interface, aluminum (Al) will be oxidized to form new oxide. Thus, the barrier layer is thicken.

$$Al \rightarrow Al^{3+}_{(ox)} + 3e \tag{1.14}$$

$$Al^{3+}_{(ox)} + O^{2-}_{(ox)} \rightarrow Al_2O_3 \tag{1.15}$$

(c) At the outer layer/inner layer interface

To maintain a constant thickness ratio of inner layer and outer layer, chemical reactions should also occur at the inner layer/outer layer interface. Here, we suggested that $OH^-_{(ox)}$ anions from the outer layer/electrolyte interface play a key role in balancing the thickness of inner layer and outer layer by reacting with oxide in inner layer.

$$Al_2O_3 + 6OH^-_{(ox)} \rightarrow 2Al(OH)_3 + 3O^{2-}_{(ox)} \tag{1.16}$$

1.5 Steady-State Anodization

It has been known that the morphology and arrangement of pores inside porous AAO membranes are governed by the chemical reactions occurred in the outer layer/electrolyte, inner layer/outer layer and inner layer/metal interfaces. During the anodization process, the dissolution of oxide is always accompanied by the formation of new oxide. The dissolution rate v_d and formation rate v_f at the interfaces are dependent on both of the anodization voltage (U) and the barrier layer thickness (t_b). For steady-state anodization, v_d and v_f will balance and the pores will grow straightly and be parallel to each other. It is worth mentioning that the potentiostatic anodization (i.e., constant voltage) is widely applied for the fabrication of porous AAO, because of the easy and precise controlling of the structural parameters. In addition, the anodization process is further divided into two types: mild anodization (MA) and hard anodization (HA).

1.5.1 *Mild Anodization*

In fact, porous AAO fabricated by conventional one-step MA processes usually have disordered pores on their surfaces, but ordered structural cells at the bottom side (i.e., barrier layer surface). Based on this result, Masuda et al. fabricated highly ordered porous AAO in 1995 for the first time by using a two-step potentiostatic MA process [22]. After that, various MA processes have been investigated and proposed for the fabrication of diverse porous AAO which can be used for a wide range of applications. At present, there are three well-known growth regimes for typical MA processes: (1) sulfuric acid (H_2SO_4) electrolyte under 25 V with a D_{int} of 63 nm [24, 26, 39], (2) oxalic acid ($H_2C_2O_4$) electrolyte under 40 V with a D_{int} of 100 nm [22, 24, 26, 40], (3) phosphoric acid (H_3PO_4) electrolyte under 195 V with a D_{int} of 500 nm [26, 41]. Figure 1.12 shows the typical SEM images of the most commonly used porous AAO membranes fabricated in 0.3 M oxalic acid electrolyte at 40 V and 0–5 °C.

Using as a template or functional material, the structural parameters of porous AAO membranes, especially the D_{int} and D_p will influence their performance directly in all the applications. Considering that the D_p can be adjusted by a further acid etching process, and the maximum D_p is dependent on the D_{int}, fabricating porous AAO membranes with tunable D_{int} in a large-range and investigating related influencing factors on D_{int} variation have attracted much attention. It has been known that the U will influence the D_{int} directly, and then the D_{int} can be easily adjusted by varying the U. In addition to the fabrication conditions mentioned above (i.e., in sulfuric, oxalic, and phosphoric acid electrolytes), many other fabrication processes have been reported. It should be noted that for a given electrolyte, there is an optimum self-ordering window for MA processes to obtain highly ordered pore arrangement. In order to widen this window, considerable efforts have

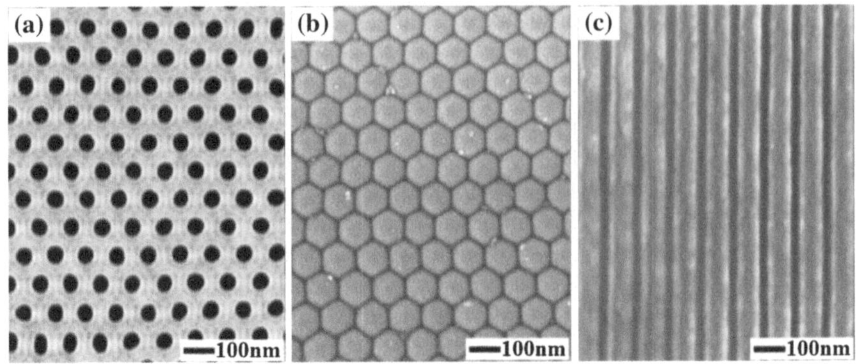

Fig. 1.12 Typical SEM images of the porous AAO membranes fabricated in 0.3 M oxalic acid electrolyte at 40 V and 0–5 °C by using the two-step potentiostatic MA process: **a** top view, **b** bottom view, **c** cross-section view

been paid, e.g., adjusting the electrolyte concentration, modifying the electrolyte with additives, varying the electrolyte systems, etc. To obtain smaller D_{int}, Zhang et al. reported a sulfuric acid based MA process, using which porous AAO membranes with D_{int} of 20–58 nm have been realized by applying lower U of 6–18 V [42]. In addition, Matsui et al. demonstrated the fabrication process of porous AAO membranes on a pre-textured Al surface. Highly ordered porous AAO membranes with a D_{int} of 13 nm has been fabricated by using 2D array of monodisperse Fe_2O_3 nanoparticles as a template [43]. Shingubara et al. used a diluted sulfuric acid electrolyte and a mixture electrolyte of sulfuric acid and oxalic acid and obtained porous AAO membranes with D_{int} of 50–98 nm under 20–48 V [44]. To obtain larger D_{int}, Sun et al. proposed a phosphoric acid based MA process, an additive of aluminum oxalate (Alox) was applied to suppress burning or breakdown of porous AAO membranes during anodization under high U. Using this method, highly ordered porous AAO membranes with continuously tunable D_{int} from 410 to 530 nm can be obtained by varying U from 180 to 230 V [45]. Moreover, considering that highly ordered porous AAO can be obtained in oxalic acid electrolyte, other organic acid electrolytes have been tried in MA processes. For example, porous AAO membranes with D_{int} of 300 nm (120 V), 500 nm (195 V), and 600 nm (240 V) can be synthesized in malonic, tartaric, and citric acid electrolytes, respectively [46, 47]. However, the resulting porous AAO usually have relatively lower ordered pore arrangement, which limits their practical applications to a large extent. Figure 1.13 shows the relationship between U and D_{int} of MA processes in commonly used electrolyte systems.

For porous AAO membranes fabricated by MA processes, it was found that D_{int} is dependent on U directly, and a nearly linear relationship of ζ_{MA} = 2.5 nm/V between D_{int} and U has been proposed (the dash line shown in Fig. 1.13). Therefore, porous AAO membranes with designated D_{int} can be fabricated by using the '2.5 nm/V' rule via MA process. In addition, for a given electrolyte, there is

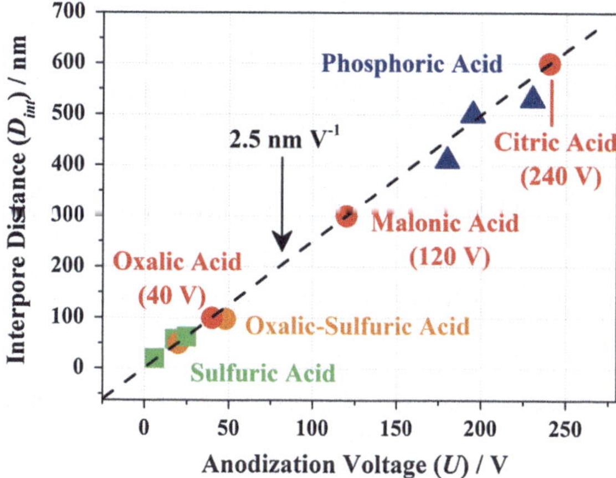

Fig. 1.13 The D_{int}-U curve of MA processes in commonly used electrolyte systems: the data marked by *green squares* were reported in [24, 26, 39, 42], by *orange circles* were reported in [44], by *red circles* were reported in [22, 24, 26, 40, 46, 47], by *blue triangles* were reported in [26, 41, 45]

always a voltage (U_e), above which the electron avalanche phenomenon can be observed. In this case, burning and breakdown of the porous AAO are most likely to occur, accompanied by an avalanche current thus resulting in the releasing of a large amount of heat [47–49]. It was found that the U_e for three commonly used electrolytes: sulfuric, oxalic, and phosphoric acid are 27, 50, and 197 V, respectively. Porous AAO membranes with best pore ordering and cell uniformity can be obtained under the U just below the U_e. It should be noted that typical self-ordering of the pores can be found at the burnt areas of porous AAO membranes, and the D_{int} in these areas are comparatively smaller than burnt-free areas, which means that the D_{int} of porous AAO can also be influenced by the J [47, 49, 50]. Considering that the J in MA processes are relatively small, their effects are usually negligible.

1.5.2 Hard Anodization

At present, although most of the porous AAO can be fabricated by MA processes which have good reproducibility and reliability, some problems still exist: these processes usually require an extremely long time to obtain highly ordered pore arrangement, and their self-ordering windows are narrow. These disadvantages limit their practical applications to a large extent. To solve the problems, HA processes have attracted much attention. Compared with MA processes, the investigations of HA processes also have a considerable long history due to the demands for surface finishing industry of aluminum and its alloys. It was found that higher U, larger J, and faster AAO growth rate are the typical characteristics of conventional HA processes. However, although various HA processes have been investigated by industry, the cell uniformity and pore ordering of the resulting porous AAO have been unsatisfactory for their applications in nanotechnology [51, 52], until the development of new HA processes. In 2005, Chu et al. reported HA processes in sulfuric acid electrolyte, and a maximum U of up to 70 V can be obtained, which is remarkably higher than the breakdown voltage (U_b, 27 V). Ideally ordered porous AAO with D_{int} of 90–130 nm under 40–70 V and 160–200 mA cm^{-2} (0.1–10 °C) have been realized by using a pre-anodized electrolyte. The J is significantly larger than that of the conventional MA processes, resulting in a much faster film growth rate of porous AAO [53]. By using an oxalic acid based electrolyte system, Lee et al. have realized new self-ordering windows of porous AAO via HA processes. Before the high voltage anodization, a thin porous AAO film has been formed on the aluminum surface, and the U was gradually increased to a high target value. Using this method, porous AAO with D_{int} of 220–300 nm under 110–150 V and 30–250 mA cm^{-2} (1–2 °C) have been fabricated [25]. Moreover, Li et al. reported HA processes in a phosphoric acid based electrolyte system. Highly ordered porous AAO with D_{int} of 320–380 nm have been fabricated under 195 V (150–400 mA cm^{-2}, −10–0 °C) by adding appropriate amount of ethanol to the electrolyte [54]. Using these new HA processes, porous AAO with fast growth rate, highly ordered pores, wide self-ordering windows can be obtained, widening their practical applications to a great extent. Figure 1.14 shows the relationship between U and D_{int} of porous AAO by HA processes in commonly used electrolyte systems. Lee et al. proposed that a ζ_{HA} of 2.0 nm/V can be established in oxalic acid electrolyte, which is different from that in MA processes [25]. Considering that the J in HA processes is remarkable larger than that in MA processes, the influence of J on the D_{int} can't be ignored. It was found that the D_{int} increases with the decrease of J, the relationships among U, J, and D_{int} have been proposed by Li et al. in phosphoric acid and sulfuric acid electrolytes, respectively.

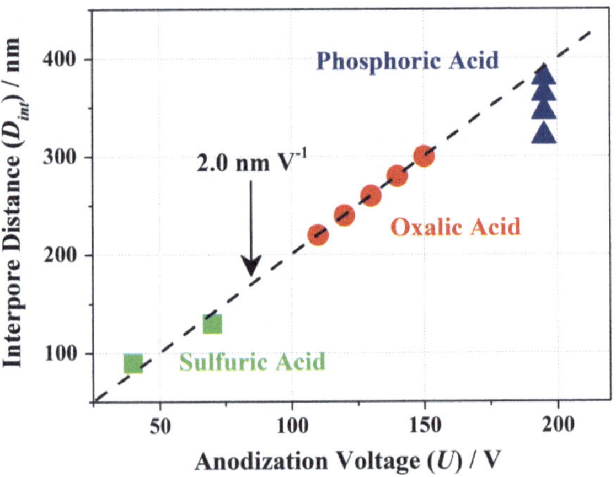

Fig. 1.14 The D_{int}-U curve of HA processes in commonly used electrolyte system: the data marked by *green squares* were reported in [53], by *red circles* were reported in [25], by *blue triangles* were reported in [54]

Their findings further confirm that the linear relationship between U and D_{int} is only a special case [54, 55]. To explain the related intrinsic mechanisms, Lee et al. assumed that a larger J will result in a higher mechanical stress at the metal/oxide interface, which may be responsible for the reduced ζ_{HA}. As a result, a smaller cell size (equal to the D_{int}) of porous AAO can be obtained to increase the surface area of the metal/oxide interface, thus adapting the high mechanical stress during HA processes [25]. Li et al. reported that the formation and dehydration of aluminum hydroxide (e.g., $Al\ (OH)_3$) during HA process can influence the cell size or D_{int} directly, and a smaller ζ_{HA} can thus be obtained due to the volume contraction accompanied by dehydration of aluminum hydroxide and high compressive stress between adjacent cells [56, 57].

1.5.3 The Maximum Anodization Voltage and the Breakdown Process

To obtain porous AAO with adjustable D_{int} in a large range, one may explore anodization processes under which the U can be adjusted in a large range. Therefore, finding proper methods to achieve high enough maximum anodization voltage (U_{max}) in various electrolyte systems and investigating related intrinsic mechanisms are important. It has been known that the commonly used U in different electrolytes are quite different. As mentioned above, the most appropriate U for MA processes in sulfuric acid, oxalic acid, and phosphoric acid electrolytes are 25, 40, and 195 V, respectively [22, 24, 26, 39–41]. For a given electrolyte system, although the U_{max} can be increased to some extent by using several methods, e.g., reducing the electrolyte concentration or lowering the reaction temperature, the practical effects of these methods are relatively limited. Therefore, a relationship of U (sulfuric) $<$ U (oxalic) $<$ U (sulfuric) can often be found in anodization processes, especially the MA processes in which the electron avalanche phenomenon does not occur.

In the anodization processes, it has been known that the J is made up of ionic current density (J_i) and electronic current density (J_e), and the J_i can be further divided into two parts: oxidation current density (J_o) and incorporation current density (J_c). The J_o represents the inwardly migrating O^{2} anions during the anodization and has a direct relationship with the formation of AAO, while the J_c can be ascribed to the migration of contaminated species (e.g., different types of acid radical anions) which makes no direct contribution to the formation of AAO [58, 59]. The relationship among these current densities was shown and illustrated in Fig. 1.15 [15].

During the anodization processes, the electrolyte anions (e.g., in commonly used electrolyte systems, SO_4^{2-}, $C_2O_4^{2-}$, and HPO_4^{2-}) will enter the barrier of porous AAO as contaminated charged particles [15, 56, 58, 59]. When a high enough U is applied during anodization processes, which is always accompanied by a high electrical field (E), some of the charged particles at the barrier layer will be ionized,

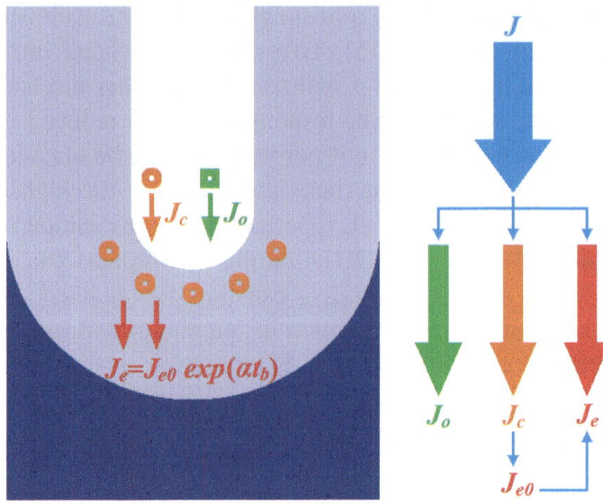

Fig. 1.15 Schematic diagram of the relationship among anodization current density (J), oxidation current density (J_o), incorporation current density (J_c), primary electronic current density (J_{e0}), and avalanche electronic current density (J_e) [15]

releasing primary electrons, i.e., primary electronic current density (J_{e0}) into the oxide conduction band according to the Poole-Frenkel mechanism [58–60]. These primary electrons can be further accelerated under a high E, thus resulting in avalanche electronic current density (J_e) by an impact ionization mechanism. The equation between J_e and J_{e0} can be illustrated as follow [15, 59]:

$$J_e = J_{e0}\exp(\alpha t_b) = J_{e0}\exp(\alpha U_d/E) \approx J_{e0}\exp(\alpha U/E) \tag{1.17}$$

where α is the impact ionization coefficient which represents the number of times of impact ionization for an electron passing a distance of 1 cm along the direction of E, U_d is the potential difference across the barrier layer. Considering that the resistance of the barrier layer is remarkably higher than that of the electrolyte, the value of U_d is approximately equal to U (1.2, 1.3, 1.17). In addition, it should be noted that the mean free path (λ) has a mutually reciprocal relationship with the α of $\alpha = 1/\lambda$, and the α increases with the increasing E [61]. When a high enough U (usually higher than the U_e) is applied, two cases may occur: (1) the avalanche J_e leads to the occurrence of breakdown of porous AAO; (2) although there is an avalanche J_e arises during anodization, the breakdown phenomenon does not occur. Taking into account that the aluminum anodization is a dynamic equilibrium process, the breakdown of porous AAO during anodization is quite different from those occurred in other dielectric mediums. It has been known that the anodic oxide is continuously formed at the barrier layer during the anodization (i.e., the repairing process), which can counteract the damaging effects caused by the breakdown process (i.e., the damaging process) to some extent. Only when the sustained damage of porous AAO can't be repaired in time in the dynamic anodization process, the breakdown phenomenon can be observed (Case 1). On the contrary, the damaged porous AAO can be effectively repaired, and then the breakdown phenomenon can be suppressed (Case 2). In case (2), the aluminum anodization will be changed from MA processes to HA processes, accompanied by avalanche J_e [15].

Compared with the MA process for a given anodization condition, the HA process always has a relatively higher U which can better promote the ionization of contaminated charged particles (i.e., the electrolyte anions) at the barrier layer, thus forming porous AAO with a smaller amount of impurities. Lee et al. reported MA processes (40 V) and HA processes (110–150 V) in 0.3 M oxalic acid electrolyte, respectively. It was found that the carbon element content in porous AAO formed by HA processes (1.8 wt%) is smaller than that fabricated by MA processes (2.4 wt%). Their results further support the above statement. In addition, the relationship between different current densities can be written as follow according to the Fig. 1.15 and (1.17) [58, 59]:

$$J = J_i + J_e = J_o + J_c + J_e = J_o + J_c + J_{e0}\exp(\alpha t_b) \approx J_o + J_c + J_{e0}\exp(\alpha U/E) \tag{1.18}$$

where the relationship between the J_c and J_o is reported as $J_c = \gamma J_o$ [58, 59, 62]. Taking into account that the J_{e0} is dependent on the acid anions of the electrolytes as we discussed above, the relationship between the J_{e0} and J_c can be written as $J_{e0} = \eta J_c$. According to the above discussion, the (1.18) can be rewritten as follow: [15, 58, 59, 62]

$$J = J_o + J_c + J_e \approx (1 + \gamma + \gamma\eta\exp(\alpha U/E))J_o = (1/\gamma + 1 + \eta\exp(\alpha U/E))J_c \tag{1.19}$$

where related investigations show that the γ and η are fractions. The former one can be determined by the electrolyte anion concentration at the barrier layer of porous AAO, while the latter one is also a fraction determined by the ability of electrolyte anions to donate electrons [15, 58, 59, 62]. According to the classical theory of breakdown, the maximum value of the avalanche J_e avoiding breakdown of anodic oxides can be calculated by the equilibrium conditions during anodization: the ionization collision rate (i.e., the damaging process) is less than or equal to the AAO formation rate (i.e., the repairing process). In addition, it has been reported that the relationship between the avalanche J_e and J_o can be written as $J_e = zJ_o$, with the condition that $z \leq 1/3$ [15, 59, 63]. Based on the above conclusions, (1.17) and (1.19), the relationship among the U_e (over which the avalanche J_e appears), $U_{\max}$ (the maximum U that can be applied before breakdown), and the breakdown voltage (U_b) can be written as follow:

$$U_e < U_{\max} < U_b \approx \frac{E}{\alpha}\ln\left(\frac{z}{\gamma\eta}\right) = \frac{E}{\alpha}\ln\left(\frac{zJ_o}{\eta J_c}\right) = \frac{E}{\alpha}\ln\left(\frac{zJ_o}{J_{e0}}\right) \tag{1.20}$$

During the anodization processes, the acid anions (e.g., SO_4^{2-}, $C_2O_4^{2-}$, and HPO_4^{2-}) exist in the electrolytes according to the following equations [19],

$$HSO_4^- \rightarrow H^+ + SO_4^{2-},\ HC_2O_4^- \rightarrow H^+ + C_2O_4^{2-},\ H_2PO_4^- \rightarrow H^+ + HPO_4^{2-} \quad (1.21)$$

and these acid anions will enter the barrier layer of porous AAO in a certain depth, replacing O^{2-} as substitution or contamination anions [17, 19]. It has been known that the secondary dissociation constant (K_{a2}) of HSO_4^-, $HC_2O_4^-$, and $H_2PO_4^-$ at 25 °C are $10^{-1.92}$, $10^{-4.27}$, and $10^{-7.20}$, respectively, and a larger K_{a2} will lead to more acid anions in the electrolytes under other conditions remain unchanged, thus resulting in a smaller U_{max}, and U_b according to the (1.20). Therefore, the experimental results which were commonly obtained of $U_{max\text{-}sulphuric} < U_{max\text{-}oxalic} < U_{max\text{-}phosphoric}$ can be reasonably explained. Furthermore, it should be noted that the U_{max} and U_b may also be modulated by adjusting the electrolyte concentration, or varying the electrolyte temperature. A lower concentration or temperature of the electrolyte will lead to fewer acid anions and a smaller J_c, thus obtaining a higher U_{max} and U_b. However, the regulating range of the U_{max} and U_b are quite narrow by using these two methods. Therefore, in the practical fabrication processes, the U_{max} and U_b are commonly adjusted by varying the electrolyte types.

1.6 Unsteady-State Anodization

In MA processes, if the U steps up suddenly, the equilibrium between the v_d and v_f will be broken. The enhancement of electrical field strength ($E \approx U/t_b$) in oxide will accelerate the migration of anions, leading to the t_b thickening. In this case, some primary stem pores stop to grow and their original areas are occupied by the pores nearby. No merging or branching but termination of pores occurs (Fig. 1.16a).

In contrast, if the U steps down suddenly, the t_b will be thinned since the electrical field strength (E) is too weak to drive the anions across the barrier layer. During this process, the dissolution reaction acts dominantly, instead of the formation reaction. The anodization does not work until the t_b is thin enough. As a result, branched pores generate from each of the primary stem pores (Fig. 1.16b) [16, 64–67].

1.6.1 Rule of Branched Channel Growth

Ideally, the porous AAO membranes grown under steady-state anodization have two elementary features:

(1) All pores align with a hexagonal close-packed structure.
(2) The interpore distance (D_{int}) is linear proportional to the U.

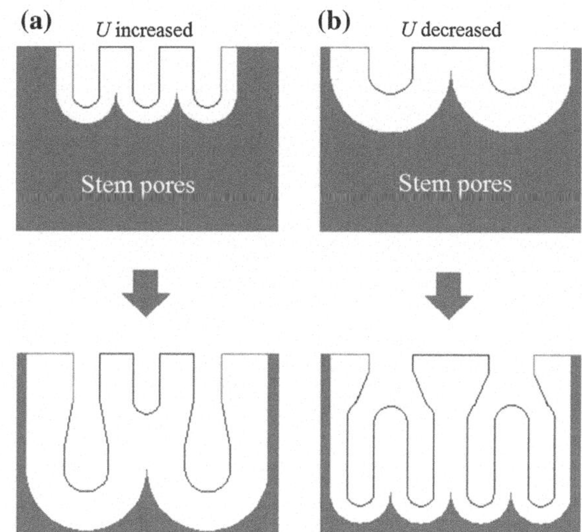

Fig. 1.16 Schematic diagrams for the pore growth by stepping up (**a**) and down (**b**) the anodization voltage (*U*)

As the *U* steps down by a factor of $\sqrt{n}$, the steady-state anodization will be ceased temporarily due to weak electrical field strength (*E*) in oxide but regain some time later when the barrier layer is thinned to a critical thickness. During this transition stage, the growth of ordered branched pores can be equivalent to the superposition of two sets of hexagonal close-packed lattice on behalf of the stem pores and the branched pores, respectively. As shown in Fig. 1.17, ordered arrangement of pores can be achieved only for the reduction of anodization voltage by the factors of $\sqrt{3}$, $\sqrt{4}$ and their common multiples and powers [68]. Otherwise, number of the branched pores can't be the same from each of the stem pores. It is impossible to grow the branched pores arbitrarily as Meng and coworkers suggested previously, $1 \prec n$, where "$\prec$" stands for the junction where the branching takes place, and n is natural numbers, standing for the number of branches produced.

To check the morphologies of $\sqrt{3}$ and $\sqrt{4}$ branched pores carefully, it is found that three types of possible pore arrangement can exist (Fig. 1.18). Although the number of branched pores from each of the stem pores is same, the branched pore arrangement can also be different. The results demonstrate that porous AAO membranes with the same branched pore morphologies are practically difficult to obtain.

1.6.2 Competitive Growth Process

As discussed above, if the anodization voltage (*U*) is lowered by factors of $\sqrt{3}$, $\sqrt{4}$ and their common multiples and powers, and the barrier layer thickness of porous AAO membranes is uniform and thin enough for anions easily migrating across under the lowered voltage, small pores will nucleate and develop in the radial direction of each stem pore bottom at first, then reach a dynamic

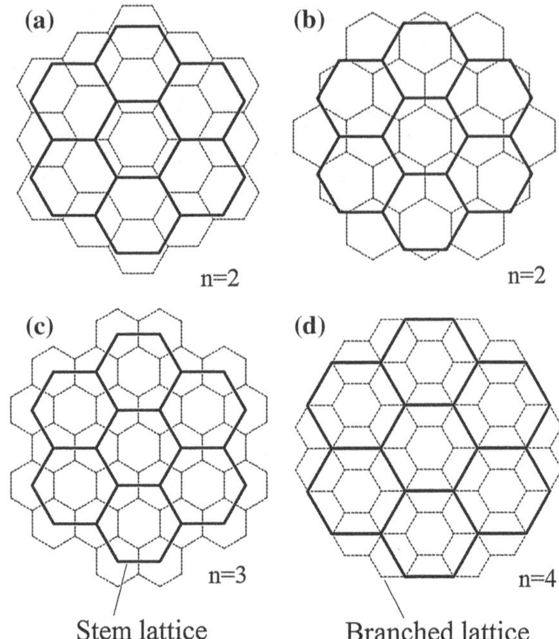

Fig. 1.17 Two sets of superposed hexagonal close-packed lattices. **a** and **b** $n = 2$, **c** $n = 3$, **d** $n = 4$

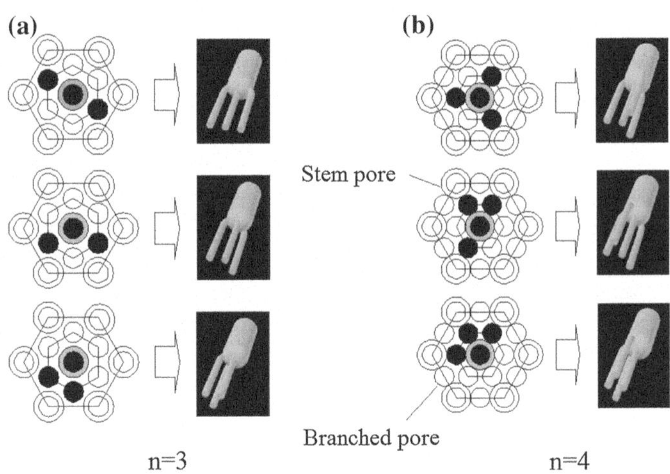

Fig. 1.18 The branched pore morphologies of **a** $1 \prec 3$ and **b** $1 \prec 4$

dissolution-formation equilibrium rapidly. Finally, the branched pores with highly ordered arrangement can be obtained (Fig. 1.19).

However, in fact, when the U steps down, the anodization current density (J) will drop to a value near zero immediately. The thinning of barrier layer in

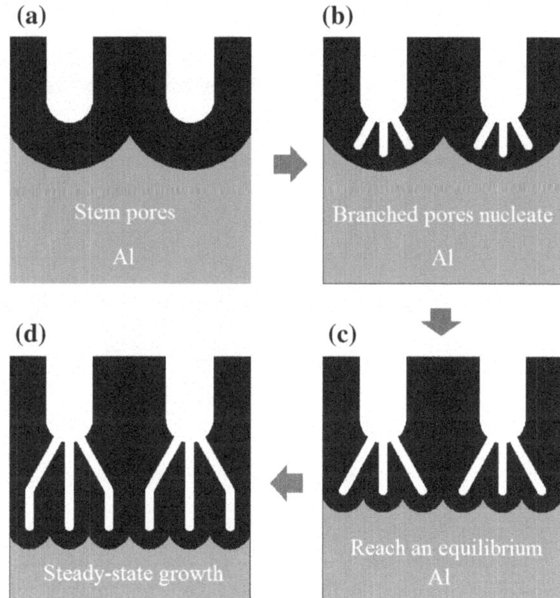

Fig. 1.19 Schematic diagrams of the growth process of controllable branched pores

electrolyte is no longer decided by the field-assisted dissolution reaction. Instead, the chemical etching reaction is dominant. At this time, the thinning rate depends not only on the acidity and temperature of solution, but also on the local field, anion content and hydrogen-bonded structure of the barrier layer, as well as any structurally or mechanically imperfections. As a result, the barrier layer thickness is non-uniform. In some particular stem pores, as the barrier layer thickness is thin enough for migration of anions across under the lowered voltage, pores branch at these sites firstly (Fig. 1.20a, b). According to the equifield-strength model proposed by Su and coworkers, the branched pores will develop along the direction of higher electrical field strength [37]. Since the pore bottom is a hemisphere, the pore growth should be in a radial direction of hemisphere as shown in Fig. 1.20c. To maintain the new barrier layer thickness (t_b) and the new interpore distance (D_{int}), the pores have to keep branching to form a larger hemispherical oxide (Fig. 1.20d) [65].

To avoid this disordered growth of branched pores, the barrier layers of porous AAO membranes are generally thinned in a specific chemical etching solution at higher temperature instead of in the electrolyte at 0 °C, such as 5 % phosphoric acid solution electrolyte at 30 °C, before the lowered voltage is applied. And the barrier layer thickness is reduced as much as possible below a critical value (t_C), where t_C can be expressed as

$$t_C = \frac{t_b}{\sqrt{n}} \tag{1.22}$$

Thus, the branched pores is preferred to extend simultaneously from each of the stem pore bottoms.

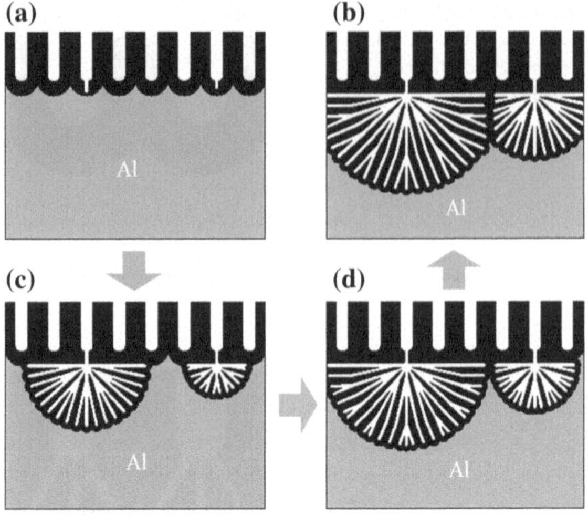

Fig. 1.20 Competitive growth process of the branched pores

To compare the *J-t* curves of porous AAO membranes anodized under the lowered voltage after the barrier layers are thinned in 5 % phosphoric acid solution electrolyte at 30 °C for 50 min and 80 min, it can be found that the former curve can be divided into three stages (Fig. 1.21a). *J* drops down to a minimum value and rises slowly in stage I, increases rapidly in stage II, and reaches a constant value in stage III. A long time-period of low current density in stage I indicates that the thinning of barrier layer is dominated by the chemical etching reaction. Non-uniform barrier layer thickness leads to a competitive growth of branched pores (Fig. 1.21b). On the contrary, no stage I appears in the latter curve, implying that the branching process starts immediately when the lowed voltage is applied (Fig. 1.21c). As a result, a morphology of ordered, upright, and uniform arrangement structure of branched pores is obtained (Fig. 1.21d) [69].

1.7 Microstructural Morphologies of Porous AAO

Obviously, the microstructural morphologies of porous AAO will influence their practical applications to a great extent. Therefore, finding proper methods for the fabrication of porous AAO with controllable microstructural morphologies is significant from the viewpoint of both scientific research and commercial applications. Since the first report of highly ordered porous AAO by Masuda et al., porous AAO with a typical "continuous porous film" morphology have been fabricated and investigated by many groups, gaining more and more attention [22, 24, 26, 39–47]. To widen the self-ordering windows of porous AAO, Chu et al. reported HA processes in sulfuric acid electrolyte, and porous AAO with a "smooth separated tubes" morphology have been realized. Similar results have also been reported by

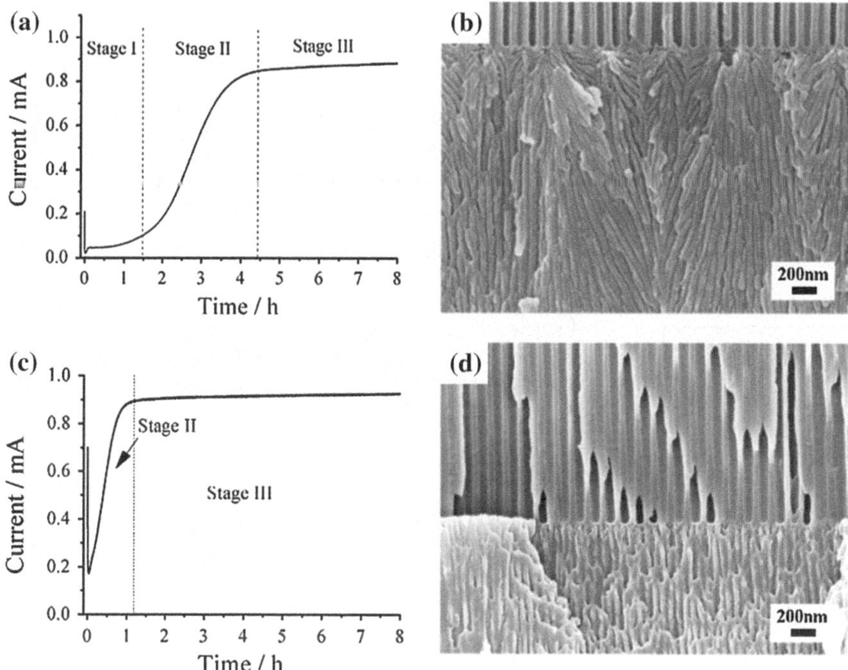

Fig. 1.21 Current-time curves and SEM images of the competitive growth process of branched pores after the chemical etching in 5 % phosphorous acid solution at 30 °C for 50 min (**a**, **b**) and 80 min (**c**, **d**)

Li et al. and Zhao et al. [35, 53, 55]. In addition, Li et al. reported the fabrication of porous AAO in an ethanol modified oxalic acid electrolyte, and the as prepared porous AAO is also made up of separated tubes, and apparent thickness variations can be observed on the tube walls [56, 57]. Therefore, it is commonly accepted that there are three typical microstructural morphologies of porous AAO, which can be formed by adjusting the anodization conditions: continuous porous film, smooth separated tubes, and tubes with wall thickness variations [22, 24, 26, 35, 39–47, 53, 55–57]. It is worth mentioning that the latter two microstructural morphologics can be further classified into two types: a gapless type and a gap type between adjacent structural cells. The former one has been reported by many groups, but the latter one has rarely been mentioned [15, 35, 53, 55–57]. Figure 1.22 shows the schematic diagrams and SEM images of porous AAO with representative microstructural morphologies.

Type I, continuous porous film (Fig. 1.22a). It has been known that the formation and dehydration of aluminum hydroxide (e.g., $Al(OH)_3$) will influence the microstructural morphologies of porous AAO to a large extent [56–58, 70]. In this type, the as-prepared porous AAO are usually formed under a relatively lower U and J. Therefore, relatively less $Al(OH)_3$ is decomposed during the anodization,

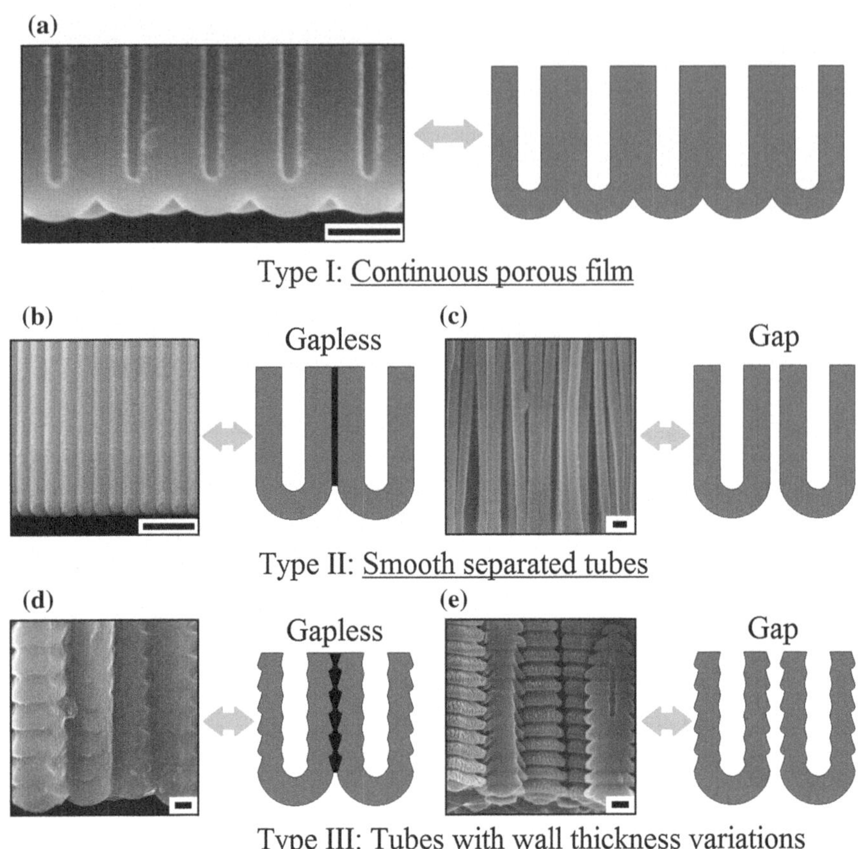

Fig. 1.22 Schematic drawings and SEM images of porous AAO with representative microstructural morphologies fabricated in different electrolytes (**a** ethanol: 0.3 M oxalic acid = 1:2, v/v, **b** H_2SO_4: 0.75 mol l^{-1} and $Al_2(SO_4)_3$: 0.15 mol l^{-1}, **c–e** ethanol: 0.3 M oxalic acid = 1:4, v/v): **a** obtained under a constant voltage of 160 V, **b** obtained under a constant current density of 500 A m^{-2}, **c** and **d** obtained under a constant voltage of 500 V with different ending J of approximately 400 and 55 A m^{-2}, respectively. *Scale bars* = 300 nm [15]

and the volume contraction due to the dehydration of $Al(OH)_3$ is slight. In this case, a comparatively larger ζ ($\sim$2.5 nm V^{-1}) can thus be obtained, corresponding to the conventional so called MA processes [22, 24, 26, 39–47]. As the increase of U, more OH^- and O^{2-} can be generated according to the (1.12), thus forming more Al $(OH)_3$ inside the porous AAO. Moreover, the higher U will promote the dehydration process of $Al(OH)_3$, thus resulting in a relatively smaller ζ under the coaction of the compressive stress between adjacent cells and the volume contraction caused by the $Al(OH)_3$ dehydration [15, 56, 57, 70, 71], corresponding to the conventional so called HA processes [25, 53–57]. In addition, it worth mentioning that in type I, most of the $Al(OH)_3$ is formed and dehydrated at the outer layer of the unit cells, and the inner layer is almost made up of high purity Al_2O_3 due to the different migration abilities of O^{2-} and OH^- [15, 56–58, 70]. Therefore, the unit cells of the as-prepared porous AAO is made up of inner layer and outer layer, and the fracture plane appears along the pores of the porous AAO [22, 24–26, 39–47, 54, 57].

Type II, smooth separated tubes (Fig. 1.22b, c). When a high enough U is applied during anodization, most of the O^{2-} is consumed and released in the form of O_2 due to the high J_e accompanied by the high U [15, 57, 58, 70]. Therefore the as-prepared porous AAO is mainly made up of $Al(OH)_3$, AlOOH, and electrolyte anions, etc., and the inner layer (i.e., high purity Al_2O_3 with relatively better chemical stability and mechanical strength) of the porous AAO often can't be observed. The dehydration of the $Al(OH)_3$ at the unit cell boundaries leads to volume contraction, thus forming separated AAO tubes [15, 53, 55–57, 70]. Considering that the thickness of the $Al(OH)_3$ layer is thinner, the volume contraction of AAO caused by the dehydration process is slight, and relatively smooth separated AAO tubes can be fabricated. In this case, the fracture plane appears along the cell boundaries [53, 55, 57].

Type III, tubes with wall thickness variations (Fig. 1.22d, e). Compared with type II, the $Al(OH)_3$ layer at the unit cell boundaries of porous AAO is thicker, and

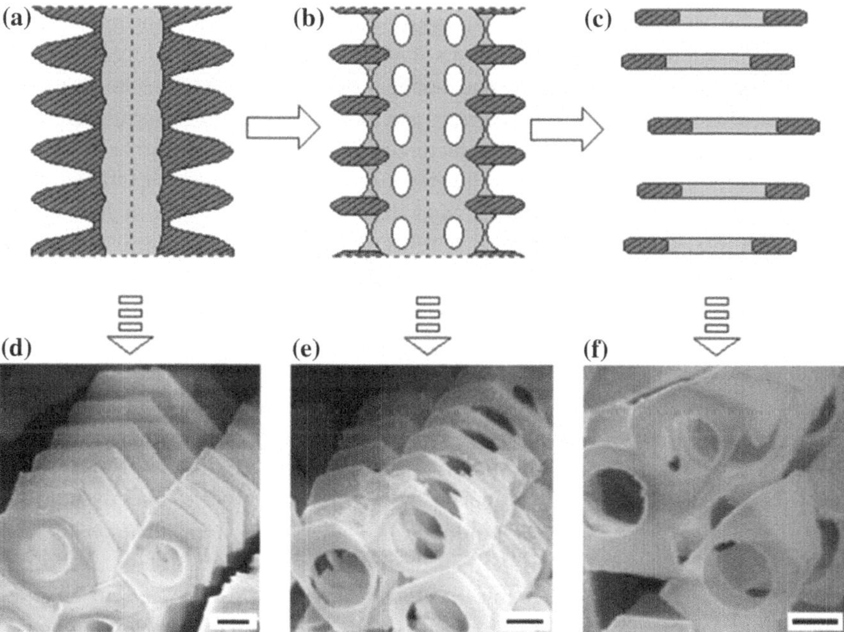

Fig. 1.23 Schematic diagrams and SEM images of novel AAO structures fabricated by further chemical etching processes. The etching time is 90 min for (**d**), 110 min for (**e**), and 140 min for (**f**), in 5 wt% H_3PO_4 solution under 45 °C. *Scale bars* = 200 nm (anodization: 500 V, 350 A m^{-2}, ethanol: 0.3 M oxalic acid = 1:2, v/v). Reproduced from [56]

the volume contraction caused by the dehydration of $Al(OH)_3$ is more obvious. Considering that the direction of volume contraction can be vertical and parallel to the growth direction of the porous AAO, separated AAO tubes with wall thickness variations can be fabricated, and the fracture plane also appears along the cell boundaries [15, 56, 57, 70].

Moreover, it can be seen from Fig. 1.22c, e that there are obvious gaps between adjacent tubes without additional acid corrosion processes. According to the related reports and investigations, the formation of these gaps may be ascribed to the coaction of the dehydration of $Al(OH)_3$ and the aggregation of oxygen gas bubbles together with cation vacancies at the unit cell junctions of the porous AAO [15, 56–58, 70, 72–75]. In addition, novel AAO structures can be obtained by further chemical etching of the porous AAO (Type III), the related schematic diagrams and SEM images were shown in Fig. 1.23 [56]. These new AAOs also have potential uses in diverse nanotechnology applications.

1.8 Conclusion

Although highly ordered porous AAO have been applied for diverse applications, e.g., as a porous template for the fabrication of various nano- or submicron materials, or as porous functional materials, investigating the intrinsic formation mechanisms and morphologies of porous AAOs is still significant for their precisely controllable fabrication processes and applications. In this chapter, the formation processes of different types (i.e., barrier type and porous type) of AAO have been reviewed in detail, and the unit cell structure of porous AAO has also been studied. The unit cells usually contain two layers: the out layer is made up of $Al(OH)_3$, AlOOH, and electrolyte anions, etc., and the inner layer consists of high purity Al_2O_3 with relatively better chemical stability and mechanical strength. The related findings and discussions will promote the precise control of the pore size of porous AAO. In addition, the intrinsic mechanisms of the mild anodization (MA), hard anodization (HA), and unsteady-state anodization processes, together with the chemical reactions during the growth of porous AAO have been investigated. Related findings will facilitate the understanding of aluminum anodization processes from qualitative to quantitative. Furthermore, the breakdown process during anodization and microstructural morphologies of porous AAO have been investigated in detailed. The J_e will influence the breakdown process directly, and the microstructural morphologies of porous AAO are mainly dependent on the formation and dehydration of $Al(OH)_3$ inside unit cells. The investigations and discussions in this chapter will enable the fabrication of porous AAOs to be more controllable, predictable and precise, and these findings may also be applied in other valve metal anodization processes.

References

1. J.W. Diggle, T.C. Downie, C.W. Goulding, Chem. Rev. **69**, 365 (1969)
2. F. Keller, M.S. Hunter, D.L. Robinson, J. Electrochem. Soc. **100**, 411 (1953)
3. G.E. Thompson, Thin Solid Films **297**, 192 (1997)
4. R.C. Furneaux, W.R. Rigby, A.P. Davidson, Nature **337**, 147 (1989)
5. G.E. Thompson, G.C. Wood, Nature **290**, 230 (1981)
6. A. Guntherschulz, H.Z. Betz, Physics **92**, 367 (1934)
7. A. Guntherschulz, H.Z. Betz, Physics **68**, 145 (1932)
8. C.-O.A. Olsson, D. Hamm, D. Landolt, J. Electrochem. Soc. **147**, 4093 (2000)
9. W. Lee, S.J. Park, Chem. Rev. **114**, 7487 (2014)
10. J.A. Davies, B. Domeij, J.P.S. Pringle, F. Brown, J. Electrochem. Soc. **112**, 675 (1965)
11. J.A. Davies, J.P.S. Pringle, R.L. Graham, F. Brown, J. Electrochem. Soc. **109**, 999 (1962)
12. J.A. Davies, B. Domeij, J. Electrochem. Soc. **110**, 849 (1963)
13. K. Shimizu, G.E. Thompson, G.C. Wood, Y. Xu, Thin Solid Films **88**, 255 (1982)
14. V.P. Parkhutik, V.I. Shershulsky, J. Phys. D Appl. Phys. **25**, 1258 (1992)
15. Y. Li, Z.Y. Ling, X. Hu, Y.S. Liu, Y. Chang, RSC Adv. **2**, 5164 (2012)
16. J.P. O'Sullivan, G.C. Wood, Proc. R. Soc. Lond., Ser. A **317**, 511 (1970)
17. K. Shimizu, K. Kobayashi, G.E. Thompson, G.C. Wood, Philos. Mag. A **66**, 643 (1992)
18. G.E. Thompson, Y. Xu, P. Skeldon, K. Shimizu, S.H. Han, G.C. Wood, Philos. Mag. B **55**, 651 (1987)
19. F. Li, L. Zhang, R.M. Metzger, Chem. Mater. **10**, 2470 (1998)
20. C. Cherki, J. Siejka, J. Electrochem. Soc. **120**, 784 (1973)
21. J. Siejka, C. Ortega, J. Electrochem. Soc. **124**, 883 (1977)
22. H. Masuda, K. Fukuda, Science **268**, 1466 (1995)
23. H. Masuda, H. Yamada, M. Satoh, H. Asoh, Appl. Phys. Lett. **71**, 2770 (1997)
24. A.P. Li, F. Muller, A. Birner, K. Nielsch, U. Gosele, J. Appl. Phys. **84**, 6023 (1998)
25. W. Lee, R. Ji, U. Gosele, K. Nielsch, Nat. Mater. **5**, 741 (2006)
26. K. Nielsch, J. Choy, K. Schwirn, R.B. Wehrspohn, U. Gosele, Nano Lett. **2**, 677 (2002)
27. F. Le Coz, L. Arurault, L. Data, Mater. Charact. **61**, 283 (2010)
28. H. Han, S.J. Park, J.S. Jang, H. Ryu, K.J. Kim, S. Baik, ACS Appl. Mater. Interfaces **5**, 3441 (2013)
29. H. Jha, T. Kikuchi, M. Sakairi, H. Takahashi, Nanotechnology **19**, 395603 (2008)
30. H. Jha, F. Schmidt-Stein, N.K. Shrestha, M. Kettering, I. Hilger, P. Schmuki, Nanotechnology **22**, 115601 (2011)
31. Y. Chang, Z.Y. Ling, Y. Li, X. Hu, Electrochim. Acta **93**, 241 (2013)
32. Y. Chang, Z.Y. Ling, Y.S. Liu, X. Hu, Y. Li, J. Mater. Chem. **22**, 7445 (2012)
33. S. Ono, H. Ichinose, N. Masuko, J. Electrochem. Soc. **138**, 3705 (1991)
34. D.D. Macdonald, J. Electrochem. Soc. **140**, L27 (1993)
35. S. Zhao, K. Chan, A. Yelon, T. Veres, Adv. Mater. **19**, 3004 (2007)
36. F. Brown, W.D. Mackintosh, J. Electrochem. Soc. **1973**, 120 (1096)
37. Z. Su, W. Zhou, Adv. Mater. **20**, 1 (2008)
38. Z. Su, M. Buhl, W. Zhou, J. Am. Chem. Soc. **131**, 8697 (2009)
39. H. Masuda, F. Hasegwa, S. Ono, J. Electrochem. Soc. **144**, L127 (1997)
40. H. Masuda, M. Satoh, Jpn. J. Appl. Phys. **35**, L126 (1996)
41. H. Masuda, K. Yada, A. Osaka, Jpn. J. Appl. Phys. **37**, L1340 (1998)
42. F. Zhang, X.H. Liu, C.F. Pan, J. Zhu, Nanotechnology **18**, 345302 (2007)
43. Y. Matsui, K. Nishio, H. Masuda, Small **2**, 522 (2002)
44. S. Shingubara, K. Morimoto, H. Sakaue, T. Takahagi, Electrochem. Solid-State Lett. **7**, E15 (2004)
45. C. Sun, J. Luo, L. Wu, J. Zhang, ACS Appl. Mater. Interfaces **2**, 1299 (2010)
46. S. Ono, M. Saito, H. Asoh, Electrochim. Acta **51**, 827 (2005)
47. S. Ono, M. Saito, M. Ishiguro, H. Asoh, J. Electrochem. Soc. **151**, B473 (2004)

48. S.Z. Chu, K. Wada, S. Inoue, M. Isogai, Y. Katsuta, A. Yasumori, J. Electrochem. Soc. **153**, B384 (2006)
49. S. Ono, M. Saito, H. Asoh, Electrochem. Solid-State Lett. **7**, B21 (2004)
50. G.C. Tu, I.T. Chen, K. Shimizu, J. Jpn. Inst. Light Met. **40**, 382 (1990)
51. P. Csokán, C.C. Sc, Electroplat. Metal Finish. **15**, 75 (1962)
52. P. Csokán, Trans. Inst. Metal Finishing **41**, 51 (1964)
53. S.Z. Chu, K. Wada, S. Inoue, M. Isogai, A. Yasumori, Adv. Mater. **17**, 2115 (2005)
54. Y.B. Li, M.J. Zheng, L. Ma, W.Z. Shen, Nanotechnology **17**, 5101 (2006)
55. Y. Li, Z.Y. Ling, S.S. Chen, J.C. Wang, Nanotechnology **19**, 225604 (2008)
56. Y. Li, Z.Y. Ling, S.S. Chen, X. Hu, X.H. He, Chem. Commun. **46**, 309 (2010)
57. Y. Li, Z.Y. Ling, X. Hu, Y.S. Liu, Y. Chang, J. Mater. Chem. **21**, 9661 (2011)
58. X.F. Zhu, Y. Song, L. Liu, C.Y. Wang, J. Zheng, H.B. Jia, X.L. Wang, Nanotechnology **20**, 475303 (2009)
59. J.M. Albella, I. Montero, J.M. Martínez-Duart, Electrochim. Acta **32**, 255 (1987)
60. I. Vrublevsky, A. Jagminas, J. Schreckenbach, W.A. Goedel, Electrochim. Acta **53**, 300 (2007)
61. J.M. Albella, I. Montero, J.M. Martínez-Duart, J. Electrochem. Soc. **131**, 1101 (1984)
62. J.M. Albella, I. Montero, J.M. Martínez-Duart, V. Parkhutik, J. Mater. Sci. **26**, 3422 (1991)
63. S. Ikonopisov, Electrochim. Acta **1977**, 22 (1077)
64. J. Li, C. Papadopoulos, J. Xu, Nature **402**, 253 (1999)
65. H. Takahashi, M. Nagayama, H. Akahori, A. Kitahara, J. Electron Microsc. **22**(2), 149 (1973)
66. J. Broughton, G. Davies, J. Membrane Sci. **106**, 89 (1995)
67. G. Meng, Y.J. Jung, A. Cao, V. Robert, M.A. Pulickel, Proc. Natl. Acad. Sci. U.S.A. **102**, 7074 (2005)
68. S.S. Chen, Z.Y. Ling, X. Hu, Y. Li, J. Mater. Chem. **19**, 1 (2009)
69. S.S. Chen, Z.Y. Ling, X. Hu, H. Yang, Y. Li, J. Mater. Chem. **20**, 1794 (2010)
70. Y. Li, Z.Y. Ling, X. Hu, Y.S. Liu, Y. Chang, Chem. Commun. **47**, 2173 (2011)
71. F. Le Coz, L. Arurault, S. Fontorbes, V. Vilar, L. Datas, P. Winterton, Surf. Interface Anal. **42**, 227 (2010)
72. X.F. Zhu, L. Liu, Y. Song, H.B. Jia, H.D. Yu, X.M. Xiao, X.L. Yang, Monatsh. Chem. **139**, 999 (2008)
73. X.F. Zhu, L. Liu, Y. Song, H.B. Jia, H.D. Yu, X.M. Xiao, X.L. Yang, Mater. Lett. **62**, 4038 (2008)
74. X.L. Yang, X.F. Zhu, H.B. Jia, T. Han, Monatsh. Chem. **140**, 595 (2009)
75. W. Lee, R. Scholz, U. Gösele, Nano Lett. **8**, 2155 (2008)

Chapter 2
Theoretical Pore Growth Models for Nanoporous Alumina

Chuan Cheng and A.H.W. Ngan

Abstract Nanoporous alumina has been extensively used in a wide range of applications, including template materials for various types of nanomaterials, high surface-area structures for energy conversation and storage, bio/chemo sensors, electronic/photonic devices, and so on. However, the formation mechanism of the nanopores and the subsequent pore growth process towards self-ordered pore arrangements have been under investigation for several decades without clear conclusions. The present models may be divided into two main groups in terms of the driving force for pore initialization, as well as the subsequent pore growth process. One group considers that the driving force is the high electric field across the oxide barrier layer at the bottom of the pore channels, which assists metal oxidation at the metal/oxide interface, and oxide dissolution at the oxide/electrolyte interface. The other group of models assumes that the driving force is mechanical stress originating from the volume expansion of the metal oxidation process. This chapter reviews the development of these models for nanoporous alumina formation, and discusses their advantages and shortcomings. A recent model proposed by us is also described, and potential directions for further development are discussed.

2.1 Introduction of Nanoporous Alumina

Nanoporous alumina, also known as anodic aluminum oxide (AAO), has attracted extensive attention both experimentally and theoretically in the past several decades, due to the unique features such as self-ordered quasi-hexagonal nanoporous structures, relative ease to control the pore size and interpore distance by anodization conditions, extremely low cost, high thermal stability, and so on [1–21]. Nowadays, AAO has been commercialized and widely used as convenient templates for non-lithographic synthesis of various nanomaterials, including nanodots

C. Cheng · A.H.W. Ngan (✉)
Department of Mechanical Engineering, The University of Hong Kong, Hong Kong, People's Republic of China
e-mail: hwngan@hku.hk

D. Losic and A. Santos (eds.), *Nanoporous Alumina*,
Springer Series in Materials Science 219, DOI 10.1007/978-3-319-20334-8_2

[22–24], nanowires [25–29], nanotubes [30–32], and many others [33–35], for applications in high density magnetic media [36–41], photonic crystals [42–49], semiconductor devices [50–58], lithium-ion batteries [59–62], solar cells [63, 64], nanocapacitors [65–69], biosensors, [70–76] and so on [77–88].

Nanoporous structured AAO can be easily fabricated by anodization of aluminum in different kinds of electrolytes, such as sulfuric acid [89], oxalic acid [90–93], phosphoric acid [94], and chromic acid [94, 95]. However, for neutral electrolytes with pH in the range of 5–7, only barrier-type alumina thin film can be formed. Here, we only focus on the former porous-type anodic alumina (AAO). Figure 2.1 illustrates the configuration of self-ordered AAO, which consists of closely packed pore channels perpendicular to the Al substrate. The pore size (D_p) and interpore distance (D_{int}) can be varied from several to hundreds of nanometers mainly by changing of anodization voltages [57, 92–94, 96]. A thin scallop-shaped oxide barrier layer exists between the porous AAO layer and Al substrate. It has been demonstrated for decades that the barrier layer thickness D_b, D_p, and D_{int} have linear relationships with anodization voltage [2, 3, 94, 97–99]. The dependence may vary slightly with temperature and acid concentration. For example, under mild anodization (MA) conditions, the voltage dependence of D_p and D_b is about 1 nm V^{-1}, and that of D_{int} is 2.5 nm V^{-1} [93, 98]. Recently, Lee et al. [92]. demonstrated that under hard anodization (HA) conditions in which the oxide growth rate is tens of micrometers per hour, the voltage dependency becomes 0.4 nm V^{-1} for the D_p, 1 nm V^{-1} for D_b, and 2 nm V^{-1} for D_{int}. A slight nonlinear relationship between D_{int} and anodization voltage was recently reported by the authors under the high acid concentration and high temperature anodization (HHA) [90]. HHA can result in much better self-ordering AAO compared with MA in a voltage range from 30 to 60 V in oxalic acid based electrolyte, and reduce the

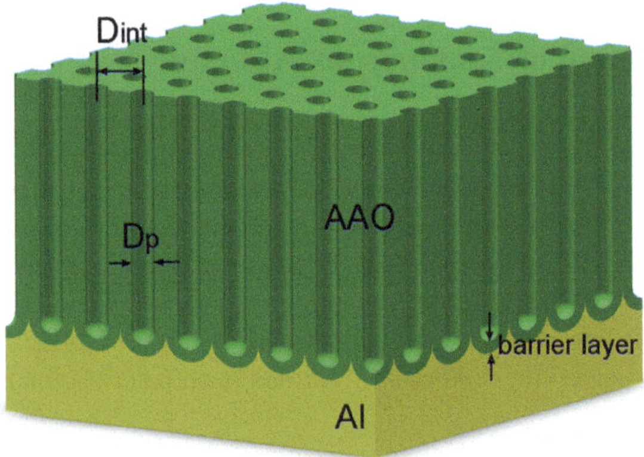

Fig. 2.1 Schematic of self-ordered anodic aluminum oxide (AAO) formed above Al substrate, with interpore distance D_{int} and pore size D_p. A scallop-shaped oxide barrier layer exits between the porous layer and the Al substrate

time needed to reach self organization of the pores from typically 2 days in MA to only 2–3 h in HHA.

From the top-view of self-ordered AAO, the pores are arranged in a quasi-hexagonal pattern. In real cases [100], the whole in-plane pattern usually exhibits local variations, with almost perfectly ordered zones separated by disordered zone boundaries. The average ordered zone size has been used as an effective factor to evaluate the ordering quality of AAO formed under different anodization conditions [90]. AAO structures with self-ordered in-plane patterns also usually possess straight pore channels in their cross-sectional view, while those with disordered in-plane patterns usually have branched channels, with frequent splitting, termination, or merging of the pore channels [57, 92–94, 96]. As a result, the aspect ratio of the pore channel, i.e. the channel length to pore diameter, can be greater than 1000 for self-ordered AAO [92, 101, 102], while the ratio may be less than 20 for disordered AAO [92, 103]. Because the in-plane porous patterns are just snap-shots during the growth of the AAO layer, it is the growth stability of pore channels during anodization which determines the self-ordering quality of AAO. In experiments, only under certain anodization conditions can self-ordered AAO with quasi-hexagonal in-plane porous patterns be fabricated, such as 25 V in 0.3 M H_2SO_4 at $\sim$0 °C with $D_{int} = 63$ nm [104], 40 V in 0.3 M $H_2C_2O_4$ at $\sim$0 °C with $D_{int} = 100$ nm [93], and 195 V in 0.3 M H_3PO_4 at $\sim$0 °C with $D_{int} = 500$ nm [105].

A general AAO growth process under constant anodization voltage condition is illustrated in Fig. 2.2, in which four typical stages in terms of AAO morphologies are involved. At the beginning stage of anodization, as shown in Fig. 2.2a, a thin and compact alumina film is quickly formed along the aluminum surface, resulting in blockage of the conductivity of the Al and a sharp decreasing of the current towards a minimum value. This process may only take several seconds to complete. Some roughness or concavity of the alumina film always exists due to the inhomogeneity of the aluminum surface, which may give rise to different oxide formation rates.

After that, as shown in Fig. 2.2b, a large amount of small pores are initiated from the concavities of the rough alumina thin film. From the calculation of electric potential distribution within alumina [106], the electric potential drop is concentrated within a concaved region. As a result, a much higher electric field intensity exists within a concavity compared with a flat region. This gives rise to a faster oxide growth rate, namely, a small concavity continues penetrating into the aluminum substrate and develops into to a pore channel. These small pores, as illustrated in Fig. 2.2b, are (i) randomly distributed; (ii) have no geometric relationship with anodization voltage (e.g. D_{int}/voltage ratio, D_p/voltage ratio); (iii) have their growth frequently terminated just at the surface region of the AAO. During the ignition of the pores, ion transportation takes place across the alumina film, mainly focusing at the pore bottom region in which the electric field is high enough to assist the ion migration. Thus, the previously blocked sample surface has ions passing through with give rise to an increase of the anodization current in Fig. 2.2e.

With anodization time increasing, as shown in Fig. 2.2c, some initial pores, which have larger depths into the Al substrate, will grow faster than their

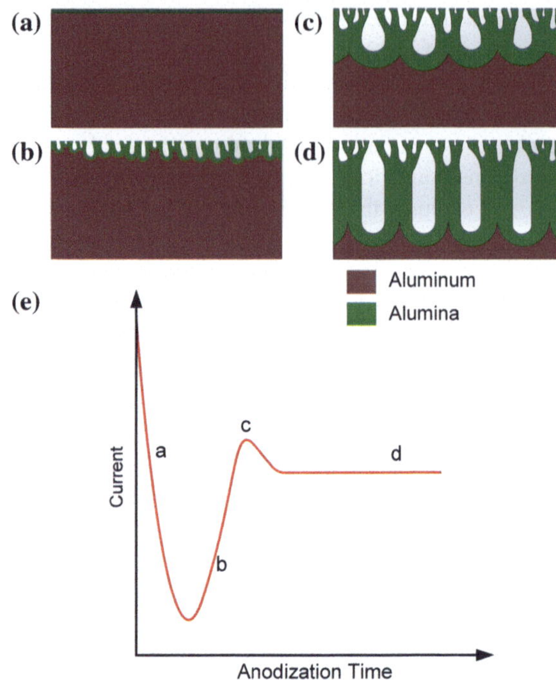

Fig. 2.2 Schematic of anodic aluminum oxide (AAO) growth process under constant voltage condition. **a–d** Morphology development with anodization time increasing; **e** a typical current against anodization time relationship, with the corresponding morphologies marked along the curve

neighboring pores. At the same time, they will expand in the horizontal, in-plane direction. As a result, water-drop shaped pore channels are developed, which have smaller pore mouths and larger pore bottoms, as shown in Fig. 2.2c. In this stage, the development of the porous structure is driven by the electric field, which continuously adjusts the barrier layer thickness in order to reach a certain electric field distribution within it. This electric field assists the ion migration across the potential barriers, in order to realize ion migration within oxide and the electrochemical reactions at the interfaces. In experiments, different anodization voltages are found to lead to different barrier layer thickness but the electric field distribution is very similar, because the driving force (electric field) for ion migration within the alumina is intrinsic. Thus, the anodization voltage directly controls the geometry of the barrier layer at the bottom of the pores. With the pore bottom penetrating into the Al substrate, a pore channel is left behind, thus the electric field indirectly controls the geometry of the pore channels with time increasing. With a water-drop shaped pore bottom in Fig. 2.2c, geometrically, the contact area between the barrier layer and aluminum substrate reaches a maximum compared with the cases in Fig. 2.2b, d. As a result, the ion current, which mainly passes through the barrier layer, reaches a maximum value, as noted in Fig. 2.2e.

After the stage of Fig. 2.2c, walls of the pores elongated along the growth direction are formed and these have much reduced electric field intensity inside them due to their length in growth direction. Thus, the pore growth is mainly concentrated at the bottom of the pores towards the substrate, and as a result, the water-drop shape of the pores is developed into a U-shape, and accordingly the current slightly decreases, as shown in Fig. 2.2e. After that, the geometry of the pore bottom tends to become stable, as a result the current reaches a steady-state value. However, this does not necessarily mean that the configuration of the pore bottom is fixed during this stage. Self origination of pores continuously takes place in terms of pore combination, splitting, and termination. The self-origination process first happens at the pore bottom and is then manifested by the trailing pore channels. It is during this stage that a disordered porous arrangement gradually develops into a much better self-ordering porous arrangement, which may take hours or days of anodization time. Thus, the self-ordering process takes place during the pore channel growth, after pores have already been initiated.

2.2 Review of Pore Growth Models

The mechanism of pore growth in AAO has been continuously investigated for decades, and currently it is still under debate [1, 95, 107–127]. In terms of the driving force for pore growth as well as the self-organization of pores towards ordering, the previous models may be divided into two types. One regards electric field as the driving force [2, 3, 94], while the other regards mechanical stress as the driving force [120, 124, 126].

2.2.1 Electric Field Assisted Pore Growth

Hoar and Mott first proposed that the formation and growth of pores in AAO was assisted by electric field [108]. They suggested that under the high electric field on the order of 1 V nm^{-1} across the oxide barrier layer, O^{2-} ions would be driven from the oxide/electrolyte interface to the metal/oxide interface for Al oxidation, while Al^{3+} ions would be driven by the electric field in the opposite direction, across the barrier layer and then ejected into the electrolyte. Ion migration in the barrier layer was proposed to take place by means of jumping from one interstitial position to another following the Cabrera-Mott equation [128]. They emphasized that space charge should not be considered within the oxide, because the process of ion migration was comparatively easy.

O'Sullivan and Wood supported the idea that electric field assisted dissolution was the reason for pore formation and growth in AAO [94]. They proposed that the thickness of oxide barrier layer was the result of a competition between oxidation and dissolution reactions at the pore bottom. The high electric field could stretch or break the Al-O bonds, thus aiding the dissolution of oxide and resulting in a faster rate than open-circuit chemical dissolution [2, 3, 94].

Nagayama and Tamura [129] demonstrated that during anodization the dissolution rate of the pore bottom was 1.04×10^{-4} cm min^{-1} under 11.9 V and 9.4 mA cm^{-2}, and this was about 10^4 times faster than the rate of 7.5×10^{-9} cm min^{-1} for the dissolution of the pores' inner surfaces, which can therefore be regarded as solely chemical dissolution. Moreover, by calculating the temperature distribution along the vertical pores, they demonstrated that the temperature rise at the pore bottom was always negligibly small at ~0.06 °C [130]. Also, Mason [131] and Li [132, 133] found that the temperature rise at the pore bottom was higher at about 25 °C, but even with this magnitude of temperature rise, the associated Joule heat would still be far insufficient to result in the observed high growth rate at the pore base. In fact Hunter and Fowle [134, 135] demonstrated that the electrolyte would have to reach boiling temperature in order for such fast growth to occur. Thus, the contribution of heat assisted dissolution of oxide should only play a minor role on pore growth during anodization [130].

Several models in different mathematical forms have been developed by regarding the electric field as the driving force for pore growth [113–117]. Parkhutik and Shershulsky proposed a model in which the electric potential in AAO obeyed Laplace equation by neglecting space charge within oxide and the electrochemical double layer along the oxide/electrolyte and metal/oxide interfaces [113]. They proposed that the movement of oxide/electrolyte interface was due to the competition between oxide formation and electric field assisted dissolution, while the metal/oxide interface should move accordingly with the oxide/electrolyte interface. This model predicts a linear relationship between D_{int} and anodization voltage, which was typically observed in experiments.

By using a linear and weakly nonlinear stability analysis, Thamida and Chang further developed the above model and predicated a critical pH value of 1.77 for the transition from barrier-type to porous-type anodic alumina [114]. Moreover, Singh et al. [115, 116] proposed a similar model by considering two situations: a long-wave instability resulting from electric field assisted dissolution, and a stabilizing effect of the Laplace pressure due to surface energy which provides a wavelength selection mechanism. They predicted that when the elastic stress in the oxide layer was significant, self-ordered pore arrays can be formed [116].

However, Thamida et al.'s, Parkhutik and Shershulsky's, and Singh et al.'s models were challenged by Friedman et al. [99]. For example, in their experiments D_{int} was independent of the electrolyte pH at constant anodization voltage [99], whereas the former two models predict that D_{int} (in nm) should vary with the pH according to $2.96V_0/(2.31–1.19\ \mathrm{pH})$, where V_0 is the constant anodization voltage. Although some of the predictions of these models do not agree with the experimental observations by Friedman et al. [99], in some aspects [98, 99], this does not necessarily mean that electric field is not the driving force for AAO growth and self-ordering, because previous models may not reflect the nature of electric-field assisted process correctly. Most recently, van Overmeere et al. [127] performed an energy-based perturbation analysis for pore growth in AAO, and they concluded that the electrostatic energy, rather than the mechanical strain energy-induced surface instability, was the main driving force for pore initiation as well as a controlling factor for pore spacing selection.

Furthermore, little efforts have been made to quantitatively investigate the electric field behavior by means of numerical simulation. For instance, in 2006, Houser and Hebert first numerically calculated the static electric potential distribution within AAO by considering the Laplace equation, as well as the Poisson equation in which the amount of space charge was artificially assumed to correlate with the distance from the pore axis [122]. To our knowledge, no real-time evolution of the pore growth process during anodization has been simulated before our recent reports [106, 136].

2.2.2 *Mechanical Stress Assisted Pore Growth*

During oxidation reaction at the oxide barrier layer of AAO, significant volume expansion may take place at the oxide/metal interface. For instance, under 100 % current efficiency without aby loss of Al^{3+}, the volume expansion ratio, namely, the Pilling-Bedworth ratio [133, 137, 138], can reach 1.64. However, it should be noted that direct loss of Al^{3+} ions (without oxide formation) can take place during anodization [139], which provides some spacing for the expanded volume and relief of the compression stress. For instance the measured ratio of the formed oxide thickness to the consumed aluminum thickness was 1.35, which still indicates a large extent of volume expansion [140]. Thus, the mechanical compression stress due to volume expansion was recently regarded as the driving force for pore growth in AAO, which was proposed to result in the plastic flow of oxide from the bottom to the walls of the pores [120, 124, 126]. In particular, recent tungsten (W) tracer experiments seemed to support the idea that the AAO formed in phosphoric acid [141, 142] and sulphuric acid [143] was due to the plastic flow of the oxide. According to their explanation, if pores were formed by electric-field assisted dissolution, W tracers at the pore base would migrate ahead of the W tracers at pore walls, but they found the inverse [141–143]. However, for the AAO formed in chromic acid [144] and alkaline borate electrolyte [145, 146], the W tracer distribution in the oxide was relatively uniform which suggests that the pore formation is due to electric-field assisted dissolution of oxide [144–146]. It is not clear why the pore formation mechanism was different in different types of electrolyte, since the electrolyte species were found not to participate in the oxide formation reaction [142].

Furthermore, by using the same electrolyte of phosphoric acid [141, 142] but different tracers [147, 148] of Nd and Hf, the tracer migration distribution within the oxide was found to be the reverse of that previously found for W tracer [141–143], and this would indicate electric-field assisted dissolution as the pore formation mechanism, just as the previously found W distribution would indicate oxide flow as the mechanism. The unexpected Nd and Hf tracer distribution was attributed to the faster migration rate of the tracer atoms compared with that of

Al ions [147, 148]. However, as noticed by Oh [135], a tracer study alone cannot yield sufficient evidence to prove oxide flow or disprove electric-field assisted dissolution as the mechanism for pore formation.

For the stress-driven self-ordering of AAO, Jessensky et al. [120, 149] proposed that repulsive forces between neighboring pores of AAO can arise during anodization due to the volume expansion. A moderate anodization voltage was found to result in a moderate magnitude of the current efficiency as well as volume expansion ratio, and only under these moderate conditions can self-ordered AAO be obtained. However, an important question is whether it is the moderate electric voltage (related to electrostatic energy) or the moderate volume expansion ratio (related to mechanical energy) which really causes the ordering in the porous structure. These two factors cannot be separated in their experiments, and so sufficient evidence has not been established to support that the main reason for self-ordering in AAO is due to the mechanical stress.

Recently, Houser and Hebert [125, 126] proposed a mathematical model for the steady state growth of AAO, in which the Al^{3+} and O^{2-} ions are transported by electrical migration and viscous flow. A good agreement between their calculation results and W tracer experimental results was obtained [141–143]. However, as discussed by Oh [135], a close examination of the boundary conditions used in this oxide flow model would show that the new oxide would be generated at the oxide/electrolyte interface (by the so-called oxygen deposition), which was inconsistent with the observation from O^{18} tracer experiments that the new oxide was only found at the metal/oxide interface [139].

2.3 A Kinetics Model for Pore Channel Growth in Nanoporous Alumina

Recently, we have developed a kinetics model for pore-channel growth as well as self-organization towards ordering in AAO [106]. This model is a further development of previous models by regarding electric field as the driving force [113–116]. The Laplacian electric potential distribution and continuity of the current density within the oxide body are considered. Both oxygen and aluminum ion current densities governed by the Cabrera-Mott equation in high-electric-field theory are formed by ion migration within the oxide as well as across the oxide/electrolyte and metal/oxide interfaces. The movements of oxide/electrolyte and metal/oxide interfaces are due to electric-field assisted oxide decomposition and metal oxidation, respectively, as governed by Faraday's law. This model has been numerically implemented by a finite element method in order to simulate the real-time evolution of the porous structure growth in two-dimensional cases corresponding to the cross-sectional view of pore channels [106, 136].

2.3.1 Electric Potential Distribution Within AAO

As has been reported by Houser and Hebert [122], during anodization, space charge within the anodic oxide may significantly influence the electric field distribution there. Although space charge was considered by Dewald [150, 151] to successfully explain the experimentally observed temperature-independent Tafel slope in the formation of barrier-type anodic tantalum oxide, Vermilyea [152] found that Dewald's consideration was unable to explain the experimental fact that the average electric field is independent of the thickness of the anodic oxide film. Thus, whether space charge should be considered during anodization still needs further investigations, and here, following Parkhutik and Shershulsky [113], Thamida and Chang [114], and Singh et al. [115, 116], we neglect space charge within the oxide. Thus, the electric potential φ within the oxide obeys the Laplace Equation:

$$\nabla^2 \varphi = 0. \tag{2.1}$$

According to Houser and Hebert [122], the potential at the oxide/electrolyte interface (typically <0.1 V) is far smaller than the anodization voltage, and so in the present model, the potential there is set to be zero. In addition, as most of the potential drop happens within the oxide body but not in the metal substrate or in the electrolyte, the potential at the metal/oxide interface is set to be the same as the anodization voltage V_0. In this chapter, we only investigate anodization under constant voltage conditions. Moreover, along the right and left edges of a simulation sample (e.g. the vertical dash dotted lines in Fig. 2.3), the Neumann boundary condition is used. Thus, the boundary conditions are summarized as

$$\begin{cases} \varphi = 0, & \text{at metal/oxide interface,} \quad (2.2) \\ \varphi = V_0, & \text{at oxide/electrolyte interface,} \quad (2.3) \\ \mathbf{n} \cdot \nabla \varphi = 0, & \text{at both edges of the sample,} \quad (2.4) \end{cases}$$

where **n** is the outward normal unit vector for each sample edge. The electric field is given as

$$\mathbf{E} = -\nabla \varphi. \tag{2.5}$$

The continuity requirement of the steady-state ion-current density **j** within the oxide bulk can be expressed as follows [113–116]

$$\nabla \cdot \mathbf{j} = 0. \tag{2.6}$$

From the above equations, we can derive the relationship between the electric field and current density along the electric-field lines across the oxide barrier layer, which will be used later. Electric-field lines are always perpendicular to equipotential contours. Consider a very small cylinder with volume V_c ($V_c \rightarrow 0$), which starts from the metal/oxide interface to the oxide/electrolyte interface along an

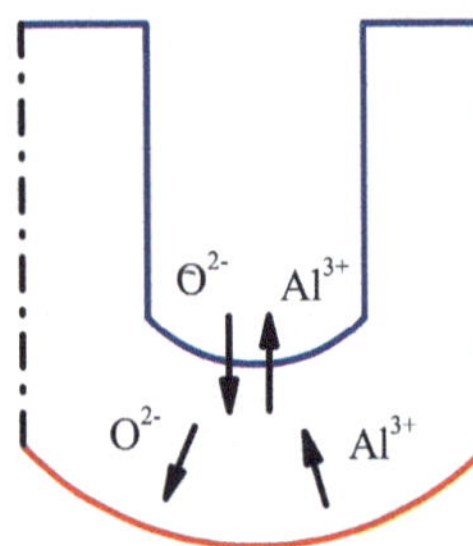

Reactions at the m/o Interface:	**Curent densities**
(R1) $Al_{(m)} \rightarrow Al^{3+}_{(ox)} + 3e^-$	Producing $\mathbf{j}_{Al,ox}\|_{m/o}$ at m/o interface, and becomes $\mathbf{j}_{Al,ox}\|_{o/e}$ at o/e interface.
(R4) $2Al_{(m)} + 3O^{2-}_{(ox)} \rightarrow Al_2O_{3\,(ox)} + 6e^-$	Supplied by $\mathbf{j}_{O,ox}\|_{o/e} = \mathbf{j}_{O,o/e} + \mathbf{j}_{O,dis}$ at o/e interface and becomes $\mathbf{j}_{O,ox}\|_{m/o}$ at m/o interface.
Reactions at the o/e Interface:	
(R2) $Al^{3+}_{(ox)} \rightarrow Al^{3+}_{(aq)}$	Supplied by $\mathbf{j}_{Al,ox}\|_{m/o}$ at m/o interface and becomes $\mathbf{j}_{Al,ox}\|_{o/e}$ at o/e interface , together with $\mathbf{j}_{Al,dis}$ from oxide decomposition at o/e interface; total value for Al^{3+} across o/e interface is $\mathbf{j}_{Al,o/e} = \mathbf{j}_{Al,ox}\|_{o/e} + \mathbf{j}_{Al,dis}$.
(R3) $Al_2O_{3\,(ox)} \rightarrow 2Al^{3+}_{(aq)} + 3O^{2-}_{(ox)}$	Producing $\mathbf{j}_{Al,dis} = \mathbf{j}_{O,dis}$, in which Al^{3+} ejected into electrolyte; O^{2-} migrating towards m/o interface.
(R5) $H_2O_{(aq)} \rightarrow 2H^+_{(aq)} + O^{2-}_{(ox)}$	Producing $\mathbf{j}_{O,o/e}$ at o/e interface, in which O^{2-} migrating towards m/o interface.

Ion Current Density Across the o/e Interface:

Al^{3+} ions ejected into electrolyte: $\mathbf{j}_{Al,o/e} = \mathbf{j}_{Al,ox}|_{o/e} + \mathbf{j}_{Al,dis}$ due to R1, R2 and R3.

O^{2-} ions supplied by electrolyte : $\mathbf{j}_{O,o/e} = \mathbf{j}_{O,ox}|_{o/e} - \mathbf{j}_{O,dis}$ due to R3, R4 and R5.

Fig. 2.3 Summary of the reactions assumed during AAO growth. o/e means oxide/electrolyte interface, and m/o means metal/oxide interface (Adapted from [106])

electric-field line across the oxide barrier layer. The top and bottom surfaces of the cylinder are elements of the oxide/electrolyte and metal/oxide interfaces with areas represented as $S_{o/e}$ and $S_{m/o}$, respectively. $S_{o/e}$ and $S_{m/o}$ are not equal because of the scalloped shape of barrier layer. The side surface S_{side} of the cylinder is along the electric-field line, so that its outward normal vector is perpendicular to the electric-field line. From (2.1) and (2.5), we get $\nabla \cdot \mathbf{E} = 0$, and with Gauss' Theorem $\oiiint_{V_c} (\nabla \cdot \mathrm{E})\, dV = \oiint_{S_c} (\mathbf{E} \cdot \mathbf{n}) dS$, we have

$$\iint_{S_{o/e}} (\mathbf{E} \cdot \mathbf{n}) dS_{o/e} + \iint_{S_{m/o}} (\mathbf{E} \cdot \mathbf{n}) dS_{m/o} + \iint_{S_{side}} (\mathbf{E} \cdot \mathbf{n}) dS_{side} = 0. \tag{2.7}$$

Since $\mathbf{E} \cdot \mathbf{n} = 0$ over S_{side}, $\mathbf{E} = \mathbf{n} E_{o/e}$ over $S_{o/e}$, and $\mathbf{E} = -\mathbf{n} E_{m/o}$ over $S_{m/o}$, where $E_{o/e}$ and $E_{m/o}$ are the electric field intensities at $S_{o/e}$ and $S_{m/o}$, respectively, and as $S_{o/e}$ and $S_{m/o}$ both tend to zero, (2.7) becomes

$$E_{o/e} S_{o/e} = E_{m/o} S_{m/o}, \tag{2.8}$$

where $S_{o/e}$ and $S_{m/o}$ are connected by the same electric-field line. By virtue of (2.6) which is of the same form as $\nabla \cdot \mathbf{E} = 0$, the above procedure can be repeated for $\mathbf{j}$ to give

$$j_{o/e}S_{o/e} = j_{m/o}S_{m/o}, \tag{2.9}$$

where $j_{o/e}$ and $j_{m/o}$ are the current density magnitudes at $S_{o/e}$ and $S_{m/o}$ respectively. From (2.8) and (2.9), we obtain

$$\frac{j_{o/e}}{j_{m/o}} = \frac{E_{o/e}}{E_{m/o}}. \tag{2.10}$$

The same derivation process actually holds for any point within the oxide bulk with electric-field intensity E_{bulk} and current density j_{bulk},

$$\frac{j_*}{j_{bulk}} = \frac{E_*}{E_{bulk}}, \tag{2.11}$$

where the subscript "$*$" represents either "o/e" or "m/o", and the oxide bulk point and the o/e (oxide/electrolyte) or m/o (metal/oxide) interface point should be connected by the same electric-field line. Equation (2.10) was first cited before by Parkhutik and Shershulsky [113], without proof, and its significance, together with that of (2.11), is as follows. For a given porous structure of AAO, the electric-field intensities can be solved directly from (2.1–2.5). After that, regardless of whether the rate-determining step of the anodization process is at the oxide/electrolyte interface, oxidc bulk, or metal/oxide interface, if we can calculate the current density at one location, e.g. the oxide/electrolyte interface, we can obtain the current density at other locations by using (2.10) and (2.11) directly. The location at which the current density is first evaluated may not necessarily be the location at which the rate-determining step occurs, but the calculated current density will be controlled by the rate-determining step through (2.11). Here, we assume that ion migration across the oxide/electrolyte interface is the rate-determining step, because the oxygen and aluminum ions are weakly bound under the effect of the high electric field [94]. It should be noted that ionic migration in the bulk oxide has been proposed previously as an alternative rate-determining step [153], but recent experiments revealed that an increase in the acid concentration of the electrolyte, which should play a role directly at the oxide/electrolyte interface, can influence the anodization process significantly, such as increasing the pore diameter [98], the current density [154], and the oxide growth rate [99]. These profound changes of the anodization process should be due to changes in the anodization conditions at the oxide/electrolyte interface, and this is the basis of the present assumption that the rate-determining step is at this interface. In the following, the current density at the oxide/electrolyte interface is derived at first, and then the current density at the metal/oxide interface is obtained from (2.10). Based on these, the interface movement equations are established from Faraday's Law.

2.3.2 *Ion Migration*

In AAO formation, Cherki and Siejka's oxygen transport study using nuclear microanalyses of O^{18} and O^{16} concluded that new oxide forms only at the metal/oxide interface but not elsewhere [139]. Also, Davies et al. [155, 156]. found that Xe^{125} and Rn^{222} tracer distributions in barrier-type anodic alumina films did not tend to broaden. These experimental observations imply that the oxidation reaction within the oxide body is negligible. On this basis we assume that the cations and anions migrating from one interface to another interface are not consumed on their way. On the oxide/electrolyte and metal/oxide interfaces, as shown in Fig. 2.3, the most possible reactions based on previous experimental observations are described in the following subsections.

2.3.2.1 Aluminum Ion Migration

Direct ejection of aluminum ions from the metal/oxide interface into the electrolyte has been indicated in many experiments, such as coating-ratio measurement and tracer experiments. The coating ratio, defined as the weight of the oxide formed to the weight of aluminum consumed in the anodization process, will be 1.89 if all of the consumed aluminum is converted into alumina, or higher if acid anions contaminate the anodic oxide, e.g. 2.2 if 14 % SO_3 contamination exists in the oxide [1]. After considering the porosity (around 10 %) [157] of the oxide due to its dissolution in the electrolyte, the coating ratio should be about 1.7 (or 1.98 if 14 % SO_3 contamination exists). However, experimentally observed values of the coating ratio are always lower. For example, Edwards and Keller [158] found that the coating ratio was smaller than about 1.46. Spooner [159] attempted to obtain a high coating ratio by increasing the current density and decreasing the dissolution rate in sulfuric acid (with SO_3 contamination in the oxide), but only 1.68 was obtained, and under other conditions the coating ratio was lower than 1.61. These imply that Al must be lost by another way beside the loss due to pore growth at the pore base assisted by the high electric field there, and this cannot be oxide dissolution loss at the pore walls or top surface, as these dissolution rates were found to be far smaller on the order of 10^{-8} cm min^{-1}, [129, 139] compared with the dissolution rate of $\sim 10^{-4}$ cm/min at the pore base [129]. A similar conclusion was reached by Cherki and Siejka from their O^{18} tracer experiments [139], which indicated that direct ejection of Al cations in the solution without formation of any oxide should take place. Recent experiments by Wu et al. [160] also support the net ejection of Al^{3+} cations across the barrier layer into the electrolyte. Thus, it is reasonable to assume that Al^{3+} cations formed at the metal/oxide interface via the reaction

$$Al_{(m)} \rightarrow Al^{3+}_{(ox)} + 3e^{-}, \qquad \text{(R1)}$$

would reach the oxide/electrolyte interface under the drive of the high electric field, and are finally ejected into the solution by the reaction

$$Al^{3+}_{(ox)} \rightarrow Al^{3+}_{(aq)}, \tag{R2}$$

without oxide formation. R1 is the only source of the aluminum ions migrating through the oxide body from the metal/oxide to oxide/electrolyte interface during the anodization process, and the density of such current is denoted as $\mathbf{j}_{Al,ox}$, where "ox" means that the corresponding aluminum ions migrate through the oxide body. The value of $\mathbf{j}_{Al,ox}$ at the metal/oxide interface is denoted as $\mathbf{j}_{Al,ox}|_{m/o}$, while that at the oxide/electrolyte interface is denoted as $\mathbf{j}_{Al,ox}|_{o/e}$.

For AAO formation, we must note that those aluminum ions which have traveled across the oxide barrier layer do not react to form new oxide at the oxide/electrolyte interface, because new oxide was found to form only at the metal/oxide interface but not at the oxide/electrolyte interface [139]. The situation in barrier-type (i.e. nonporous-type) anodic alumina film formation is, however, different, since new oxide was found to form at both the metal/oxide and oxide/electrolyte interfaces [155, 161]. In other words, a net aluminum current passes through the oxide barrier layer in both cases of barrier-type and porous-type alumina formation, but whether the aluminum ions reaching the oxide/electrolyte interface can form new oxide there would determine the type of alumina finally formed. We surmise that the acid concentration or the pH of the electrolyte would determine the fate of the aluminum ions migrated to the oxide/electrolyte interface, and in the model development below, this effect will be incorporated (see (2.15) later).

In addition to the direct ejection of aluminum ions, dissolution of the old oxide to form pores should take place at the oxide/electrolyte interface, which is thought to be also electric-field assisted, because of the extremely fast dissolution rate at the pore base ($\sim 10^{-4}$ cm min^{-1}) compared with the rate at the pore walls ($\sim 10^{-8}$ cm min^{-1}) as found in experiments [129, 139]. Such a great difference in the dissolution rates should be mainly due to the large difference in electric-field intensities between these two locations. Furthermore, Siejka and Ortega's O^{18} tracer experiments [162] showed that oxygen loss during pore formation is negligible, which would contradict the dissolution reaction $Al_2O_{3(ox)} + 6H^+_{(aq)} \rightarrow 2Al^{3+}_{(aq)} + 3H_2O_{(aq)}$ assumed in some previous studies [114, 133], since this reaction would involve the loss of oxygen from the oxide into the electrolyte. Instead, the old oxide at the pore base is likely to be consumed by the following decomposition reaction: [162]

$$Al_2O_{3(ox)} \rightarrow 2Al^{3+}_{(aq)} + 3O^{2-}_{(ox)}, \tag{R3}$$

in which the product oxygen remains in the oxide body, and is then driven by the high electric field to reach the metal/oxide interface to form new oxide there. Thus, the so-called "electric-field assisted dissolution of oxide" referred to by some previous researchers [2, 94, 110] is interpreted here as the electric-field assisted decomposition of oxide at the oxide/electrolyte interface. Let $\mathbf{j}_{O,dis}$ and $\mathbf{j}_{Al,dis}$ denote the oxygen ion and aluminum ion current density due to R3 at the oxide/electrolyte interface. Their values are equal, but the corresponding ion movements are in opposite directions, i.e.

$$\mathbf{j}_{Al,dis} = \mathbf{j}_{O,dis}. \tag{2.12}$$

The experimentally established electric-field assisted ejection of aluminum ions into the electrolyte, the current density of which is denoted as $\mathbf{j}_{Al,o/e}$ hereafter, is contributed by aluminum ions produced by oxide decomposition at the o/e interface (of current density $\mathbf{j}_{Al,dis}$ in R3), as well as ions migrated from the metal/oxide interface (of current density $\mathbf{j}_{Al,ox}|_{o/e}$ in R2), i.e.

$$\mathbf{j}_{Al,o/e} = \mathbf{j}_{Al,ox}\big|_{o/e} + \mathbf{j}_{Al,dis}. \tag{2.13}$$

At the oxide/electrolyte interface, although the aluminum ions ejected into the electrolyte come from two sources, the actual ejection process which is reaction R2 has no difference from an electrolyte point of view. Physically, this process is governed by the high-field theory [4, 128, 163] in which the aluminum ions are assumed to jump across a potential barrier W_{Al} at the oxide/electrolyte interface, the effective value of which is reduced by an amount $\alpha_{Al} a_{Al} q_{Al} E_{o/e}$ in the jumping direction along the electric field $E_{o/e}$, and increased by $(1 - \alpha_{Al}) a_{Al} q_{Al} E_{o/e}$ in the opposite direction. Thus, the $\mathbf{j}_{Al,o/e}$ can be expressed as the Cabrera-Mott equation [128],

$$\mathbf{j}_{Al,o/e} = \left\{ n_{Al} q_{Al} \nu_{Al} \exp\left(-\frac{W_{Al} - \alpha_{Al} q_{Al} a_{Al} E_{o/e}}{kT} \right) - n_{Al} q_{Al} \nu_{Al} \exp\left[-\frac{W_{Al} + (1 - \alpha_{Al}) q_{Al} a_{Al} E_{o/e}}{kT} \right] \right\} \hat{\mathbf{E}}_{o/e} \tag{2.14}$$

where n_{Al} is the surface density of mobile aluminum ions at the oxide/electrolyte interface which is dependent on electric intensity [4], q_{Al} is the charge of one aluminum ion, ν_{Al} is the vibration frequency of aluminum ions, α_{Al} is a transfer coefficient related to the symmetry of the potential barrier (e.g. if the potential barrier is symmetrical, then $\alpha_{Al} = 0.5$), a_{Al} is the jump distance (twice the activation distance) of aluminum ions, $\mathbf{E}_{o/e}$ is the electric field at oxide/electrolyte interface, $E_{o/e} = |\mathbf{E}_{o/e}|$ is the electric field intensity, $\hat{\mathbf{E}}_{o/e}$ is the unit vector $\mathbf{E}_{o/e}/E_{o/e}$, k is the Boltzmann constant, and T is the absolute temperature. The Cabrera-Mott equation above contains terms that describe jumps in both the forward and backward directions, but in practice, the backward current density (the second term in (2.14)) is far smaller than the forward one (the first term in (2.14)) [163], and so to save computation time only the forward current was considered in the present numerical simulations. Furthermore, following Diggle [110] and Vermilyea [164], to describe the fact that the dissolution process is strongly influenced by the acid concentration C_{H^+}, the current density is scaled by the factor $(C_{H^+})^{\eta}$, where $\eta = \alpha/\varsigma \in [0, 1]$ is the ratio of the number of protons α involved in the dissolution process to the stoichiometric number ς appropriate to the dissolution mechanism [110]. This power term $(C_{H^+})^{\eta}$ was also used in previous reports [113–116]. Diggle [110] stated that only the current of the ion species involved in the rate determining process should be scaled by $(C_{H^+})^{\eta}$, and here, we believe that aluminum ions rather than oxygen ions are more likely the rate controlling species, since, as discussed above, aluminum ions need to jump across a high potential barrier at the oxide/electrolyte interface to enter the electrolyte, while oxygen ions migrate within the oxide body towards the metal/oxide interface, and such migration can take place along some easy paths such as microchannels [139, 165, 166] or by vacancy motion [162]. Thus, after neglecting the backward current density and scaling the current density by the acid concentration in (2.14), the total aluminum ion current which goes into the electrolyte is given as

$$\mathbf{j}_{Al,o/e} = n_{Al} A_{Al} \exp(k_{Al} E_{o/e}) \hat{\mathbf{E}}_{o/e}, \tag{2.15}$$

where $A_{Al} = C_{H^+}^{\eta} q_{Al} v_{Al} \exp\left(-W_{Al}/kT\right)$ and $k_{Al} = \alpha_{Al} q_{Al} a_{Al}/kT$.

2.3.2.2 Oxygen Ion Migration

According to Cherki and Siejka's oxygen transport study [139], new oxide is only formed at the metal/oxide interface, but not at the electrolyte/barrier layer interface or at the outer surface of the porous film. This means that O^{2-} ions have to migrate from the oxide/electrolyte interface to the metal/oxide interface across the barrier layer under the high electric field. Once the oxygen ions reach the metal/oxide interface, the following reaction may take place:

$$2Al_{(m)} + 3O^{2-}_{(ox)} \rightarrow Al_2O_{3(ox)} + 6e^-. \tag{R4}$$

R4 accounts for the entire migration of oxygen ions through the oxide body, the current density of which is denoted as $\mathbf{j}_{O,ox}$, where "ox" again means that the current goes through the oxide body, and the local value of $\mathbf{j}_{O,ox}$ at the metal/oxide interface is denoted as $\mathbf{j}_{O,ox}|_{m/o}$, while that at the oxide/electrolyte interface is denoted as $\mathbf{j}_{O,ox}|_{o/e}$. In turn $\mathbf{j}_{O,ox}|_{o/e}$ is contributed by two sources of oxygen ions: one is from water decomposition at the oxide/electrolyte interface [167]

$$H_2O_{(aq)} \rightarrow 2H^+_{(aq)} + O^{2-}_{(ox)}. \tag{R5}$$

the current density of which is denoted as $\mathbf{j}_{O,o/e}$, and the other source is from decomposition of old oxide at the oxide/electrolyte interface by reaction R3, the current density of which is $\mathbf{j}_{O,dis}$ which is equal to $\mathbf{j}_{Al,dis}$ (2.12). Thus,

$$\mathbf{j}_{O,ox}\big|_{o/e} = \mathbf{j}_{O,o/e} + \mathbf{j}_{O,dis}. \tag{2.16}$$

As stated before, after oxide decomposition according to R3, the product aluminum ions will jump across the oxide/electrolyte interface to enter the electrolyte, while the oxygen ions will not cross that potential barrier but will migrate towards the metal/oxide interface by some easy paths. Thus, only those oxygen ions coming from water decomposition (with current density $\mathbf{j}_{O,o/e}$ from R5) need to jump across the potential barrier at the oxide/electrolyte interface, and this current density should also follow the Cabrera-Mott equation [128]. By neglecting the backward current density which is small, $\mathbf{j}_{O,o/e}$ is given as

$$\mathbf{j}_{O,o/e} = n_O A_O \exp(k_O E_{o/e}) \hat{\mathbf{E}}_{o/e}, \tag{2.17}$$

where $A_O = q_O v_O \exp(-W_O/kT)$ and $k_O = \alpha_O q_O a_O/kT$, and the parameters in these expressions have similar meanings as in (2.14) albeit now for oxygen ions. From (2.15) and (2.16), the total ion current density which will go across the oxide/electrolyte interface is

$$\mathbf{j}_{total,o/e} = \mathbf{j}_{Al,o/e} + \mathbf{j}_{O,o/e} = \left[n_{Al} A_{Al} \exp\left(k_{Al} E_{o/e}\right) + n_O A_O \exp\left(k_O E_{o/e}\right)\right] \hat{\mathbf{E}}_{o/e}. \tag{2.18}$$

It should be emphasized that in the present model, the oxide body is assumed to be the channel for ion migration, and ions are assumed not able to accumulate or be neutralized. As mentioned above, we also assume that the jump of ions across the oxide/electrolyte interface is the rate determining step for their entire migration across the oxide body, where the oxygen and aluminum ions are weakly bound under the effect of the high electric field, in accordance with O'Sullivan and Wood's electric-field assisted dissolution theory [94].

2.3.2.3 Relationship Between Aluminum Ion Current Density and Oxygen Ion Current Density Within the Oxide Body

According to the discussion in Sects. 2.3.2.1 and 2.3.2.2, continuous growth of porous alumina depends on the outward migration of aluminum ions (with current density $\mathbf{j}_{Al,ox}$) and inward migration of oxygen ions (with current density $\mathbf{j}_{O,ox}$) across the oxide barrier layer. We propose that these two current densities should have a fixed relationship because of the following reason. During anodization, many experiments have proven that the metal substrate and the oxide barrier layer are in good contact with each other [94], although the theoretical volume expansion ratio (the Pilling-Bedworth ratio) [133, 137, 138] equals to 1.7 at the metal/oxide interface. This implies that the oxygen ions must be provided with enough spaces at the metal/oxide interface to form new oxide without influencing the close contact between metal and oxide. These spaces can only be due to the ejected aluminum ions from the metal/oxide interface which will migrate across the oxide barrier layer. As the volume expansion accompanying the oxidation reaction at the metal/oxide interface is fixed under a certain anodization condition, the required spaces to accommodate such volume expansion for maintaining good metal-oxide contact is then fixed, and so the ratio between the outward amount of aluminum ion current density and the inward amount of oxygen ion current density,

$$\beta = \frac{j_{Al,ox}\big|_{m/o}}{j_{O,ox}\big|_{m/o}} = \frac{j_{Al,ox}\big|_{o/e}}{j_{O,ox}\big|_{o/e}}, \tag{2.19}$$

should also be fixed during anodization, where $j_{Al,ox}|_{m/o} = |\mathbf{j}_{Al,ox}|_{m/o}|$, $j_{O,ox}|_{m/o} = |\mathbf{j}_{O,ox}|_{m/o}|$, $j_{Al,ox}|_{o/e} = |\mathbf{j}_{Al,ox}|_{o/e}|$, and $j_{O,ox}|_{o/e} = |\mathbf{j}_{O,ox}|_{o/e}|$. In (2.19), "ox" means that the corresponding ions migrate across the oxide, and $|_{m/o}$ and $|_{o/e}$ mean that the values of the corresponding current densities are at the metal/oxide or oxide/electrolyte interfaces, respectively. In achieving the second step in (2.19), a special case of (2.10), namely,

$$j_{,ox}\big|_{m/o} = j_{,ox}\big|_{o/e}\frac{E_{m/o}}{E_{o/e}}. \tag{2.20}$$

which links the current densities $\mathbf{j}_{,ox}|_{m/o}$ and $\mathbf{j}_{,ox}|_{o/e}$ at two points on the metal/oxide and oxide/electrolyte interfaces connected by the same electric-field line, is used for both ion species, noting that the electric-field intensities at the two points $E_{m/o} = |\mathbf{E}_{m/o}|$ and $E_{o/e} = |\mathbf{E}_{o/e}|$ are common for both species. From (2.12), (2.13), (2.16) and (2.19), and noting that $\mathbf{j}_{Al,ox}|_{o/e}$, $\mathbf{j}_{O,ox}|_{o/e}$, $\mathbf{j}_{Al,o/e}$, $\mathbf{j}_{O,o/e}$, $\mathbf{j}_{Al,dis}$, and $\mathbf{j}_{O,dis}$ have the same direction $\hat{\mathbf{E}}_{o/e}$ at a given point on the oxide/electrolyte interface,

$$\mathbf{j}_{Al,dis} = \frac{j_{Al,o/e} - \beta j_{O,o/e}}{1+\beta}\hat{\mathbf{E}}_{o/e}, \tag{2.21}$$

where $j_{Al,dis} = |\mathbf{j}_{Al,dis}|$. Strictly speaking, under different anodization conditions such as voltage, electrolyte type, concentration, or substrate grain orientation, β may change a little because the volume expansion ratio may change, but the change is expected to be small as the oxide density is usually around 3 g/cm^3 from experiments [4, 168]. As a typical condition, we set β to be 3/7 in accordance with Siejka and Ortega's experimental results [162]. It should also be noted that the β defined in (2.19) is not the same as the current efficiency $\mu = j_{O,o/e}/(j_{O,o/e} + j_{Al,o/e})$, and so a constant β does not mean that the current efficiency is also a constant.

2.3.3 Interface Movement Equations

From Faraday's law [1], the change in volume V of the oxide caused by a passed charge Q carried by ions is

$$V = \frac{MQ}{zF\rho} = \frac{MAjt}{zF\rho}, \tag{2.22}$$

where M is the molecular weight of oxide Al_xO_y, $z = xy$, ρ is the oxide density, j is the amount of current density corresponding to the reaction, A is the area of oxide surface, t is time and F is Faraday's constant. Since the oxide thickness is given by $D = V/A$, the moving velocity $\mathbf{v}$ of a given point at the interface is proportional to the current density as

$$\mathbf{v} = -\frac{dD}{dt}\hat{\mathbf{E}} = -\frac{M}{zF\rho}j\hat{\mathbf{E}}. \tag{2.23}$$

where $\hat{\mathbf{E}} = \mathbf{E}/E$ is the unit vector of the electric field at that given point on the interface. Equation (2.23) is not only suitable for the metal/oxide interface where the oxidation reaction R4 takes place but is also suitable for the oxide/electrolyte interface movement where the oxide decomposition reaction R3 takes place. The moving velocity direction is in the opposite direction of the electric field at a given point on the interface. More specifically, at the oxide/electrolyte interface, the interface movement velocity is $\mathbf{v}_{o/e} = -\mathbf{j}_{Al,dis}M/zF\rho$, and substituting in (2.21), and replacing $\mathbf{j}_{Al,o/e}$ and $\mathbf{j}_{O,o/e}$ by (2.15) and (2.17), respectively, we obtain

$$\mathbf{v}_{o/e} = -\frac{M}{zF\rho(1+\beta)}[n_{Al}A_{Al}\exp(k_{Al}E_{o/e}) - \beta n_o A_o \exp(k_o E_{o/e})]\hat{\mathbf{E}}_{o/e}. \tag{2.24}$$

Similarly, the metal/oxide interface movement velocity is $\mathbf{v}_{m/o} = -\mathbf{j}_{O,ox}|_{m/o}M/zF\rho$, and from (2.12), (2.16), (2.15), (2.17), (2.20) and (2.21), this is given as

$$\mathbf{v}_{m/o} = -\frac{M}{zF\rho(1+\beta)}\frac{E_{m/o}}{E_{o/e}}\left[n_{Al}A_{Al}\exp\left(k_{Al}E_{o/e}\right) + n_O A_O \exp\left(k_O E_{o/e}\right)\right]\hat{\mathbf{E}}_{m/o}. \tag{2.25}$$

where $\hat{\mathbf{E}}_{m/o} = \mathbf{E}_{m/o}/E_{m/o}$. In (2.25), as in (2.20), the two electric-field intensities $E_{m/o}$ and $E_{o/e}$ are those at two points on the metal/oxide and oxide/electrolyte interfaces connected by a given electric-field line. It should also be noted that, although (2.25) is for the velocity of the metal/oxide interface, the present formalism is such that the parameters n_{Al}, n_O, A_{Al}, A_O, k_{Al} and k_O all refer the oxide/electrolyte interface where the rate-determining energy barrier exists.

The density of mobile ions n_{Al} and n_O on the oxide/electrolyte interface should depend on the electric field. For instance, from pulse experiments [4], the relative change of the mobile ion density depends exponentially on the electric-field intensity by a factor less than 10. A cutoff electric-field intensity $E_{cutoff} = 1.1$ V nm^{-1} was predicated, above which all ions become mobile [4], which means all ions have the possibility to jump over the potential barrier to realize migration according to the Cabrera-Mott equation [1, 4, 128]. To reflect such experimental results, here, we set the relative change of the mobile ion density to depend exponentially on the electric-field intensity as

$$n_{Al} = n_{Al}^0 \exp\left[\ln(\lambda) - \ln(\lambda)\frac{E_{o/e}}{E_{cutoff}}\right] \tag{2.26}$$

for Al^{3+} ions [106], where n_{Al}^0 is the number of Al^{3+} ions when all of them are mobile and $\lambda = 0.2$. For O^{2-} ions the same relationship was used. Other λ values, such as 0.1 and 0.5, has also been checked, but no significant difference was found in the simulation results compared to the case of $\lambda = 0.2$, because at the pore bottom the electric field intensity is always around 1 V nm^{-1}.

At the oxide/electrolyte interface, Valand and Heusler experimentally established the following relation between the O^{2-} current density $j_{O,o/e}$ and the Al^{3+} current density $j_{Al,o/e}$ [169]:

$$\left(\frac{\partial \ln j_{O,o/e}}{\partial \ln j_{Al,o/e}}\right)_{pH} = 1.38(\pm 0.14), \tag{2.27}$$

where the slope 1.38 is independent of the pH of the electrolyte from 0 to 11. By substituting (2.15) and (2.17) into (2.27), we can set a relationship of $k_O/k_{Al} = 1.5$ for the simulation.

In order to reduce the complexity involved with the number of parameters in the interface movement equations of (2.24) and (2.25), we adopt the following reduced parameters in our model:

$$B_{Al} = n_{Al}^0 A_{Al} = n_{Al}^0 C_{H^+}^{\eta} q_{Al} v_{Al} \exp(-W_{Al}/kT), \tag{2.28}$$

$$B_O = n_O^0 A_O = n_O^0 q_O v_O \exp(-W_O/kT). \tag{2.29}$$

By substituting in (2.26–2.29), (2.24) and (2.25) become

$$\begin{aligned} \mathbf{v}_{o/e} = &-\frac{M}{zF\rho(1+\beta)} \exp\left[\ln(\lambda) - \ln(\lambda)\frac{E_{o/e}}{E_{cutoff}}\right] \\ &\times \left[B_{Al}\exp(k_{Al}E_{o/e}) - \beta B_O \exp(k_O E_{o/e})\right]\hat{\mathbf{E}}_{o/e}, \end{aligned} \tag{2.30}$$

$$\mathbf{v}_{m/o} = -\frac{M}{zF\rho(1+\beta)}\frac{E_{m/o}}{E_{o/e}}\exp\left[\ln(\lambda) - \ln(\lambda)\frac{E_{o/e}}{E_{cutoff}}\right] \times \left[B_{Al}\exp\left(k_{Al}E_{o/e}\right) + B_O\exp\left(k_O E_{o/e}\right)\right]\hat{\mathbf{E}}_{m/o}. \quad (2.31)$$

The values of B_{Al} and B_O were estimated based on each of the parameters involved, which can produce an oxide growth rate on the order of 1 nm s^{-1} at the pore bottom as is commonly observed in mild anodization experiments [92, 93]. For instance, B_{Al} and B_O can be set within the ranges of [0.12, 1.5] A m^{-2} and [0.024, 0.12] A m^{-2}, respectively. An example value of B_{Al} = 1 A m^{-2} can be obtained by using physically reasonable values for the various parameters [4], such as charge density $n^0_{Al}q_{Al} = 1800\,\mathrm{C\,cm^{-3}}$, vibration frequency ν = 10^{12} s^{-1}, temperature T = 275 K, pH = 1, η = 1 and potential barrier W_{Al} = 1.105 eV.

2.4 Simulation Results and Discussion

The above model has been numerically realized by the finite element method to simulate the real-time evolution of pore growth during anodization [170]. Although the model may be applicable for three-dimensional simulations, for the sake of computational simplicity, only two-dimensional (2-D) simulations corresponding the cross-sectional views of pore channels were conducted [106, 136]. The pore channels start to grow from a pre-patterned configuration, in which small concavities have already existed on the alumina surface. These concavities may represent the defects on the sample surface. Also, according to the simulations as well as experiments, pre-patterns cannot determine the final configuration of AAO after enough anodization time when the self-organization condition has been established, after which the configuration is controlled by the anodization conditions or simulation parameters [106, 136]. Details of the simulation results and experimental comparisons have been reported elsewhere [102, 106, 136], and here, we just provide some typical examples.

Figure 2.4 shows how two small initial cavities in a pre-patterned alumina surface will evolve into steady-state U-shaped pore channels. At the anodizaiton time t = 0, the initial configuration has two concavities with diameter 20 nm. The interpore distance is 200 nm, satisfying the stable interpore distance to voltage ratio of ~2.5 nm V^{-1} typically found in mild anodization experiments [92, 93]. The barrier layer is flat along the oxide/metal interface, and has a thickness of 20 nm at the pore bottom region. Because of the Neumann boundary condition in (4) for the left and right edges of the sample, the present simulation represents an infinite series of repeating units. The simulation parameters are listed in the caption of Fig. 2.4.

With time increased to 50 s in Fig. 2.4, the oxide growth accelerates around the pore bottom regions, due to the concentrated electric-potential distribution there which gives rise to much higher electric-field intensity compared to the pore walls and the top surface of the sample. The initial pores develop into water-drop shaped

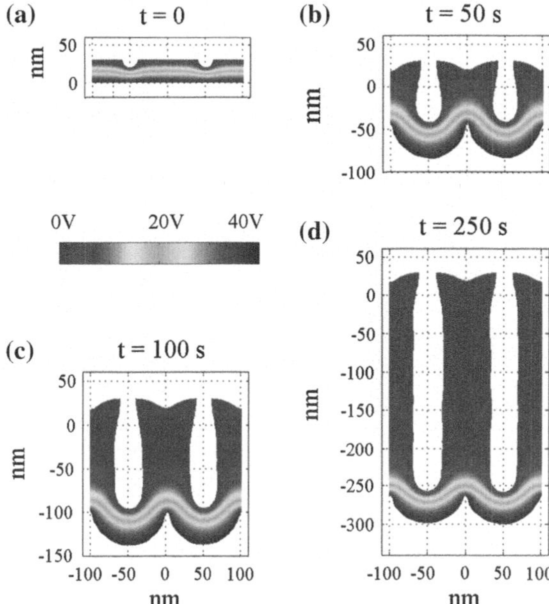

Fig. 2.4 Simulated growth process of porous structure in AAO, starting from a pre-patterned initial configuration at $t = 0$ with an initial barrier layer thickness 20 nm. The simulation parameters are 40 V; $B_O = 0.072$ A m^{-2}; $B_{Al} = 0.72$ A m^{-2}, $k_O/k_{Al} = 1.5$; $k_O = 3.8$ nm V^{-1}; $\beta = 3/7$. The simulated porous structures at anodization time points of $t = 50$, 100 and 250 s are shown (Adapted from [170])

pore channels, which were typically observed in experiments at the beginning of anodization, appears. Also, a scallop-shape of the oxide barrier layer frequently observed in experiments [171, 172] is formed from $t = 50$ s onward [171, 172]. With time increased to 250 s, U-shaped pore channels corresponding to the usual steady-state configuration of AAO are developed. From (2.30) and (2.31), the interface velocity exponentially depends on the electric field along the interface. In a domain with finger-like features connected by a thin common base the Laplacian electric-potential distribution has to exhibit concentrated potential drop within the base region, and this leads to the faster oxide growth rate at the bottom of the pores driven by the electric field. In addition, from the steady-state configuration in Fig. 2.4, the average barrier layer thickness along the two pore axes is 41.7 nm, which is in accordance with the stable barrier-layer thickness-to-voltage ratio of ~ 1 nm V^{-1} found in experiments [92, 172, 173].

Figure 2.5 shows the results of another set of simulations which were different from Fig. 2.4 in terms of the initial configurations, but the same in terms of other simulation parameters. Among the three initial configurations simulated, only that in Fig. 2.5b has the initial interpore distance-to-voltage ratio satisfied the 2.5 nm V^{-1} self-ordering condition found in mild anodization experiments [92, 93]. In Fig. 2.5a, c, the initial interpore distance-to-voltage ratio are 5.0 and 1.7, respectively. In Fig. 2.5b, stable growth of the pore channels takes place, with the initial interpore distance-to-voltage ratio maintained, at all times within the simulation. However, for both Fig. 2.5a, c, the growth of the pore channels is unstable, with branching of the single pore happening in the former, and termination of two of the three pores in latter. However, self-organization takes place, and after an

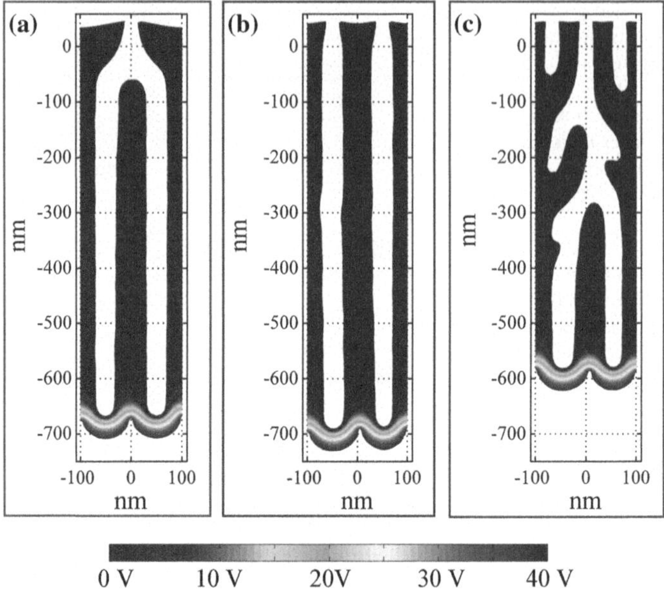

Fig. 2.5 AAO porous structures after 700 s anodization time. The simulation cell initially contains **a** *1* initial pore; **b** *2* initial pores; **c** *3* initial pores. Each initial configuration is centro-symmetric. Except the number of initial pores, the initial pore dimensions, as well as anodization conditions (voltage, B_{Al} and B_O), are the same as in Fig. 2.4 (40 V; $B_O = 0.072$ A m^{-2}; $B_{Al} = 0.72$ A m^{-2}, $k_O/k_{Al} = 1.5$; $k_O = 3.8$ nm V^{-1}; $\beta = 3/7$) (Adapted from [106])

enough long anodization time (here, 700 s), the final configurations of both Fig. 2.5a, c reach to the same interpore distance-to-voltage ratio as in Fig. 2.5b. This indicates that the final configuration of pore channels does not depend on the initial configuration, but is determined by the simulation parameters, which have been set the same for these three simulations. Figure 2.5 also confirms that our simulation can reproduce the typical experimental result of interpore distance-to-voltage ratio of ~2.5 nm V^{-1}.

Figure 2.6 shows the current density against time relationship corresponding to the pore channel growth process in Fig. 2.5. The evolution trend is much similar to the illustration in Fig. 2.2. However, for the case of two initial pores in Fig. 2.6b, much shorter time of ~75 s is required to reach the steady-state current density (~20 A m^{-2}) compared to the other two cases. Due to the incommensurate interpore spacing in the pre-pattern with the applied voltage, the case with one initial pore takes ~150 s and that with three initial pores takes ~500 s to reach the steady state. In experiments, for instance, if a pre-pattern has already been formed by a first-step anodization, then during a second-step anodization the current density needs less time to reach steady state than the first anodization process [133, 174].

It should be noted that the simulation parameters used in both Figs. 2.4 and 2.5 are located within a window in which stable, self-ordered pore growth always occurs [106]. If the parameters are selected from outside this window, then, as

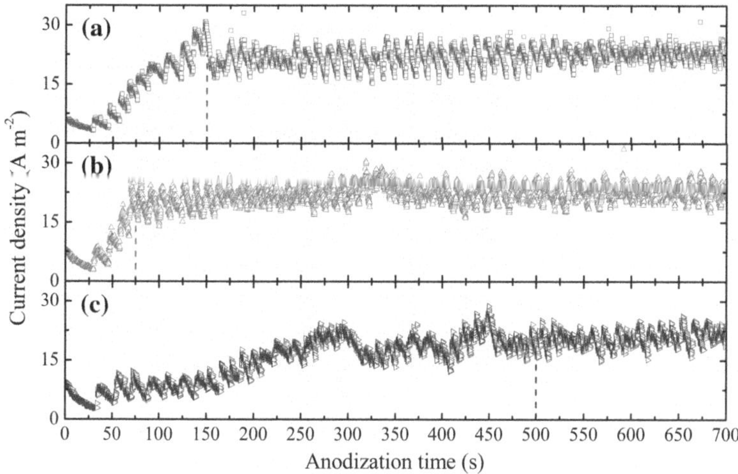

Fig. 2.6 Plot of current density against time corresponding to the anodization process with **a** 1 initial pore in Fig. 2.5 (**a**); **b** 2 initial pores in Fig. 2.5 (**b**); **c** 3 initial pores in Fig. 2.5 (**c**) (Adapted from [106])

shown in our previous reports [106], the pore-channels cannot maintain straight, but frequently branch or terminate during their growth. This window of simulation parameters corresponding to stable pore growth should correspond to the experimentally established window of processing conditions under which self-ordered AAO formation occurs.

For instance, Fig. 2.7 shows how the same initial configuration (Fig. 2.7a) can develop in different stable or unstable manners under different values of $B_{Al,}$. From Fig. 2.7b–e, by increasing $B_{Al,}$ only, the configuration at prolonged simulation times can vary from a barrier type in Fig. 2.7b, to an unstable porous structure in Fig. 2.7c, stable structure in Fig. 2.7d, and back to unstable porous structure in Fig. 2.7e. From (2.28), B_{Al} is proportional to the H^+ concentration C_{H^+} in the electrolyte. For comparison with experiments, therefore, we can regard the change of B_{Al} as only due to the change of C_{H^+}, which can be easily varied in experiments by controlling the acid concentration. Good agreements between the simulation results in Fig. 2.7 and experimental results by changing acid concentration only have been demonstrated, the details of which can be found in our previous report [136].

The situation depicted in Fig. 2.7 is only one typical trend with B_{Al} increasing, as shown in Fig. 2.8. By varying both B_O and B_{Al} while keeping other simulation parameters the same as Fig. 2.6, a map is plotted which shows the growth stability of the same initial porous structure under different parameter values. Figure 2.8 shows that, under a certain B_O value, the structural transformation trend is from barrier-type to unstable porous to stable porous and unstable porous again with B_{Al} increasing. Figure 2.8 indicates that the stable pore growth region is very narrow, and this explains why the experimental processing window for self ordering is very narrow [91–93].

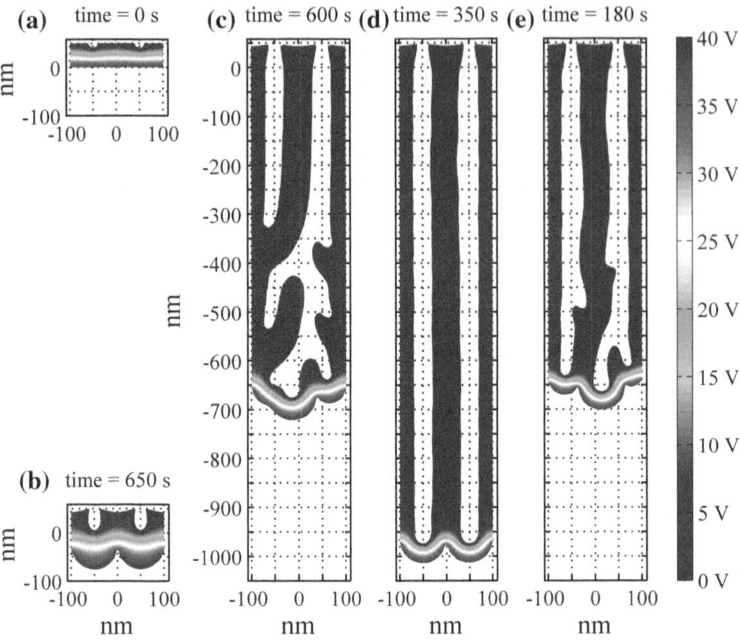

Fig. 2.7 Simulation of pore channel growth in AAO starting from (**a**) the same initial configuration (t = 0 s), but under different values of B_{Al} ($\propto C_{H^+}$): **b** 0.3, **c** 0.96, **d** 1.32, and **e** 1.56 A m^{-2}. Other parameters are kept the same (40 V, B_O = 0.096 A m^{-2}, k_O/k_{Al} = 1.5, k_O = 4.2 nm V^{-1}, β = 3/7) (Adapted from [136])

Fig. 2.8 Map of B_O and B_{Al} conditions for different types of nanoporous alumina structures to occur. (40 V, k_O/k_{Al} = 1.5, k_O = 4.2 nm V^{-1}, β = 3/7) (Adapted from [136])

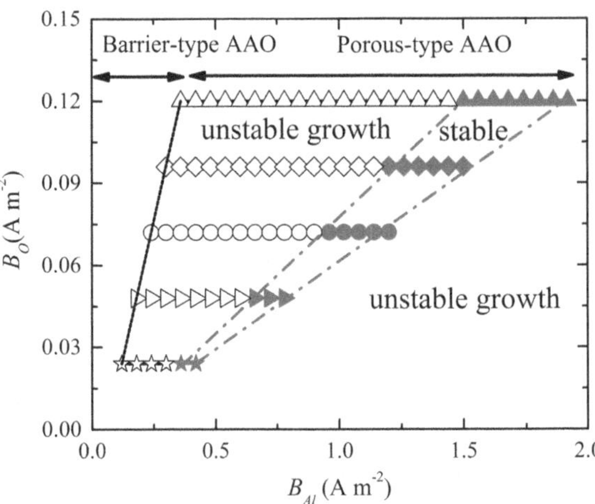

2.5 Outlook

In the model presented in Sect. 2.3, space charge within the AAO body and the double-layer effects along both the oxide/electrolyte and metal/oxide interfaces have been neglected. If space charge is considered, the Poisson equation for electric-field distribution in AAO should be used. The question is how the space charge can distribute within AAO. In addition, volume expansion during oxidation reaction may produce a significant compression stress at the oxide barrier layer. To incorporate the effects of the stress field into the current electric-field based model may be an important step that would end the current debate in terms of the driving force. Furthermore, in real cases, due to the contamination of electrolyte ions within the outside layer of the AAO pore channels, both electric-potential distribution and oxide density may vary. A complete model should also consider the electrolyte type dependent effects. The above factors may be correlated with each other and require future investigations.

2.6 Conclusions

Different growth mechanisms for nanoporous alumina have been reviewed. In terms of driving force, the previous models may be summarized into two groups, namely, electric-field assisted, and mechanical-stress induced mechanisms. A kinetics model recently developed by the authors has been presented. In our model, pore growth is driven by electric-field assisted oxide decomposition at the oxide/electrolyte interface and oxide formation at the metal/oxide interface. Numerical simulation of two-dimensional pore-channel growth in nanoporous alumina with pre-patterned initial configurations has been performed by the finite element method. To the best of our knowledge, this is the first successful attempt to numerically simulate the real-time evolution process of porous alumina growth, starting from non-steady-state initial porous configurations to reach steady-state configurations. This model can capture typical features observed in experiments including pore-channel growth and self-organization processes towards ordering, which supports that electric field can be the key driving force for pore growth in nanoporous alumina.

Acknowledgments The work described here was funded by a grant from the Research Grants Council of the Hong Kong Special Administrative Government, P.R. China (project code: 17206114).

References

1. J.W. Diggle, T.C. Downie, C.W. Goulding, Chem. Rev. **69**, 365 (1969)
2. G.C. Wood, in *Oxide and Oxide Films*, vol. 2, ed. by J.W. Diggle (Marcel Dekker, New York, 1973), p. 167
3. G.E. Thompson, G.C. Wood, in *Treatise on Materials Science and Technology*, vol. 23, ed. by J.C. Scully (Academic Press, New York, 1983), p. 205

4. M.M. Lohrengel, Mater. Sci. Eng., R **11**, 243 (1993)
5. H. Chik, J.M. Xu, Mater. Sci. Eng., R **43**, 103 (2004)
6. B. Sakintuna, Y. Yurum, Ind. Eng. Chem. Res. **44**, 2893 (2005)
7. H.T. Wang, J.F. Yao, Ind. Eng. Chem. Res. **45**, 6393 (2006)
8. F. Cheng, Z. Tao, J. Liang, J. Chen, Chem. Mater. **20**, 667 (2008)
9. C. Bae, H. Yoo, S. Kim, K. Lee, J. Kim, M.A. Sung, H. Shin, Chem. Mater. **20**, 756 (2008)
10. Y. Lei, S.K. Yang, M.H. Wu, G. Wilde, Chem. Soc. Rev. **40**, 1247 (2011)
11. K. Okada, T. Isobe, K. Katsumata, Y. Kameshima, A. Nakajima, K.J.D. MacKenzie, Sci. Technol. Adv. Mater. **12**, 064701 (2011)
12. H.M. Chen, R.S. Liu, J. Phys. Chem. C **115**, 3513 (2011)
13. B. Platschek, A. Keilbach, T. Bein, Adv. Mater. **23**, 2395 (2011)
14. A. de la Escosura-Muniz, A. Merkoci, ACS Nano **6**, 7556 (2012)
15. Q. Hao, T. Qiu, P.K. Chu, Prog. Surf. Sci. **87**, 23 (2012)
16. C.J. Ingham, J. ter Maat, W.M. de Vos, Biotechnol. Adv. **30**, 1089 (2012)
17. J. Martin, J. Maiz, J. Sacristan, C. Mijangos, Polymer **53**, 1149 (2012)
18. C. Nicolini, N. Bragazzi, E. Pechkova, Adv. Drug. Delivery Rev. **64**, 1522 (2012)
19. C.R. Simovski, P.A. Belov, A.V. Atrashchenko, Y.S. Kivshar, Adv. Mater. **24**, 4229 (2012)
20. J.T. Zhang, C.M. Li, Chem. Soc. Rev. **41**, 7016 (2012)
21. M.E. Warkiani, A.A.S. Bhagat, B.L. Khoo, J. Han, C.T. Lim, H.Q. Gong, A.G. Fane, ACS Nano **7**, 1882 (2013)
22. G. Tai, K. Wang, Z. Sun, J. Yin, S.M. Ng, J. Zhou, F. Yan, C.W. Leung, K.H. Wong, W. Guo, et al., J. Phys. Chem. C **116,** 532 (2012)
23. T. Kondo, H. Masuda, K. Nishio, J. Phys. Chem. C **117**, 2531 (2013)
24. M.H. Wu, L.Y. Wen, Y. Lei, S. Ostendorp, K. Chen, G. Wilde, Small **6**, 695 (2010)
25. X.Y. Zhang, G.H. Wen, Y.F. Chan, R.K. Zheng, X.X. Zhang, N. Wang, Appl. Phys. Lett. **83**, 3341 (2003)
26. D. Borissov, S. Isik-Uppenkamp, M. Rohwerder, J. Phys. Chem. C **113**, 3133 (2009)
27. K.G. Biswas, H.E. Matbouly, V. Rawat, J.L. Schroeder, T.D. Sands, Appl. Phys. Lett. **95**, 073108 (2009)
28. X.D. Li, G.W. Meng, Q.L. Xu, M.G. Kong, X.G. Zhu, Z.Q. Chu, A.P. Li, Nano Lett. **11**, 1704 (2011)
29. J. Kim, H. Han, Y.H. Kim, S.H. Choi, J.C. Kim, W. Lee, ACS Nano **5**, 3222 (2011)
30. S.L. Sung, S.H. Tsai, S.H. Tseng, F.K. Chiang, X.W. Liu, H.C. Shih, Appl. Phys. Lett. **74**, 197 (1999)
31. H. Chun, M.G. Hahm, Y. Homma, R. Meritz, K. Kuramochi, L. Menon, L. Ci, P.M. Ajayan, Y.J. Jung, ACS Nano **3**, 1274 (2009)
32. S.H. Jeong, H.Y. Hwang, K.H. Lee, Y. Jeong, Appl. Phys. Lett. **78**, 2052 (2001)
33. S.A. Knaack, M. Redden, M. Onellion, Am. J. Phys. **72**, 856 (2004)
34. J. Zou, X. Qi, L. Tan, B.J.H. Stadler, Appl. Phys. Lett. **89**, 093106 (2006)
35. S. Hong, T. Kang, D. Choi, Y. Choi, L.P. Lee, ACS Nano **6**, 5803 (2012)
36. Y.T. Chong, M.Y.E. Yau, Y. Yang, M. Zacharias, D. Gorlitz, K. Nielsch, J. Bachmann, J. Appl. Phys. **110**, 043930 (2011)
37. M. Koohbor, S. Soltanian, M. Najafi, P. Servati, Mater. Chem. Phys. **131**, 728 (2012)
38. X.W. Wang, Z.H. Yuan, B.C. Fang, Mater. Chem. Phys. **125**, 1 (2011)
39. T. Kim, L. He, J.R. Morales, W.P. Beyermann, C.J. Bardeen, Nanotechnology **22**, 455704 (2011)
40. V. Vega, T. Bohnert, S. Martens, M. Waleczek, J.M. Montero-Moreno, D. Gorlitz, V.M. Prida, K. Nielsch, Nanotechnology **23** 465709 (2012)
41. D.C. Leitao, J. Ventura, C.T. Sousa, A.M. Pereira, J. B. Sousa, M. Vazquez, J.P. Araujo, Phys. Rev. B **84**, 014410 (2011)
42. A. Santos, J.M. Montero-Moreno, J. Bachmann, K. Nielsch, P. Formentin, J. Ferre-Borrull, J. Pallares, L.F. Marsal, A.C.S. Appl, Mater. Interfaces **3**, 1925 (2011)
43. J.H. Fang, I. Aharonovich, I. Leychenko, K. Ostrikov, P.G. Spizzirri, S. Rubanov, S. Prawer, Cryst. Growth Design **12**, 2917 (2012)

44. X. Sheng, J.F. Liu, N. Coronel, A.M. Agarwal, J. Michel, L.C. Kimerling, I.E.E.E. Photo, Tech. Lett. **22**, 1394 (2010)
45. P. Yan, G.T. Fei, G.L. Shang, B. Wu, L.D. Zhang, J. Mater. Chem. C **1**, 1659 (2013)
46. Y. Su, G.T. Fei, Y. Zhang, P. Yan, H. Li, G.L. Shang, L.D. Zhang, Mater. Lett. **65**, 2693 (2011)
47. X.L.Q. Wang, D.X. Zhang, H.J. Zhang, Y. Ma, J.Z. Jiang, Nanotechnology **22**, 849–854 (2011)
48. I. Maksymov, J. Ferre-Borrull, J. Pallares, L.F. Marsal, Photon. Nanostruc. Fund. Applic. **10**, 459 (2012)
49. A. Sato, Y. Pennec, T. Yanagishita, H. Masuda, W. Knoll, B. Djafari-Rouhani, G. Fytas, New J. Phys. **14** (2012)
50. X.D. Li, G.W. Meng, S.Y. Qin, Q.L. Xu, Z.Q. Chu, X.G. Zhu, M.G. Kong, A.P. Li, ACS Nano **6**, 831 (2012)
51. F.M. Han, G.W. Meng, Q.L. Xu, X.G. Zhu, X.L. Zhao, B.S. Chen, X.D. Li, D.C. Yang, Z.Q. Chu, M.G. Kong, Angew. Chem. Int. Ed. **50**, 2036 (2011)
52. G.D. Sulka, A. Brzozka, L.F. Liu, Electrochim. Acta **56**, 4972 (2011)
53. I.E. Rauda, R. Senter, S.H. Tolbert, J. Mater. Chem. C **1**, 1423 (2013)
54. B. Benfedda, L. Hamadou, N. Benbrahim, A. Kadri, E. Chainet, F. Charlot, J. Electrochem. Soc. **159**, C372 (2012)
55. M.K. Date, B.C. Chiu, C.H. Liu, Y.Z. Chen, Y.C. Wang, H.Y. Tuan, Y.L. Chueh, Mater. Chem. Phys. **138**, 5 (2013)
56. W. Liu, X.D. Wang, R. Xu, X.F. Wang, K.F. Cheng, H.L. Ma, F.H. Yang, J.M. Li, Mater. Sci. Semicond. Proc. **16**, 160 (2013)
57. Y. Xiang, A. Keilbach, L.M. Codinachs, K. Nielsch, G. Abstreiter, A.F.I. Morral, T. Bein, Nano Lett. **10**, 1341 (2010)
58. S.L. Brock, I.U. Arachchige, J.L. Mohanan, Science **307**, 397 (2005)
59. M. Tian, W. Wang, Y.J. Wei, R.G. Yang, J. Power Sour. **211**, 46 (2012)
60. M. Tian, W. Wang, S.H. Lee, Y.C. Lee, R.G. Yang, J. Power Sour. **196**, 10207 (2011)
61. G. Ferrara, L. Damen, C. Arbizzani, R. Inguanta, S. Piazza, C. Sunseri, M. Mastragostino, J. Power Sour. **196**, 1469 (2011)
62. W. Wang, M. Tian, A. Abdulagatov, S.M. George, Y.C. Lee, R.G. Yang, Nano Lett. **12**, 655 (2012)
63. Y.C. Tsao, T. Sondergaard, E. Skovsen, L. Gurevich, K. Pedersen, T.G. Pedersen, Opt. Exp. **21**, A84 (2013)
64. D. Chen, W. Zhao, T.P. Russell, ACS Nano **6**, 1479 (2012)
65. L.C. Haspert, S.B. Lee, G.W. Rubloff, ACS Nano **6**, 3528 (2012)
66. T. Xue, X. Wang, J.M. Lee, J. Power Sour. **201**, 382 (2012)
67. L.J. Li, B. Zhu, S.J. Ding, H.L. Lu, Q.Q. Sun, A.Q. Jiang, D.W. Zhang, C.X. Zhu, Nanoscale Res. Lett. **7** (2012)
68. W. Lee, H. Han, A. Lotnyk, M.A. Schubert, S. Senz, M. Alexe, D. Hesse, S. Baik, U. Gösele, Nat. Nanotechnol. **3**, 402 (2008)
69. J. Jiang, Y. Li, J. Liu, X. Huang, C. Yuan, X.W. Lou, Adv. Mater. **24**, 5166 (2012)
70. K. Hotta, A. Yamaguchi, N. Teramae, ACS Nano **6**, 1541 (2012)
71. K. Vasilev, Z. Poh, K. Kant, J. Chan, A. Michelmore, D. Losic, Biomaterials **31**, 532 (2010)
72. J.B. Li, Y. Yu, X.N. Peng, Z.J. Yang, Z.K. Zhou, L. Zhou, J. Appl. Phys. **111**, 123110 (2012)
73. K. Hotta, A. Yamaguchi, N. Teramae, J. Phys. Chem. C **116**, 23533 (2012)
74. A. Santos, M. Alba, M.M. Rahman, P. Formentin, J. Ferre-Borrull, J. Pallares, L.F. Marsal, Nanoscale Res. Lett. **7**, 228 (2012)
75. S.H. Yeom, O.G. Kim, B.H. Kang, K.J. Kim, H. Yuan, D.H. Kwon, H.R. Kim, S.W. Kang, Opt. Exp. **19**, 22882 (2011)
76. C. Nicolini, T. Bezerra, E. Pechkova, Nanomedicine **7**, 1117 (2012)
77. J.T. Chen, W.L. Chen, P.W. Fan, ACS Macro Lett. **1**, 41 (2012)
78. A.Y.Y. Ho, L.P. Yeo, Y.C. Lam, I. Rodriguez, ACS Nano **5**, 1897 (2011)

79. Y. Wang, L. Tong, M. Steinhart, ACS Nano **5**, 1928 (2011)
80. T. Yanagishita, R. Fujimura, K. Nishio, H. Masuda, Chem. Lett. **39**, 188 (2010)
81. N. Suzuki, M. Imura, Y. Nemoto, X.F. Jiang, Y. Yamauchi, Cryst. Eng. Comm. **13**, 40 (2011)
82. J.T. Chen, C.W. Lee, M.H. Chi, I.C. Yao, Macromol. Rap. Comm. **34**, 348 (2013)
83. K. Kwon, C.W. Park, D. Kim, Sensor. Actuat. A-Phys. **175**, 108 (2012)
84. Y.H. Ma, G.Q. Zhan, M. Ma, X. Wang, C.Y. Li, Bioelectrochemistry **84**, 6 (2012)
85. J.R. Deneault, X.Y. Xiao, T.S. Kang, J.S. Wang, C.M. Wai, G.J. Brown, M.F. Durstock, ChemPhysChem **13**, 256 (2012)
86. R.M. Michell, A.T. Lorenzo, A.J. Muller, M.C. Lin, H.L. Chen, I. Blaszczyk-Lezak, J. Martin, C. Mijangos, Macromolecules **45**, 1517 (2012)
87. Y. Suzuki, H. Duran, M. Steinhart, H.J. Butt, G. Floudas, Soft Matter **9**, 2621 (2013)
88. H. Wu, Z.H. Su, A. Takahara, Soft Matter **7**, 1868 (2011)
89. K. Shwirn, W. Lee, R. Hillebrand, M. Steinhart, K. Nielsch, U. Gösele, ACS Nano **2**, 302 (2008)
90. C. Cheng, A.H.W. Ngan, Nanotechnology **24**, 215602 (2013)
91. C.K.Y. Ng, A.H.W. Ngan, Chem. Mater. **23**, 5264 (2011)
92. W. Lee, R. Ji, U. Gösele, K. Nielsch, Nat. Mater. **5**, 741 (2006)
93. H. Masuda, K. Fukuda, Science **268**, 1466 (1995)
94. J.P. O'Sullivan, G.C. Wood, Proc. R. Soc. London, Ser. A **317**, 511 (1970)
95. G.E. Thompson, G.C. Wood, Nature **290**, 230 (1981)
96. E. Moyen, L. Santinacci, L. Masson, W. Wulfhekel, M. Hanbucken, Adv. Mater. **24**, 5094 (2012)
97. F. Keller, M.S. Hunter, D.L. Robinson, J. Electrochem. Soc. **100**, 441 (1953)
98. N.Q. Zhao, X.X. Jiang, C.S. Shi, J.J. Li, Z.G. Zhao, X.W. Du, J. Mater. Sci. **42**, 3878 (2007)
99. A.L. Friedman, D. Brittain, L. Menon, J. Chem. Phys. **127**, 154717 (2007)
100. C. Cheng, K.Y. Ng, A.H.W. Ngan, AIP Adv. **1**, 042113 (2011)
101. A.P. Li, F. Müller, A. Birner, K. Nielsch, U. Gosele, J. Vac. Sci. Technol., A **17**, 1428 (1999)
102. C. Cheng, K.Y. Ng, N.R. Aluru, A.H.W. Ngan, J. Appl. Phys. **113**, 204903 (2013)
103. A.L. Friedman, L. Menon, J. Appl. Phys. **101**, 084310 (2007)
104. H. Masuda, F. Hasegwa, S. Ono, J. Electrochem. Soc. **144**, L127 (1997)
105. H. Masuda, K. Yada, A. Osada, Jpn. J. Appl. Phys. **37**, L1340 (1998)
106. C. Cheng, A.H.W. Ngan, Electrochim. Acta **56**, 9998 (2011)
107. S. Anderson, J. Appl. Phys. **15**, 477 (1944)
108. T.P. Hoar, N.F. Mott, J. Phys. Chem. Solids **9**, 97 (1959)
109. T.P. Hoar, in *Modern Aspects of Electrochemistry*, ed. by J.O.M. Bockris (Butterworths Scientific Publications, London, 1959), p. 263
110. J.W. Diggle, in *Oxide and Oxide Films*, vol. 2, ed. by J.W. Diggle (Marcel Dekker, New York, 1973), p. 281
111. G.E. Thompson, R.C. Furneaux, G.C. Wood, J.A. Richardson, J.S. Goode, Nature **271**, 433 (1978)
112. G.E. Thompson, Thin Solid Films **297**, 192 (1997)
113. V.P. Parkhutik, V.I. Shershulsky, J. Phys. D Appl. Phys. **25**, 1258 (1992)
114. S.K. Thamida, H.C. Chang, Chaos **12**, 240 (2002)
115. G.K. Singh, A.A. Golovin, I.S. Aranson, V.M. Vinokur, Europhys. Lett. **70**, 836 (2005)
116. G.K. Singh, A.A. Golovin, I.S. Aranson, Phys. Rev. B **73**, 205422 (2006)
117. C. Sample, A.A. Golovin, Phys. Rev. B **74**, 041606 (2006)
118. G. Patermarakis, J. Chandrinos, K. Masavetas, J. Solid State Electrochem. **11**, 1191 (2007)
119. G. Patermarakis, K. Moussoutzanis, Electrochim. Acta **54**, 2434 (2009)
120. O. Jessensky, F. Müller, U. Gosele, Appl. Phys. Lett. **72**, 1173 (1998)
121. A.P. Li, F. Müller, A. Birner, K. Nielsch, U. Gösele, J. Appl. Phys. **84**, 6023 (1998)
122. J.E. Houser, K.R. Hebert, J. Electrochem. Soc. **153**, B566 (2006)
123. J.E. Houser, K.R. Hebert, Phys. Stat. Sol. (a) **205**, 2396 (2008)
124. K.R. Hebert, J.E. Houser, J. Electrochem. Soc. **156**, C275 (2009)

125. K.R. Hebert, S.P. Albu, I. Paramasivam, P. Schmuki, Nat. Mater. **11**, 162 (2012)
126. J.E. Houser, K.R. Hebert, Nat. Mater. **8**, 415 (2009)
127. Q. Van Overmeere, F. Blaffart, J. Proost, Electrochem. Commun. **12**, 1174 (2010)
128. N. Cabrera, N.F. Mott, Rep. Prog. Phys. **12**, 163 (1949)
129. M. Nagayama, K. Tamura, Electrochim. Acta **12**, 1097 (1967)
130. M. Nagayama, K. Tamura, Electrochim. Acta **13**, 1773 (1968)
131. R.B. Mason, J. Electrochem. Soc. **102**, 671 (1955)
132. F. Li, Ph.D. Thesis, University of Alabama (1998)
133. F. Li, L. Zhang, R.M. Metzger, Chem. Mater. **10**, 2470 (1998)
134. M.S. Hunter, P. Fowle, J. Electrochem. Soc. **101**, 514 (1954)
135. J. Oh, Ph.D. Thesis, Massachusetts Institute of Technology (2010)
136. C. Cheng, A.H.W. Ngan, J. Phys. Chem. C **117**, 12183 (2013)
137. N.B. Pilling, R.E. Bedworth, J. Inst. Metals **29**, 529 (1923)
138. R.E. Smallman, A.H.W. Ngan, *Physical Metallurgy and Advanced Materials* (Elsevier, Amsterdam, 2007)
139. C. Cherki, J. Siejka, J. Electrochem. Soc. **120**, 784 (1973)
140. S.J. Garcia-Vergara, L. Iglesias-Rubianes, C.E. Blanco-Pinzon, P. Skeldon, G.E. Thompson, P. Campestrini, Proc. R. Soc. A **462**, 2345 (2006)
141. S.J. Garcia-Vergara, P. Skeldon, G.E. Thompson, H. Habazaki, Electrochim. Acta **52**, 681 (2006)
142. P. Skeldon, G.E. Thompson, S.J. Garcia-Vergara, L. Iglesias-Rubianes, G.E. Blanco-Pinzon, Electrochem. Solid State Lett. **9**, B47 (2006)
143. S.J. Garcia-Vergara, P. Skeldon, G.E. Thompson, H. Habazaki, Corro. Sci. **49**, 3772 (2007)
144. S.J. Garcia-Vergara, P. Skeldon, G.E. Thompson, H. Habazaki, Surf. Inerface Anal. **39**, 860 (2007)
145. S.J. Garcia-Vergara, P. Skeldon, G.E. Thompson, H. Habazaki, Thin Solid Films **515**, 5418 (2007)
146. S.J. Garcia-Vergara, P. Skeldon, G.E. Thompson, T. Hashimoto, H. Habazaki, J. Electrochem. Soc. **154**, C540 (2007)
147. S.J. Garcia-Vergara, P. Skeldon, G.E. Thompson, H. Habazaki, Corro. Sci. **50**, 3179 (2008)
148. S.J. Garcia-Vergara, T. Hashimoto, P. Skeldon, G.E. Thompson, H. Habazaki, Electrochim. Acta **54**, 3662 (2009)
149. O. Jessensky, F. Müller, U. Gösele, J. Electrochem. Soc. **145**, 3735 (1998)
150. J.F. Dewald, Acta. Met. **2**, 340 (1954)
151. J.F. Dewald, J. Electrochem. Soc. **102**, 1 (1955)
152. D.A. Vermilyea, Acta. Met. **1**, 282 (1953)
153. G. Patermarakis, J. Electroanal. Chem. **635**, 39 (2009)
154. S. Ono, M. Saito, M. Ishiguro, H. Asoh, J. Electrochem. Soc. **151**, B473 (2004)
155. J.A. Davies, B. Domeij, J.P.S. Pringle, F. Brown, J. Electrochem. Soc. **112**, 675 (1965)
156. J.A. Davies, B. Domeij, J. Electrochem. Soc. **110**, 849 (1963)
157. K. Nielsch, J. Choi, K. Schwim, R.B. Wehrspohn, U. Gösele, Nano Lett. **2**, 677 (2002)
158. J.D. Edwards, F. Keller, Trans. Electrochem. Soc. **79**, 180 (1940)
159. R.C. Spooner, J. Electrochem. Soc. **102**, 156 (1955)
160. Z. Wu, C. Richter, L. Menon, J. Electrochem. Soc. **154**, E8 (2007)
161. J.L. Whitton, J. Electrochem. Soc. **115**, 58 (1968)
162. J. Siejka, C. Ortega, J. Electrochem. Soc. **124**, 883 (1977)
163. L. Young, *Anodic Oxide Films* (Academic Press, London, 1961)
164. D.A. Vermilyea, J. Electrochem. Soc. **113**, 1067 (1966)
165. G.A.J. Dorsey, J. Electrochem. Soc. **113**, 169 (1966)
166. L. Vecchia, G. Piazzesi, F. Siniscalco, Electrochim. Metal **2**, 71 (1967)
167. J. Siejka, J.P. Nadai, G. Amsel, J. Electrochem. Soc. **118**, 727 (1970)
168. S. Lee, H.S. White, J. Electrochem. Soc. **151**, B479 (2004)
169. T. Valand, K.E. Heusler, J. Electroanal. Chem. **149**, 71 (1983)
170. C. Cheng, Ph.D. Thesis, The University of Hong Kong (2013)

171. C. Cheng, A.H.W. Ngan, J. Appl. Phys. **113**, 184903 (2013)
172. Z. Su, W. Zhou, Adv. Mater. **20**, 3663 (2008)
173. K. Nishio, T. Yanagishita, S. Hatakeyama, H. Maegawa, H. Masuda, J. Vac. Sci. Technol., A **B26**, L10 (2008)
174. J.M. Montero-Moreno, M. Sarret, C. Muller, J. Electrochem. Soc. **154**, C169 (2007)

Chapter 3
Synthesis of Nanoporous Anodic Alumina by Anodic Oxidation of Low Purity Aluminum Substrates

Leszek Zaraska, Ewa Wierzbicka, Elżbieta Kurowska-Tabor and Grzegorz D. Sulka

Abstract The aim of this chapter is to present some recent findings on the fabrication of anodic aluminum oxide (AAO) layers by anodization of low purity aluminum substrates. The use of low purity, technical aluminum alloys, instead of high purity substrates, can significantly reduce the cost of AAO fabrication, however, this can also affect the structure and properties of as produced alumina layers. Here, we focused on the comparison of oxide layer growth on substrates with different Al contents, as well as on the new procedures used for the synthesis of well-ordered nanoporous oxides from technical aluminum alloys. Some applications of the formed nanoporous AAO layers are presented.

3.1 Introduction

During the recent years nanomaterials have received significant attention due to many unique chemical and physical properties that make them promising materials for various modern applications. As a result, nanotechnology is now one of the most popular sciences which have opened a lot of new perspectives in almost all scientific disciplines [1]. Up to now, the main problem that limits the practical application of nanomaterials is a relatively high cost of nanofabrication [2]. That is why considerable research efforts focus on the design and control of novel nanostructures via innovative synthetic strategies. A big attention has been paid to development a simple and inexpensive method of nanofabrication that can be used not only in a laboratory scale but also could have some potential for a commercial-scale implementation. Among numerous strategies reported in the literature, one of the most popular method for fabricating nanostructured oxide layers on the metal surface is controlled anodic oxidation (anodization) of metals under

L. Zaraska · E. Wierzbicka · E. Kurowska-Tabor · G.D. Sulka (✉)
Department of Physical Chemistry & Electrochemistry, Faculty of Chemistry,
Jagiellonian University in Krakow, Ingardena 3, 30-060 Krakow, Poland
e-mail: sulka@chemia.uj.edu.pl

D. Losic and A. Santos (eds.), *Nanoporous Alumina*,
Springer Series in Materials Science 219, DOI 10.1007/978-3-319-20334-8_3

specified conditions [3]. Anodization, widely recognized as an easy way to synthesize nanoporous alumina [4, 5], and nanoporous or nanotubular titania [6–8] have been recently employed to fabricate nanostructured oxides also on the surface of other metals such as: Nb, Hf, W, V, Zr, Cr, Fe, Ni, Sn, and so on [3].

Over the past decades, electrochemical fabrication of anodic aluminum oxide (AAO) layers has become one of the most important and widespread method that allows for fabrication of arrays of well ordered, close-packed nanostructures [4, 5, 9]. A tremendous number of publications on fabrication and potential applications of this kind of structures have been already published. Moreover, due to the regular pore arrangement and precisely controlled dimensions of nanopores, porous anodic alumina is now a very popular and widely used template for fabrication of other nanostructured materials. Recently, many of nanostructured materials with different morphologies (e.g., nanowires, nanotubes, nanodots) have been fabricated by deposition of various substrates including metals, semiconductors, oxides and polymers into the pores of alumina membranes (for details see our previous work [9]). Moreover, a through-hole AAO membrane can be successfully used as an etching mask for pattern transfer into other substrates [10, 11].

Two different groups of methods for fabrication of nanoporous alumina can be distinguished. The first one, called prepatterned-guided anodization is based on anodization of the substrate which is pre-textured prior to anodizing by imprinting of desired structure on the Al surface. A big advantage of this approach is possibility to obtain ideally ordered hexagonal or even square and triangle structures of nanopores [12]. Unfortunately, the size of as obtained, ideal structure is limited to the dimensions of the master mold used for pre-patterning. In addition, fabrication of the mold is often time consuming and expensive. On the other hand, a two-step anodization based only on self-organization, proposed originally by Masuda et al., results always in non-perfectly ordered structures with defects in the hexagonal arrangement of pores [13]. However, when the optimized anodizing conditions are applied, an arrays of hexagonally arranged nanopores with a satisfactory degree of pore order can be achieved. The method involves the chemical etching of oxide layer formed during first anodization, and subsequent re-anodization of the sample under the same conditions. A great advantage of this method is that the surface area of the sample is now even non-limited. Moreover, no sophisticated and expensive technologies are required. A schematic representation of the hexagonal array of anodic alumina is shown in Fig. 3.1.

As can be seen, a typical porous alumina layer consists of hexagonal cells with nanopores at their centers. In the ideal structure, uniform and continuous pores are parallel to each other and perpendicular to the substrate. At the pore bottoms, compact dielectric oxide layer, called "barrier layer", is formed. Nanoporous oxide layers are often characterized by such parameters as pore diameter (D_p), interpore distance (cell diameter, D_c), wall thickness (W), "barrier layer" thickness (B) and thickness of the oxide layer (H). From the practical point of view, it is sometimes beneficial to give the porosity of the structure (σ) defined as a ratio of surface occupied by the pores to the total surface of the sample, and pore density (ρ) being a number of pores that can be found on specified area.

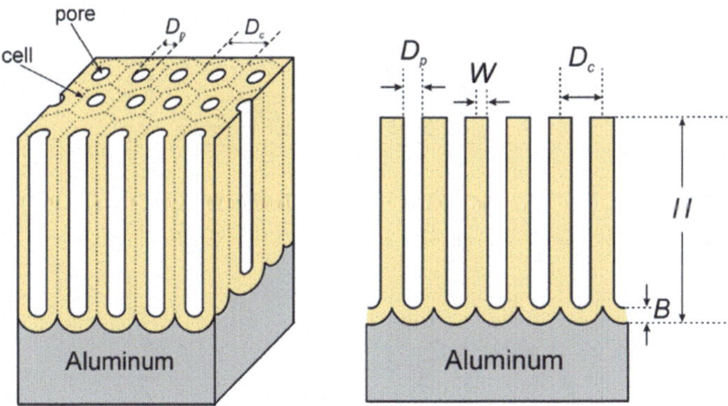

Fig. 3.1 An ideal, hexagonal structure of nanoporous anodic alumina

Nowadays, nanoporous alumina layers are synthesized mainly by current- or voltage-controlled (more often) two-step anodization carried out in sulfuric [14–16], oxalic [17–21] and phosphoric acid [22, 23] solutions. However, some publications on the fabrication of porous alumina by anodization in other inorganic and organic electrolytes can be also found [24–32].

It should be mentioned that all structural features of anodic alumina strongly depend on anodizing conditions including type and concentration of the electrolyte, applied potential/current density, temperature and duration of the process [4]. Moreover, the degree of nanopore order also depends on processing parameters, especially on anodizing potential. It is widely known that the best hexagonal pore order can be achieved only at a certain value of the anodizing potential, known as self-ordering regime. For anodizing carried out in sulfuric, oxalic and phosphoric acid, this voltage is about 25, 40, and 195 V, respectively [4]. From the practical point of view and further applications of such kind of alumina layers, the most important issue is that the structural features of AAO templates can be fine-tuned by adjusting anodization conditions.

Although many different techniques were proposed for the fabrication of well ordered AAO, the most of them employed high-purity aluminum (min. 99.99 %) as a substrate for anodization. Unfortunately, the relatively high cost of high-purity Al foils limits the potential application of porous alumina membranes in the large-scale template synthesis of nanomaterials. Therefore, the use of low purity, commercially-available technical aluminum alloys with a lower Al content instead of high purity substrates can significantly reduce the cost of AAO fabrication. On the other hand, many different alloying elements such as Si, Fe, Mg, Mn, Cu etc. can be found in technical Al alloys, and proportions and distributions of these impurities depend on the type of alloy. Examples of aluminum alloys often used as starting materials for anodization are collected in Table 3.1, together with their rough composition.

Alloying elements exhibit very often completely different behaviors during anodization. This can strongly affect the formation of nanoporous alumina and

Table 3.1 Chemical composition (in wt%) of several types of aluminum alloys used for anodization

Alloy	Alloy composition									
	Si	Fe	Cu	Mn	Mg	V	Cr	Zn	Ti	Al
1070	**≤0.20**	**≤0.25**	≤0.04	≤0.03	≤0.03	≤0.05	–	≤0.04	≤0.03	≥99.70
1050	**≤0.25**	**≤0.40**	≤0.05	≤0.05	≤0.05	≤0.05	–	–	–	≥99.5
1100	**≤0.95**		0.05–0.20	≤0.05	–	–	–	≤0.10	–	≥99.0
2024	≤0.50	≤0.50	**3.8–4.9**	0.30–0.90	**1.20–1.80**	–	≤0.10	≤0.25	≤0.15	90.7–94.7
3003	≤0.60	≤0.70	0.05–0.20	1.00–1.50	–	–	–	0.10	–	96.7–99.0
6111	**0.60–1.10**	≤0.40	**0.50–0.90**	0.10–0.45	**0.50–1.00**	–	≤0.10	≤0.15	≤0.1	95.6–98.3
5005	**≤0.3**	**≤0.7**	≤0.2	≤0.2	**0.50–1.10**	–	≤0.1	≤0.25	–	≤97
5083	≤0.40	≤0.40	≤0.10	**0.40–1.00**	**4.00–4.90**	–	0.05–0.25	≤0.25	≤0.15	92.4–95.6
6060	**0.30–0.60**	0.10–0.30	≤0.10	≤0.10	**0.35–0.60**	–	≤0.05	≤0.15	≤0.10	≤97.8
6061	**0.40–0.80**	**≤0.70**	0.15–0.40	≤0.15	**0.80–1.20**	–	0.04–0.35	≤0.25	≤0.15	95.8–98.6
6082	**0.70–1.30**	≤0.50	≤0.10	**0.40–1.00**	**0.60–1.20**	–	≤0.25	≤0.20	≤0.10	95.2–98.3
7175	≤0.15	≤0.20	**1.2–2.0**	≤0.10	**2.10–2.90**	–	0.18–0.28	**5.10–6.10**	≤0.10	88–91.4

The main alloying elements are highlighted in *bold*

results in nanostructures with different morphologies. So, the synthesis of nanoporous AAO membranes with a desired morphology and satisfactory degree of nanopore order from technical aluminum is not a trivial issue and requires specific conditions, sometimes completely different from those typically applied for a pure aluminum anodization.

In this chapter we present some recent findings on anodizing of low purity aluminum substrates. A special emphasis is put on the comparison of an oxide growth on different substrates as well as on new procedures used for the synthesis of highly-ordered nanoporous templates from technical Al and their possible applications.

3.2 Synthesis of Porous Alumina on Low Purity Al Substrates

3.2.1 Al Substrate Pre-treatment

It was proved that the state of Al surface could be also a crucial parameter in the fabrication of self-ordered anodic alumina layers, especially when technical substrates are used for anodizing. It is widely known that the AAO structures with the best hexagonal pore arrangement can be obtained only when the surface roughness of the Al foil is reduced to minimum. That is why the process of surface pretreatment should be carefully optimized to obtain high quality alumina layers. When low purity Al is a starting material for anodization, electrochemical polishing as an entire process of the surface pretreatment is often not enough to obtain desired surface morphology. Moreover, the existence of even small amount of alloying elements on the surface could also result in a formation of defects in the anodic oxide structure (see below). Therefore, not only a native alumina layer should be removed prior to anodizing but also the number of intermetallic compounds present on the surface must be reduced. Several strategies of the surface pretreatment have been already reported.

For instance, Saenz de Miera et al. have found that after the surface pretreatment, consisting of etching in 0.5 M NaOH at 313 K and desmutting for 15 s in 30 vol% HNO_3, the intermetallic compounds in the AA2024-T3 alloy (mainly Al–Cu and Al–Cu–Fe phases) occupied about 2 % of the surface (Fig. 3.2a) [33]. On the other hand, a reduced amount of intermetallic compounds ($\sim$1.4 %) was found on the surface of the AA7075-T6 alloy after desmutting (Fig. 3.2b). A 90 s anodizing in 0.4 M H_2SO_4 results in almost complete removal of S-phase (Al_2CuMg) as well as Al–Cu particles (see Fig. 3.2c, d), leaving evident cavities in the surface. Also, a size reduction of Al–Cu–Fe phases can be noticed, while Si-rich particles remain almost unaffected [33].

In another report, Montero-Moreno et al. proposed a five-step pretreatment procedure before anodization of the AA1050 alloy which includes alkaline

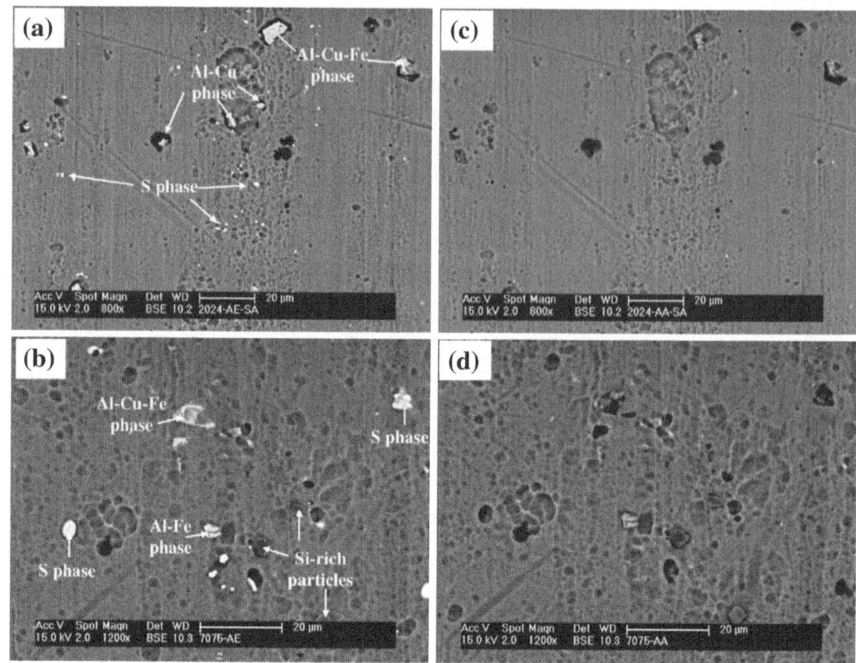

Fig. 3.2 SEM images of **a** and **c** AA2024-T3 and **b** and **d** AA7075-T6 alloys after etching/desmutting (**a** and **b**) and after subsequent anodizing in sulfuric acid (**c** and **d**). Reprinted with permission from [33]

degreasing, alkaline etching in a NaOH/sodium gluconate solution, desmutting in HNO_3, galvanostatic electropolishing in a H_3PO_4/H_2SO_4 mixture and, finally, acid etching in a H_3PO_4/CrO_3 solution [34]. The first three steps were applied in order to clean the surface and reduce the amount of intermetallic species on the surface. The alkaline etching resulted in the formation of precipitates which were then dissolved during desmutting. As a result of these steps an evident decrease of surface roughness was observed. However, the surface was found to be still too rough to obtain well ordered alumina layers, even after applying the two-step anodizing procedure. Only when further electrochemical polishing and acid etching were performed after desmutting, the surface roughness was low enough to enable the fabrication of well-ordered hexagonal nanoporous oxide structures. It should be mentioned that the electropolishing parameters, especially temperature, have to be accurately controlled and optimized [34]. For instance, the optimal temperature for the galvanostatic electropolishing carried out in a mixture of H_3PO_4 and H_2SO_4 (3:2) at 15 A dm^{-2} is 77.5 °C [34]. These conditions allow for formation of the dense and compact barrier layer during early stages of the first anodizing step, and its more uniform and homogeneous breakdown. In case when a non-compact barrier layer is formed during first few seconds of anodizing, the formation of well-ordered alumina is almost impossible.

On the other hand, Yu et al. reported that a suitable pretreatment of the AA1050 substrate by electrochemical polishing, carried out in a mixture of ethanol and perchloric acid followed by chemical polishing, can result in the reduction of the surface roughness to ≤5 nm. This allows for obtaining a quite uniform alumina film in the mm^2 scale [35]. However, it should be mentioned that electropolishing conditions, especially temperature, potential, electropolishing time and electrolyte composition (perchloric acid content in the $C_2H_5OH–HClO_4$ mixture) strongly influence the surface roughness of the AA1050 alloy. In addition, a precise control of an average surface roughness can be achieved by adjusting electropolishing parameters. Surprisingly, it was proved that the structural features of anodic alumina layers formed by anodizing of AA1050 in a mixture of sulfuric and oxalic acid, depend on the surface roughness. It was found that the average pore size of AAO decreases from about 90–60 nm when the surface roughness of Al substrate increases from 3 to 30 nm [35].

In our recent work, we also studied the effect of electropolishing procedure on the surface morphology of the AA1050 substrate [36]. The process was carried out in a mixture of perchloric acid (60 wt%) and ethanol (1:4 vol.) at 20 V and 10 °C for 1 min. The SEM images of Al surface before and after electrochemical polishing are shown in Fig. 3.3a, b (low magnification) and 3c (high magnification), respectively. It can be seen that just after electropolishing, the surface of Al

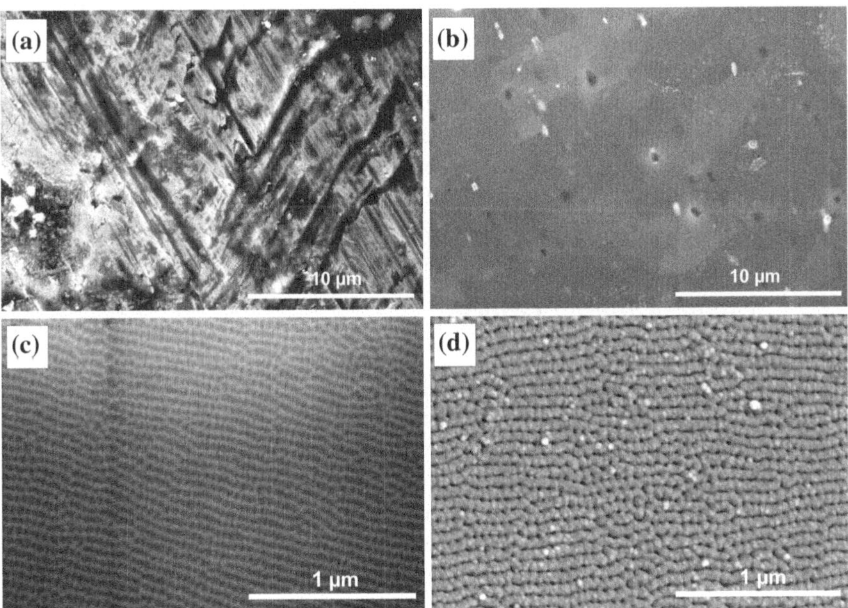

Fig. 3.3 SEM images of the AA1050 alloy surface before (**a**) after electropolishing (**b**, **c**) and the oxide structure obtained after 1 h of first anodization carried out in 0.3 M oxalic acid at 20 °C under 45 V

substrate is almost flat, however, some pits and bumps being a direct result of the local allocation of impurities can be easily recognized. As can be seen in the high magnification image (Fig. 3.3c), an ordered structures of nanostripes or nanopores are formed on the Al surface during electrochemical polishing. According to the literature, the formation of such kinds of nanopatterns is probably a result of the preferential adsorption of alcohol molecules on the surface ridges [37, 38]. It should be mentioned that the type of nanopattern (porous or cellular) formed during electropolishing depends on the polishing parameters, especially applied potential and duration of the process. In addition, the size of created patterns increases with the applied potential [39]. We also confirmed that the pattern formed on the Al surface by electropolishing is then replicated at the initial stages of first anodizing step (compare Fig. 3.3c, d), so the morphology of initial nanoporous oxide is strongly determined by the morphology of aluminum substrate used as a starting material for anodization [36]. It was also reported that the significant improvement of nanostructured oxide morphology can be also achieved by annealing of impure Al foils prior to anodizing that could be attributed to the reduction of amount of grain boundaries and smoothing of the Al surface during annealing [40]. Surprisingly, according to Lo et al., when both annealing and electropolishing are applied for the sample pre-treatment, nanostructured oxides with a destroyed pore arrangement are obtained [40].

In a typical two-step anodizing procedure, the oxide layer formed during the first anodization should be completely removed prior to second anodizing step. It is very important to optimize also the duration of oxide stripping in order to achieve a completely clean Al surface. On the other hand, too long exposition of the sample to the aggressive environment could result in damage of the aluminum substrate. For instance, Montero-Moreno et al. have optimized the time needed to remove the oxide layer formed during the first anodization of AA1050 alloy. For chemical etching in the H_3PO_4:CrO_3 3.5 %:2 % solution at 55 °C, it was suggested that the time of removal in min should twice the thickness of the alumina layer in μm [34].

3.2.2 *Anodic Alumina Growth*

3.2.2.1 Current/Potential Versus Time Behavior and Growth Rates

Both potentiostatic [e.g., 36, 41–43] and galvanostatic anodizations [44–46] were applied for the fabrication of nanoporous alumina layers on the surface of low-purity Al substrates. A typical voltage versus time curve recorded during anodization under a constant current regime is shown in Fig. 3.4 and compared with a typical current-time dependence obtained for the potentiostatic process [44]. Three main stages that correspond to the mechanism of oxide growth can be noticed. At the beginning of the process, the growth of compact oxide layer on the

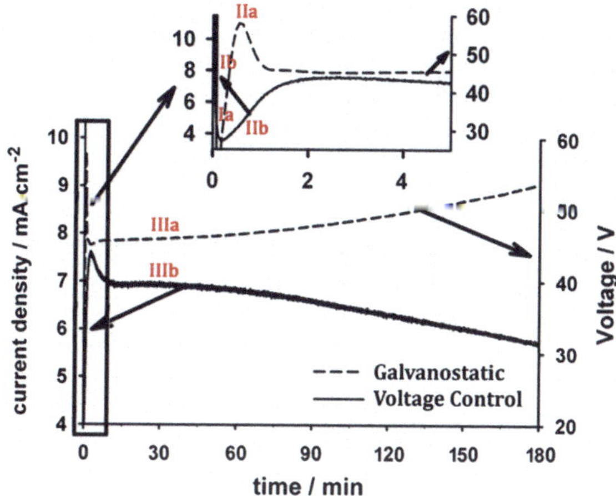

Fig. 3.4 Current density and potential evolution during the first anodizing step carried out at constant voltage or current. The different stages of anodization are indicated: *Ia*, *IIa*, *IIIa* belong to constant current; and *Ib*, *IIb*, *IIIb* belong to constant voltage. Reprinted with permission from [44]

Al surface is observed, resulting in a significant increase of the potential or decrease of the current density for the galvanostatic and potentiostatic process, respectively (stage I). Further continuation of the anodization results in a pore nucleation at surface concavities (stage II) that is responsible for a decrease of recorded potential or increase of current density. Finally, the voltage or current plateau is reached (stage III) as a result of the steady-state growth of nanoporous oxide. During this stage, the thickness of compact barrier layer at the pore bottoms remains constant (the rate of oxide formation at the metal–oxide interface and oxide dissolution at the oxide–electrolyte interface are the same), while the porous layer thickens at a constant rate. It is also noticeable, that after certain time a significant increase of recorded potential or decrease of current density can be observed for galvanostatic and potentiostatic anodizations, respectively. This can be explained in terms of the diffusion limited electrochemical oxidation of aluminum at the pore bottoms. When the nanoporous alumina layer is thick enough, the diffusion of the reactant species from the bulk electrolyte to the pore bottoms can be strongly limited. As a consequence, the composition of the electrolyte at the pore bottoms gradually changes that results in voltage/current changes. When potentiostatic anodizations are performed, the gradual decrease of current density results in a decrease of the oxide growth rate (for further details see our previous work [47]), but all the structural features of anodic alumina, dependent mainly on the applied potential, remain unchanged. An alternative situation is observed when the anodization is carried out at the constant current regime. The local pH changes at the pore bottoms together with increasing pore length result in decrease of the oxide etching rate and, consequently, increase of the barrier layer thickness. As a result, the measured potential

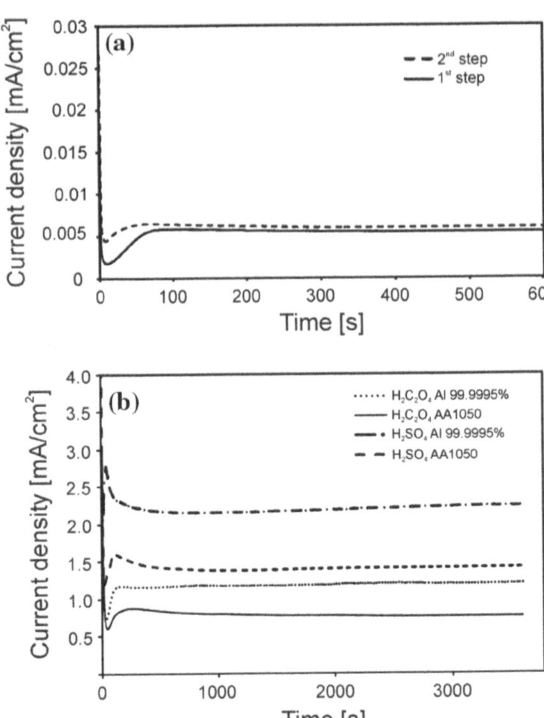

Fig. 3.5 Current density versus time curves recorded during the first and second step of the AA1050 alloy anodization (**a**) and during the second anodizing step of different substrates (**b**)

increases. Moreover, as the interpore distance (cell diameter, D_c) is determined mainly by the anodizing potential (see (3.1)) [48, 49]

$$D_c = -1.7 + 2.81 \cdot U \tag{3.1}$$

the progressive increase of the interpore distance could be observed [44].

The current density versus time curves recorded during the first and second anodizations of AA1050 alloy in 0.3 M oxalic acid at 45 V and 20 °C are shown in Fig. 3.5a. As can be seen, the current density minimum observed during the first anodization is significantly deeper than that recorded in the case of second anodizing step. It is well known that during the first anodization, a periodic structure of concaves is formed on the Al surface. During prolonged anodization of as pre-textured aluminum the lowest resistance and the highest electric field are observed at the bottom of each concave. Hence, the pore nucleation occurs easier, so the lower rate of the barrier oxide growth and more rapid increase in current density are observed [36]. This dependence observed in our group for AA1050 is in excellent agreement with the results obtained e.g., by Li et al. [50], and Wang et al. [51] for anodization of high-purity substrates. Current-time behaviors recorded during second steps of anodization of the AA1050 alloy and high-purity aluminum at 25 V for 0.3 M H_2SO_4 and at 40 V for 0.3 M $H_2C_2O_4$ are shown in Fig. 3.5b.

Table 3.2 Nanoporous oxide growth rates (μm h^{-1}) for different anodizing conditions

Starting material	Anodizing conditions		
	0.3 M H_2SO_4, 25 V, 1 °C	0.3 M $H_2C_2O_4$, 40 V, 1 °C	0.3 M $H_2C_2O_4$, 45 V, 20 °C
AA1050	2.32 ± 0.19	1.61 ± 0.15	10.83 ± 1.36
Al 99.999 %	4.11 ± 0.22	2.20 ± 0.08	14.87 ± 1.91

As can be seen, significantly lower current density values were recorded for anodizations of low-purity substrates. Moreover, an increase as well as a local maximum of current density, being an exact evidence of transformation of the compact layer into porous oxide, followed by the pore rearrangement are achieved at the later time than typically, when high-purity Al is anodized. It means that the formation of pores and their reorganization are significantly retarded on the AA1050 surface. This phenomenon could be attributed to the presence of alloying elements that could affect the cooperative migration of ions through the anodic layer (e.g., Si, Mg and Cu) or the ionic resistivity of the oxide film (e.g., Fe or Mn) [41, 52–55].

The lower current densities during the steady state growth of oxide on the AA1050 alloy are also a reason of lower average oxide growth rates observed on technical Al substrate when compared to high purity aluminum (see Table 3.2). This phenomenon was observed in our group for different sets of anodizing conditions.

Recently, Voon et al. have investigated the effect of Mn content in the Al substrate on the anodizing behavior and the structure of anodic alumina layers formed by potentiostatic anodization in oxalic acid electrolytes [56]. The main conclusions drawn from this study are in general agreement with the results described above, obtained in our group [41, 47]. It was proved that an increase of Mn content in the alloy significantly reduced the steady-state current density values and retarded the pore nucleation process. It is well known that at the beginning of anodization of the Al–Mg alloy, a preferential oxidation of aluminum occurs resulting in the formation of the amorphous oxide layer, while manganese is enriched in the alloy matrix just beneath the growing film [57]. As a result, availability of Al for oxidation is reduced, and the pore nucleation and organization are retarded. As a consequence, the local minimum of the current density as well as the steady state current density are achieved within a longer anodizing time. The lower current densities result also in lower oxide growth rates [56].

On the other hand, no significant effect of the alloying elements such as Mg and Ti on the oxide growth rates in nitric acid electrolyte was found. The similar thicknesses of oxide layers formed on the surface of pure Al and binary alloys were observed. However, the efficiency of the porous alumina formation in this electrolyte is significantly reduced (40 %) due to the reactive nature of HNO_3 to alumina [58].

Tsangaraki-Kaplanoglou et al. have analyzed and compared the anodizing behavior of two different aluminum alloys, AA5083 and AA6111 anodized at constant potential in H_2SO_4 electrolyte. It was found that during anodization of the AA6111 alloy, the stage of pore rearrangement, before the steady-state current stage, is almost completely missing as a result of enrichment of the alloying elements at the metal/oxide interface [59]. On the other hand, when the AA5080 alloy was anodized, no significant difference between the oxide formation on this substrate and high purity Al was found, however, the higher oxide growth rates were observed for the technical alloy [59].

3.2.2.2 Structure of the Nanoporous Oxide

Pore Arrangement

As we mentioned before, the fabrication of nanoporous alumina via anodization of low-purity aluminum with satisfactory nanopore order is not a trivial issue. In most cases, the structure of AAO layers, especially degree of nanopore order as well as the number of defects in the pore arrangement strongly depend on the presence of impurities in the starting material. In 2000, Dasquet et al. reported the one step anodization of two different alloys of various Al contents carried out in different electrolytes [60, 61]. It was found that when the AA1050 alloy with a higher Al content (99.5 wt%) was used as a starting material, the oxide layer exhibits quite well defined morphology, with an array of parallel pores perpendicular to the substrate and the structural features of these layers (e.g., pore diameter, barrier layer thickness, wall thickness, and oxide layer thickness) were dependent on the type of electrolyte [60]. On the other hand, when the 2024 alloy (<95 wt% Al) was used as a substrate for anodization, the "sponge" like structures with non unidirectional pores were obtained independently of the electrolyte used [60].

The fabrication of anodic alumina from the commercially available impure Al foil was also reported by Lo et al., and compared with the oxide layers obtained from the high-purity substrates [40]. Although chemical composition of anodic oxide films and, surprisingly, anodization current behavior for the impure and ultrapure substrates were similar, the morphology of obtained AAO films were quite different. The nanoporous oxides with a good hexagonal pore arrangement were formed by two-step anodization of the high-purity substrate in oxalic acid electrolyte at 40 V that is widely recognized as the self-ordering regime. On the contrary, the porous alumina layers obtained from low-purity aluminum exhibited no planar pore ordering. Moreover, much smaller pores with inconstant and non-circular shapes were obtained, and many of them were merged with neighboring pores. The specific pore lines with spacing between 60 and 120 nm were noticeable (even after second anodizing step) at some areas of the alumina layer (see Fig. 3.6a). These lines follow the direction of rolling in the aluminum foil used for anodizing. As we mentioned in the previous section, some improvement of the nanostructured oxide morphology was achieved by annealing of impure Al foils prior to anodizing. The SEM image of such kind of the nanoporous structure with a quite uniform pore diameter (about 50 nm) and interpore distance (about 80 nm),

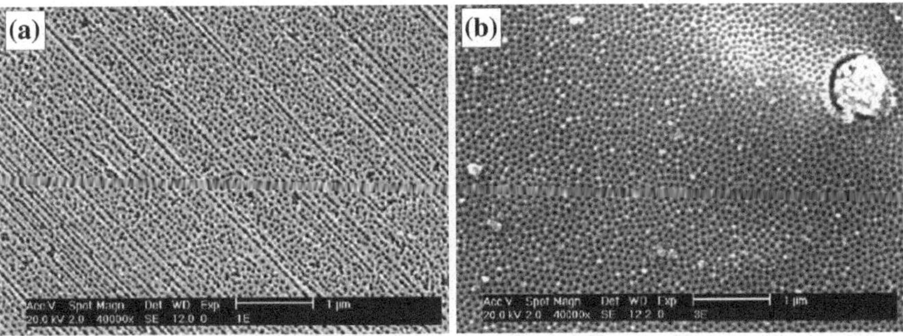

Fig. 3.6 SEM images of AAO layers obtained by anodizing of the impure aluminum foil in 0.3 M oxalic acid after 4 h of the second anodization. The used substrate was subjected to **a** no treatment, and **b** annealing prior to anodizing. Reprinted with permission from [40]

and noticeable only short range hexagonal pore order on the surface of the annealed substrate is shown in Fig. 3.6b.

Very recently, Kim et al. have also examined the effect of aluminum purity on the pores formation by comparison of the morphologies of anodic oxide layers obtained by two-step anodizing of the high purity (99.999 %) an low purity (99.8 %) substrates in oxalic acid and phosphoric acid electrolytes [62]. No direct effect of the substrate composition on the average interpore distance (D_c) was found, since the pore spacing is directly proportional to the anodizing potential (see further sections). On the other hand, less uniform structures with slightly smaller pores were observed when the low purity substrate was used as a starting material. In addition, the number of defective pores, including multiple and branched pores significantly increases when low purity Al is used. Moreover, anodizing of the Al 99.8 % foil in the phosphoric acid solution at 190 V (very close to the self-ordering regime) results in formation of the characteristic mesh-like nanoporous structure. This finding is in excellent agreement with the results obtained previously by our group—an example of such kind of structure synthesized by two-step anodizing of the AA1050 alloy (99.5 % Al) in 0.1 M H_3PO_4 at 190 V is shown in Fig. 3.7 together with the SEM images of nanoporous oxide obtained by anodizing of the high purity substrate. As can be seen, a completely irregular structure of nanopores with double and triple sub-channels underneath was formed on the surface of low purity aluminum. On the contrary, the surface of anodic oxide obtained on pure Al exhibits much less pore defects (see Fig. 3.7b). Moreover, straight nanochannels having no branches and sub-pores can be easily observed (see Fig. 3.7d). The formation of such kinds of disordered structures can be attributed to the fact that the presence of impurities in a given region of the metal/oxide interface influences the local rate of oxide growth that can result in the non-uniform oxide layer. Moreover, impurities can also disturb the direction of the electric field during anodization that results in the formation of branched or meshed pore structures. Such kind of the oxide layer with the non-uniform pore arrangement and high surface area can be beneficial in some practical applications, e.g., gas adsorption (see the last section).

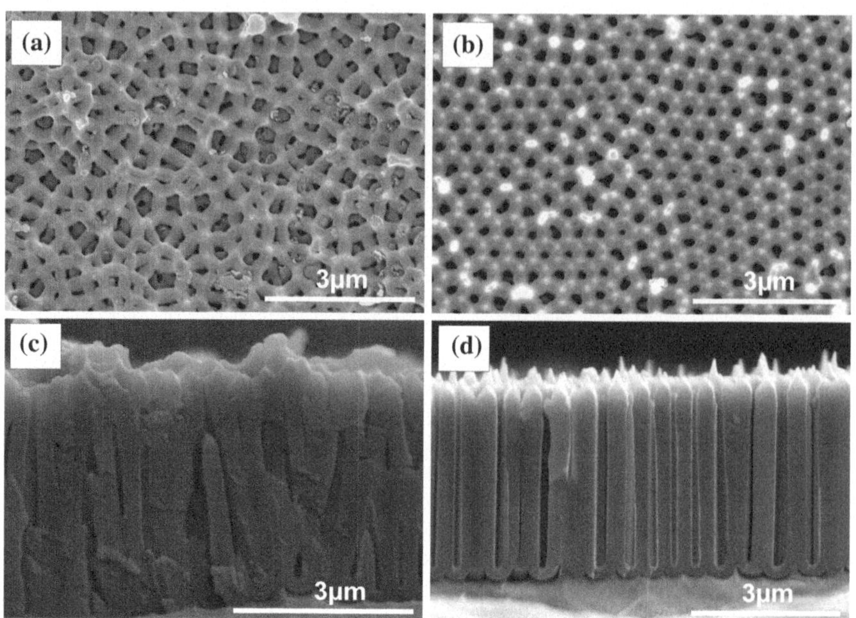

Fig. 3.7 SEM images of nanoporous alumina formed by two-step anodization of the AA1050 alloy (**a**, **c**) and high purity Al (**b**, **d**) in 0.1 M H_3PO_4 at 190 V and 0 °C

In our recent work we performed a qualitative and quantitative comparison of morphology of nanoporous alumina structures grown on the substrates with different Al contents by two-step anodization in different electrolytes [41]. Figure 3.8 shows the SEM micrographs of the samples obtained by anodization of high purity (99.9995 %) aluminum (Figs. 3.8a, c) and AA1050 alloy (Fig. 3.8b, d) in sulfuric acid (Fig. 3.8a, b) and oxalic acid (Fig. 3.8c, d) electrolytes. As can be seen, the two-step anodization of high purity Al under the self-ordering regime results in the formation of nanoporous alumina layers with the almost ideal, hexagonal pore arrangement and high uniformity of pore sizes and pore-to-pore distances, independently of the anodizing electrolyte (Fig. 3.8a, c). On the contrary, when AA1050 was used as a starting material for anodization, the AAO layers with much worse degree of pore ordering and many defects in the hexagonal arrangement of pores were formed (Fig. 3.8b, d). In addition, the networks of pores connected by narrow channels can be noticed in the structures synthesized in a sulfuric acid solution (Fig. 3.8b). The presence of this kind of defects can significantly decrease the mechanical stability of AAO membranes that strongly limits their further practical applications, e.g., as templates for nanofabrication.

Basing on fast Fourier transform (FFT) images taken from the SEM micrographs we also performed detailed, quantitative inspection of the hexagonal pore arrangement in as obtained nanoporous oxides. It is well known, that in the case of ideal, long-range hexagonal order of pores the FFT image consists of six spots

Fig. 3.8 SEM micrographs of anodized high purity aluminum (**a** and **c**), and AA1050 alloy (**b** and **d**) in 0.3 M sulfuric acid (**a** and **b**) and 0.3 M oxalic acid (**c** and **d**). The duration of the second anodizing step and applied voltage was 8 h and 25 V for sulfuric acid or 4 h and 40 V for oxalic acid

distributed on the edges of the hexagon. Any disturbance in the hexagonal pore arrangement results in the ring or even disk shape forms in the FFT pattern. The values of regularity ratio, defined as a ratio of the maximum intensity of the FFT radius profile to the width of the peek at half maximum (for further details of this procedure see our previous works [18, 41, 63]), were calculated for AAO samples obtained from substrates with different Al contents. The significantly larger values of this parameter were obtained for the anodized high purity substrate when compared to those obtained for the anodized AA1050 alloy. It confirms that the worse nanopore ordering is achieved in the anodic alumina layers formed on technical alloys. The same conclusions were drawn from the defect maps, known as Delaunay triangulations, constructed on the basis of pore centers coordinates by using executable paper [64, 65] powered by GridSpace 2 Virtual Laboratory [66]. Over 40 % of defective (non six fold coordinated) pores were found in the structures formed from the AA1050 alloy [36], while a typical percentage of defects of about 20 % is obtained for the samples anodized from the high-purity substrate [21].

Moreover, we confirmed that the higher average circularity of pores or better uniformity of the structure is always observed in the porous anodic alumina films formed by two-step anodizing of the high purity substrate [41].

On the contrary, no significant effect of substrate purity on degree of pore order was found by Michalska-Domańska et al. for the anodized AA1050 alloy in 0.3 M sulfuric acid in a mixture of water and ethylene glycol (3:2, v/v) at various anodizing potentials (from 15 to 35 V) [67]. The FFT-based regularity ratios, calculated on the basis of FFT images of AAOs formed on the surface of AA1050 alloy, were only slightly lower than those obtained for the oxide layers grown on the high-purity substrate. This would suggest that the presence of glycol in the electrolyte could be a quite promising method for improvement of quality of porous alumina obtained from the cheap, technical alloys. This phenomenon could be related with the reduction of anodization current, oxide growth rate and heat produced during the reaction after addition of glycol to the system. However, in our opinion, further systemic studies are required to confirm and fully understand this effect.

On the other hand, Vonn et al. reported that the presence of Mn can strongly influence the uniformity of the oxide structure [56]. In general, the higher Mn content in the substrate, the lower regularity of as synthesized oxide. For the alloys with Mn content <1.0 wt%, it is possible to achieve the hexagonal arrangement of pores, however, their shape and size are less uniform than those obtained from the high-purity substrate [56]. Moreover, a number of pits appear on the surface of AAO layers formed from the substrate with a higher Mn content. The presence of this kind of defects can be attributed to the incorporation of intermetallic particles to the oxide structure (see next section) that results in non-uniform oxidation rates and volume expansion between the intermetallic particles and the aluminum alloy, what in consequence lead to the deformation of neighboring pores. A further increase of the Mn content (up to 2.0 wt%) in the starting material disturbs the hexagonal pore arrangement to a greater extend. As a result, the sizes of the pits and cracks in the oxide structure also increase [56].

Formation of Defects and Cracks in the Oxide Structure

The formation of some characteristic defects in the structure of anodic alumina films obtained by anodization of low purity aluminum alloys was already reported and it was shown that such kind of the specific morphology depends mainly on the substrate composition and nature of alloying elements. The formation of characteristic cracks and hillocks in the nanoporous oxide structure was widely reported for anodizing of the low purity (AA1050) alloy in sulfuric acid electrolytes. Gaston-Garcia et al. attributed the appearance of this kind of defects to the presence of impurities, mainly Fe and Si, in the anodized alloy [68, 69] which initiates the local burning of the sample. On the contrary, Michalska-Domańska et al. have found that the oxide morphology with the number of hillocks on the surface can be formed not only during anodization of technical Al alloys, but also on the surface of high purity aluminum. Careful inspection revealed that the amount of formed hillocks increased significantly when anodizing potential was raised to 30 and 35 V (above the self-ordering regime for H_2SO_4) [67]. Moreover, an increased content of sulfur on the central part of hillocks was found by EDS and XPS analyses that would suggest that a local "burning" of the sample significantly increases the rate of incorporation of sulfate anions to the oxide layer. In our opinion, the presence of impurities in the AA1050 alloy increases the probability of sample "burning" also at lower potentials (e.g., 25 V), so the oxide structures with hillocks and holes can appear much more often on the anodized technical alloy than on the high-purity substrate.

Chung et al. gave a possible explanation for the presence of the characteristic concavities of various sizes that appear very often on the alumina surface formed from the low-purity substrate [70]. Some examples of such kind of oxide structures obtained by our group can be seen in Fig. 3.9 together with the proposed schematic representation of the formation of these defects. It was suggested that the impurities with a higher electrical resistivity can be responsible for the increased generation of

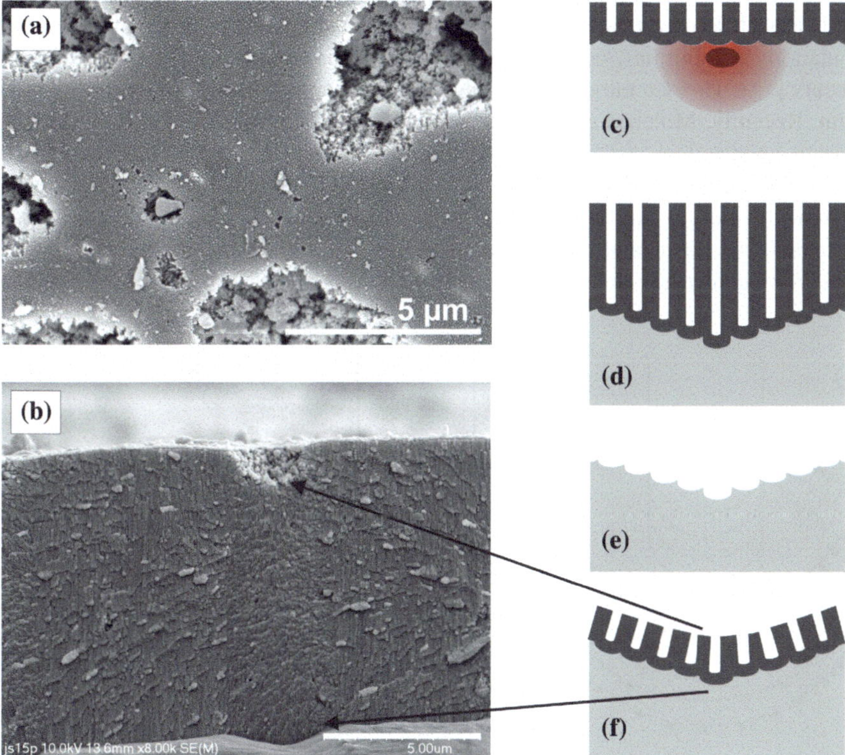

Fig. 3.9 SEM images of the surface (**a**) and cross section (**b**) of the porous alumina layer obtained during two-step anodization of the AA1050 alloy in 0.3 M H_2SO_4 at 25 V and 0 °C, together with the schematic representation of the cavity formation in the AAO structure (**c**–**f**)

Joule heat, so the formation and dissolution rates of porous alumina are accelerated around the impurity (see Fig. 3.9c). As a result, a lot of hemisphere cavities are formed on the substrate surface (see Fig. 3.9d, e). During the second anodization, nanopores grow perpendicular to the surface at the curved metal/oxide interface that can results in branching and bending of nanopores (see Fig. 3.9f) [71].

The effect of Cu impurities on the morphology of porous oxide layers formed by anodizing of the AA2024 alloy in sulfuric acid electrolyte was discussed by Curioni et al. [72]. It was proved that the presence of alloying elements, both in a solid solution and as a second-phase material, has a significant effect on the morphology of the oxide layer. In general, a less regular morphology is developed above the Al alloy matrix (see also previous section), however, the formation of some characteristic morphologies near the second-phase particles was confirmed and attributed to the oxygen evolution occurring during Cu oxidation at the alloy/film interface [72]. This is in general agreement with the results obtained previously by Garcia-Vergara et al. who also ascribed the altered morphology of nanoporous oxide formed on the surface of Al–Cu alloys to generation of oxygen bubbles within the alumina that affects the distribution of stress in the oxide film and the process of pore formation [73].

It is well known that during electropolishing of aluminum, a significant accumulation of impurities, especially Cu in a $\sim$2 nm-thick layer below the surface, occurs [74]. This Cu enrichment strongly affects the subsequent growth of porous film. Recently, Molchan et al. proposed a mechanism of gas-filled void formation during anodization carried out in the phosphoric acid electrolyte (see Fig. 3.10) [75]. Briefly, it is well known that the formation of alumina cells develops hemispherical voids at the metal/oxide interface in which the neighboring cells are bounded by hexagonal cell sides. During anodizing, the movement of the metal/oxide interface along the local electric field is observed, and the speed of this

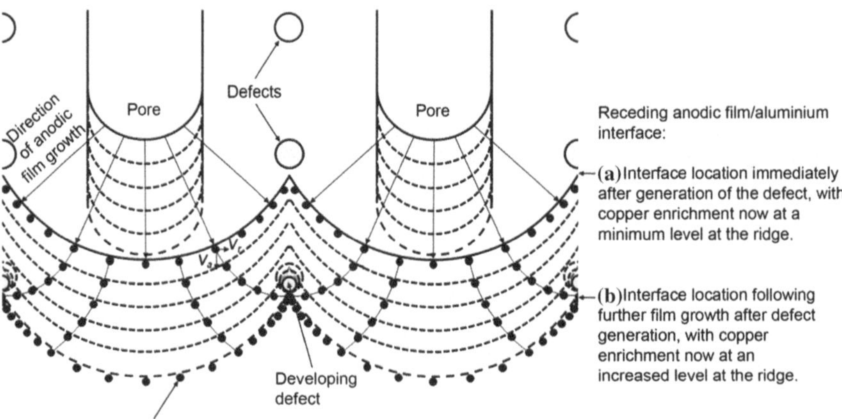

Fig. 3.10 Schematic representation of the formation of copper-rich regions at ridges at the metal/film interface and the subsequent generation of defects. Reprinted with permission from [75]

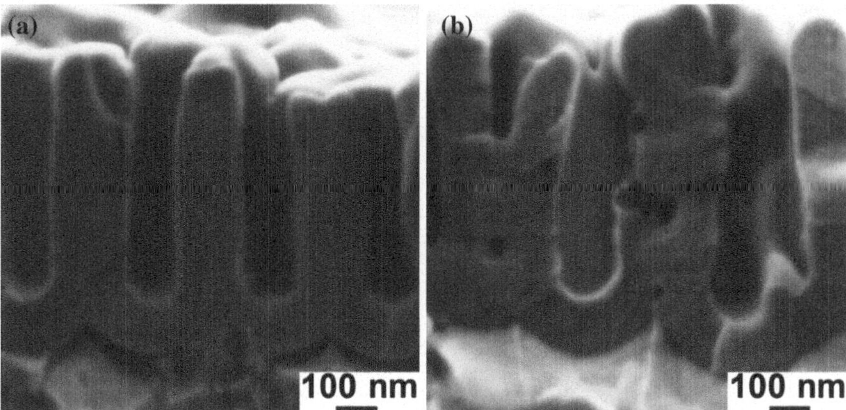

Fig. 3.11 SEM image of AAO obtained at 120 V in 0.3 M H_3PO_4 on magnetron sputtered Al-99.999 % (**a**) and Al–0.05 % Cu alloy (**b**). Reprinted with permission from [75]

migration is determined mainly by the local current density. The movement is directed to the cell boundary at which the accumulation of metal particles occurs. The Cu particles accumulated mainly at the triple-point junctions where neighboring cells meet, are then oxidized that results in the generation of oxygen gas and formation of the oxygen-filled bubbles within the cell boundaries. It should be mentioned that the rate of Cu supply to the triple points strongly depends on such parameters as the oxide growth rate, size of the cell and Al purity. In addition, under appropriate conditions oxygen bubbles can be responsible for the generation of tubular branches inside the pores. The evidence for a crucial role of Cu impurities in the formation of this kind of defects is shown in Fig. 3.11. As can be seen, no defects were observed in the porous alumina film grown on the surface of high purity Al, while a lot of small holes appeared in the oxide film formed on the Al–Cu alloy. On the other hand, it is believed that the controlled formation of defects together with the chemical pore widening can result in well ordered AAO structures with cylindrical nanopores interconnected by branched voids. Such kind of the three-dimensional network in the porous anodic alumina structures were successfully obtained by Chu et al., from various industrial pure Al materials [76]. For instance, when a low purity (99.0 % Al) substrate is used as a starting material, an array of large (c.a. 250 nm in diameter) pores perpendicular to the surface with a lot of small (50–80 nm in diameter) transverse pores with a regular spacing (220 nm) across the pore walls could be formed by anodizing in phosphoric acid electrolyte. What is more, the transverse pores connect all "regular" pores into a network. The formation of the transverse pores was attributed mainly to the chemical dissolution of impurities in the Al alloy during anodizing in a corrosive environment. The SEM images of both types of porous alumina layers formed on a low purity and high purity substrate are shown in Fig. 3.12.

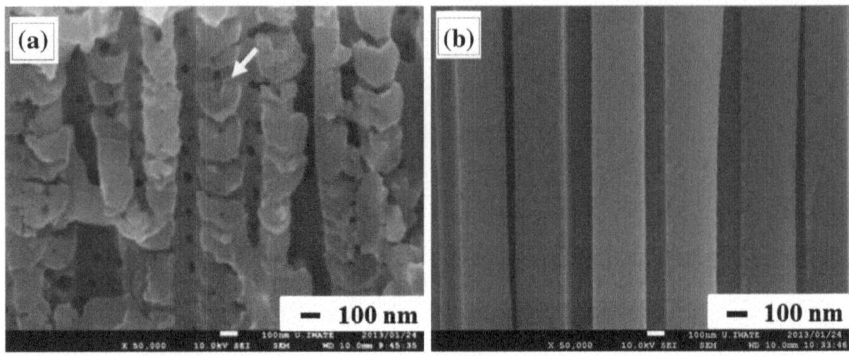

Fig. 3.12 Cross-sectional FE-SEM images of AAO layer synthesized in a H_3PO_4 solution at 160 V from Al sheets of (**a**) Al-99.0 % and (**b**) Al-99.999 %. Reprinted with permission from [76]

According to Fratila-Apachitei et al., during anodization of various aluminum alloys, an accumulation of Si species in the barrier layer occurs [53]. Furthermore, the presence of Si and Fe impurities at the metal/oxide interface significantly reduces the oxide growth rate. The local accumulation of impurities on the alloy surface influences the local oxidation rates and, thus, the morphology of anodic oxide [53–55, 77]. So, a specific microstructure of anodic oxide formed on the surface of the low-purity alloy (e.g., AA1050) can be explained in terms of the increased content of principal alloying elements (Fe and Si) in the barrier oxide layer in comparison to the alloy that results in the increased roughness of the metal/oxide interface. In addition, the occluded Si or Fe–Si particles are oxidized during anodizing that is accompanied by a significant increase in the volume fraction. This non-uniform oxidation results in the formation of cracks and flaws on the surface. What is more, some branched and deflected pores are formed around occluded particles [41, 53, 55].

The formation of characteristic defects in the anodic oxide structure due to the incorporation of intermetallic particles into the alumina layer was confirmed by Montero-Moreno et al. for the AA1050 alloy [44] and by Walmsey et al. for the AA6061 alloy [78]. It is well known that alloying elements can aggregate and form intermetallic compounds such as Al–Fe–Si, Al–Fe and Al–Si. This kind of species exhibits significantly different electrochemical behavior than pure Al. Applying a suitable surface pre-treatment procedure can reduce the amount of intermetallic particles on the alloy surface. However, during anodization, underneath particles can be incorporated into the oxide film, resulting in the formation of defects. It was found that the rate of oxide growth around the impurities is significantly lower that disrupts the homogeneous pore formation. Consequently, the formation of hole-like defects at the bottom of the oxide layer is observed (see Fig. 3.13c). This kind of defects formed in the metal-oxide interface during first anodizing step, are then replicated during second anodizing. The resulting oxide film has a number of local protrusions that can be easily recognized on the top of AAO (see Fig. 3.13a).

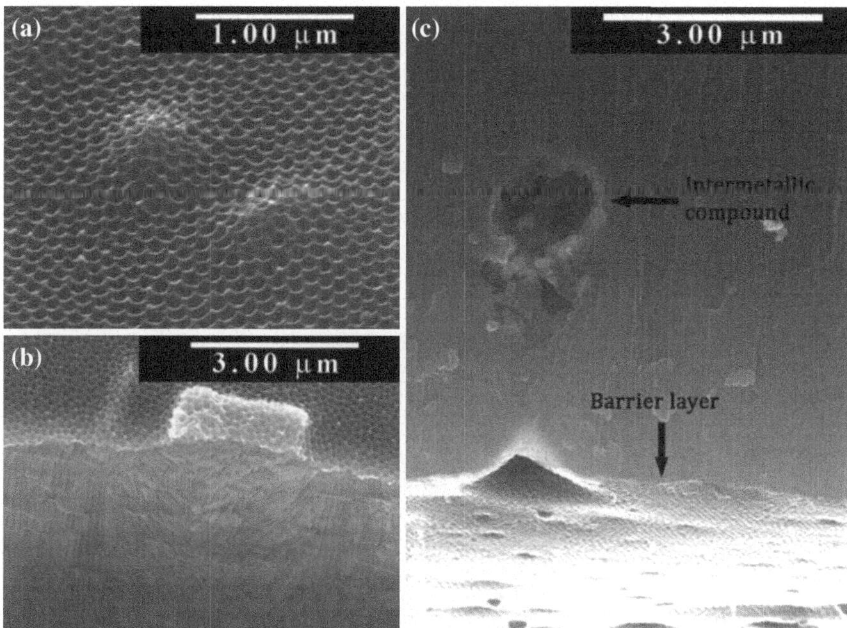

Fig. 3.13 **a** Top view of the porous AAO layer after a 60 min second anodization. The intermetallic compound from the first anodizing step resulted in the formation of the characteristic protrusions on the oxide surface. **b** Cross-section of an intermetallic compound on the surface of the porous AAO layer after a two-step anodization. **c** Cross-section of the AAO layer after the Al substrate was removed. The effect of intermetallic compound presence on the barrier layer side is clearly visible. Reprinted with permission from [44]

The incorporation of intermetallic particles to AAO is schematically illustrated in Fig. 3.14. As can be seen, when the intermetallic particle reaches the metal/oxide interface pores placed over the particle are terminated. With further continuation of the anodization, these terminated pores are replaced by branched and deviated neighboring pores. This deviation is caused by a natural tendency of pores to grow perpendicular to the Al surface. Over time, this local disorder caused by incorporated intermetallic compound is completely recovered, and a normal anodization with parallel pore arrangement occurs at this location. This phenomenon was also confirmed by our group (the SEM image of such kind of structure is shown in Fig. 3.15). The cavities with intermetallic compounds are clearly visible in the SEM cross-section of the AAO layer formed by anodization of the AA1050 substrate. It is obvious that the amount of particles incorporated in AAO and, consequently, the amount of defects within the oxide structure increases with increasing anodizing time. Montero-Moreno et al. suggested that the optimum order in the oxide structure is observed after 60 min of anodization in 0.3 M oxalic acid at the constant current density of 10 mA cm^{-2} [44]. Na et al. have successfully grown the nanoporous alumina layers on the surface of AA1050 alloy by one step anodization in a

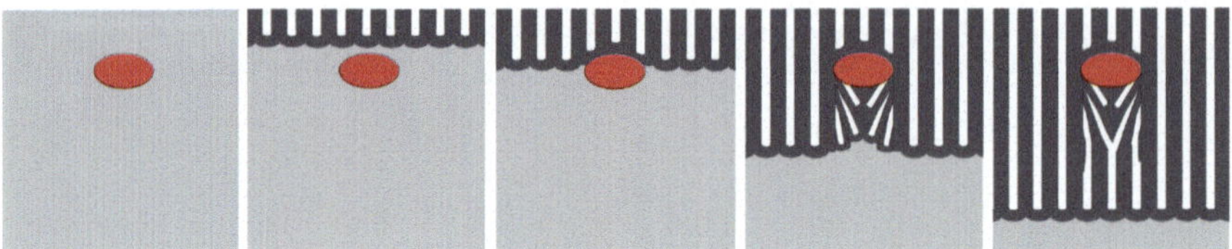

Fig. 3.14 Schematic representation of incorporation of intermetallic particle to porous alumina layer

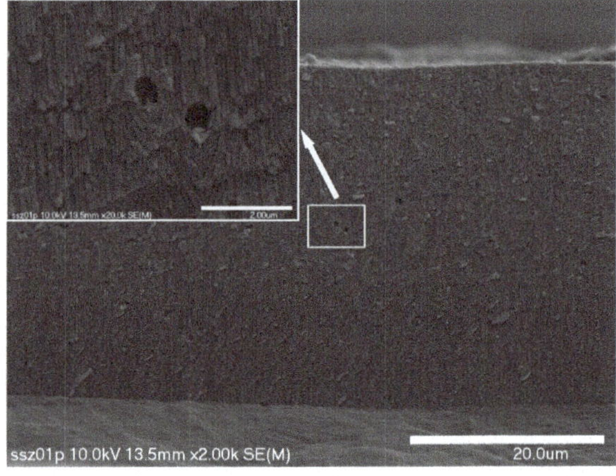

Fig. 3.15 SEM cross-sectional view of the AAO layer formed by anodization of the AA1050 substrate in 0.3 M $H_2C_2O_4$ at 45 V and 20 °C

diluted sulfuric acid electrolyte under the constant current density of 3 A dm^{-2} at different temperatures [46]. The AAO layers with a thickness of 105–284 μm were successfully obtained, however, when the anodizing time was longer than 4 h, a lot of cracks were observed in the cross sectional images. Moreover, the fraction of cracks increases significantly with anodizing time and electrolyte temperature. The crack-free oxide layers were obtained at the anodizing time of 2 h, irrespective of electrolyte temperature [46].

Although the presence of Mg and Ti slightly influences the oxide growth rate (see previous section), it has a significant influence on the morphology of oxide. Mg^{2+} ions incorporate into the alumina film that results in a high-population of localized voids near the alloy/oxide interface [58].

Structural Features of AAO Layers

It is well known that anodizing parameters, such as type and concentration of the electrolyte, applied anodizing potential or current density, temperature, duration of the process, dimensions of the electrodes, stirring speed, or even the geometry of electrochemical cell, can influence the structural features of anodic alumina layers.

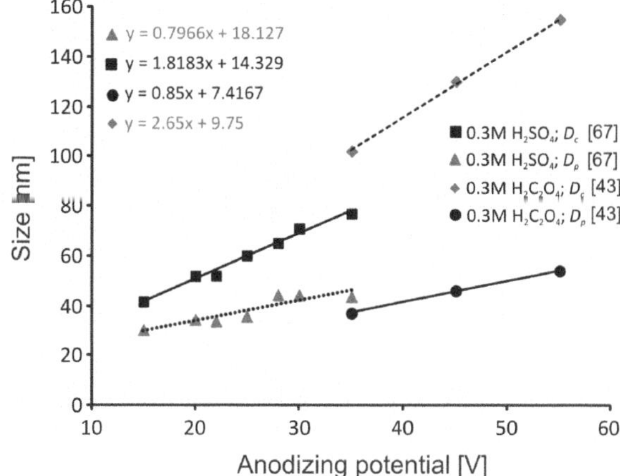

Fig. 3.16 The effect of anodizing potential on the pore diameter (D_p) and cell diameter (interpore distance, D_c) observed for potentiostatic anodizations of the AA1050 alloy in oxalic and sulfuric acid electrolytes

These effects were well established for the high-purity Al anodization (see for example our previous works [4, 16, 21]). However, as we mentioned before, the use of low-purity technical alloys as starting materials for anodization can result in quite different morphologies. Recently, the effect of anodizing conditions on the structural features of AAO layers formed from low-purity Al was extensively studied [e.g., 41, 79, 80].

For instance, Montero-Moreno et al. reported the fabrication of uniform, ordered and quite homogeneous porous alumina layers on the AA1050 alloy by two-step anodizing in oxalic acid [43]. The effect of anodizing voltage on the structural features of as obtained layers was investigated [43]. As expected, the linear relations between the applied potential and pore diameter or cell size were found for the samples produced by so-called "symmetric anodizing" (the same potential was used for both anodizing steps)-see Fig. 3.16. The similar linear relationships between these parameters, observed also for the anodizations carried out in other electrolytes (e.g., sulfuric acid), are also shown in Fig. 3.16 [67]. On the other hand, when "asymmetric anodizing" was performed (different potentials were used in anodizing steps), the cell size was found to be related only with the potential applied during the first anodizing step. However, little influence of the potential applied during the second anodizing step on the oxide layer morphology was noticed. In addition, when the first anodizing potential was higher than the second one, a characteristic structure with more than one pore per cell was formed. An example of this kind of structure is shown in Fig. 3.17 [43].

The same authors have extensively studied the effect of anodizing conditions on the nanoporous AAO structures obtained by a two-step constant-current anodization of the AA1050 alloy [44]. For the galvanostatic anodization of this substrate, the self-ordering phenomenon was observed only when the final potential corresponding to the steady-state oxide growth ($U_{plateau}$) was about 45 V. This potential value is well known as the self-ordering regime for oxalic acid electrolytes [4, 21]. What is interesting, the value of $U_{plateau}$ was successfully achieved for different

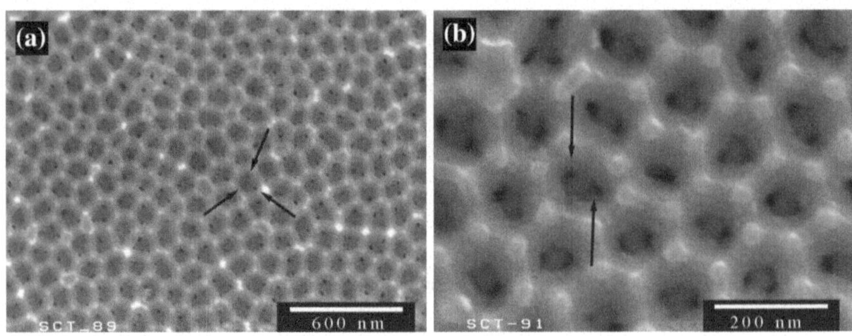

Fig. 3.17 SEM images of the anodic alumina layer after two-step asymmetric anodizing. The *arrows* show more than 1 pore per cell when U_{1AN} is higher than U_{2AN}: **a** U_{1AN} = 55 V, U_{2AN} = 35 V; **b** U_{1AN} = 55 V, U_{2AN} = 45 V. (U_{1AN}—potential applied during the 1st anodizing step, U_{2AN}—potential applied during the 2nd anodizing step). Reprinted with permission from [43]

Table 3.3 The effect of anodizing conditions on the structural features of porous alumina formed by galvanostatic anodization of the AA1050 alloy in oxalic acid

j (mA cm^{-2})	T (°C)	C (mol dm^{-3})	D_c (nm)	$U_{plateau}$ (V)	H (μm)
3.1	10	0.30	119 ± 8	44	5.4
6.6	20	0.30	128 ± 10	45	11.5
10	20	0.60	123 ± 8	46	16.8
10	20	0.30	159 ± 16	53	15.5
10	10	0.60	167 ± 19	60	16.9
10	10	0.30	202 ± 33	67	16.7
10	25	0.30	122 ± 8	43	16.4

j current density, *T* temperature, *c* electrolyte concentration, D_c interpore distance, $U_{plateau}$ final, stable value of anodizing potential recorded during galvanostatic anodization, *H* oxide layer thickness. Duration of both anodizing steps was 60 min. Reprinted with permission from [44]

current densities (3–10 mA cm^{-2}) by careful adjusting other experimental conditions (especially concentration and temperature of the electrolyte). A general rule was established, that higher current densities require higher concentrations or temperatures of the electrolyte to achieve the desired cell voltage. As can be seen in Table 3.3, the anodic alumina layers obtained at the same U_{plateau} exhibited very similar interpore distances, despite the different experimental conditions. In addition, the thickness of the oxide layer depends mainly on the total charge that passed through the system.

The possibility to adjust self-ordering as a function of temperature and electrolyte concentration for each current density introduced a degree of freedom for optimization of the process to obtain the highest possible growth rates [44].

On the other hand, Aerts et al. reported results on the effect of anodizing temperature on the growth of anodic oxide on the surface of AA1050 alloy. Firstly, it was reported, that the pore diameter and porosity of the nanoporous oxide layer formed by anodizing in the H_2SO_4 based electrolyte increase with increasing temperature of the solution [81]. In addition, the same research group has recently found that during the galvanostatic anodization the electrochemical behavior of such kind of Al electrode depends rather on the temperature of the electrode than electrolyte. A normal porous oxide layer can be successfully obtained during anodizing at even high electrolyte temperatures if the electrode is cooled efficiently. However, anodizing in a cool electrolyte at the high electrode temperature results in the formation of the nanoporous AAO structures with a partially collapsed outer part of the oxide layer which was in contact with a warm electrolyte [45]. What is more, the electrode temperature also influences the oxide growth rate much more than the temperature of the electrolyte [45, 82, 83].

Very recently, Vonn et al. studied the effect of anodizing temperature on the growth of nanoporous oxides on the surface of Al–0.5 wt% Mn substrate during the one-step anodization carried out in 0.5 M $H_2C_2O_4$ under 50 V for 60 min at the temperature range of 5–25 °C [84]. The AAO structures with the hexagonal pore arrangement were obtained only for the temperatures up to 15 °C. Moreover, with increasing temperature of anodizing an increase of the oxide growth rate and current density values were observed. However, the current efficiency decreased when the anodizations were carried out at elevated temperatures. This was attributed mainly to the enhanced oxide dissolution in acidic electrolyte. It was found that anodizing temperature does not affect interpore distance of formed AAOs [84].

A composition of the electrolyte used for anodization can also strongly influence the morphology of as obtained oxides. For instance, Chung et al. have observed that an oxide layer thickness and pore diameter increase with increasing concentration of the electrolyte and anodizing time [85]. At higher electrolyte concentrations, the conductivity of the whole cell increases that results in a higher current density and growth rate. On the other hand, the AAO structures with denser pore distributions are formed when more concentrated electrolytes are used for anodizing [85].

Vonn and Derman also proved that the concentration of oxalic acid electrolyte is a crucial parameter for the fabrication of well ordered nanoporous alumina by anodizing of Al–0.5 wt% Mn alloys [86]. The use of electrolyte with too low (0.1 M) or too high (0.7 M) concentrations does not result in AAO layers with a hexagonal pore arrangement. The high-quality, hexagonal nanoporous AAO structures can be only achieved in moderate concentrations of $H_2C_2O_4$ (i.e. 0.3–0.5 M) [86].

The addition of some other substances to the electrolyte could also affect the process of oxide growth as well as the morphology of obtained AAO layers. For instance, Bai et al. obtained well-ordered nanoporous structures by anodization in the electrolyte with a commercially available Al protecting agent [87]. On the other hand, the addition of tartaric acid to the sulfuric acid electrolyte reduced the rate of anodizing by about 10–20 % [88]. The presence of this additive could results in a significant reduction of porosity of the anodic film obtained on the surface of the AA2024 alloy [89].

Michalska-Domańska et al. proposed anodization of the AA1050 alloy (99.5 wt % Al) in a mixture of water and ethylene glycol (3:2 v/v) containing 0.3 M sulfuric acid at various anodizing potentials (from 15 to 35 V) [67]. Surprisingly, as obtained nanoporous oxide layers exhibited quite large pore diameters (>30 nm), especially when compared to those previously formed in the same electrolyte without ethylene glycol [41, 63]. Moreover, the addition of ethylene glycol to the electrolyte results in the reduced current, oxide growth rate and heat produced during the reaction. It should be mentioned that in case when anodizations are carried out in sulfuric acid without any additives, raising the anodizing potential above 25 V often leads to local "burning" of the sample and complete destruction of the nanoporous oxide layer. The addition of ethylene glycol allows for anodization at higher potentials (even up to 40 V). Since the pore diameter is linearly dependent on the applied potential, the anodization at higher voltages must lead to larger pores [67].

Due to the fact that all discussed parameters affect the morphology of anodic oxide formed during anodizing of technical alloys, comprehensive studies have to be performed to optimize the procedure of fabrication of AAO with a desired morphology. Such complex investigations were carried out by Sanz et al. for anodizing in sulfuric acid [90] and oxalic acid [91]. The obtained results are collected in Fig. 3.18. As can be seen, all anodizing conditions, i.e. electrolyte concentration, current density and temperature strongly affect the structural features of AAO layers. For instance, it is clearly visible that an average interpore distance decreases significantly with increasing anodizing temperature (Fig. 3.18a), while the opposite trend is observed for the pore diameter (Fig. 3.18b). On the other hand, the increase in current density results in the increase of both D_c (Fig. 3.18d) and D_p (Fig. 3.18e). The AAO structures with the highest pore density (number of pores

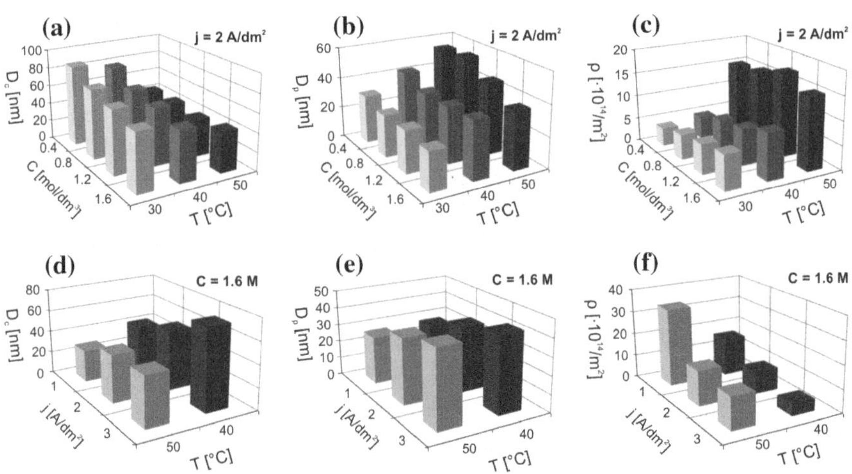

Fig. 3.18 The effect of anodizing conditions such as electrolyte concentration (**a–c**), current density (**d–f**) and temperature on pore diameter (D_p), interpore distance (D_c) and pore density (ρ) of the porous anodic alumina layer formed in oxalic acid by galvanostatic anodization of technical Al. All values taken from the [91]

per surface unit) can be obtained when the anodization is carried out at high temperature and low current density (Fig. 3.18c, f).

The presence of impurities in the starting material can influence not only the uniformity and degree of nanopore order, but also in some cases, the structural features of nanoporous alumina layers. For instance, in our recent work, we demonstrated that the average pore diameter, porosity and pore density of the AAO membranes formed from the AA1050 alloy are slightly smaller than those obtained for the high purity Al foil. An opposite trend was found for the other parameters, such as wall thickness and barrier layer thickness—the higher values were observed for the oxides layers synthesized on the technical alloy surface [41]. The similar results were obtained recently by Michalska-Domańska et al. [67]. Vonn et al. have investigated the effect of Mn content in the starting material on the structural features of anodic aluminum oxide. A significant decrease in the pore diameter and interpore distance was observed with increasing the Mn content in the Al substrate. The authors attributed this phenomenon to the enrichment of Mn species at the metal/oxide interface that disturbs the local electric field during pore nucleation. However, the detailed mechanism of this phenomenon requires further investigations [56].

On the contrary, for anodization performed in phosphoric acid at 160 V Chu et al. reported that the Al content in the range between 99.0–99.999 % does not affect the interpore distance and the barrier layer thickness. These results indicate that the insulating nature of the anodic films formed on different substrates is quite similar [76].

The structural features of as synthesized nanoporous alumina layers can also depend on some other factors. For instance, we found that the sample size could influence the morphology of anodic oxide formed via two-step anodization of the AA1050 alloy in oxalic acid electrolyte. We observed that the pore diameter, interpore distance, and porosity increase slightly with increasing surface area of the aluminum sample exposed to the anodizing electrolyte. On the other hand, a slight decrease in pore density and cell wall thickness was observed when the surface area of the sample was increased [36].

Very recently, Stępniowski et al. have proved that even the cold rolling of the AA1050 foil prior to anodizing can affect not only the uniformity and regularity of AAO layers formed on the surface, but also such parameters as interpore distance and pore density [92].

3.2.2.3 Some Less Common Procedures of Anodization of Low Purity Al Alloys

Besides the common one or two-step anodization carried out at constant potential (potentiostatic) or constant current density (galvanostatic), some other anodizing procedures have been applied for anodizing of technical Al substrates.

For instance, Bai et al. proposed for the AA1050 alloy a new, one-step anodizing procedure that allows for the fabrication of highly-ordered nanoporous AAO layers with uniform and well-controlled pore diameters [87]. The electrolysis was carried out in a mixture of sulfuric acid and oxalic acid with an addition of the commercially available Al protection agent (BH-1 from Yang-Chi Co., Taiwan). Initially, the galvanostatic anodization was performed until the desired voltage was reached. Then, the anodization mode was changed to potentiostatic regime. The applied procedure results in the formation of nanoporous alumina films with pore diameters ranging from 18 to 150 nm. The pore size depends not only on the final anodizing voltage, but also on the concentration of sulfuric acid. Surprisingly, no significant effect of other anodizing parameters such as bath temperature, concentration of oxalic acid and agitation rate on the pore diameter was found. It should be mentioned that the variation in the pore diameter for every sample was relatively small (i.e., ≤10 nm). However, the fabrication of highly ordered AAO films on the surface of AA1050 alloy by using this procedure is almost impossible without using the protection agent. An interesting approach to the fabrication of porous alumina layers on the surface of commercial purity Al (99 %) was proposed and extensively investigated by Chung et al. [70, 85, 93–98]. The authors have compared morphologies of the oxide layers formed by the conventional, potentiostatic method and two different pulse anodizations. The simple pulse anodization (PA) is typically performed by applying a sequence of anodic pulses without any potential applied between the pulses. On the other hand, a hybrid pulse anodization (HPA) is based on applying a sequence of anodic (positive) and cathodic (negative) pulses. In every cycle, two positive potential pulses are separated by a slight negative potential pulse in order to suppress the anodic current between positive pulses (see Fig. 3.19).

It was found that well ordered nanoporous layers could be obtained by using both pulse and pulse reverse methods. The pore arrangement was better when compared to AAO films obtained by the conventional potentiostatic anodization. This improvement in pore order was attributed mainly to the enhanced cooling effect and a less effective generation of Joule's heat during the pulse anodic oxide

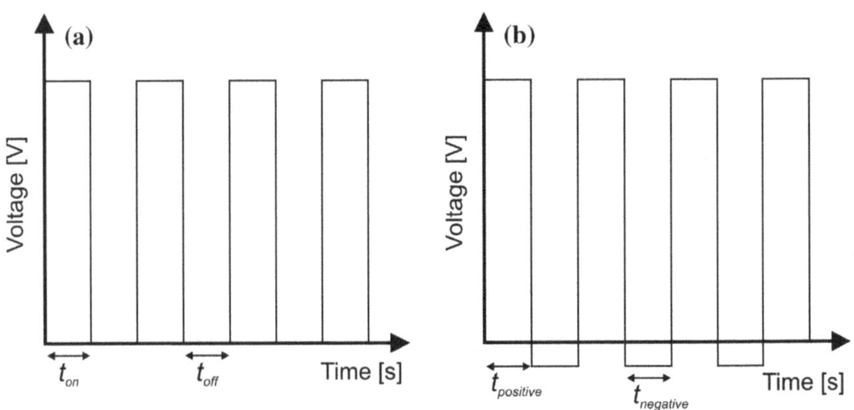

Fig. 3.19 Schematic diagram of **a** simple potential pulses and **b** bipolar potential pulses applied during anodization

growth [95, 97]. It was also reported, that pulse reverse anodizing results in the formation of the porous alumina films with smaller pores and higher pore densities. In addition, the AAO films formed by pulse reverse anodizing have a higher pore density and smaller pore size compared to that produced by the pulse voltage procedure [85]. Moreover, an extension of both positive and negative pulse periods during hybrid pulse anodization leads to an increase in the uniformity and quality of as synthesized nanoporous oxide [94, 95].

A different procedure for the anodic oxide layer formation on the surface of AA2024 alloy was proposed by Iglesias-Rubianes et al. [88]. The anodizations were performed in sulfuric acid based electrolytes by stepping the applied potential up by 3 V increments every 65 s. When the final value of 18 V was reached the constant potential was maintained for 1800 s. The anodic oxide films with a layered morphology in cross-section were produced as a result of the formation of embryo cells of alumina at the metal/film interface. It was confirmed that the thickness of each individual layer was proportional to the anodizing potential. Moreover, the pore formation is accompanied by the increase of Cu content in the alloy and incorporation of copper species into the oxide film. In addition, generated oxygen gas influences integrity of the oxide layer [88].

Recently, Hu et al. reported the fabrication of nanoporous AAO films from the AA2024 alloy deposited on Si wafers by anodizing in 20 wt% H_2SO_4 at low potentials (<20 V) [99]. When conventional potentiostatic anodizations were performed the authors observed stripping of the oxide films, especially at higher anodizing potentials. Therefore, an alternative approach called a current-limited voltage-controlled procedure was proposed. The idea of this protocol is to maintain the constant potential until a maximum current limit is reached. Then, the anodization is continued at the constant current density in the galvanostatic mode [99]. Thus, the rate of anodic oxidation could be precisely controlled. It allows to observe the oxide film evolution during anodizing. Moreover, the rate of pore growth is slow enough to prevent from stripping the alumina off form the underlying metal film during anodic oxidation. By using this method the nanopores with an average diameter from 5 to 15 nm, depending mainly on the applied potential, were achieved. The results indicated that the pore size and interpore distance are fixed in the first 10 s of anodization. Surprisingly, it was found that the barrier layer is only slightly denser than the porous outer layer, that indicates a considerable porosity in the barrier layer. In addition, the degree of hydration of as synthesized oxide is also dependent on anodizing potential and decreases linearly with increasing applied potential [99].

An interesting method for the fabrication of porous alumina on the surface of low-purity (99.7 %) aluminum was reported in 2005 by Chen et al. [100]. The proposed process consists of one anodizing step followed by a long-term heat treatment and pore widening. The results suggest that the disordered nanopore structure formed during anodizing is rearranged into the ordered array as a consequence of self-diffusion inside alumina occurring during long-term heat treatment (self-repairing process). The examples of anodic alumina before and after thermal

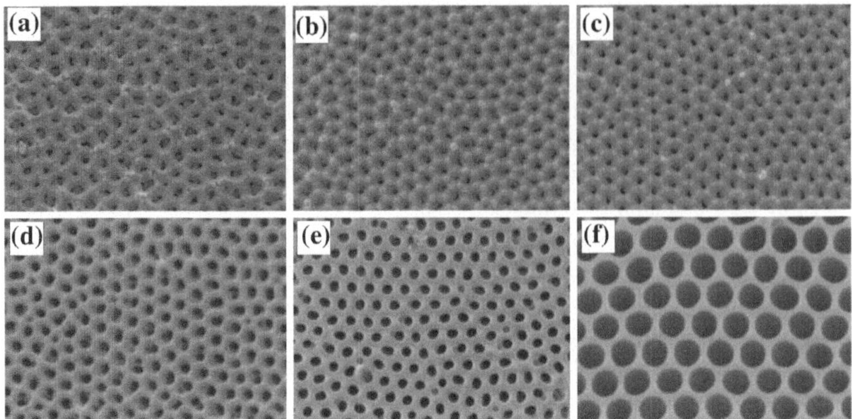

Fig. 3.20 SEM images of AAO after anodization in 0.3 M oxalic acid at 25 °C at 40 V before heat treatment (**a**); after of 4 h heat treatment at 600 °C (**b**); after annealing and 5 min (**c**), 50 min (**d**), 70 min (**e**) and 85 min (**f**) of pore widening. Reprinted with permission from [100]

post-treatment at 600 °C are shown in Fig. 3.20a, b, respectively. As can be seen, the self-diffusion process results in the formation of more uniform structure with less number of sub-pores. After 4 h of annealing, the average pore diameter is reduced of about 10 nm. Moreover, further improvement in the nanopore order was achieved by pore widening. The SEM images of nanoporous AAO layers for various durations of pore widening are shown in Fig. 3.20c, d, e. As can be seen in Fig. 3.20e, a quite uniform and well ordered nanoporous structure with an average pore diameter of 75 nm is obtained after 85 min of pore widening [100]. Unfortunately, it should be mentioned that according to our knowledge the occurrence of the self-repairing process was not confirmed by other research groups [40].

3.2.2.4 Miscellaneous Properties of AAO Layers Obtained from Technical Al

Due to many potential applications (see the last section), the properties of anodic porous alumina layers formed during anodization of low purity Al were extensively investigated. For instance, the thermal stability of nanoporous alumina formed on the surface of AA1050 alloy by one step anodization performed in oxalic acid was studied by Fernandez-Romero et al. [42]. It was proved that the as-obtained amorphous oxide membranes can be converted to polycrystalline by annealing at above 750 °C, and the pore array is maintained during annealing even up to 1100 ° C. However, it should be mentioned, that there was observed a slight change in the morphology of oxide during thermal treatment. In addition, a significant difference in crystallization of the porous and barrier oxide films was confirmed. The porous layer crystallizes mainly in the γ-Al_2O_3 phase during annealing at the temperatures of 900–1100 °C. On the other hand, the barrier layer is converted to the α-Al_2O_3 phase at 1100 °C [42]. It was also confirmed that the presence of barrier layer provides thermal stability to the alumina membranes and prevents them against bending or cracking during annealing. So, the removal of barrier layer before annealing significantly reduces the thermal stability of as obtained AAO membranes. This should be taken into consideration, especially when further applications of nanoporous alumina layers require their stability to thermal shocks [42].

Mechanical properties of the anodic oxide layer formed on the surface of AA1050 alloy by potentiostatic anodization carried out at various temperatures in a H_2SO_4 based electrolyte were examined by Aerts et al. [81]. In general, a significant decrease in microhardness and wear resistance was observed for the oxide layers formed at elevated temperatures. It was attributed to the higher porosities of AAO layers obtained at this conditions.

Rocca et al. synthesized nanoporous alumina layers on the surface of AA7175 alloy via one step anodizing in sulfuric acid and studied the evolution of the nanoporous structure during sealing process [101, 102]. Using various techniques, it was confirmed that the amorphous oxide structure is constituted by 2/3 of aluminum in tetrahedral coordination, 1/3 in octahedral coordination and sulfate species. During contact with hot water, the hydrolysis of AlO_4 into AlO_6 cluster occurs accompanied by a partial release of sulfate ions. The diffusion properties in alumina pores were also examined and correlated with the oxide composition and the pore size. It was confirmed that due to the acidic nature of the tetrahedral aluminum atoms, the oxide surface is positively charged. This fact plays a crucial role in diffusion of ionic species within the pores (diffusion of anions inside the channels is favorable). On the contrary, a rapid diffusion of water molecules along aluminum oxide nanopores was observed [101, 102].

Optical properties of the AAO films formed on the surface of AA1050 alloy by galvanostatic anodization carried out in sulfuric acid electrolytes were investigated by Shih et al. [103]. It was proved that the thickness of AAO film and pore density are main factors that determine the surface reflectance of the oxide layer. An increase in refractive index and decrease in extinction coefficient were observed for increasing pore density. The structures with the highest pore densities were obtained at high temperatures and at relatively low current densities, that indicates an excellent agreement with the results observed by Sanz et al. [91] (see above). Very recently, Canulescu et al. have extensively studied the optical properties and electronic structure of anodic oxides formed by anodizing of Ti–Al alloys with different Ti concentrations (from 2 to 20 at.%). The anodizations were carried out at the constant voltage of 20 V in 20 wt% H_2SO_4 at 18 °C until the observed current density was stabilized at 0.1 mA cm^{-2}. It was found that the band gap energy of Ti-alloyed alumina decreases with increasing Ti concentration [104].

3.3 Practical Applications of AAO from Low Purity Substrates

3.3.1 Template-Assisted Fabrication of Nanowire Arrays

Anodic alumina membranes are widely used as templates for fabrication of various nanostructures such as nanowires, nanotubes or nanodots [9]. Electrodeposition is a common method of depositing of metals inside the channels of alumina layers and it is often used for the fabrication of metallic nanowire arrays [9]. The use of nanoporous membranes formed from low-purity substrates as hard templates is not very popular due to (i) the worse degree of pore order, (ii) the presence of voids and cracks in the oxide structure, and (iii) lower mechanical strength of these membranes in comparison with those obtained from high-purity Al. However, some works on the successful fabrication of metallic nanowire arrays by electrodeposition into nanoporous AAO membranes formed from technical alloys were already published. The examples include Au, Ag, Sn [105], Pd [36] and Ni [106, 107] nanowire arrays. Two main strategies for using AAO membranes as templates for electrodeposition of nanowires have been developed. The first one is based on the through-hole alumina membranes prepared by a suitable post-treatment procedure. This procedure includes the removal of the remaining metallic substrate and selective etching of the dielectric barrier layer from the pore bottoms [18, 36, 47, 105]. The second strategy involves direct AC electrodeposition of metals inside the pores of AAO membranes subjected to electrochemical reduction of the barrier layer thickness [106]. In this approach the removal of the remaining Al substrate is not required, hence the mechanical stability of the template is significantly higher.

Although the fabrication of through-hole nanoporous AAO membranes from commercially available technical Al alloys is difficult, mainly due to their low mechanical stability, we proved that it is possible [36, 105]. The anodic alumina layers were formed by two step anodization of the AA1050 alloy in 0.3 M oxalic acid at 45 V and 20 °C. The duration of the first and second step was 1 and 4 h, respectively. After the second anodization the remaining Al was dissolved in a saturated $HgCl_2$ solution [105] or in a saturated $CuSO_4$ containing 4 wt% of HCl [36]. In order to obtain the through-hole membranes, the pore opening process was performed in 5 wt% phosphoric acid at 25 °C. It should be mentioned that the rate of chemical etching of the barrier layer strongly depends on the electrolyte concentration and temperature, therefore, the process of barrier layer removal should be carefully optimized. The SEM images of bottom side of the AAO membranes subjected to the dissolution of the barrier layer are shown in Fig. 3.21. As can be seen, after 30 min of dissolution (Fig. 3.21a) the pore bottoms are still closed. A complete removal of the barrier layer is achieved after 50 min of pore opening and further etching results in gradual increase of the pore size. After 90 min of immersion, an average pore diameter observed at the bottom side of the membrane

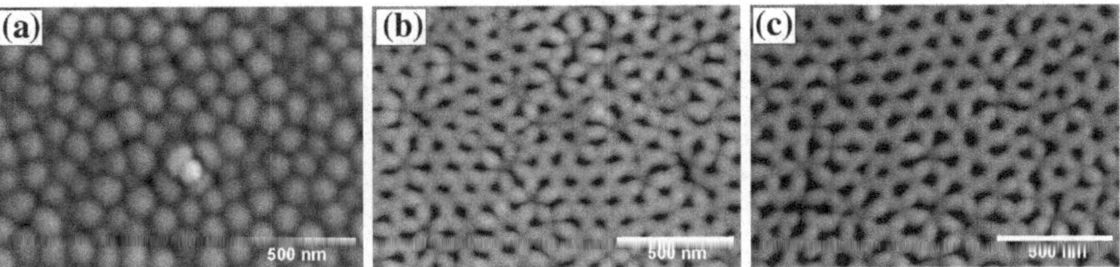

Fig. 3.21 SEM images of the bottom-side of the membranes after various times of immersion in 5 wt% H_3PO_4: 30 min (**a**), 50 min (**b**), and 90 min (**c**)

was about 50 nm. However, contrary to the membranes obtained in the analogical process from the high purity aluminum foil (compare with [18, 47]), uniformity in the pore distribution is much worse and the pore shape is far from a perfect circle. In addition, the pore diameters in AAO membranes fabricated from high purity substrates are higher than those obtained from the AA1050 alloy. It was attributed to an accumulation of alloying elements (especially Si) in the barrier layer during anodization that results in a reduced rate of oxide etching under the same conditions of temperature and concentration of phosphoric acid (for further details see [105]).

The as prepared AAO templates were successfully used for the fabrication of Au, Ag, Sn [105] and Pd [36] nanowire arrays. It should be mentioned that the AAO membranes were easy to handle and their mechanical stability allows for manipulation without a serious damage. The SEM images of various nanowire arrays are shown in Fig. 3.22. The average diameter of individual nanowire is between 80 and 90 nm.

Nanoporous alumina layers produced by anodization of the AA1050 alloy were also used as templates for pulse electrodeposition of Ni nanowires [106]. The method employed two-step anodizing in 0.3 M oxalic acid electrolyte at 45 V and 20 °C, followed by an electrochemical thinning of the barrier layer. This additional process was based on the stepwise reduction of the current during anodization. The current was reduced in five steps, each time by half, and the duration of each step was doubled. As a result, branched tree-shaped structures were created at the pore bottoms. Subsequently, pulse electrodeposition of Ni was applied to form the uniform metal nanowires with the tree-shaped endings, being an exact replication of the pores. In addition, using AAO templates obtained by asymmetric two-step anodizing, the fabrication of Ni nanowires with the diameter of about 25 nm, that is much lower than expected for templates anodized in oxalic acid, was demonstrated [106]. The preparation of the AAO template was based on a stepwise reduction of anodizing potential in two or three steps of 10 V during the first anodization, followed by chemical etching of oxide and a subsequent second anodization carried out at the final potential of the first anodization.

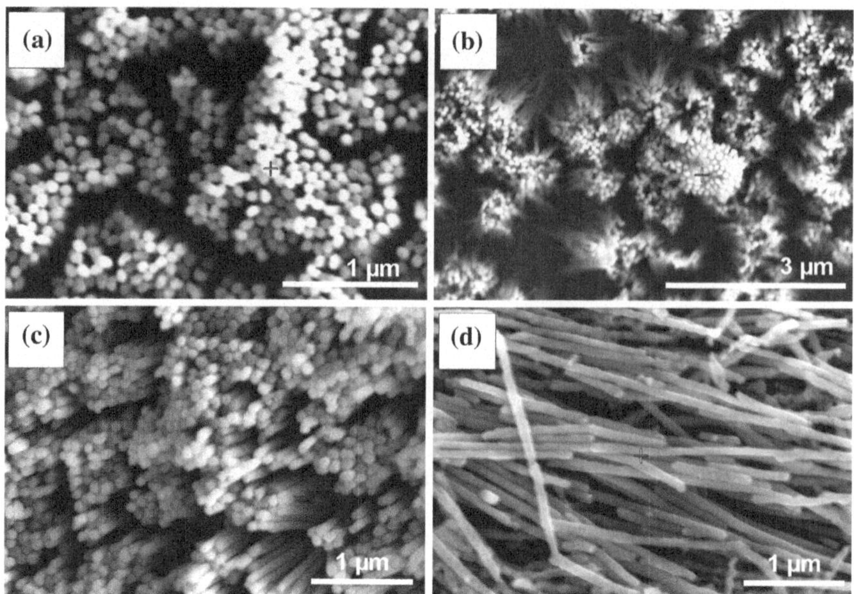

Fig. 3.22 SEM images of Au (**a**), Ag (**b**), Pd (**c**) and Sn (**d**) nanowire arrays obtained by electrodeposition inside the porous AAO templates followed by dissolution of Al_2O_3

3.3.2 Nanoporous Capsules for Biofiltration and Drug Delivery

In 2003, Gong et al. reported the fabrication of nanoporous alumina capsules (Fig. 3.23a) by two-step anodization of the $Al_{98.6}Mn_{1.2}Cu_{0.12}$ alloy pipe in oxalic acid electrolyte followed by the removal of remaining Al and pore opening/widening procedure [108]. The application of such capsules for controlled drug delivery was also presented. The authors successfully obtained the nanoporous oxide capsules with various pore sizes which could be precisely controlled by appropriate adjusting of anodizing potential. The curves of fluorescein release from the capsules of different pore sizes (55, 40 and 25 nm) synthesized by anodization under different potentials (40, 30 and 20 V, respectively) are shown in Fig. 3.23b. As can be seen, the release rate increases nearly seven times with increasing pore size from 25 to 55 nm. Moreover, the capsules with a given pore size are completely permeable to small molecules, but prevents the diffusion of large molecules. These observations suggest that indeed the capsules can be used in biofiltration [108].

Similar tubular nanoporous alumina films were obtained by Belwalker et al. by two-step anodizing of the AA3003 alloy tubes in oxalic and sulfuric acid electrolytes [109]. The authors showed that the nanoporous membranes with quite uniform pores and a narrow pore size distribution can be easily obtained. It was observed

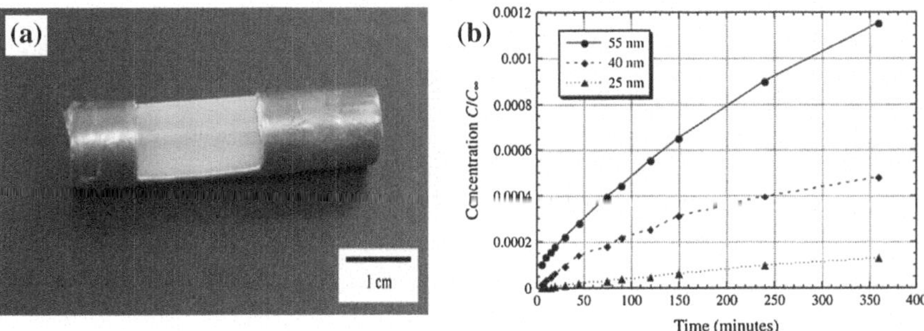

Fig. 3.23 Photograph of the completed tubular AAO capsule (**a**) together with normalized release curves of fluorescein from the capsules of different pore diameters. Reprinted with permission from [108]

that an average pore diameter increases with increasing applied voltage and decreasing concentration of electrolyte. On the contrary, an interpore distance (cell diameter) was found to be a function of the applied voltage only. A well known linear relation between these parameters was confirmed. Finally, it was suggested that the increase in acid concentration results in increasing wall thickness of tubular membrane that improves significantly its mechanical handling. This is especially important from the practical point of view [109].

Recently, Kasi et al. reported the fabrication of AAO tubular membranes by a conventional two-step anodization of technical alloys in oxalic acid and demonstrated their potential application in hemodialysis [110]. The effect of purity of the substrate on the membrane structure was studied. The membranes formed from the alloy with a lower Al content (99.35 %) exhibited poor hexagonal pore order and a lower uniformity in pore size when compared to those obtained from the higher purity substrates (99.58 %). Moreover, the presence of impurities results in the formation of micro size cracks and voids. Fortunately, this kind of impurities and defects do not penetrate deeply inside the membrane. As we mentioned in the previous sections, the "self-repairing" of the structure can be attributed to the self organization. When the micro size impurity blocks the growth of alumina, then the neighboring nanopores start to bend and branch in order to fulfill the gap. As a result, the micro size hole is not created. Concluding, this kind of membranes can be still used in hemodialysis for both diffusive and convective filtration [110].

3.3.3 *Coloring of the Anodic Film*

Coloring of porous anodic alumina films by simple AC electrodeposition of metals is widely reported. Recently, this kind of layers has been a subject of great interest due to their possible implementation, e.g., as light-absorbing coatings in solar cells.

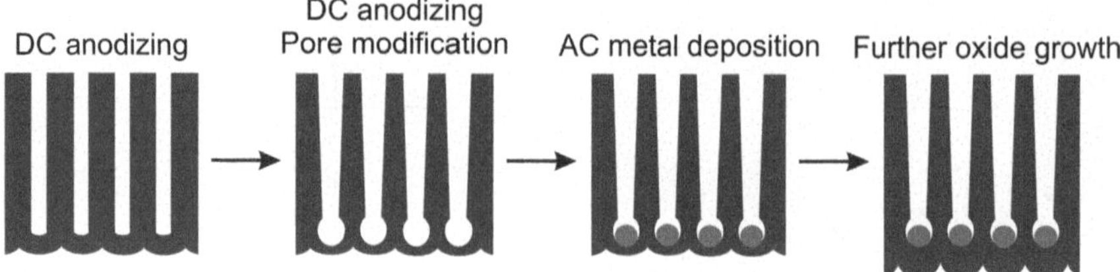

Fig. 3.24 Schematic representation of the formation of electro-colored anodized aluminum

For example, Kanazirski et al. reported the fabrication of porous AAO matrices with the thickness of about 20 μm and porosity of 15 % by simple one step galvanostatic anodization of the technical Al alloy (99.5 % Al). The synthesized porous films were electrochemically colored in three types of electrolytes containing $Ni(SO_3NH_2)_2$ and $CuSO_4$ [111]. The obtained anodic films exhibited high degree of light absorbance combined with good decorative properties.

De Graeve et al. reported the fabrication of electro-colored anodic alumina on the surface of AA5005 alloy [112]. A schematic representation of the applied procedure is shown in Fig. 3.24. At first, an alloy sample was anodized galvanostatically in a sulfuric acid solution to form the conventional porous film structure. The second anodizing step was performed in the same solution or in phosphoric acid in order to modify the pore-base structure (Fig. 3.24b). Further AC electrodeposition of tin or nickel at the pore bottoms resulted in the formation of grey or clear spectral colors depending on the amount of metal deposited as well as on the growth of pores underneath the primary oxide layer during the electro-coloring step. Finally, the porous oxide layer can be sealed by immersing in boiling water for a specific period of time [112].

Shih et al. reported electrochemical coloring of the anodized AA1100 alloy and examined the effect of coloring conditions such as electrolyte concentration, temperature, applied potential and duration of the coloring process on the properties of as obtained films [113]. The films with the darkest shade of black exhibiting a minimum visible light reflectivity (3 %) were obtained in the electrolyte containing 35 g dm^{-3} $CuSO_4$, 10 g dm^{-3} H_2SO_4 and 2 g dm^{-3} Na_2SO_4 at 40 °C at the alternating current voltage of 30 V. It was also found, that coloring parameters, especially temperature have a significant impact on the anti-corrosion properties of these layers [113].

On the other hand, Goueffon et al. extensively studied the formation and properties of black anodic films on aluminum alloys which are often used in space industry due to their promising thermo-optical properties [114–116]. The AA7175 alloy was anodized galvanostatically in the sulfuric acid electrolyte. Then, the coloring step based on the insertion of CoS precipitates into the pores was carried out followed by the sealing of pores. The effects of anodizing parameters and post-treatment procedures on the morphology and mechanical properties of obtained oxide layers were studied in detail. It was found that anodizing conditions, such as temperature, electrolyte concentration, current density and time of anodization, strongly affect the morphology of oxide coatings, especially their thickness and porosity. It was shown that with increasing porosity of anodic layer a significant deterioration of its mechanical properties occurs. On the other hand, the quantity of CoS precipitates was increased with the AAO porosity, however, the precipitated dyes were mainly allocated near the oxide surface. The authors proved that the reduction of initial porosity could be a good way to limit the risk of crazing and flaking without affecting thermo-optical properties of colored layers [116].

3.3.4 Catalyst Supports

Nanoporous alumina layers formed during anodizing of low-purity alloys can be also used as catalyst supports. In this case, a cracked and disordered morphology of anodized oxide can be even beneficial due to the high surface area of oxide layers. For instance, Sanz et al. reported galvanostatic anodization of the AA1002 alloy in the sulfuric acid electrolyte as an effective way for the fabrication of alumina layer over the Al monolith that served as a support for powdered CeO_2 and Au–CeO_2 catalysts [90]. It was shown that the texture and morphology of obtained alumina layers strongly depend on anodizing conditions, especially time, temperature, current density and concentration of the electrolyte, as well as on the applied post-treatment procedure. The AAO layer with a characteristic cracked morphology was successfully obtained under extreme anodizing conditions (30 °C, 50 min, 2 A dm^{-2} and 2.6 M of H_2SO_4). It was also suggested that if the surface cracks are larger in size than the catalyst particles, the adhesion of catalyst coating to the AAO support is significantly enhanced due to mechanical anchoring and interlocking of the active particles. A catalytic activity of as prepared structured catalytic reactors toward oxidation of CO was also proved [90].

On the other hand, Ganley et al. proposed the Ru-impregnated anodic alumina-based microreactors for the production of hydrogen from ammonia. A series of about 300 × 300 μm posts were created on the surface of the AA1100 alloy by using microelectric discharge machining followed by anodization. The resultant AAO structures were about 60 μm thick with an average pore size of about 50 nm. Finally, Ru was dispersed on the alumina posts by a wet impregnation method. It was reported that the arrays of monolithic anodized aluminum posts can achieve a 95 % decomposition of anhydrous ammonia at 650 °C [117]. Moreover, further optimization of the anodization procedure accompanied by applying hydrothermal treatment can significantly increase a specific surface area of alumina which serves as the Ru or Ni metal catalyst support [118].

3.3.5 Fabrication of Superhydrophobic Surfaces

Recently, Peng and co-workers reported the fabrication of hierarchical alumina superhydrophobic surfaces by simple one-step galvanostatic anodization of silica gel coated Al plates (Al 97.17 %) in oxalic acid electrolyte [119]. Depending on the anodizing time various morphologies with different properties were obtained. For instance, the hierarchical morphology such as nanowire bunches-on-nanopores exhibits an extremely slippery property with a sliding angle of 1°. On the other hand, the nanowire pyramids-on-nanopores structure shows strong water adhesion with the adhesive ability to support 15 μl inverted water droplet. Good mechanical durability and resistivity to the ice water, hot water, high temperature, organic solvent, and oil contamination make this kind of materials very promising for many important applications such as microdroplet transport, microfluidic devices and self-cleaning surfaces [119].

The fabrication of the superhydrophobic surface on the 5A06 alloy by galvanostatic anodization in a sulfuric and oxalic acid mixture with an addition of sodium chloride, followed by deposition of the polymer (e.g., polypropylene and polypropylene with a silane coupling agent) coating on the AAO surface was recently reported by Liu et al. [120]. It was found that anodizing conditions strongly influence the superhydrophobicity of the AAO surface. Under the optimal conditions, the best superhydrophobic surface with the contact angle of 162° and the sliding angle of 2° was successfully obtained [120].

3.3.6 Large Scale Fabrication of Nanostructured Low-Cost Aluminum Foil

A very promising method for the nanotexturing of low-cost aluminum foils so-called the roll-to-roll anodization was proposed by Lee et al. together with several examples of practical applications of such nanostructures [121]. In general, a stock aluminum foil travels through a series of rollers to the reaction bath filled with a suitable solution. The temperature of solution as well as rolling speed might be easily controlled by the user. An appropriate potential is applied directly to the Al roll, while two graphite rods serve as the counter electrodes. The roll-to-roll process consists of four main steps (see Fig. 3.25): (i) Al electropolishing to make the substrate smoother; (ii) first anodization that results in the formation of a quite irregular nanoporous oxide layer; (iii) chemical etching of alumina created during the previous step that results in nanobowl Al surface texturing (see below); (iv) second anodization (optional) to form well-ordered nanoporous structures with desired dimensions. It should be mentioned, that each process is carried out

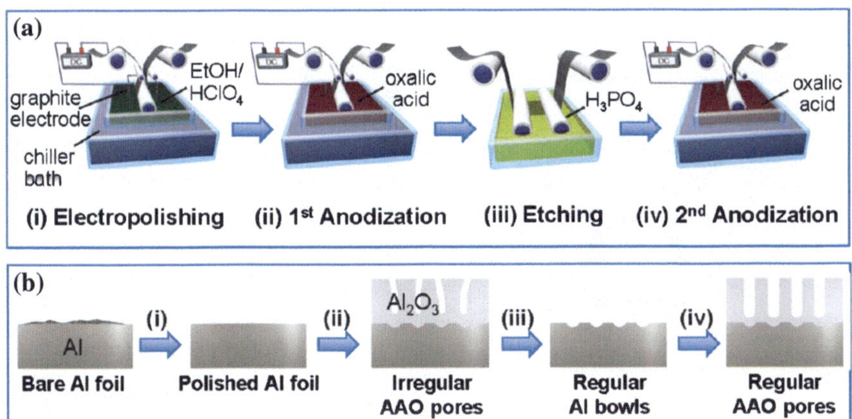

Fig. 3.25 Schematic diagrams of **a** the roll-to-roll anodization procedure and **b** the resulting surface structure after (*i*) electropolishing, (*ii*) first anodization, (*iii*) AAO wet etching, and (*iv*) second anodization steps used for the fabrication of various surface textures. Reprinted with permission from [121]

sequentially by using the same system. Only the electrolyte solution and rolling speed are changed. The optical image of the nanotextured Al roll after anodization and chemical etching is shown in Fig. 3.26a. The schematic representation together with the SEM image of the nanobowl-like Al surface after the removal of anodic oxide formed during the first anodization step is presented in Fig. 3.26c, d, respectively. A dense array of nanoconcaves across the Al surface can be easily seen. Moreover, the size and spacing between nanobowls can be controlled by adjusting the appropriate anodizing conditions. For instance, the anodization at 60 V in the oxalic acid electrolyte followed by alumina wet etching results in formation of the nanobowls with an average diameter of about 95 nm, depth of 75 nm and distance between concaves of about 150 nm. Further continuation of the anodization at the same voltage results in the formation of porous alumina with the similar pore diameter and interpore distance as were obtained for the nanobowls (Fig. 3.26e, f). What is more, the oxide layer thickness is a function of rolling speed and the corresponding anodization time (Fig. 3.26b). Figure 3.26b shows the window of the applicable experimental conditions, marked with the arrows, which results in the generation of array of pores over the entire length of the Al foil. It should be mentioned that, according to the authors, the presented concept can be broaden to larger-scale systems in order to increase the area of the nanostructured surfaces. Moreover, some examples of practical use of this nanotextured surface, e.g., formation of CdS nanocrystals, Ge nanoparticles and nanopillars, polyurethane nanostructures were also presented [121]. It is believed that the technology

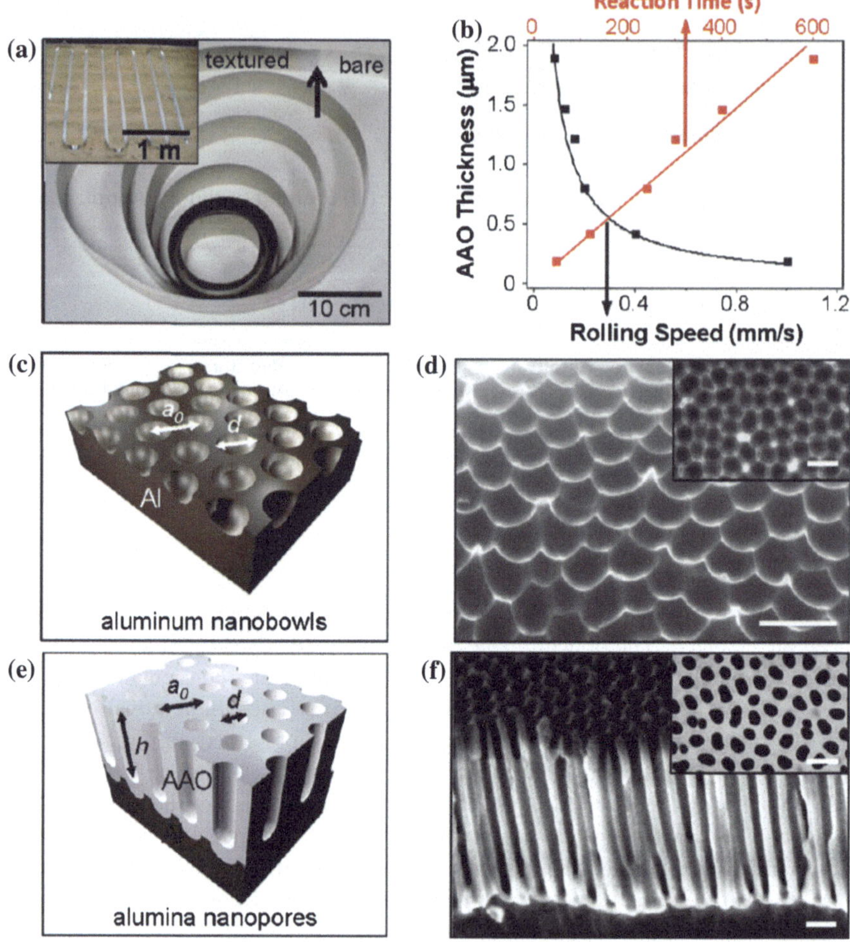

Fig. 3.26 **a** Optical image of a large-area Al roll (length of >10 m) prepared by the roll-to-roll anodization and etching processes (the *inset* shows the same Al foil in an unrolled configuration). **b** The thickness of the anodized alumina as a function of rolling speed (*bottom axis*) and reaction time (*top axis*). **c** Schematic and **d** tilted-view SEM image of the Al foil with nanobowl surface patterns. **e** Schematic and **f** tilted-view SEM image of alumina nanopores. The scale bars are 200 nm. Reprinted with permission from [121]

proposed by the authors can be used for the synthesis of supports for efficient solar cells, self-cleaning surfaces, supercapacitors and in many other practical applications [121].

Acknowledgments Some of the research presented here was supported by the Polish Ministry of Science and High Education (grant no. N N507 481237 and the European COST action D41 Inorganic oxide surfaces and interfaces). A part of the research was carried out with the equipment purchased thanks to the financial support of the European Regional Development Fund in the framework of the Polish Innovation Economy Operational Program (contract no. POIG.02.01.00-12-023/08). Some part of work was made possible with assistance of the PL-Grid project, contract number: POIG.02.03.00-00-007/08-00, website: www.plgrid.pl. The project is co-funded by the European Regional Development Fund as part of the Innovative Economy program. The SEM imaging was performed in the Laboratory of Field Emission Scanning Electron Microscopy and Microanalysis at the Institute of Geological Sciences, Jagiellonian University, Poland.

References

1. B. Bhushan (ed.), *Springer Handbook of Nanotechnology*, 2nd edn. (Springer-Verlag, Berlin, 2007)
2. S. Chen (ed.), *Nanomanufacturing* (American Scientific Publishers, New York, 2010)
3. A. Ghicov, P. Schmuki, Self-ordering electrochemistry: a review on growth and functionality of TiO_2 nanotubes and other self-aligned $MO_{(x)}$ structures. Chem. Commun. **2009**, 2791–2808 (2009)
4. G.D. Sulka, Highly ordered anodic porous alumina formation by self-organised anodising and template-assisted fabrication of nanostructured materials, in *Nanostructured materials in electrochemistry*, ed. by A. Eftekhari (Wiley, Weinheim, 2008), pp. 1–116
5. W. Lee, S.-J. Park, Porous anodic aluminum oxide: anodization and templated synthesis of functional nanostructures. Chem. Rev. **114**, 7487–7556 (2014)
6. G.D. Sulka, J. Kapusta-Kołodziej, A. Brzózka et al., Anodic growth of TiO_2 nanopore arrays at various temperatures. Electrochim. Acta **104**, 526–535 (2013)
7. K. Lee, A. Mazare, P. Schmuki, One-dimensional titanium dioxide nanomaterials: nanotubes. Chem. Rev. **114**, 9385–9454 (2014)
8. C.A. Grimes, G.K. Mor, *TiO_2 Nanotube Arrays: Synthesis, Properties, and Applications* (Springer, Dordrecht, Heidelberg, London, New York, 2009)
9. G.D. Sulka, L. Zaraska, W.J. Stępniowski, Anodic porous alumina as a template for nanofabrication, in *Encyclopedia of Nanoscience and Nanotechnology*, vol. 11, 2nd edn., ed. by H.S. Nalwa (American Scientific Publishers, California, 2011), pp. 261–349
10. H. Masuda, K. Yasui, Y. Sakamoto et al., Ideally ordered anodic porous alumina mask prepared by imprinting of vacuum-evaporated Al on Si. Jpn. Appl. Phys. **40**, L1267–L1269 (2001)
11. H. Masuda, K. Yasui, M. Watanabe et al., Fabrication of through-hole diamond membranes by plasma etching using anodic porous alumina mask. Electrochem. Solid-State Lett. **4**, G101–G103 (2001)
12. H. Masuda, H. Asoh, M. Watanabe et al., Square and triangular nanohole array architectures in anodic alumina. Adv. Mater. **13**, 189–192 (2001)
13. H. Masuda, K. Fukuda, Ordered metal nanohole arrays made by a two-step replication of honeycomb structures of anodic alumina. Science **268**, 1466–1468 (1995)
14. G.D. Sulka, S. Stroobants, V. Moshchalkov et al., Synthesis of well-ordered nanopores by anodising aluminium foils in sulphuric acid. J. Electrochem. Soc. **149**, D97–D103 (2002)
15. G.D. Sulka, M. Jaskuła, Defect analysis in self-ordered nanopore structures grown by anodization of aluminium at various temperatures. J. Nanosci. Nanotechnol. **6**, 3803–3811 (2006)
16. G.D. Sulka, K.G. Parkoła, Anodizing potential influence on well-ordered nanostructures formed by anodization of aluminium in sulphuric acid. Thin Solid Films **515**, 338–345 (2006)
17. G.D. Sulka, W.J. Stępniowski, Structural features of self-organized nanopore arrays formed by anodization of aluminum in oxalic acid at relatively high temperatures. Electrochim. Acta **54**, 3683–3691 (2009)

18. G.D. Sulka, A. Brzózka, L. Zaraska et al., Through-hole membranes of nanoporous alumina as templates for fabricating silver and tin nanowire arrays. Electrochim. Acta **55**, 4368–4376 (2010)
19. L. Zaraska, W.J. Stępniowski, E. Ciepiela et al., The effect of anodizing temperature on structural features and hexagonal arrangement of nanopores in alumina synthesized by two-step anodizing in oxalic acid. Thin Solid Films **534**, 155–161 (2013)
20. L. Zaraska, W.J. Stępniowski, G.D. Sulka et al., Analysis of nanopore arrangement and structural features of anodic alumina layers formed by two-step anodizing in oxalic acid using the dedicated executable software. Appl. Phys. A **114**, 571–577 (2014)
21. L. Zaraska, W.J. Stępniowski, M. Jaskuła et al., Analysis of nanopore arrangement of porous alumina layers formed by anodizing in oxalic acid at relatively high temperatures. Appl. Surf. Sci. **305**, 650–657 (2014)
22. L. Zaraska, G.D. Sulka, M. Jaskuła, The effect of n-alcohols on porous anodic alumin formed by self-organized two-step anodizing of aluminum in phosphoric acid. Surf. Coat. Technol. **204**, 1729–1737 (2010)
23. L. Zaraska, G.D. Sulka, M. Jaskuła, Properties of nanostructures obtained by anodization of aluminum in phosphoric acid at moderate potentials. J. Phys: Conf. Ser. **146**, 012020 (2009)
24. T. Kikuchi, O. Nishinaga, S. Natsui et al., Self-ordering behavior of anodic porous alumina via selenic acid anodizing. Electrochim. Acta **137**, 728–735 (2014)
25. W.J. Stępniowski, M. Norek, M. Michalska-Domańska et al., Fabrication of anodic aluminum oxide with incorporated chromate ions. Appl. Surf. Sci. **259**, 324–330 (2012)
26. T. Kikuchi, T. Yamamoto, S. Natsui et al., Fabrication of anodic porous alumina by squaric acid anodizing. Electrochim. Acta **123**, 14–22 (2014)
27. S. Ono, M. Saito, H. Asoh, Self-ordering of anodic porous alumina formed in organic acid electrolytes. Electrochim. Acta **51**, 827–833 (2005)
28. X. Chen, D. Yu, L. Cao et al., Fabrication of ordered porous anodic alumina with ultra-large interpore distances using ultrahigh voltages. Mater. Res. Bull. **57**, 116–120 (2014)
29. T. Kikuchi, T. Yamamoto, R.O. Suzuki, Growth behavior of anodic porous alumina formed in malic acid solution. Appl. Surf. Sci. **284**, 907–913 (2013)
30. T. Kikuchi, D. Nakajima, J. Kawashima et al., Fabrication of anodic porous alumina via anodizing in cyclic oxocarbon acids. Appl. Surf. Sci. **313**, 276–285 (2014)
31. S. Stojadinovic, R. Vasilic, I. Belca et al., Structural and luminescence characterization of porous anodic oxide films on aluminum formed in sulfamic acid solution. Appl. Surf. Sci. **255**, 2845–2850 (2008)
32. T. Kikuchi, O. Nishinaga, S. Natsui et al., Fabrication of anodic nanoporous alumina via acetylenedicarboxylic acid anodizing. ECS Electrochem. Lett. **7**, C25–C28 (2014)
33. M. Saenz de Miera, M. Curioni, P. Skeldon et al., Modelling the anodizing behaviour of aluminium alloys in sulphuric acid through alloy analogues. Corros. Sci. **50**, 3410–3415 (2008)
34. J.M. Montero-Moreno, M. Sarret, C. Müller, Influence of the aluminum surface on the final results of a two-step anodizing. Surf. Coat. Technol. **201**, 6352–6357 (2007)
35. C.U. Yu, C.C. Hu, A. Bai et al., Pore-size dependence of AAO films on surface roughness of Al-1050 sheets controlled by electropolishing coupled with fractional factorial design. Surf. Coat. Technol. **201**, 7259–7265 (2007)
36. L. Zaraska, E. Kurowska, G.D. Sulka et al., The effect of anode surface area on nanoporous oxide formation during anodizing of low purity aluminum (AA1050) alloy. J. Solid State Electrochem. **18**, 361–368 (2014)
37. L.B. Kong, Y. Huang, Y. Guo et al., A facile approach to preparation of nanostripes on the electropolished aluminum surface. Mater. Lett. **59**, 1656–1659 (2005)
38. V.V. Yuzhakov, S.C. Chang, A.E. Miller, Pattern formation during electropolishing. Phys. Rev. B **56**, 12608–12624 (1997)
39. N. Wang, W.D. Zhang, J.P. Xu et al., Fabrication of anodic aluminum oxide templates with small interpore distances. Chin. Phys. Lett. **27**, 066801 (2010)

40. D. Lo, R.A. Budiman, Fabrication and characterization of porous anodic alumina films from impure aluminum foils. J. Electrochem. Soc. **154**, C60–C66 (2007)
41. L. Zaraska, G.D. Sulka, J. Szeremeta et al., Porous anodic alumina formed by anodization of aluminum alloy (AA1050) and high purity aluminum. Electrochim. Acta **55**, 4377–4386 (2010)
42. L. Fernandez-Romero, J.M. Montero-Moreno, E. Pellicer et al., Assessment of the thermal stability of anodic alumina membranes at high temperatures. Mater. Chem. Phys. **111**, 542–547 (2008)
43. J.M. Montero-Moreno, M. Sarret, C. Müller, Some considerations on the influence of voltage in potentiostatic two-step anodizing of AA1050. J. Electrochem. Soc. **154**, C169–C174 (2007)
44. J.M. Montero-Moreno, M. Sarret, C. Müller, Self-ordered porous alumina by two-step anodizing at constant current: behaviour and evolution of the structure. Micropor. Mesopor. Mat. **136**, 68–74 (2010)
45. T. Aerts, J.B. Jorcin, I. De Graeve et al., Comparison between the influence of applied electrode and electrolyte temperatures on porous anodizing of aluminium. Electrochim. Acta **55**, 3957–3965 (2010)
46. H.C. Na, T.J. Sung, S.H. Yoon et al., Formation of unidirectional nanoporous structures in thickly anodized aluminum oxide layer. Trans. Nonferrous Met. Soc. China **19**, 1013–1017 (2009)
47. L. Zaraska, G.D. Sulka, M. Jaskuła, Anodic alumina membranes with defined pore diameters and thicknesses obtained by adjusting the anodizing duration and pore opening/widening time. J. Solid State Electrochem. **15**, 2427–2436 (2011)
48. A.P. Li, F. Muller, A. Birner et al., Hexagonal pore arrays with a 50–420 nm interpore distance formed by self-organization in anodic alumina. J. Appl. Phys. **84**, 6023–6026 (1998)
49. K. Ebihara, H. Takahashi, M. Nagayama, Structure and density of anodic oxide films formed on aluminum in oxalic acid solutions. J. Met. Finish. Soc. Jpn. **34**, 548–553 (1983)
50. F. Li, L. Zhang, R.M. Metzger, On the growth of highly ordered pores in anodized aluminum oxide. Chem. Mater. **10**, 2470–2480 (1998)
51. N. Wang, J. Xu, W. Zhang et al., Initial stage of pore formation process in anodic aluminum oxide template. J. Solid State Electrochem. **14**, 1377–1382 (2010)
52. H. Habazaki, K. Shimitzu, P. Skeldon et al., Formation of amorphous anodic oxide films of controlled composition on aluminium alloys. Thin Solid Films **300**, 131–137 (1997)
53. L.E. Fratila-Apachitei, F.D. Tichelaar, G.E. Thompson et al., A transmission electron microscopy study of hard anodic oxide layers on AlSi(Cu) alloys. Electrochim. Acta **49**, 3169–3177 (2004)
54. L.E. Fratila-Apachitei, H. Terryn, P. Skeldon et al., Influence of substrate microstructure on the growth of anodic oxide layers. Electrochim. Acta **49**, 1127–1140 (2004)
55. L.E. Fratila-Apachitei, J. Duszczyk, L. Katgerman, Voltage transients and morphology of AlSi(Cu) anodic oxide layers formed in H_2SO_4 at low temperature. Surf. Coat. Technol. **157**, 80–94 (2002)
56. C.H. Voon, M.N. Derman, U. Hashim, Effect of manganese content on the fabrication of porous anodic alumina. J. Nanomater. **2012**, 752926 (2012)
57. A.C. Crossland, G.E. Thompson, C.J.E. Smith et al., Formation of manganese-rich layers during anodizing of Al-Mn alloys. Corros. Sci. **41**, 2053–2069 (1999)
58. J.F. Garcia-Garcia, E.F. Koroleva, G.E. Thompson et al., Anodic film formation on binary Al–Mg and Al–Ti alloys in nitric acid. Surf. Interface Anal. **42**, 258–263 (2010)
59. I. Tsangaraki-Kaplanoglou, S. Theohari, T. Dimogerontakis et al., Effect of alloy types on the anodizing process of aluminum. Surf. Coat. Technol. **200**, 2634–2641 (2006)
60. J.P. Dasquet, D. Caillard, E. Conforto et al., Investigation of the anodic oxide layer on 1050 and 2024T3 aluminium alloys by electron microscopy and electrochemical impedance spectroscopy. Thin Solid Films **371**, 183–190 (2000)
61. J.P. Dasquet, J.P. Bonino, D. Caillard et al., Zinc impregnation of the anodic oxidation layer of 1050 and 2024 aluminium alloys. J. Appl. Electrochem. **30**, 845–853 (2000)

62. B. Kim, J.S. Lee, Effect of aluminum purity on the pore formation of porous anodic alumina. Bull. Korean Chem. Soc. **35**, 349–352 (2014)
63. G.D. Sulka, K.G. Parkoła, Temperature influence on well-ordered nanopore structures grown by anodization of aluminium in sulphuric acid. Electrochim. Acta **52**, 1880–1888 (2007)
64. E. Ciepiela, L. Zaraska, G.D. Sulka, GridSpace2 Virtual Laboratory case study: Implementation of algorithms for quantitative analysis of grain morphology in self-assembled hexagonal lattices according to Hillebrand method. Lect. Notes Comput. Sci. **7136**, 240–251 (2012)
65. E. Ciepiela, L. Zaraska, G.D. Sulka, Implementation of algorithms of quantitative analysis of the grain morphology in self-assembled hexagonal lattices according to Hillebrand method. http://gs2.cyfronet.pl/epapers/hillebrand-grains/. Accessed Oct 2014 (2011)
66. P. Nowakowski, E. Ciepiela, D. Harężlak et al., The collage authoring environment. Procedia Comput. Sci. **4**, 608–617 (2011)
67. M. Michalska-Domańska, M. Norek, W.J. Stepniowski et al., Fabrication of high quality anodic aluminum oxide (AAO) on low purity aluminum—a comparative study with the AAO produced on high purity aluminum. Electrochim. Acta **105**, 424–432 (2013)
68. B. Gastón-García, E. García-Lecina, J.A. Díez et al., Local burning phenomena in sulfuric acid anodizing: analysis of porous anodic alumina layers on AA1050. Electrochem. Solid State Lett. **11**, C33–C35 (2010)
69. J.J. Roa, B. Gastón-García, E. García-Lecina et al., Mechanical properties at nanometric scale of alumina layers formed in sulphuric acid anodizing under burning conditions. Ceram. Int. **38**, 1627–1633 (2012)
70. C.K. Chung, M.W. Liao, C.T. Lee et al., Anodization of nanoporous alumina on impurityinduced hemisphere curved surface of aluminum at room temperature. Nanoscale Res. Lett. **6**, 596–601 (2011)
71. A. Yin, R.S. Guico, J. Xu, Fabrication of anodic aluminium oxide templates on curved surfaces. Nanotechnology **18**, 035304 (2007)
72. M. Curioni, M. Saenz de Miera, P. Skeldon et al., Macroscopic and local filming behavior of AA2024 T3 aluminum alloy during anodizing in sulfuric acid electrolyte. J. Electrochem. Soc. **155**, C387–C395 (2008)
73. S.J. Garcia-Vergara, K. El Khazmi, P. Skeldon et al., Influence of copper on the morphology of porous anodic alumina. Corros. Sci. **48**, 2937–2946 (2006)
74. C.E. Caicedo-Martinez, E. Koroleva, P. Skeldon et al., Behavior of impurity and minor alloying elements during surface treatments of aluminum. J. Electrochem. Soc. **149**, B139–B145 (2002)
75. I.S. Molchan, T.V. Molchan, N.V. Gaponenko et al., Impurity-driven defect generation in porous anodic alumina. Electrochem. Commun. **12**, 693–696 (2010)
76. S.-Z. Kure-Chu, K. Osaka, H. Yashiro et al., Controllable fabrication of networked three-dimensional nanoporous anodic alumina films on low-purity Al materials. J. Electrochem. Soc. **162**, C24–C34 (2015)
77. A.K. Mukhopadhyay, A.K. Sharma, Influence of Fe-bearing particles and nature of electrolyte on the hard anodizing behaviour of AA 7075 extrusion products. Surf. Coat. Technol. **92**, 212–220 (1997)
78. J.C. Walmsley, C.J. Simensen, A. Bjørgum et al., The structure and impurities of hard DC anodic layers on AA6060 aluminium alloy. J. Adhesion **84**, 543–561 (2008)
79. M. Schneider, K. Kremmer, S.K. Weidmann, W. Fürbeth, Interplay between parameter variation and oxide structure of a modified PAA process. Surf. Interface Anal. **45**, 1503–1509 (2013)
80. M. Schneider, K. Kremmer, The effect of bath aging on the microstructure of anodic oxide layers on AA1050. Surf. Coat. Technol. **246**, 64–70 (2014)
81. T. Aerts, T. Dimogerontakis, I. De Graeve et al., Influence of the anodizing temperature on the porosity and the mechanical properties of the porous anodic oxide film. Surf. Coat. Technol. **201**, 7310–7317 (2007)

82. T. Aerts, I. De Graeve, H. Terryn, Control of the electrode temperature for electrochemical studies: A new approach illustrated on porous anodizing of aluminium. Electrochem. Commun. **11**, 2292–2295 (2009)
83. T. Aerts, E. Tourwé, R. Pintelon et al., Modelling of the porous anodizing of aluminium: generation of experimental input data and optimization of the considered model. Surf. Coat. Technol. **205**, 4388–4396 (2011)
84. C.H. Voon, M.N. Derman, U. Hashim et al., Effect of temperature of oxalic acid on the fabrication of porous anodic alumina from Al-Mn alloys. J. Nanomater. **2013**, 167047 (2013)
85. C.K. Chung, R.X. Zhou, T.Y. Liu et al., Hybrid pulse anodization for the fabrication of porous anodic alumina films from commercial purity (99 %) aluminum at room temperature. Nanotechnology **20**, 055301 (2009)
86. C.H. Voon, M.N. Derman, Effect of electrolyte concentration on the growth of porous anodic aluminium oxide (AAO) on Al-Mn alloys. Adv. Mater. Res. **626**, 610–614 (2013)
87. A. Bai, C.C. Hu, Y.F. Yang et al., Pore diameter control of anodic aluminum oxide with ordered array of nanopores. Electrochim. Acta **53**, 2258–2264 (2008)
88. L. Iglesias-Rubianes, S.J. Garcia-Vergara, P. Skeldon et al., Cyclic oxidation processes during anodizing of Al–Cu alloys. Electrochim. Acta **52**, 7148–7157 (2007)
89. G. Boisier, N. Pébère, C. Druez et al., FESEM and EIS study of sealed AA2024 T3 anodized in sulfuric acid electrolytes: influence of tartaric acid. J. Electrochem. Soc. **155**, C521–C529 (2008)
90. O. Sanz, T.L.M. Martínez, F.J. Echave et al., Aluminium anodisation for Au-CeO_2/Al_2O_3-Al monoliths preparation. Chem. Eng. J. **151**, 324–332 (2009)
91. O. Sanz, F.J. Echave, J.A. Odriozola et al., Aluminum anodization in oxalic acid: controlling the texture of Al_2O_3/Al monoliths for catalytic applications. Ind. Eng. Chem. Res. **50**, 2117–2125 (2011)
92. W.J. Stępniowski, M. Michalska-Domańska, M. Norek et al., Anodization of cold deformed technical purity aluminum (AA1050) in oxalic acid. Surf. Coat. Technol. **238**, 268–274 (2014)
93. C.K. Chung, T.Y. Liu, W.T. Chang, Effect of oxalic acid concentration on the formation of anodic aluminum oxide using pulse anodization at room temperature. Microsyst. Technol. **16**, 1451–1456 (2010)
94. C.K. Chung, W.T. Chang, M.W. Liao et al., Effect of pulse voltage and aluminum purity on the characteristics of anodic aluminum oxide using hybrid pulse anodization at room temperature. Thin Solid Films **519**, 4754–4758 (2011)
95. C.K. Chung, W.T. Chang, M.W. Liao et al., Fabrication of enhanced anodic aluminum oxide performance at room temperatures using hybrid pulse anodization with effective cooling. Electrochim. Acta **56**, 6489–6497 (2011)
96. C.K. Chung, M.W. Liao, H.C. Chang et al., Effects of temperature and voltage mode on nanoporous anodic aluminum oxide films by one-step anodization. Thin Solid Films **520**, 1554–1558 (2011)
97. C.K. Chung, W.T. Chang, M.W. Liao et al., Improvement of pore distribution uniformity of nanoporous anodic aluminum oxide with pulse reverse voltage on low-and-high-purity aluminum foils. Mater. Lett. **88**, 104–107 (2012)
98. C.K. Chung, M.W. Liao, H.C. Chang et al., On characteristics of pore size distribution in hybrid pulse anodized high-aspect-ratio aluminum oxide with Taguchi method. Microsyst. Technol. **19**, 387–393 (2013)
99. N. Hu, X. Dong, X. He et al., Interfacial morphology of low-voltage anodic aluminium oxide. J. Appl. Crystallogr. **46**, 1386–1396 (2013)
100. C.C. Chen, J.H. Chen, C.G. Chao Post-treatment method of producing ordered array of anodic aluminum oxide using general purity commercial (99.7 %) aluminum. Jpn. J. Appl. Phys. **44**, 1529–1533 (2005)
101. E. Rocca, D. Vantelon, A. Gehin et al., Chemical reactivity of self-organized alumina nanopores in aqueous medium. Acta Mater. **59**, 962–970 (2011)

102. E. Rocca, D. Vantelon, S. Reguer et al., Structural evolution in nanoporous anodic aluminium oxide. Mater. Chem. Phys. **134**, 905–911 (2012)
103. T.S. Shih, P.S. Wei, Y.S. Huang, Optical properties of anodic aluminum oxide films on Al1050 alloys. Surf. Coat. Technol. **202**, 3298–3305 (2008)
104. S. Canulescu, K. Rechendorff, V.N. Borca et al., Band gap structure modification of amorphous anodic Al oxide film by Ti-alloying. Appl. Phys. Lett. **104**, 121910 (2014)
105. L. Zaraska, G.D. Sulka, M. Jaskuła, Porous anodic alumina membranes formed by anodization of AA1050 alloy as templates for fabrication of metallic nanowire arrays. Surf. Coat. Technol. **205**, 2432–2437 (2010)
106. J.M. Montero-Moreno, M. Belenguer, M. Sarret et al., Production of alumina templates suitable for electrodeposition of nanostructures using stepped techniques. Electrochim. Acta **54**, 2529–2535 (2009)
107. M.T. Safarzadeh, A. Arab, S.M.A. Boutorabi, The effects of anodizing condition and post treatment on the growth of nickel nanowires using anodic aluminum oxide. Iran. J. Mater. Sci. Eng. **7**, 12–18 (2010)
108. D. Gong, V. Yadavalli, M. Paulose et al., Controlled molecular release using nanoporous alumina capsules. Biomed. Microdevices **5**, 75–80 (2003)
109. A. Belwalkar, E. Grasing, W. Van Geertruyden et al., Effect of processing parameters on pore structure and thickness of anodic aluminum oxide (AAO) tubular membranes. J. Membrane Sci. **319**, 192–198 (2008)
110. A.K. Kasi, J.K. Kasi, M. Hasan et al., Fabrication of low cost anodic aluminum oxide (AAO) tubular membrane and their application for hemodialysis. Adv. Mater. Res. **550–553**, 2040–2045 (2012)
111. I. Kanazirski, C. Girginov, A. Girginov, Electrolytic colouring of anodic alumina films in $NiSO_4$, $CuSO_4$ and ($NiSO_4$ + $CuSO_4$) containing electrolytes. Adv. Nat. Sci. Theory Appl. **1**, 45–51 (2012)
112. I. De Graeve, P. Laha, V. Goossens et al., Colour simulation and prediction of complex nano-structured metal oxide films test case: Analysis and modeling of electro-coloured anodized aluminium. Surf. Coat. Technol. **205**, 4349–4354 (2011)
113. H.-H. Shih, Y-Ch. Huang, Study on the black electrolytic coloring of anodized aluminum in cupric sulfate. J. Mater. Process. Technol. **208**, 24–28 (2008)
114. Y. Goueffon, C. Mabru, M. Labarrère et al., Black anodic films on 7175 aluminium alloy for space applications. Surf. Coat. Technol. **204**, 1013–1017 (2009)
115. Y. Goueffon, C. Mabru, M. Labarrère et al., Investigations into the coefficient of thermal expansion of porous films prepared on AA7175 T7351 by anodizing in sulphuric acid electrolyte. Surf. Coat. Technol. **205**, 2643–2648 (2010)
116. Y. Goueffon, L. Arurault, S. Fontorbes et al., Chemical characteristics, mechanical and thermo-optical properties of black anodic films prepared on 7175 aluminium alloy for space applications. Mater. Chem. Phys. **120**, 636–642 (2010)
117. J.C. Ganley, E.G. Seebauer, R.I. Masel, Porous anodic alumina microreactors for production of hydrogen from ammonia. AIChE J. **50**, 829–834 (2004)
118. J.C. Ganley, K.L. Riechmann, E.G. Seebauer et al., Porous anodic alumina optimized as a catalyst support for microreactors. J. Catal. **227**, 26–32 (2004)
119. S. Peng, D. Tian, X. Miao et al., Designing robust alumina nanowires-on-nanopores structures: Superhydrophobic surfaces with slippery or sticky water adhesion. J. Colloid Interface Sci. **409**, 18–24 (2013)
120. W. Liu, Y. Luo, L. Sun et al., Fabrication of the superhydrophobic surface on aluminum alloy by anodizing and polymeric coating. Appl. Surf. Sci. **264**, 872–878 (2013)
121. M.H. Lee, N. Lim, D.J. Ruebusch et al., Roll-to-roll anodization and etching of aluminum foils for high-throughput surface nanotexturing. Nano Lett. **11**, 3425–3430 (2011)

Chapter 4
Structural Engineering of Porous Anodic Aluminum Oxide (AAO) and Applications

Woo Lee

Abstract Porous anodic aluminum oxide (AAO) films can be conveniently produced by anodization of aluminum. Porous oxide layer formed on aluminum contains a large number of mutually parallel pores. Each cylindrical nanopore and its surrounding oxide constitute a hexagonal cell aligned normal to the metal surface. Under proper conditions, the oxide cells are self-organized to form a hexagonally close-packed structure. The novel and tunable structural features of porous AAOs have been intensively exploited for templated synthesis of a variety of functional nanostructures and also for fabrication of nanodevices. On the other hand, porous AAOs with modulated pores may provide an additional degree of freedom in templated synthesis. In addition, they can be used as model systems for systematically investigating structure-property relations of nanostructured materials. Based on the anodization techniques developed recently, one can fabricate porous AAOs with tailor-made internal pore structures. This chapter is devoted to conveying the most recent advances in structural engineering of porous AAOs and nanotechnology applications. In order to provide context, a brief description is given of the general structure and fundamental electrochemical processes associated with pore formation. Subsequently, two common anodizing techniques (i.e., mild and hard anodizations) that have been explored for nanotechnology applications are discussed. Next, various nanostructuring approaches for custom-designed porous AAOs are reviewed. The chapter covers the properties of porous AAOs derived from structural engineering and their applications to various nanotechnology researches and finally presents the challenges and future prospects.

W. Lee (✉)
Department of Physics, Hankuk University of Foreign Studies (HUFS), Yongin, Daejeon 449-791, Korea
e-mail: woolee@hufs.ac.kr

D. Losic and A. Santos (eds.), *Nanoporous Alumina*,
Springer Series in Materials Science 219, DOI 10.1007/978-3-319-20334-8_4

4.1 Introduction

Anodization of aluminum is an electrochemical oxidation process employed for the formation of protective and decorative coating on aluminum surfaces. In general, the anodic aluminum oxide (AAO) films form with two different morphologies depending on the anodizing electrolytes and conditions; (i) compact nonporous barrier-type AAO films from neutral electrolytes in which the anodic oxide is practically insoluble and (ii) porous-type AAO films from acid electrolytes in which the anodic oxide is slightly soluble [1]. Since the early 1920's, AAO films on aluminum, especially porous AAO films, have received substantial attention in industry due to their diverse applications in electrical insulation, corrosion protection, decoration, and improvement of mechanical properties of aluminum products (e.g., machine parts, architectural items, vehicles, electronic gadgets, outdoor products, etc) [2]. With the discovery of self-organized formation of ordered pore in 1995 [3], the porous-type anodizing processes have been drawing increasing attention in nanotechnology research.

Porous AAO formed on aluminum contains a large number of mutually parallel pores. Each cylindrical pore and its surrounding oxide constitute a hexagonal cell aligned normal to the aluminum surface. Under properly chosen conditions, the oxide cells are self-organized to form a hexagonally close-packed structure, like a honeycomb. Self-ordered porous AAOs with uniform pore diameter in the range of 10–400 nm and with pore density in the range of 10^8–10^{10} pores cm^{-1} can be conveniently prepared by simply subjecting aluminum under anodic polarization [4]. The unique and tunable structural features, combined with their excellent thermal stability, chemical inertness, and bio-compatibility, and easy modification of surface chemistry, make porous AAOs not only as ideal templates for synthesizing a variety of functional nanostructures (e.g., nanodots, nanowires, and nanotubes), but also as excellent platform materials for exploring various advanced devices for molecule separation, cell adhesion and culture, drug delivery, chemical/biological sensing, high performance catalyst, information storage, energy storage and harvest, etc. [4–9]. Moreover, base on the recent advances, the pore structure of AAO can be designed and engineered to a variety of geometries (e.g., sharp featured non-circular pores, symmetrically or asymmetrically modulated pores, funnel-like pores, hierarchical three-dimensional architectures of pores, etc.) [4]. The structural engineering of porous AAO can result in the generation of optically active structures by imparting a range of tailored optical properties, such as selective light transmission, reflection, enhancement, confinement, or guide. Therefore, the capability of structural engineering, coupled with the aforementioned characteristics, would make porous AAOs even more versatile as template or platform materials for advanced applications. In fact, utilization of structurally engineered porous AAOs for label-free, real-time, ultra-sensitive chemical and biosensing applications has become an emerging research field during the last few years [10–13].

This chapter conveys the most recent advances in structural engineering of porous AAOs and nanotechnology applications of the resulting porous AAOs. In order to provide context, first a brief description will be given of the general structure and fundamental electrochemical processes associated with pore formation. Subsequently, two common anodizing techniques (i.e., mild anodization (HA) and hard anodizations (HA)) that have been explored for nanotechnology applications will be discussed. Next, various nanostructuring approaches for the fabrication of custom-designed porous AAO architectures will be reviewed. These include micromachining of porous AAO, anodization of pre-textured aluminum, multistep anodization with controlled wet-chemical oxide etching after each anodizing step, programmed potential reduction, pulse anodization (PA) and cyclic anodization (CA) that selectively combine MA and HA processing conditions. The properties of porous AAOs derived from structural engineering and their applications to various nanotechnology researches will be also discussed. Finally, the challenges and future prospects of the field will be given.

4.2 Structure of Porous Anodic Aluminum Oxide (AAO) and Its Formation

Porous anodic aluminum oxide (AAO) film grown on aluminum has some structural analogy with a honeycomb (Fig. 4.1). It consists of a thin barrier oxide layer in contact with aluminum, and an overlying, relatively thick, porous oxide film containing mutually parallel cylindrical pores extending from the barrier oxide layer to the film surface. The thickness of porous AAO film can be varied from a few tens of nanometers up to hundreds of micrometers by controlling anodizing time. Each cylindrical pore and its surrounding oxide region constitute a hexagonal cell oriented normal to the metal surface. Under proper conditions, the oxide cells self-organize into hexagonal close-packed arrangement. Pore diameter and density of self-ordered porous AAOs are tunable in wide ranges by properly choosing anodizing conditions: pore diameter = 10–400 nm and pore density = 10^8–10^{10} pores cm^{-2}. The structure of self-ordered porous AAO can be defined by several structural parameters, such as interpore distance (D_{int}), pore diameter (D_p), barrier layer thickness (t_b), pore wall thickness (t_w), pore density (ρ_p), and porosity (P). These structural parameters have the following relationships [4]:

$$D_{int} = D_p + 2t_w \tag{4.1}$$

$$\rho_p = \left(\frac{2}{\sqrt{3}D_{int}^2}\right) \times 10^{14}\,\mathrm{cm}^{-2} \tag{4.2}$$

$$P(\%) = \left(\frac{\pi}{2\sqrt{3}}\right)\left(\frac{D_p}{D_{int}}\right)^2 \times 100 \tag{4.3}$$

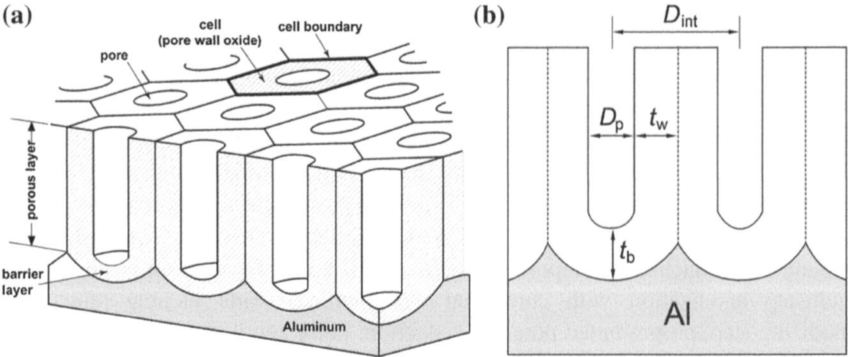

Fig. 4.1 Schematic structure of **a** porous anodic aluminum oxide (AAO) on Al substrate and **b** cross-sectional view; D_{int} = interpore distance, D_p = pore diameter, t_w = pore wall thickness, and t_b = barrier layer thickness

Studies to date have indicated that these structural parameters are dependent on the anodizing conditions, such as the type of electrolyte, anodizing potential (U), current density (j), temperature (T), etc.

During the formation of AAO films, the current (j) in the film is associated with the movement of charged ions in the barrier oxide layer, and can be described by the high-field conduction theory [14, 15]:

$$j = j_0 \exp(\beta E) = j_0 \exp(\beta U/t_b) \tag{4.4}$$

where j_0 and β are the constants, and U/t_b is the electric field (E, typically 10^6–10^7 Vcm^{-1}) impressed on the barrier layer with thickness t_b, the inverse of which is called "anodizing ratio (i.e., $AR = t_b/U$)" and describes the potential dependence of the barrier layer thickness. As such, the high-field conduction theory describes the field-driven movement of charged ions in barrier oxide. It is now well accepted that both Al^{3+} cations and oxygen-carrying anions (e.g., O^{2-}/OH^-) are mobile within the barrier oxide: Al^{3+} ions migrate outwardly toward the oxide/electrolyte (o/e) interface, while O^{2-}/OH^- ions move inwardly toward the metal/oxide (m/o) interface [16–20]. In the case of barrier-type AAO formation, the cooperative counter movement of Al^{3+} and O^{2-}/OH^- ions results in simultaneous formation of anodic oxide both at the o/e- and m/o-interfaces: at 100 % current efficiency, about 40 and 60 % of the total anodic oxide thickness is formed at the o/e- and m/o-interfaces, respectively [16–19]. On the other hand, in the case of porous AAO formation, anodic oxide forms at the m/o-interface by the inward migration of O^{2-}/OH^- anions, while the outward migrating Al^{3+} cations are lost to the electrolyte through field-assisted direct ejection mechanism without contributing to the oxide formation [21–23]. Thus, current efficiency (η) in porous AAO formation is lower than that of barrier-type AAO formation.

Porous AAO can be prepared either under a constant potential (i.e., potentiostatic) or constant current (i.e., galvanostatic) condition using an electrochemical

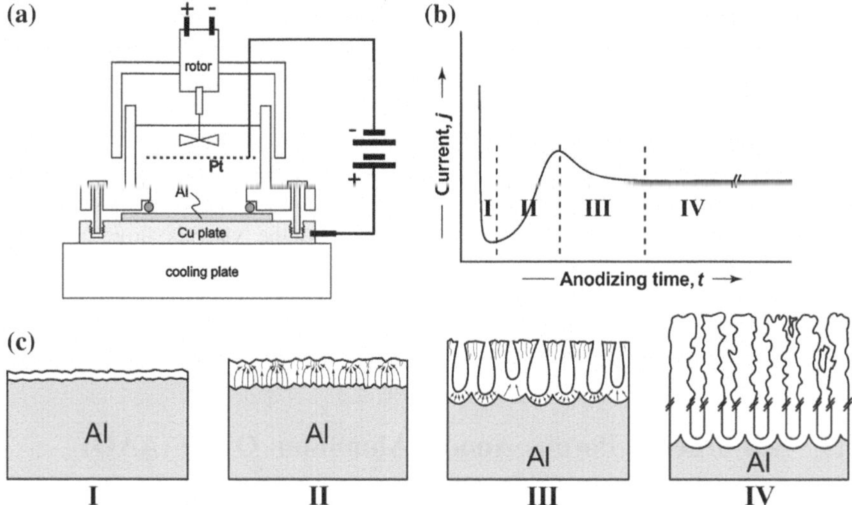

Fig. 4.2 Schematic diagram of **a** an electrochemical cell for typical anodization experiment, **b** a current (*j*)-time (*t*) transient during a potentiostatic anodization, and **c** the kinetics of porous AAO growth, corresponding to the anodization stages I–IV in panel (**b**)

setup shown in Fig. 4.2a. In general, potentiostatic anodization is widely employed for the fabrication of self-ordered porous AAO, because of the linear relation between the applied potential (U) and the structural parameters (i.e., D_p, D_{int}, and t_b) of the resulting AAO. Figure 4.2b, c illustrate a typical current (j)-time (t) transient in a potentiostatic anodization and the stages of pore structure development. Upon applying anodic potential, a thin compact barrier-type anodic oxide starts to grow on the surface of aluminum (stage I). When the thickness of compact barrier oxide reaches a certain value, current (j) drops rapidly to hit the minimum value. Electric field (E) may concentrate on local defects, impurities, or pits on the growing oxide surfaces to develop paths for electrolyte penetration, which serve as pore nucleation sites. Local field enhancement at the penetration paths facilitates field-assisted oxide decomposition, and leads to eventual development of incipient pores. Accordingly, further anodization leads to gradual increase in current (j) to a local maximum due to facile diffusion of electrolyte (stage II). At this stage, the pores are not uniform in size with random spatial distribution and undergo persistent merging with neighboring incipient pores to develop major pores (stage III). After that, current (j) reaches a steady-state value after passing an overshoot. During this state, gradual rearrangement of pores occurs over time until the field (E) across the barrier layers becomes the same for every pores (stage IV).

Several models have been explored to explain the formation of porous AAO. For steady-state pore growth, the velocity of the m/o- and o/e-interfaces should be balanced, keeping the barrier layer thickness (t_b) constant. This balance has long been attributed to an equilibrium between the oxide formation at the m/o-interface and the removal of oxide at the o/e-interface either by Joule's heat-induced chemical dissolution [24–28] or by field-assisted oxide dissolution [26, 29–33]. But, recent studies have indicated that the dissolution-based pore formation model is operative for the initial stage of pore formation [34–36], whereas it does not adequately account for the steady-state pore formation. For steady-state pore growth, there is an increasing number of results from experiments and theoretical modeling, disproving the field-assisted oxide dissolution [23, 37–43]. The experimental evidences have been derived from the W-tracer studies by Skeldon and coworkers, who suggested that pores grow due to the viscous flow of oxide materials from the pore base toward the cell boundary under high electric field and growth stresses [23, 37–40]. The reader interested in this matter is referred to the recent review article given in [4].

4.3 Self-ordered Porous Anodic Aluminum Oxide (AAO)

4.3.1 Mild Anodization (MA)

Since early 1920s, various anodizing processes have been developed and extensively utilized by industry for the purpose of surface protection of aluminum products. But porous AAOs produced by industrial processes, represented by hard anodization (HA), are characterized by disordered structures with non-uniform pore shape, size, and spatial ordering. Therefore, the porous-type anodization processes have not been implemented in the nanotechnology research, until the discovery of self-ordered pore formation under mild anodization (MA) condition in 1995 [3]. MA is characterized by slow formation of porous AAOs due to the one- or two-orders of magnitude lower current density (j), compared to HA process. In their seminal work, Masuda and Fukuda found that long-term mild anodization (MA) of aluminum in 0.3 M oxalic acid ($H_2C_2O_4$) at 40 V leads to self-ordering of pores as a result of the gradual rearrangement of the initially disordered pores [3]. Based on this finding, Masuda and Satoh developed the so-called “two-step anodization” process, by which porous AAOs with highly ordered pore arrangement could be obtained [44].

Typical two-step anodizing process consists of (i) the first long-term anodization followed by removal of the resulting porous AAO layer with disordered pores in its top part (Fig. 4.3a) to obtain textured aluminum with arrays of almost hemispherical concaves (Fig. 4.3b) and (ii) the subsequent second-step anodization with the textured aluminum at the same condition used for the first-step anodization to obtain highly ordered pore arrangement (Fig. 4.3c). Typical porous AAOs formed by two-step MA process exhibit a polydomain structure, in which each domain contains hexagonally ordered defect-free pores and contacts with neighboring domains forming boundaries along which defect pores or imperfections in pore arrangement are present (Fig. 4.3d). The lateral size of individual domains increases with the anodizing time, but is limited to several micrometers.

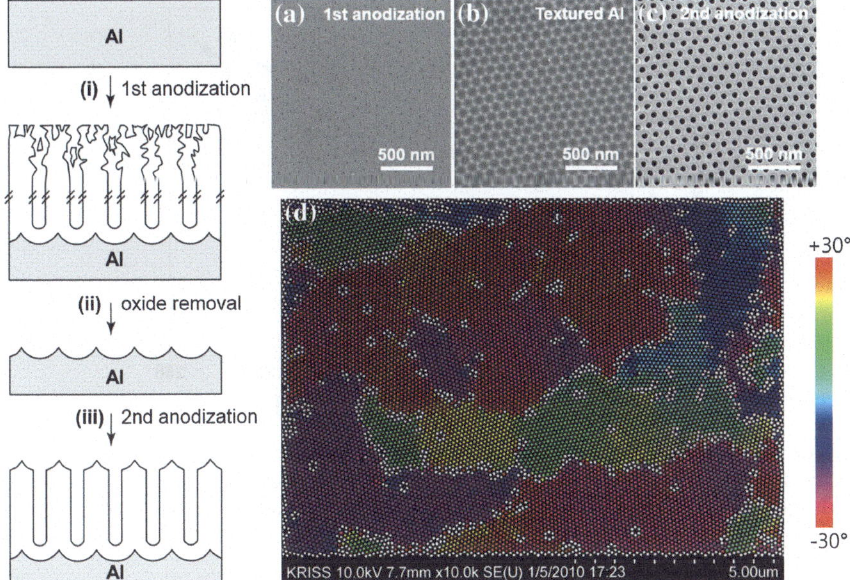

Fig. 4.3 *Left* A schematic procedure of a conventional two-step mild anodization (MA) for self-ordered porous anodic aluminum oxide (AAO): (i) the first long-term anodization step, (ii) removal of disordered porous AAO, and (iii) the second anodization step at the identical condition to the first one. (**a–c**) Representative plane-view SEM images of the samples of the respective steps. **d** A color-coded SEM image of AAO formed by two-step MA using 0.3 M oxalic acid at 40 V, showing a poly-domain structure. An area with the same color consists of a domain. The pores are color-coded on the basis of the average angle to the six nearest neighbors. Pores that have no apparent hexagonal coordination (i.e., defect pores) are marked with *white*

The development of the two-step process inspired and led to an overwhelming number of studies in attempts to fabricate ordered porous AAOs with different pore sizes and densities, to improve the spatial arrangement of pores, and to understand the mechanism of self-organized growth of pores. Particular efforts have been made to find the optimum pore ordering conditions for H_2SO_4, $H_2C_2O_4$, and H_3PO_4 solutions, the electrolyte systems that were investigated by Keller et al. in the early 1950s [24]. Studies to date have indicated that for a given anodizing electrolyte there exists a narrow MA processing window (known as the “self-ordering regime”) leading self-organized growth of ordered pores (see Fig. 4.4): (i) sulfuric acid (0.3 M H_2SO_4) at U = 25 V for an interpore distance (D_{int}^{MA}) = 65 nm [45, 46]; (ii) oxalic acid (0.3 M $H_2C_2O_4$) at 40 V for D_{int}^{MA} = 103 nm [27, 44–46]; (iii) selenic acid (0.3 M H_2SeO_4) at 48 V for D_{int}^{MA} = 112 nm [47]; and (iv) phosphoric acid (0.3 M H_3PO_4) at 195 V for D_{int}^{MA} = 500 nm [48, 49]. Considerable efforts have been devoted to exploring new self-ordering regimes in a wider range of interpore distance (D_{int}^{MA}). Most studies in this direction have been made by properly tuning the aforementioned three popular pore-forming acid electrolytes or by seeking new electrolyte systems. Shingubara et al. reported that anodization of aluminum in a

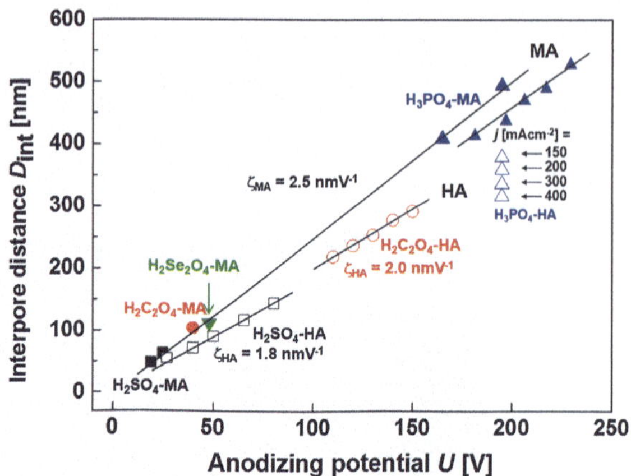

Fig. 4.4 Self-ordering regimes in MA (*filled symbols*) and HA (*open symbols*) by using H_2SO_4 (*black*), $H_2C_2O_4$ (*red*), H_2SeO_4 (*green*), and H_3PO_4 (*blue symbols*). The *black solid lines* are the linear regressions of the data with proportionality constants of ζ_{MA} = 2.5 nm V^{-1} and ζ_{HA} = 1.8–2.0 nm V^{-1}. Reprinted with permission from [4]. Copyright 2014, The American Chemical Society

mixture solution of 0.3 M H_2SO_4 and 0.3 M $H_2C_2O_4$ (v/v = 1:1) yields self-ordered porous AAO with D_{int}^{MA} = 73 nm at U = 37 V [50]. Sun and coworkers have recently reported that the interpore distance can be continuously tuned from D_{int}^{MA} = 410 to 530 nm by anodizing aluminum in a mixture solution of H_3PO_4 and aluminum oxalate ($(AlC_2O_4)_2C_2O_4$) at U = 180 to 230 V [51]. About 50 years ago, Kape reported that organic acids containing at least two carboxyl (–COOH) group can act as anodizing electrolytes [52–54]. These acids include malonic, citric, tartaric, succinic, maleic, phthalic acid, and their mixtures of sulfuric or oxalic acid. Ono et al. have reported that porous AAOs with D_{int}^{MA} = 300–600 nm can be obtained by MA in some of these acid electrolytes: D_{int}^{MA} = 300 nm for 5 M malonic acid at 120 V, D_{int}^{MA} = 500 nm for 3 M tartaric acid at 195 V, and D_{int}^{MA} = 600 nm for 2 M citric acid at 240 V [55, 56]. Table 4.1 summarizes the self-ordering regimes for mild anodization (MA) of aluminum in some of pore-forming acid electrolytes.

It has been widely accepted that structural parameters of porous AAO formed under self-ordering regimes depend mainly on anodizing potential (U). The interpore distance (D_{int}^{MA}) and barrier layer thickness (t_b^{MA}) increase linearly with anodizing potential (U) with proportionality constants ζ_{MA} = 2.5 nm V^{-1} for D_{int}^{MA} and AR_{MA} (so-called "anodizing ratio") = 1.2 nm V^{-1} for t_b^{MA} [46, 49, 58]. Pore diameter (D_p^{MA}) has also been known to increase with anodizing potential. However, recent studies have indicated that the potential dependence of D_p^{MA} is not as sensitive as the interplay between the current density and the temperature, concentration and nature of the anodizing electrolyte [47, 55, 59–61]. For MA-AAOs, Nielsch et al. proposed that self-ordering of pores requires a porosity (P_{MA}) of about 10 %, independent of the specific anodizing conditions by assuming D_{int}^{MA} = 2.5 nm V^{-1} [49].

Table 4.1 Self-ordering regimes for mild anodization (MA) of aluminum

Electrolyte	Self-ordering voltage, (U) (V)	Interpore distance, $\left(D_{int}^{MA}\right)$ (nm)	$\zeta = D_{int}^{MA}/U$ (nm V^{-1})	References
0.3 M sulfuric acid H_2SO_4	19–25	50–65	2.60	[45, 46]
0.3 M oxalic acid $(HCOO)_2$	40	103	2.57	[27, 44–46]
0.3–3 M selenic acid H_2SeO_4	42–48	95–112	2.26–2.33	[47, 57]
0.3 M phosphoric acid H_3PO_4	160–195	420–500	2.63–2.56	[46, 48, 49]
5 M malonic acid HOOC–CH_2–COOH	120	300	2.50	[56]
3 M tartaric acid HOOC–$(CHOH)_2$–COOH	195	500	2.56	[56]
2 M citric acid HOOC–$CH_2COH(COOH)CH_2$–COOH	240	600	2.50	[55]

This requirement dictates a linear increase of the pore diameter (D_p^{MA}) with the anodizing potential at a rate of $\zeta_p = 0.83$ nm V^{-1}, irrespective of anodizing conditions. On the other hand, there are a number of recent works, reporting that the self-ordering of pores can occur at other porosity (P_{MA}) values ranging from 0.8 to 30 % depending on the MA conditions [47, 55, 59–61]. For example, Nishinaga et al. have recently reported fabrication of porous AAO with porosity $P_{MA} = 0.8$ % ($D_p^{MA} = 10.4$ nm and $D_{int}^{MA} = 112$ nm) by anodization of aluminum in 0.3 M H_2SeO_4 (0 °C) at 48 V [47]. The reported results clearly indicate that the anodizing potential is not the only parameter determining the pore diameter.

4.3.2 *Hard Anodization (HA)*

For a given anodizing electrolyte, there is a breakdown potential (U_B), above which stable anodization is difficult to maintain due to the commencement of local thickening, burning, and cracking of the growing oxide: $U_B = 27$, 50, and 197 V for sulfuric, oxalic, and phosphoric acid, respectively. It has been known that breakdown of anodic oxide occurs due to the catastrophic local current flow and consequent Joule heating [4, 62–64]. Chu et al. reported that the best self-ordering of pores occurs when anodization is conducted at potentials just below the breakdown potential (U_B) [64]. Ono et al. have found that the locally thickened areas formed by breakdown exhibit domains of highly ordered cell arrangement [55, 65]. Further, they found that the interpore distance (D_{int}) and the barrier layer thickness (t_b) of porous AAO at the burnt areas are smaller than those at the burnt-free areas, which is consistent with the earlier results reported by Tu et al. [66]. All these results imply that at a given anodizing potential (U) current density (j) may determine the spatial ordering of pores and the structural parameters (i.e., cell size D_{int} and the barrier layer thickness t_b) of porous AAOs.

Industrial hard anodization (HA) have been typically conducted at potentials higher than the breakdown value (U_B) [67–70]. HA-AAO films can be produced at an efficient rate (typically 50–100 μm h^{-1}) due to the high anodizing current at an increased anodizing potential [71–74]. The current density (j) in HA process is typically one or two orders of magnitude higher than that of conventional MA processes. Thus, Joule's heat (Q) during HA is two or four orders of magnitude greater than the ordinary MA processes: $Q = Ujt = R_b j^2 t$, where t is the anodization time and R_b is the resistance of the barrier layer. The excessive heat should be properly dissipated, otherwise it promotes dissolution and breakdown of the growing anodic oxide. The porous AAO films formed by industrial HA process are characterized by uneven surfaces with numerous cracks and non-uniform pores due to the local breakdown of anodic oxide, and thus are not suited to nanotechnology applications. Various attempts have been made to overcome the problems associated with the local breakdown events during HA. The researches in this direction include appropriate modifications of the three major pore-forming acid electrolytes (i.e., H_2SO_4, $H_2C_2O_4$, and H_3PO_4), exploration of new anodizing electrolytes, and efficient removal of the reaction heat.

Chu et al. reported that the breakdown potential (U_B = 27 V) in a sulfuric acid-based anodization can be increased up to 70 V by ageing the electrolyte after a long-term pre-anodizing at 10–20 ampere hours per liter [64, 75]. The authors showed that self-ordered porous AAOs with D_{int} = 90–130 nm at U = 40–70 V and j = 160–200 mA cm^{-2} at 0.1–10 °C. However, porous AAOs formed by this process exhibited poor mechanical stability due to weak cell junction strength, which would greatly limit their practical applications to current nanotechnology research. Chu et al. claimed that the concentration of Al^{3+} ions dissolved in an aged solution plays a key role in the stable growth of porous AAOs at high potentials ($U > U_B$) and current densities (j). But, later studies by Schwirn et al. have indicated that stable anodization at high potentials ($U > U_B$) is actually determined by the initial limiting current density (j_{limit}), not by the solution state [76, 77].

Lee et al. have shown that the self-ordering regimes can be extended by performing HA of aluminum in oxalic acid [78]. By introducing a thin (*ca.* 400 nm) porous oxide layer onto an aluminum substrate and effectively removing the reaction heat, they were able to prevent the breakdown of oxide films and grow mechanically stable ordered porous AAOs at anodizing potentials of U > 100 V [78]. Their HA process established a new self-ordering regime with widely tuneable interpore distances: D_{int}^{HA} = 220–300 nm at U = 110–150 V and j = 30–250 mA cm^{-2}. They found that the porosity (P_{HA}) of HA-AAOs is 3.3–3.4 %, which is about one-third of the porosity value ($P_{MA} \approx 10$ %) that was proposed as a requirement for self-ordered MA-AAOs by Nielsch et al. [49]. The experimental method has also been applied to sulfuric and malonic acid-based HAs [77, 79].

Li et al. have reported that large reaction heat during HA can be effectively removed by adding ethanol (C_2H_5OH) to anodizing electrolytes, and thus can achieve stable anodization at high potential ($U > U_B$) and current density (j) [80, 81]. The added ethanol served not only as an agent for lowering the freezing point of the electrolyte down to −10 °C, but also as a coolant for the removal of reaction heat through its vaporization from the pore bases. The process allowed the fabrication of self-ordered porous AAOs with various interpore distances; D_{int}^{HA} = 70–140 nm for H_2SO_4-HA at 30–80 V, 225–450 nm for $H_2C_2O_4$-HA at 100–180 V, and 320–380 nm for H_3PO_4-HA at 195 V [80, 81].

Figure 4.4 shows the self-ordering potentials (U) and corresponding interpore distances (D_{int}) of porous AAOs formed by MA and HA in three major pore-forming electrolytes (i.e., H_2SO_4, $H_2C_2O_4$, and H_3PO_4). HA-AAOs exhibit a reduced potential (U) dependence of the interpore distance (D_{int}) with a proportionality constant ζ_{HA} = 1.8–2.1 nm V^{-1} [75, 77–83], compared to self-ordered MA-AAOs (i.e., ζ_{MA} = 2.5 nm V^{-1}). At a fixed anodizing potential, the interpore distance (D_{int}^{HA}) of HA-AAOs decreases with the current density (j), indicating that anodizing potential (U) is not the only parameter determining the cell size of HA-AAOs [77–80]. On the other hand, the barrier layer thickness (t_b) of HA-AAOs increases at a rate of AR_{HA} = 0.6–1.0 nm V^{-1} with respect to anodizing potential (U) depending on the current density (j) [64, 77, 78, 80], which is smaller than $AR_{MA} \approx 1.2$ nm V^{-1} for MA-AAO [24, 30, 84]. Lee et al. found that pore diameter (D_p) of HA-AAOs increases with current density (j) under potentiostatic conditions (i.e., U = fixed) (Fig. 4.5) [85]. Based on this observation, they suggested structural engineering of porous AAOs through appropriate control of the current density (j) during potentiostatic anodization.

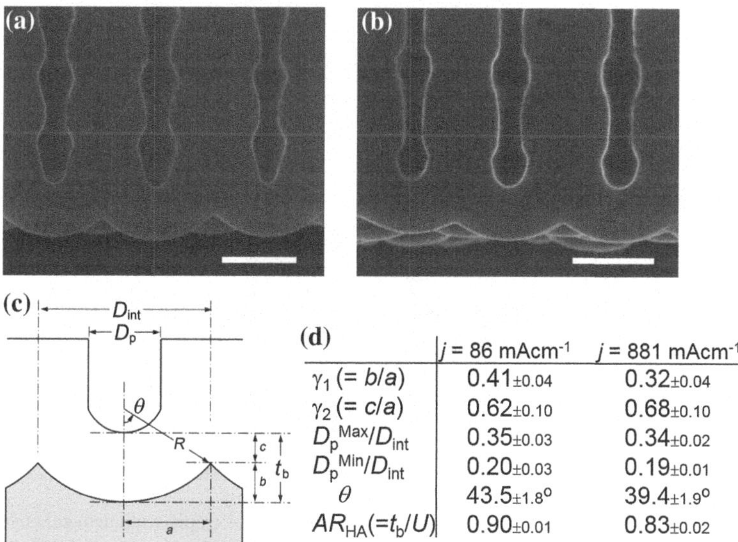

	j = 86 mAcm^{-1}	j = 881 mAcm^{-1}
γ_1 (= b/a)	0.41±0.04	0.32±0.04
γ_2 (= c/a)	0.62±0.10	0.68±0.10
D_p^{Max}/D_{int}	0.35±0.03	0.34±0.02
D_p^{Min}/D_{int}	0.20±0.03	0.19±0.01
θ	43.5±1.8°	39.4±1.9°
AR_{HA}(=t_b/U)	0.90±0.01	0.83±0.02

Fig. 4.5 Cross-section SEM micrographs of AAOs prepared from two separate anodisation experiments (U = 200 V), whose reaction were terminated near j = 86 mA cm^{-2} and j = 881 mA cm^{-2} in oscillating currents. Scale bars = 250 nm. **c** A schematic cross-section of AAO on Al. **d** The parameters defining the geometry of the pore bottom. Adapted with permission from [85]. Copyright 2010, WILEY-VCH Verlag GmbH & Co. KGaA, Weinheim

4.4 Structural Engineering of Porous AAO

4.4.1 Microstructuring of Porous AAO

The unique structural characteristics of the porous AAO make it not only an ideal template material for the preparation of arrays of nanodots, wires and tubes, but also an excellent platform material for the development of integrated nanodevices [4]. The elongated cylindrical feature of oxide pores can be exploited for the micro-structuring of porous AAO films by anisotropic etching of the pore wall oxide. Tan et al. [86] demonstrated first the fabrication of high-aspect-ratio microstructures of porous AAOs by combining a conventional lithographic process and a wet-chemical etching process. For micromachining of porous AAO films, the same general process steps can be applied (Fig. 4.6a) [86–89]. First, a thin masking layer is patterned on the surface of AAO to selectively block the pores. Next, the resulting sample is immersed into an appropriate etching solution to selectively remove anodic oxide of the unmasked areas. During this process step, lateral etching of pore wall oxide takes place simultaneously along the entire length of the every unmasked pore, resulting in vertical side walls and negligible undercutting at the mask edges (Fig. 4.6b, c). Multilevel micromachining of porous AAO is also possible, as shown in Fig. 4.6d. After the first level microstructuring of AAO, the

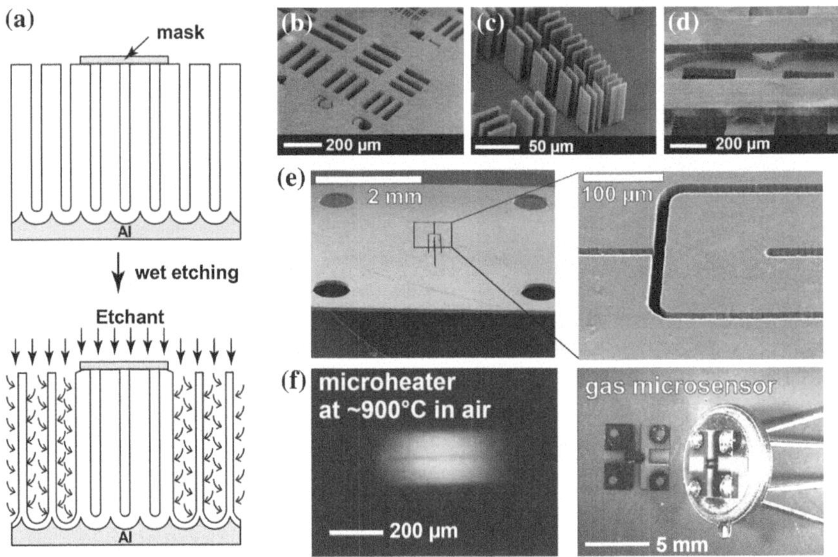

Fig. 4.6 **a** Scheme for the micromaching of porous AAO by anisotropic wet-chemical etching. **b–d** SEM micrographs of micromachined porous AAOs. **e** SEM micrographs of micromachined sensor substrate. **f** Device prototypes fabricated from micromachined porous AAO: microheater operating in air (*left*) and gas microsensor (*right*). Panels (**b–f**) adapted from [88]. Copyright 2000, American Society of Mechanical Engineers, ASME

remaining aluminum substrate can be further anodized. Afterwards, the newly formed porous AAO is patterned and etched similar to the first level micromachining.

Tan et al. [86] found that the time required for the complete removal of the unmasked areas is nearly independent of the thickness of porous AAO. This observation indicates that the process is governed by the oxide etching reaction, rather than the vertical diffusion of etching solution along the long cylindrical nanopores. Otherwise, the microstructures formed by etching would exhibit tilted sidewalls due to the etch rate difference between the top and bottom of pores. In other words, the critical thickness that must be etched corresponds to the pore wall thickness, rather than the film thickness of porous AAO.

In general, the pore wall oxide of as-prepared porous AAOs is amorphous and contaminated with the electrolyte-derived anionic impurity (see Sect. 4.4.3.1). Thus, porous AAOs exhibit poor chemical stability against both acid and base. But an amorphous AAO can be converted to polycrystalline alumina by proper heat treatment at high temperatures in air. The resulting porous alumina ceramic exhibits markedly improved hardness and chemical stability against acid, base, and other corrosive chemicals. Therefore, micromachined porous AAO can be used an excellent platform material for developing microelectromechanical system (MEMS) devices that operate even in high temperatures or harsh environments (Fig. 4.6e, f) [88, 89].

4.4.2 Control of the Arrangement and Shape of the Pores

As discussed in Sect. 4.3, self-ordered porous AAOs exhibit a poly-domain structure, in which each domain contains hexagonally arranged nanopores of an identical orientation (see Fig. 4.3d). The lateral size of ordered pore domains is typically limited to several micrometers. For advanced applications, on the other hand, porous AAOs with a single domain configuration of nanopores over macroscopic areas ($\sim cm^2$) are required. In addition, porous AAOs with non-hexagonal arrangement of pores or sharp-featured non-circular pore cross-sections may have extend application potential in biomedical sensors based on localized surface plasmon resonance (LSPR) or surface-enhanced Raman spectroscopy (SERS). The works in this direction was pioneered by Masuda and coworkers. They first reported the fabrication of porous AAOs with a single domain configuration of pores over a few mm^2 area [90]. The key of their method is to generate an array of shallow depressions on the surface of aluminum by mechanical nanoindentation with a hard SiC imprint stamp prior to anodization (Fig. 4.7a). Each surface depression on aluminum defined the position of pore growth during anodization, and thus led to a perfect hexagonal arrangement of pores within the stamped area (Fig. 4.7b). Masuda et al. further demonstrated fabrication of porous AAOs with square- or triangle-shaped pore openings in square or triangular arrangements (Fig. 4.7c, d) [91].

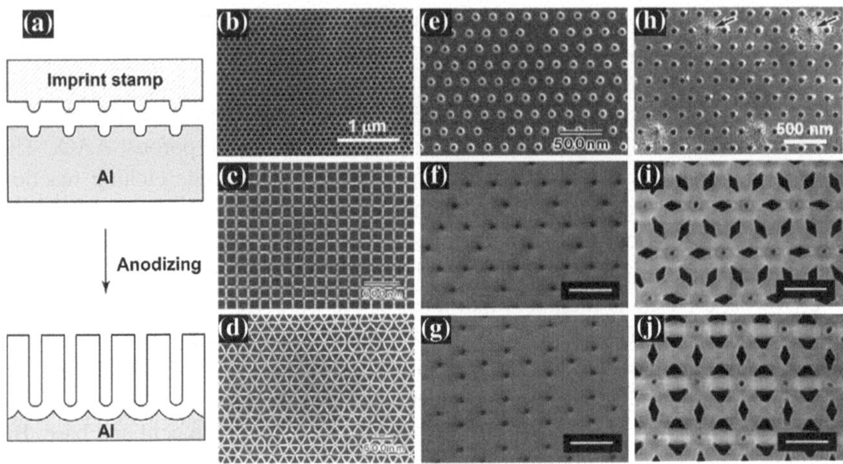

Fig. 4.7 **a** Schematic process for the preparation of an ideally ordered porous AAO by anodization of pre-patterned aluminum. **b–d** SEM micrographs of porous AAOs with different arrangements and shapes of pores: **b** circular, **c** square, and **d** triangular pore openings. **e–j** SEM images showing self-compensated pore formation; **e–g** surface pre-patterns in non-equilibrium tessellation and **h–j** the respective porous AAOs formed by anodization. SEM images shown in (**f**, **g**, **i**, **j**) shows formation of porous AAOs with (**f**, **i**) hybrid circular-diamond and (**g**, **j**) hybrid cicular-diamond-triangle pore cross-sections; the scale bars are 500 nm. Panel (**b**) adapted with permission from [90]. Copyright 1997, AIP Publishing LLC; Panels (**c**, **d**) adapted with permission from [91]. Copyright 2001, WILEY-VCH Verlag GmbH & Co. KGaA, Weinheim; Panels (**e**, **h**) reprinted with permission from [105], Copyright 2001, AIP Publishing LLC; Panels (**f**, **g**, **i**, **j**) reprinted with permission from [106]. Copyright 2008, AIP Publishing LLC

To date, various strategies for surface pre-patterning of aluminum have been developed in order to generate surface pre-patterns either on bulk aluminum substrate or on thin aluminum layer deposited on fragile substrates (e.g., glass or Si wafer). These include imprinting with optical diffraction grating [92], lithographically fabricated Si_3N_4 [93] or Ni stamps [94], imprinting with self-assembled mono- or multi-layer of microbeads [95, 96], surface patterning by focused ion-beam (FIB) lithography [97–99], interference lithography [100], holographic lithography [101], block-copolymer lithography [102], step-and-flash imprint lithography (SFIL) [103], and soft lithography utilizing elastomeric poly-dimethylsiloxane (PDMS) stamp [104]. Each of these techniques has its advantages and disadvantages. The choice of the technique should be made in consideration of the process cost, the fragility of the substrate, the required size of single domain, etc. The reader interested in this matter is referred to the recent review article given in [4].

Masuda and coworkers found that in a pre-pattern on aluminum the deficiency sites of the pattern are compensated during anodization in order to achieve the closest packing configuration of the cylindrical cells (Fig. 4.7e, f) [105]. The size of the pores formed at the deficiency sites was smaller than that of pores formed at the patterned sites, which was attributed to the different pore growth rates at the patterned and deficient sites [105]. Since the pore formation at the deficiency sites is associated with the establishment of the equilibrium tessellation arrangement of oxide cells, the anodizing potential (U) should be chosen to be the value that satisfies the potential (U)-interpore distance (D_{int}) relation required for the self-ordering of pores (see Sect. 4.3). By exploiting the self-compensation ability in AAO growth, Smith et al. have demonstrated the fabrication of porous AAOs with hybrid circular-diamond and circular-triangular-diamond pore cross-sections [106]. From anodizing experiments with Al surface pre-patterns in non-equilibrium tessellation arrangements (Fig. 4.7f, g), the authors found that the arrangement of the patterned and deficiency sites determines the cell geometry of the resulting porous AAO. On the other hand, the cell geometry was found to direct the shape of pores: circular, diamond, and triangular pores respectively in regular, elongated, and partially compressed hexagonal cells (Fig. 4.7i, j). It was suggested that a coupled effect between the thick pore wall oxide and the longer segment length of cell-boundary bands would account for the formation of diamond and triangular shaped pores [106].

4.4.3 Engineering of the Internal Pore Structure

4.4.3.1 Pore Widening

Pore wall oxide of as-prepared AAOs is amorphous and contaminated with varying amounts of anions derived from acid electrolyte. The incorporation of acid anions occurs via inward migrations under an electric field (E) during the anodization. The amount of incorporated acid anions and their spatial distribution in AAO depend on the type and concentration of electrolytes, anodizing potential (U), current density (j), and temperature (T) [28, 107–112]. The incorporated electrolyte anions influence the chemical, optical, and mechanical properties of porous AAOs. For instance, oxalate ions and the oxygen vacancies in AAOs formed in $H_2C_2O_4$ have been known to be the origin of the blue photoluminescences (PL) [81, 113–117]. In general, the pore wall oxide of AAOs exhibit a duplex structure in terms of chemical composition [108, 118–120]. The outer oxide layer next to the pores is contaminated with acid anions, yet the inner oxide layer is comparatively pure.

The duplex nature of the pore wall oxide can be conveniently evidenced by performing pore widening experiments. For a given set of pore widening conditions (i.e., temperature and concentration of an etchant solution), the rate of oxide etching is dependent on the chemical composition of the pore wall oxide [112]. Figure 4.8 shows the systematic increase in pore diameter (D_p) with the pore widening time (t_{etch}) for porous AAOs formed in H_2SO_4, $H_2C_2O_4$, and H_3PO_4. As shown in the figure, all D_p versus t_{etch} plots have the inflection points, at which the slopes of the curves change. Pore wall oxide in the early stage is etched at higher rates (*ca.* 1 nm min^{-1}) than those (*ca.* 0.37 nm min^{-1}) in the later stage. The retarded rate of etching in the later stage can be attributed to the relatively pure nature of the inner pore wall oxide, compared to the less dense outer pore wall oxide due to the incorporation of acid anions. From the estimated inflection points, one can easily deduce the relative thickness ratios of the anion contaminated outer pore wall (t_o) to the more compact inner pore wall (t_i); they are summarized as an inset table in Fig. 4.8. The relative thickness of the outer pore wall (i.e., $t_o/(t_i + t_o)$) increases in the order:

$$H_2SO_4 - AAO(0.70) > H_2C_2O_4 - AAO(0.67) > H_3PO_4 - AAO(0.64)$$

From the view point of material's optical property, the pore wall oxide of AAOs have a trilayer structure [113, 115, 116]: (i) outer oxide layer contaminated with electrolyte anions (F-PL centers at 2.66 eV), (ii) middle oxide layer with singly ionized oxygen vacancies (F^+-PL centers at 3.06 eV), and (iii) inner oxide layer (relatively pure alumina without PL centers). Yamamoto et al. [113] observed that pore widening treatment on porous AAO results in blue shift of the emission peak. They noted that the outer oxide layer screens the excitation light and the emission light from the middle oxide layer. Therefore, PL intensity increased initially with pore widening time (t_{etch}), but decreased upon oxide etching for an extended period of time due to the dissolution of the middle oxide layer.

Huang et al. reported that thin porous AAOs can act as a Fabry-Pérot optical cavity constituted by the air-AAO-Al, producing oscillating reflectance and PL spectra due to multiple-reflection interferences of the propagating waves [121]. From the fact that the band gap of alumina is much bigger than the photon energy of the considered PL spectral range (typically, 2.25–3.54 eV) and the absorption coefficient is very small, the authors deduced the optical thickness ($2nL_{AAO}$) of porous AAOs [121]:

$$2nL_{AAO} = hc/\Delta E \tag{4.5}$$

where n and L_{AAO} are respectively the refractive index and the thickness of a porous AAO, h and c are the Planck's constant (6.626×10^{-34} Js) and the speed of light (2.998×10^8 ms^{-1}), and ΔE is the mode spacing (i.e., the oscillation period) that determines the number of PL oscillations. The above equation predicts that the number of oscillations in a PL spectrum increases with the thickness (L_{AAO}) of porous AAO. From the laser-excited PL spectra of porous AAOs with different thicknesses, the authors estimated the refractive index of porous AAO to be $n = 1.66$ [121]. The refractive index (n) was about 2 % deviation to the effective refractive index ($n_{eff} = 1.62$) calculated from the Bruggeman's equation (4.6) assuming 8 % porosity (P) of porous AAOs [122]:

$$(1 - P)\frac{n^2_{Al_2O_3} - n^2_{eff}}{n^2_{Al_2O_3} + 2n^2_{eff}} + P\frac{1 - n^2_{eff}}{1 + 2n^2_{eff}} = 0 \tag{4.6}$$

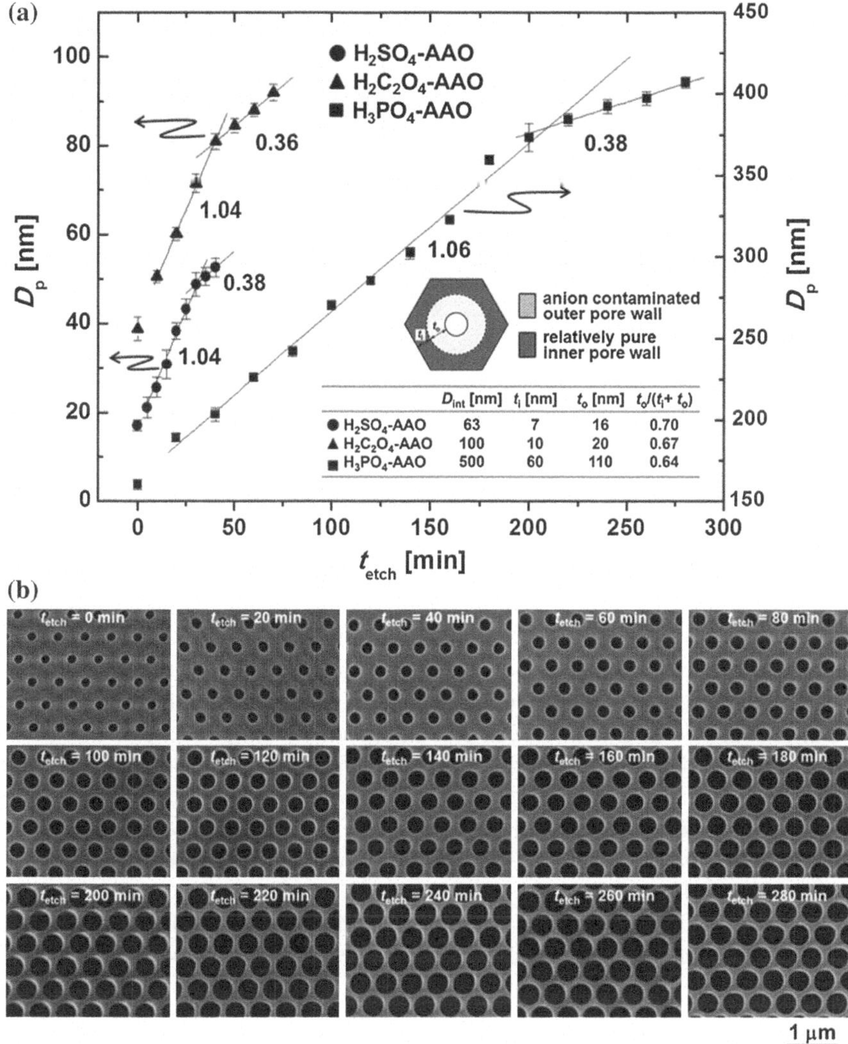

Fig. 4.8 The evolution of pore diameter (D_p) as a function of time (t_{etch}) upon wet-chemical etching of porous AAOs formed in 0.3 M sulfuric acid at 25 V (H_2SO_4-AAO), 0.3 M oxalic acid at 40 V ($H_2C_2O_4$-AAO), and 1 wt% H_3PO_4 at 195 V (H_3PO_4-AAO). Wet-chemical etchings of pore wall oxide were performed in 5 wt% H_3PO_4 at 29 °C. The numbers in each plot are the slopes (in nm min^{-1}) of the corresponding linear fits. The table presented as an inset summarizes the relative thickness of the anion contaminated outer pore wall oxide layer (t_o) to the relatively pure inner pore wall oxide layer (t_i), which were determined from the inflection points of the D_p versus t_{etch} plots for three kinds of porous AAOs. **b** SEM images showing systematic increase in pore diameter (D_p) as a function of time (t_{etch}) upon wet-chemical etching in 5 wt% H_3PO_4 (29 °C). Data for $H_2C_2O_4$-AAO were adapted from [112]

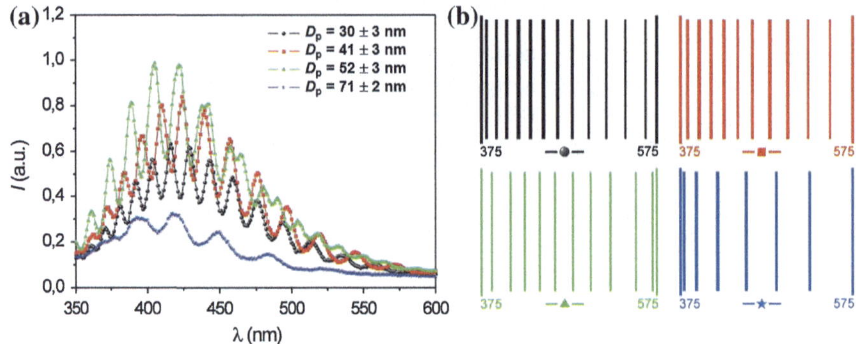

Fig. 4.9 **a** Photoluminescence (PL) spectra (λ_{ex} = 320 nm) of porous AAOs with the same thickness (t_{AAO} = 5.0 ± 0.1 mm), yet different pore diameters (D_p). **b** The corresponding barcode patterns after encoding the measured PL oscillations in (**a**). Adapted with permission from [123]. Copyright 2012 Wiley-VCH Verlag & Co. KGaA, Weinheim

where $n_{Al_2O_3}$ is the refractive index of alumina (1.67). Since the effective refractive index (n_{eff}) is a function of porosity (P), PL oscillation details can be tuned by changing the pore diameter (D_p) of porous AAO through pore widening treatment.

Santos and coworkers [123, 124] put forward the Fabry-Pérot effect to develop an optical encoding system based on the PL spectra of porous AAOs in the UV-Visible range (Fig. 4.9). Each PL oscillation corresponds to a bar in a barcode. The intensity and position of each PL oscillation are related to the width and position of each bar. As discussed above, the number, intensity and position of the PL oscillations depend on the thickness (L_{AAO}), pore diameter (D_p), and the nature of effective medium. By exploiting these features, the authors demonstrated that their barcode system can be used as optical biosensors for solvochromic oxazine molecules and glucose [123].

In an air-AAO-Al system, the wavelength of the peak maxima in a white light interference spectrum is governed by the Fabry-Pérot relationship:

$$2n_{eff}L_{AAO} = m\lambda \tag{4.7}$$

where λ is the wavelength of maximum constructive interference for spectral fringe of order m. The binding of molecules to the pore wall oxide gives rise to a shift of the characteristic interference pattern due to a change in the effective optical thickness (EOT, $2n_{eff}L_{AAO}$) that can be determined by the Fourier transform of the reflectivity spectrum. This analysis, so-called "the reflectometric interference spectroscopy (RIfS)" has been used by Sailor group for the investigation of gas adsorption characteristics within AAOs and also for label-free detection of biomolecules [125–127]. More recently, Losic group has extended RIfS technique extensively for real-time detection of biologically relevant metal cations, volatile sulfur compound (VSC), and molecules [124, 128–131].

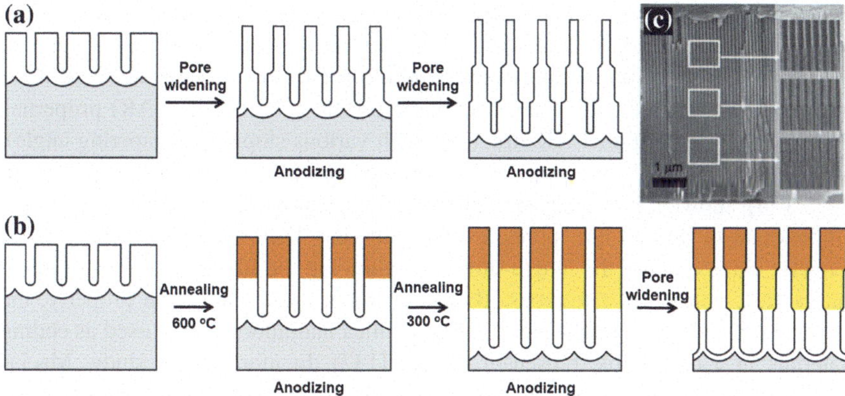

Fig. 4.10 Schematic procedures for the fabrication of porous AAOs with **a** normal and **b** inverted funnel-like pores: **a** combination of anodization steps and pore widening treatments in a sequential fashion and **b** combination of anodization steps and annealing treatments and final pore widening treatment. In (**b**), AAO strata after annealing at 600 and 300 °C are represented by *orange* and *yellow* colors, respectively. **c** Cross-sectional SEM micrograph of porous AAO with funnel-like pores prepared by the process (**a**). Panel (**c**) was reprinted with permission from [134]. Copyright 2007 The American Chemical Society

Pore widening treatment allows one to systematically increase the pore diameter (D_p) without affecting the interpore distance (D_{int}) of porous AAOs. This feature can be exploited in investigating the size dependence of physical properties (e.g., magnetic interaction) of low-dimensional nanostructures that can be fabricated by AAO template-based synthesis. Pore widening also offers a simple way of engineering the internal pore structure. For example, porous AAOs with funnel-like pores can be prepared by repeating the process comprised of anodizing and the subsequent pore widening treatment (Fig. 4.10a), in which the anodizing time and pore widening duration determine the length of each segment and pore diameter, respectively [132–134]. Li et al. have recently reported tailoring of the shape of funnel-like pores in AAOs [135]. The authors successfully demonstrated the fabrication of linear cones, whorl-embedded cones, funnels, pencils, parabolas and trumpets-like pore structures by controlling the anodizing time, pore widening time and cyclic time. Fabrication of inverted funnel-like pore structure has also been demonstrated recently by Santos et al. [136]. The approach was based on the different etching rate of pore wall oxide with the annealing temperature (Fig. 4.10b). Combination of two or more anodization steps and annealing at different temperatures after each anodizing step produced porous AAO strata having different chemical stabilities against etching solution. Pore widening treatment on the resulting porous AAO resulted in cylindrical pores with increasing pore diameter from top to bottom.

Funnel-shaped AAO pore structures have been used not only as sensing platforms, but also as templates to replicate optical nanostructures. Nagaura et al. fabricated porous AAO with low aspect ratio funnel-like pores and utilized it for replicating ordered arrays of Ni nanocones by electroless Ni deposition [137, 138]. With similar approach, Yanagishita et al. replicated ordered arrays of tapered polymer pillars from funnel-like pores by photo-imprinting process and used it for investigating the effect of the tapering angle on the antireflection (AR) properties [139]. Among the polymer AR structures with various slopes (i.e., tapering angle), those with a gradually changing slop exhibited the lowest reflectance. He et al. [132, 134] showed that each segment of silica nanotubes prepared from AAO templates with funnel-like pores exhibit different light reflectance due to the diameter change along the length direction, allowing the identification of different shapes (i.e., codes) of nanotubes. Based on this optical reflectance property, the authors demonstrated that differently shaped silica nanotubes can be used as coding materials in a dispersible biosensor system [132]. In more recent study, Macias et al. showed that porous AAOs with funnel-like pores can be used for RIfS-based label-free optical biosensing applications [140]. The authors prepared porous AAOs with two porous layers of different pore diameters (i.e., a top layer with large pores and a bottom layer with smaller pores (see Fig. 4.11a, b). The pore diameters of the

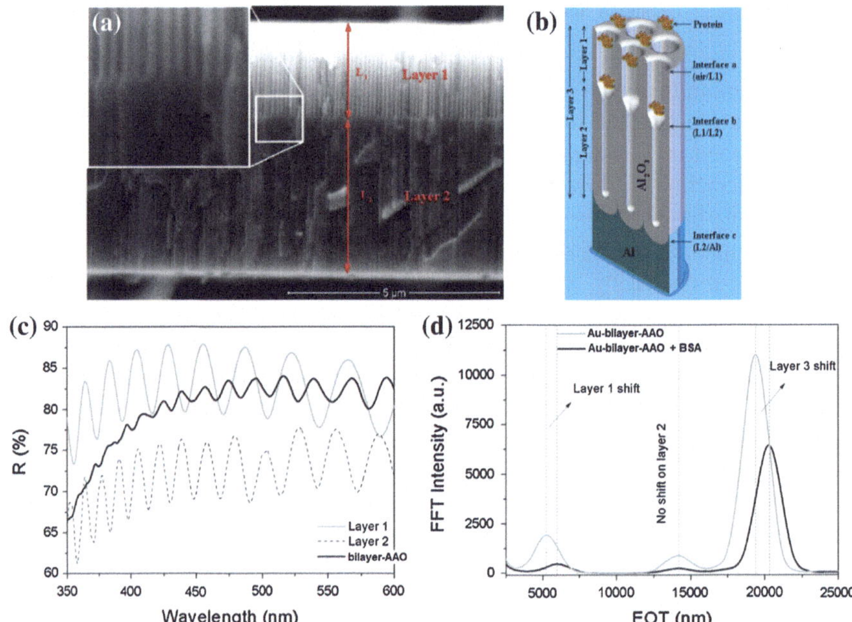

Fig. 4.11 **a** Cross-sectional SEM image of bilayered AAO. **b** Schematic of the sensing and self-referencing procedure, where the protein to be sensed can infiltrate Layer 1, but not Layer 2. Layer 3 = Layer 2 + Layer 2. **c** Reflectance spectra obtained from two single AAO layers and from a bilayer-AAO consisted of Layer 1 and 2: Layer 1 = 1.9-mm-thick and D_p = 68 nm (41 % porosity), Layer 2 = 4.7 mm-thick and D_p = 33 nm (9 % porosity). **d** FFT plot of the optical response of the Au coated-bilayer AAO before and after the introduction of the BSA protein. Adapted with permission from [140]. Copyright 2013, The American Chemical Society

top and bottom layers were designed for analyte (i.e., bovine serum albumin protein, BSA) to be infiltrated only into the top layer (i.e., L_1 in Fig. 4.11a), so that the bottom layer (i.e., L_2 in Fig. 4.11a) can serve as a self-reference, accounting for changes in the effective optical thickness that are caused by the medium of analyte. However, bilayered AAO exhibited mon-periodic Fabry-Pérot oscillations with much reduced amplitude compared to L_1- and L_2-only AAOs due to the low refractive index contrast (Fig. 4.11c), which caused difficulties in the direct analysis of the top L_1 layer. In order to enhance the refractive index contrast between the top and both the bottom layer and the top incidence medium, a thin layer of Au was deposited on top of the porous AAO. This resulted in the appearance of the fast Fourier transform (FFT) peak corresponding to the top layer, otherwise barely visible, and in the amplified intensity of the peaks corresponding to the top layer as well as the whole bilayer structure (i.e., $L_1 + L_2$ in Fig. 4.11a). The amplified response translated upon analyte infiltration into a more sensitive determination of the change in the effective optical thickness (EOT) and in a much bigger change in the FFT peak amplitude ratios (Fig. 4.11d).

4.4.3.2 Potential Reduction

As discussed in Sect. 4.3, anodizing potential (U) determines the steady-state morphology of porous AAO. In other words, the barrier layer thickness (t_b), pore diameter (D_p), and interpore distance (D_{int}) of AAO are mainly determined by the potential (U). At a steady-state, the barrier layer thickness (t_b) remains constant due to the dynamic balance of the movement rates of the metal/oxide interface and the oxide/electrolyte interface. Thus, the current density (j) and the electric field ($E = U/t_b$) across the barrier layer are constant. When the anodizing potential (U_1) is reduced abruptly to the potential (U_2), the anodizing current (j) drops to a very small value due to the sudden decrease of the electric field (E), and after a certain period of delay it slowly increases to a steady value dictated by U_2. This phenomenon has been known as "current recovery" and attributed to the thinning of the barrier layer accompanied by branching of pores [30, 141, 142]. Since the movements of ions (i.e., Al^{3+} and O^{2-}/OH^-) within the barrier oxide are governed by the electric field, the time required for complete current recovery increases with decreasing the field strength at the moment of potential drop [143, 144]. Porous AAO layer grown under a newly established steady-state exhibits reduced barrier layer thickness, pore size and spacing (i.e., increased pore density or porosity).

Furneaux et al. [145] reported that effective elimination of the barrier layer can be achieved by reducing anodizing potential in a series of small steps near to zero (Fig. 4.12). The authors suggested that the stepwise potential reduction maintains the electric field at a relatively high level and causes uniform thinning of the barrier layer. When the barrier layer was thin enough, porous AAO film was separated from the underlying Al substrate due to the dissolution of thinned barrier oxide by

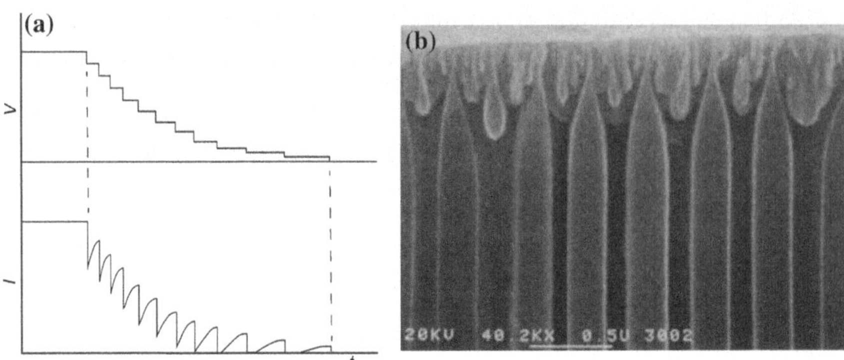

Fig. 4.12 **a** Schematic plots of voltage (V) and current (I) against time during the stepwise potential reduction process, where potential is stepped down, when the rate of current rise had fallen to 75 % of its maximum value. **b** Cross-section SEM image of porous AAO after the stepwise potential reduction process. Adapted with permission from [145]. Copyright 1989, Nature

the electrolyte. The resulting porous AAO film had an asymmetric membrane structure: larger straight pores extending through the bulk of membrane are interconnected with smaller branched pores which formed with a thin skin layer during the stepwise potential reduction process. Since the seminal work of Furneaux et al. [145], thinning of the barrier layer via stepwise potential reduction has been extensively exploited for template-based synthesis of various nanomaterials. It has been reported that the thinning may effectively reduce the potential barrier for the electrons to tunnel through the barrier layer, and thus allows alternating current (AC) electrodeposition of metallic nanowires into the pores of AAO that is attached on aluminum [146–149]. Free-standing AAO membranes obtained by stepwise potential reduction process have also been utilized for the preparation of metallic nanowires or multisegmented nanotubes by direct current (DC) electrodeposition [150, 151]. More recently, Jeon et al. [152] have demonstrated that AAOs with asymmetric membrane structure can be used as filters in enriching various fragile viruses in stocks without losing viral activity. For hepatitis c virus (HCV), the authors reported enrichment efficiency can be as high as 91 %, which is four times higher than that ($\sim$22 %) of conventional centrifuge-based technique. The origin of the efficiency enhancement was attributed to the perfect filtration of the viron particles without damage, whereas significant damage to the virons occurs during ultra-high speed centrifugation [152].

As mentioned above, potential reduction gives rise to branching of the primary pore channels. In general, pore branching is accompanied by random merging or dying of the newly generating pores, especially at the early stage of anodization at a reduced potential. Thus, it is rather difficult to define the pore structure at the oxide layer interconnecting the stem and branched pores. Considerable attention has been paid to control the pore branching behavior, since porous AAOs with tailor-made pore branches could be promising template materials for the synthesis various functional nanostructures with controlled levels of morphological complexity.

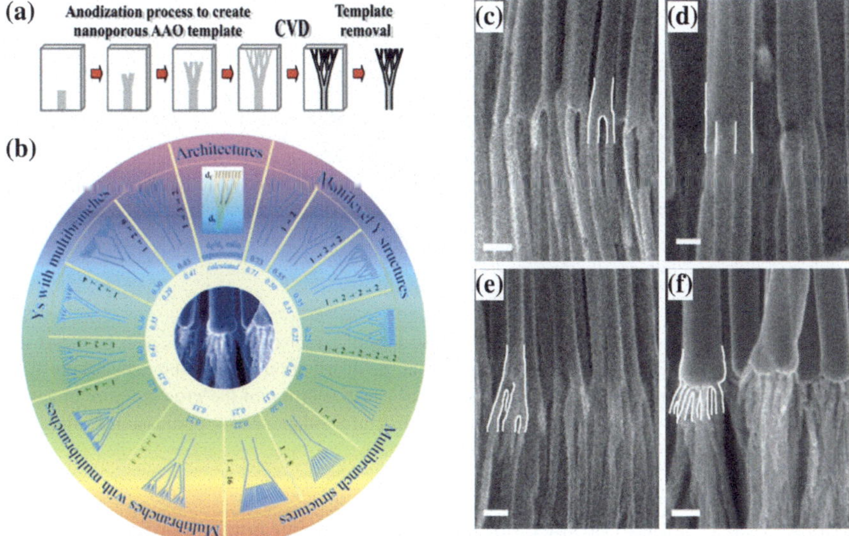

Fig. 4.13 **a** Schematic procedure for the preparation of carbon nanotubes (CNTs) having several hierarchical levels of multiple branching from AAO template with controlled porous architectures, where porous AAO is fabricated by consecutive steps of anodization, and then utilized as template to replicate CNTs by chemical vapor deposition (CVD). **b** Schematic showing CNT architectures that can be prepared from AAO templates with multiply branched pores. **c**–**f** SEM images of multiply branched CNTs showing the primary stem abruptly dividing into n = 2, 3, 4 and 16 branches, respectively (scale bars = 100 nm). Adapted with permission from [157]. Copyright 2005, The National Academy of Sciences of the USA

The work in this direction was first made by Li et al., who fabricated porous AAO with Y-branched pore channels by reducing anodizing potential by a factor of $1/\sqrt{2}$ [153]. Y-branched AAO pores have been utilized as templates for the preparation of Y-Junction carbon nanotubes (CNT) and metallic nanowires [153–156]. Later, Meng et al. expanded this work in order to fabricate multiply connected and hierarchically branched pores inside AAOs, of which hierarchical levels of multiple branching were represented by the notation $1 \prec n \prec m$, where "$\prec$" denotes the junction where the branching occurs, and "n or m" denotes the number of branched pores (Fig. 4.13) [157]. After the anodization for the stem pores in oxalic acid, the authors thinned the barrier layer prior to reducing the anodizing potential by a factor of $1/\sqrt{n}$. If the potential is lower than 25 V for pore branching, sulfuric acid was used instead of oxalic acid. Meng et al. claimed that the number and frequency of pore branching, dimensions, and the overall architecture of pore channels can be precisely controlled through properly designed potential reduction scheme [157].

In the works of Li et al. and Meng et al [153, 157], the potential reduction factors were chosen in consideration of (i) the linear dependence of the steady-state oxide cell dimension on the anodizing potential (i.e., $D_{\text{int}} = \zeta U$) and (ii) the invariance of the oxide cell area of a stem pore after branching; reducing the anodizing potential

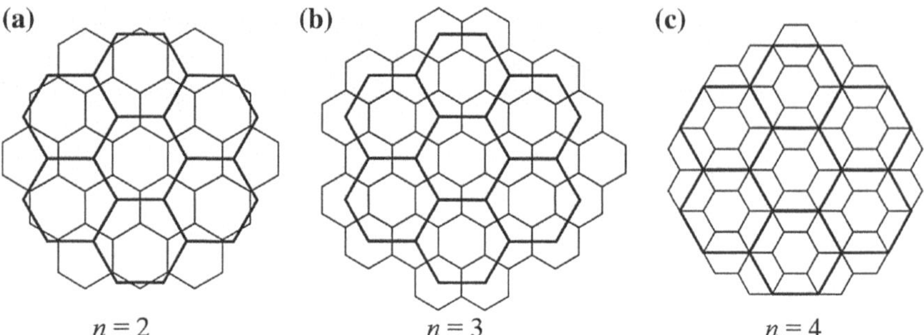

Fig. 4.14 Superposed hexagonal oxide cells. The area of larger oxide cells is **a** 2, **b** 3, and **c** 4 times that of the smaller cells. Panel **a** illustrates that the distribution of the Y-branched pores (i.e., $1 \prec 2$) cannot be close-packed. The extra space will cause splitting of the growing pore channels and result in a disordered and uncontrollable pore structure. Adapted with permission from [158]. Copyright 2009, The Royal Society of Chemistry

by a factor of $1/\sqrt{n}$ will produce n times as many smaller pores in order to maintain the original area of the oxide cell. However, this equal-area model for the growth of Y- or n-branched pore channels was argued by Shuoshuo et al. [158]. The authors claimed that it is impossible to control arbitrarily the growth of branched pore channels as a $1 \prec n$ structure, because the growth of some n-branched pores (e.g., $n = 2$; Y-branched pores) could not be a steady process and thus such branched pores could not be obtained uniform in a large scale. Based on theoretical consideration on the geometric arrangement of the branched pores, Shuoshuo et al. proposed that close-packed hexagonal arrangement of oxide cells can be achieved only for the potential reduction by the factors of $1/\sqrt{3}$, $1/\sqrt{4}$ and their common multiples and powers (i.e., $n = 3, 4, 9, 12, \ldots$) (see Fig. 4.14); otherwise, the number of branched pores cannot be grown proportionally from each of the stem pores [158].

Non-steady-state formation of branched pores by potential reduction was systematically investigated by Ho and coworkers [159]. The authors fabricated three-dimensional (3D) two and three-tiered branched AAOs by stepping down anodizing potential for each anodizing step while thinning of the barrier layer is performed after each step; AAOs with the first tier having an average pore diameter of 285 nm branching into four 125 nm sub-pores in the second-tier and four 55 nm sub-pores in the third tier (Fig. 4.15a, b). They observed that the ratio of interpore distance (D_{int}) to the anodizing potential (U) of the second and subsequent tiers does not follow the characteristic constant value of $\zeta = 2.5$ nm V^{-1} for steady-state mild anodization (MA), but varies with an exponential function from $\zeta = 1.77$ to 1.26 nm V^{-1} when the potential ratio of the second- to the first-tier anodization changes from 0.5 to 0.96 (Fig. 4.15c). This difference was attributed to the constricted formation of the sub-pores by the boundaries of the preceding pores. On the other hand, the average number of sub-pores that can be generated within each preceding pore was observed to be decreased from 4.27 to 1.28 for the corresponding range of potential ratio (Fig. 4.15d).

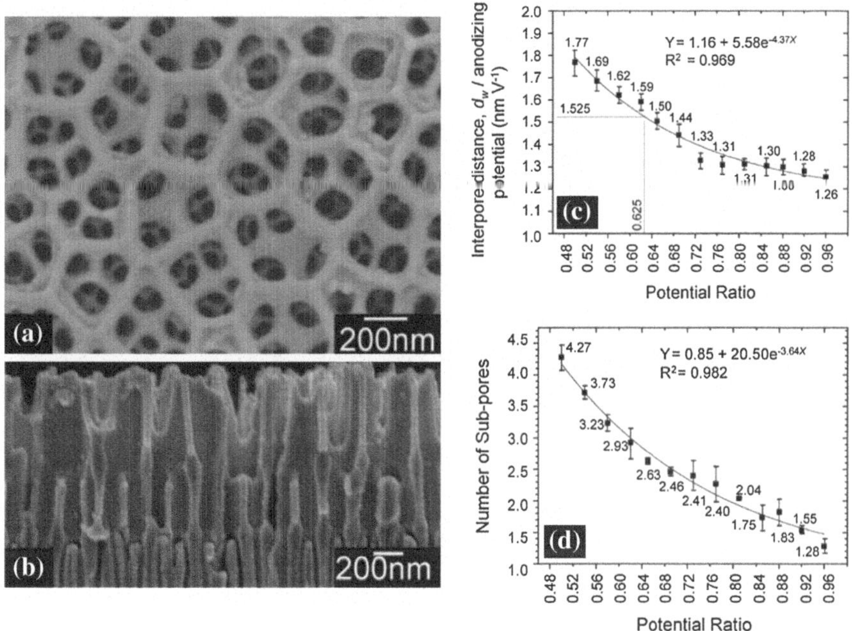

Fig. 4.15 **a** Top and **b** cross-sectional SEM images of AAOs with three-tiered branched pores prepared by multiple anodizations in 0.3 M H_3PO_4 (130 V), 0.15 M $H_2C_2O_4$ (80 V) and 0.15 M $H_2C_2O_4$ (80 V) followed by thinning of the barrier layer using 5 wt% H_3PO_4. **c** Ratio of the sub-pores' interpore distance and anodizing potential ratio. **d** Number of sub-pores within each preceding pore versus anodizing potential ratio. Reprinted with permission from [159]. Copyright 2008 Wiley-VCH Verlag GmbH & Co. KGaA, Weinheim

4.4.3.3 Pulse Anodization (PA)

As discussed in Sect. 4.3.2, porous AAOs formed by hard anodization (HA) have one-third lower porosity (P) than those produced by mild anodization (MA) (i.e., $P_{HA} \approx 3$ % for HA and $P_{MA} \approx 10$ % for MA). Based on this experimental finding, Lee et al. [78] first demonstrated fabrication of highly ordered porous AAOs with periodically modulated pore diameters along the pore axes by properly combining MA and HA processes (Fig. 4.16). An ideally ordered porous AAO was first prepared by conducting MA in 0.4 M H_3PO_4 (10 °C) at 110 V. After that, the resulting sample was further anodized in an HA condition using 0.015 M $H_2C_2O_4$ (0.5 °C) at 137 V. Porous AAOs with periodically modulated pore diameters were obtained by repeating the above two anodization processes. The anodizing potentials (U) for MA and HA were chosen in consideration of the potential (U) and interpore distance (D_{int}) relations of the respective anodizing processes; $\zeta_{MA} = 2.5$ nm V^{-1} and $\zeta_{HA} = 2.0$ nm V^{-1}. With a similar approach, Pitzschel et al. fabricated porous AAOs with modulated pore diameters using the same electrolytes yet under different anodizing conditions [160]. The resulting porous AAOs were used to prepare concentric $SiO_2/Fe_3O_4/SiO_2$ nanotubes by atomic layer deposition

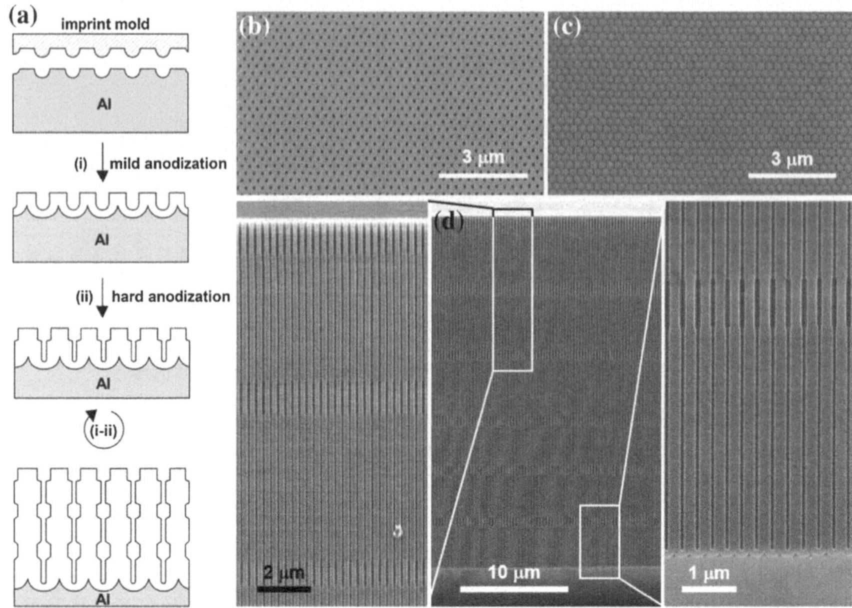

Fig. 4.16 **a** Scheme for the fabrication of porous AAO with modulated pore diameters by a combination of MA and HA. **b**, **c** SEM images of the *top* and *bottom* surface view of the porous AAO ($D_{\text{int}} = 275$ nm) formed by the process in (**a**). **d–f** SEM images showing the cross-section view of the porous AAO with periodically modulated pore diameters. Reproduced with permission from [78]. Copyright 2006 Nature Publishing Group

(ALD) in order to investigate the effect of diameter modulations on the magnetic properties. In aforementioned approach, each step for pore modulation required a tedious manual change of the anodizing electrolytes and temperatures in order to satisfy both MA and HA processing conditions. Lee et al. have reported that this drawback can be overcome by performing pulse anodization (PA) of aluminum under potentiostatic conditions in sulfuric or oxalic acid electrolytes, demonstrating continuous structural and compositional engineering of porous AAOs (Fig. 4.17) [144].

In a typical PA process, a low potential (U_{MA}) and a high potential (U_{HA}) is alternately pulsed to achieve MA and HA conditions, respectively. A representative current-time (*j-t*) transient during PA of aluminum is shown in Fig. 4.17b. Current profile during a MA potential pulses is characterized by “current recovery” behavior (see the inset of Fig. 4.17b), similarly to the current evolution after potential reduction as discussed in Sect. 4.4.3.2: initially high current drops abruptly, hits a minimum value after a capacitive decay for a short time, and then gradually increases to reach a steady-state value after passing an overshoot. On the other hand, upon applying an HA pulse the current density (*j*) increases steeply for a short period of time and then decreases exponentially, which is the typical anodization

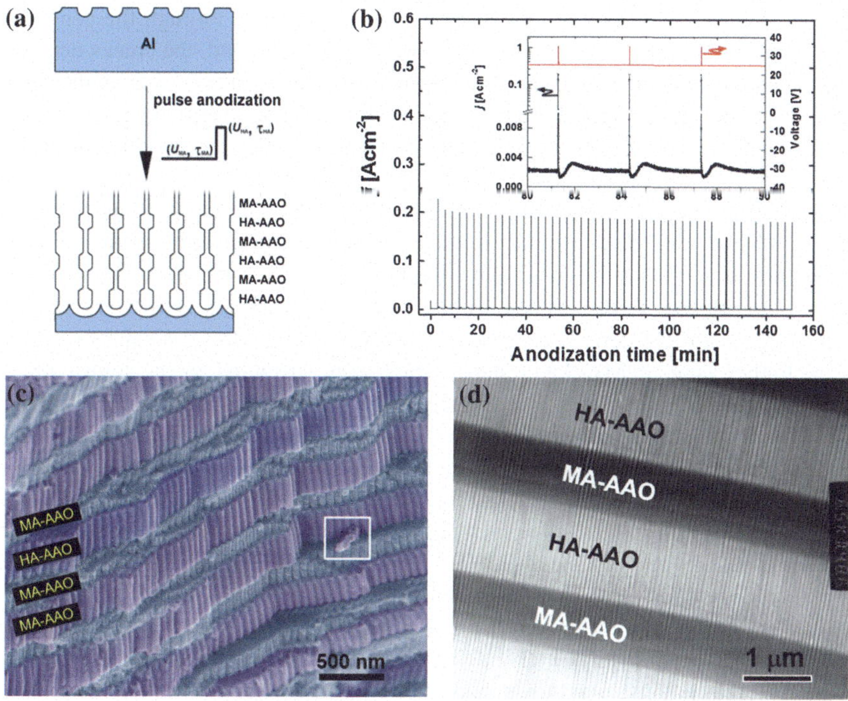

Fig. 4.17 **a** Scheme for the preparation of porous AAOs with modulated pore diameters by pulse anodization (PA). **b** A typical current-time (j-t) transient during PA of aluminium; pulses consisting of an MA pulse ($U_{MA} = 25$ V and $\tau_{MA} = 180$ s) followed by an HA pulse ($U_{HA} = 35$ V and $\tau_{HA} = 0.1$ s) were applied. *Inset* An enlarged j-t curve at $t = 80$–90 min. **c** False-colored SEM image of AAO formed by H_2SO_4-PA. AAO slabs formed by MA and HA pulses are indicated by MA-AAO and HA-AAO, respectively. A white rectangle in (**c**) highlights alumina nanotubes (so-called "Keller-Hunter-Robinson cells") that were separated from the nearby HA-AAO slab. **d** Cross-sectional TEM image of AAO formed by pulse anodization using 0.3 M H_2SO_4 ($U_{MA} = 25$ V, $\tau_{MA} = 180$ s, $U_{HA} = 37$ V, $\tau_{HA} = 1$ s), showing image contrast in MA- and HA-slabs due to the difference in pore diameter. Reproduced with permission from [144]. Copyright 2008 Nature Publishing Group

kinetics of potentiostatic HA of aluminum [78]. As shown in Fig. 4.17b, current density (j) changes periodically to the values dictated by pulsed potentials (i.e., j_{MA} for U_{MA} and j_{HA} for U_{HA}). As a results of periodic variations of current amplitude, the resulting porous AAO exhibits a layered structure with alternate stacking of MA-AAO slab with a smaller pore diameter and HA-AAO slab with a larger pore diameter (Fig. 4.17c, d). The thickness of each oxide slab is determined by the pulse durations (i.e., τ_{MA} for MA-pulse and τ_{HA} for HA-pulse). As a characteristic feature of porous AAOs formed by pulse anodization in sulfuric acid (i.e., H_2SO_4-PA), MA-AAO and HA-AAO slabs exhibit different fracture behaviors against external stresses (Fig. 4.17c). For MA-AAO slabs, cracks propagate through the center of pores from pore to pore. For HA-AAO slabs, on the other hand, the cracks develop along the cell boundaries and not through the center of the pores, indicating fairly weak cell junction strength. Cleavage through the cell boundaries in AAO formed by HA pulses resulted even in single Al_2O_3 nanotubes or bundles of tubes (so-called "Keller-Hunter-Robinson cells") [24, 161, 162].

Apart from periodic modulations of pore diameter, Lee et al. found that the change of the current amplitude during PA results in periodic compositional modulation of the resulting AAO along the pore axes. Their microscopic elemental analysis of porous AAO formed by sulfuric acid-based PA indicated that the content of electrolyte-derived impurities (mostly SO_4^{2-}) in HA-AAO slabs is about 88 % higher than in MA-AAO slabs, which was attributed to the high current density (j_{HA}) during HA-pulsing (typically, $j_{HA} \approx 10–10^2 \times j_{MA}$) [144]. In general, anodic oxide contaminated with anionic impurities is labile to chemical attack of an oxide etchant (e.g., 5 wt% H_3PO_4). Therefore, a periodic compositional modulation in pulse anodized AAO may result in an etching contrast between MA- and HA-AAO pore wall oxides, allowing the fabrication of porous AAOs, in which the difference in the pore diameters of HA- and MA-AAO is pronounced than in as-prepared samples. Upon wet-chemical etching of pore wall oxide, pulse anodized AAO exhibits a structural color that is tunable by controlling the pulse durations (i.e., τ_{MA} and τ_{HA}). By taking advantage of poor chemical stability of HA-AAO slabs against an oxide etchant, Lee et al. [144] could completely separate the entire MA-AAO slabs from a single as as-prepared porous AAO by selectively etching HA-AAO slabs (Fig. 4.18). The thickness of MA-AAO slabs could be varied from a few hundred nanometer to several micrometers by changing MA pulse duration (τ_{MA}). This demonstration indicated that PA of aluminum and wet-chemical etching of the resulting porous AAO could be a simple, continuous, and economic way for the mass production of porous AAO sheets.

As mentioned above, when the anodizing potential is changed from a higher U_{HA} to a lower U_{MA}, the current density drops abruptly from j_{HA} to a minimum value and then increases gradually to a value (j_{MA}) corresponding to U_{MA} (i.e., current recovery). It was reported that the time required for a complete recovery of anodizing current depends on the chemical nature of the barrier oxide (i.e., the content of anionic impurities), the electrolyte temperature, and the potential difference between U_{HA} and U_{MA} (i.e., the electric field strength at the moment of potential drop): For three popular anodizing electrolytes, it increases with the order $H_2SO_4 < H_2C_2O_4 < H_3PO_4$ [143, 144]. Unlike H_2SO_4-based PA, PA of aluminum in $H_2C_2O_4$ or H_3PO_4 electrolyte is difficult to achieve continuously within a reasonable period of time because of the retarded current recovery, especially at a low temperature and a large potential difference [143, 144]. In order to solve the problem associated with the slow current recovery, Lee and Kim increased gradually anodizing potential prior to pulsing a high anodizing potential [143]. By

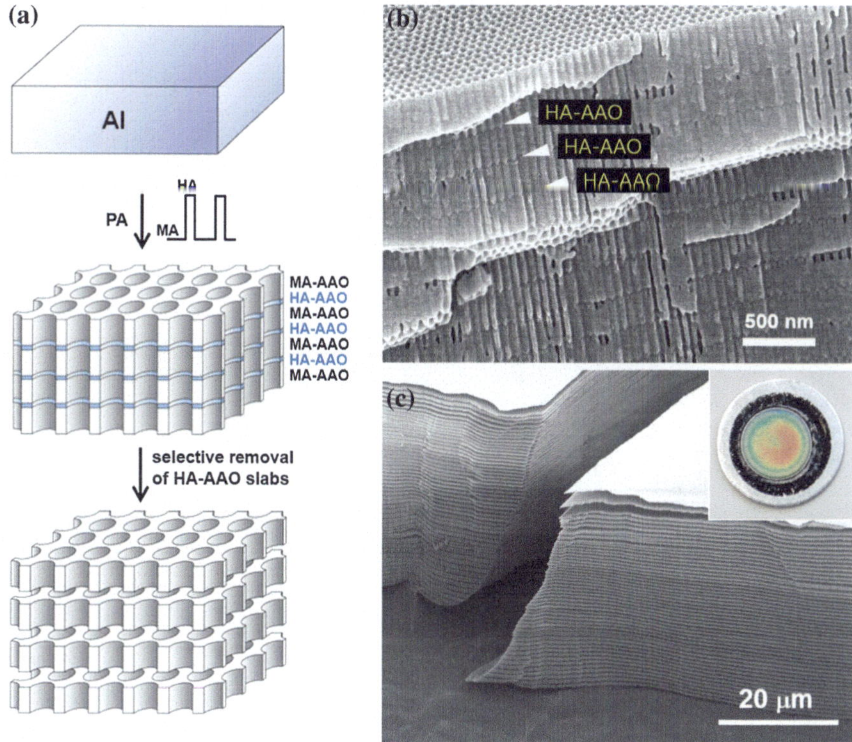

Fig. 4.18 **a** Scheme for the mass production of porous AAO membranes by a combination of pulse anodization (PA) and wet-chemical etching. **b**, **c** SEM images of porous AAO formed by PA and stacks of MA-AAO slabs after selective removal of HA-AAO slabs by wet-chemical etching. The inset of **c** shows a photograph of pulse anodized AAO on aluminum after wet-chemical etching treatment. SEM images were reproduced with permission from [144]. Copyright 2008 Nature Publishing Group

employing potential pulses with specifically designed periods and amplitudes, they successfully demonstrated engineering of the internal pore geometry (Fig. 4.19), enabling fabrication of optically active 3D porous architectures and also expanding the degree of freedom not only in AAO-based synthesis of functional nanowires and nanotubes with modulated diameters.

Structural modulation of porous AAOs can also be achieved by galvanostatic PA, where current pulses satisfying MA and HA conditions are periodically applied. As mentioned above, HA-AAO slabs formed by H_2SO_4-PA shows weak junction strength between cells. By exploiting this feature, Lee et al. [163] developed a convenient route for the mass preparation of uniform anodic alumina nanotubes (AANTs) with prescribed lengths. They applied periodically galvanic MA and HA pulses to achieve continuous modulations of pore diameter and also to weaken the

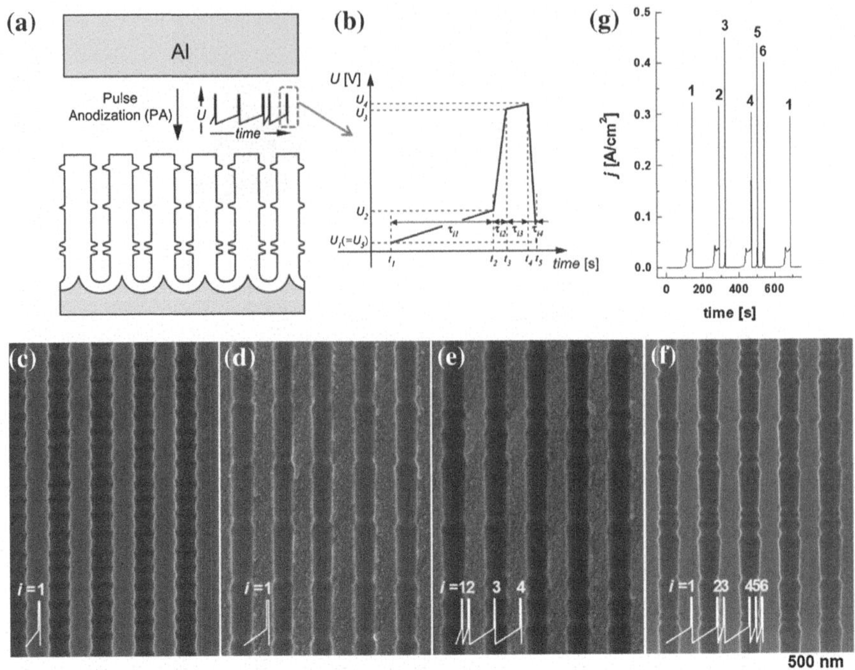

Fig. 4.19 Schematics showing **a** the experimental process for the fabrication of AAO with tailor-made pore structures by pulse anodization (PA) and **b** a generalized form of a potential pulse employed in pulse anodizations. U_j and τ_{ij} define the repeating unit of potential waves, where U_j = the potential at the time t_j with $U_1 = U_5$ (j = 1–4), $\tau_{ij} = t_{j+1} - t_j$, i = the pulse number (i = 1, 2, 3, …). Representative SEM image of porous AAOs with modulated pores prepared by pulse anodization; **c** τ_{11} = 36 s, **d** τ_{11} = 144 s, **e** $\tau_{11} = \tau_{21}$ = 36 s, $\tau_{31} = t_{41}$ = 144 s, and **f** $\tau_{11} = \tau_{21}$ = 144 s, τ_{31} = 36 s, τ_{41} = 144 s, $\tau_{51} = \tau_{61}$ = 36 s. Other parameters were fixed at U_1 = 80 V, U_2 = 140 V, $U_3 = U_4$ = 160 V, $\tau_{i2} = \tau_{i4}$ = 0 s, τ_{i3} = 0.2 s. The repeating units of potential pulses are shown as insets in the respective images. **g** Current (j)-time (t) transient for the sample shown in (**f**). From [143]. Copyright 2010 IOP Publishing

junction strength between oxide cells. AANTs, the length of which is determined by the HA-pulse duration (τ_{HA}), could be obtained by immersing the resulting porous AAO into an acidified $CuCl_2$ solution, followed by ultrasonic treatment in order to release individual oxide nanotubes from the as-anodized sample (Fig. 4.20). Uniform AANTs with controlled length may merit various applications because of material's excellent physicochemical properties, including a large effective surface area, high dielectric constant, chemical inertness, and biocompatibility [164–167]. Recently, Logic group have demonstrated that AANTs formed by galvanostatic PA can be used as new drug delivery nanocarriers by using tumor necrosis factor-related apoptosis-inducing ligand (Apo2L/TRAIL) as a model drug Fig. 4.21) [168].

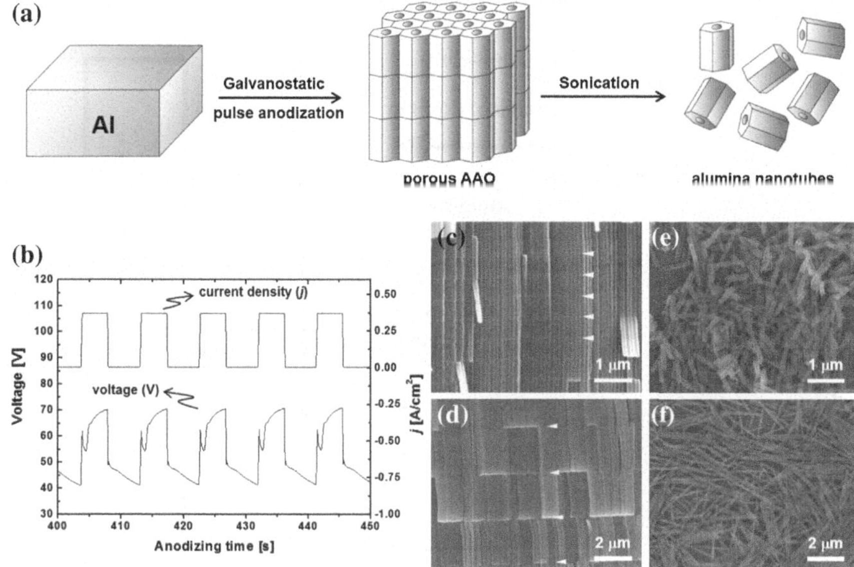

Fig. 4.20 **a** Scheme for the mass production of Al_2O_3 nanotubes with prescribed lengths. **b** A typical current (j)-voltage (V) relation during galvanostatic pulse anodization (PA) of aluminum. **c**, **d** Cross-sectional SEM micrographs of AAOs formed by galvanostatic PA: $\tau_{HA} = 2$ s for (**c**) and $\tau_{HA} = 6$ s for (**d**). Other pulse parameters were fixed at $j_{HA} = 368.42$ mA cm^{-2}, $j_{MA} = 3.16$ mA cm^{-2} and $\tau_{MA} = 5$ s. **e**, **f** SEM micrographs of Al_2O_3 nanotubes obtained from the sample shown in panel (**c**) and (**d**), respectively. Reprinted with permission from [163]. Copyright 2008 The American Chemical Society

Their study have shown that as-prepared AANTs exhibit ultra-high pro-apoptotic protein Apo2L/TRAIL loading capacity (104 ± 14.4 μg mg^{-1}), and provide a successful release with considerable local concentration (>24 μg) at in vitro simulated physiological condition. The effectiveness of AANTs-Apol2L/TRAIL nanocarriers was confirmed by effective reduction in viability of MDA-MB231-TXSA breast cancer cell, which was attributed to the cell apoptosis induced by high concentration of Apo2L/TRAIL loaded in AANTs.

An approach for pore modulations without combination of MA and HA processes has been reported by Santos et al. [169]. The process, so-called "discontinuous anodization (DA)", uses potential pulses under MA regime and thermally activated oxide dissolution condition [169]. In this method, self-ordered porous AAO was first grown by anodizing a textured aluminum in 0.3 M H_3PO_4 (1 °C) at 170 V. After 6 min, anodizing was stopped and the electrolyte temperature was increased to 35 °C. After that, DA process was started. DA consisted of applying consecutive MA pulses ($U_{MA} = 170$ V and $\tau_{MA} = 1$ min) separated each other by a designated relax period ($U_{relax} = 0$ V and $\tau_{relax} = 3, 6, 9,$ or 12 min). The modulated pores of the resulting AAOs changed from a slender node-like, ellipsoidal, and

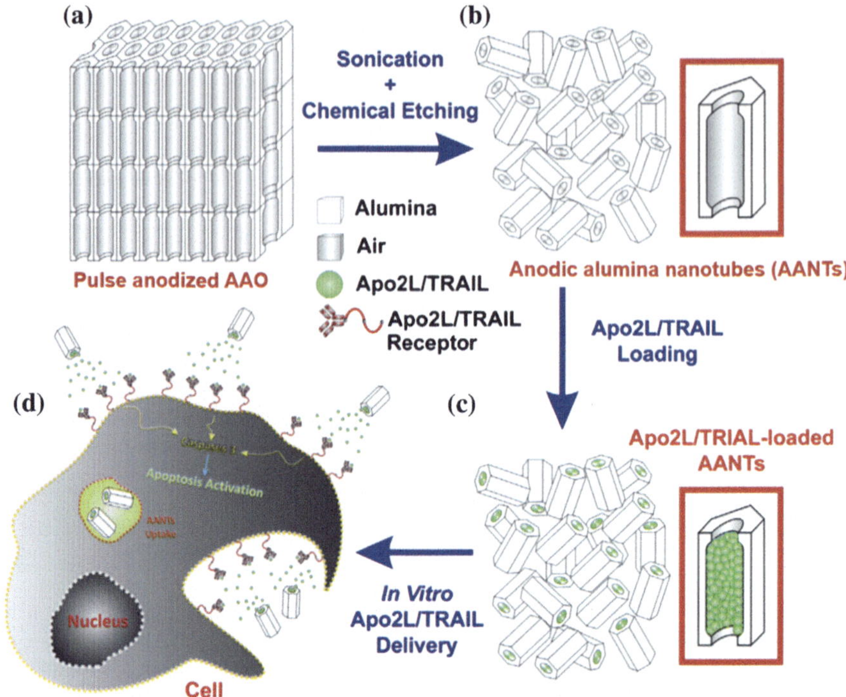

Fig. 4.21 Schematic illustration showing application of anodic alumina nanotubes (AANTs) for in vitro drug delivery of tumor necrosis factor-related apoptosis-inducing ligand (Apo2L/TRAIL). Adapted with permission from [168]. Copyright 2014 Elsevier Ltd

asymmetric serrated morphology depending on the relax period. It was proposed that the generation of the pore modulations is associated with the formation of a gel-like layer (e.g., $Al(OH)_3$) during the period of relax and its subsequent detachment during the beginning of each potential pulse [169]. The proposed mechanism appears to explain the experimental observations. But it should be supported by the evidence that can confirm the presence of gel-like layer on the entire surfaces of the pores. It was reported that gel-like material on pore wall surface is very soluble in acid electrolytes even at room temperature [85]. Gel-like layer on the pore wall surface may be formed by re-deposition of dissoluted oxide species within the pores, which typically occurs upon drying of an improperly rinsed porous AAO. Santos et al. [169] attributed the sharp increase in anodizing current (j) at the beginning of each potential pulse to the detachment of the gel-like layer from the pore base. But, an alternative interpretation can be made for this observation as follow: During the period of relax, acidic dissolution of pore wall oxide would occur due to the high electrolyte temperature, thus pores are enlarged. Simultaneously, the barrier layer is thinned. As a consequence, anodizing current (*j*) increases steeply upon applying MA pulses but decreases almost exponentially to steady-state value as the pores grow with their equilibrium morphology. The reported maximum current density (*j*) upon MA pulses was about 40 mA cm^{-2}, which is about 10 times higher than that of typical H_3PO_4-MA at comparable temperature. The high current density (*j*) at the beginning of each potential pulse would account for observed pore modulation. In fact, it has been reported that at a given anodizing potential (i.e., U = constant) pore diameter (D_p) increases with current density (*j*) [85].

4.4.3.4 Cyclic Anodization (CA)

Losic et al. put forward the concept of structural engineering of internal pore structures of AAO by combining MA and HA [170]. They developed a new anodization process, termed "cyclic anodization (CA)", which is based on slow and oscillatory changes of the anodizing conditions from MA to HA regimes. The authors used periodically oscillatory current signals with different profiles, amplitudes, and periods in order to achieve pore modulations in porous AAOs (Fig. 4.22). By applying current signals of different cyclic parameters, porous AAOs with different pore shapes (circular- or ratchet-type), lengths, periodicity and gradients were prepared. Microscopic investigation on the samples indicated that the internal pore geometry is directed by the characteristics (i.e., profile, period, and amplitude) of the applied cyclic signal. The authors pointed out that the transitional anodization (TA) mode, the transition from MA to HA regime, is important for structural engineering of pores in their CA process; the minimum current of the applied cycle is responsible for the formation of the smallest pores (MA), the slope in the current directs the main pore shape (TA), and the maximum current is responsible for the formation of the largest pores. The authors further demonstrated fabrication of porous AAOs with distinctive, hierarchical internal pore structures by employing multiprofiled current signals; for example, three different successive CA steps, beginning with first cycle having a gradually increasing amplitude of current, then a double-profiled cycle, and lastly a series of triangular galvanic cycles. In a separate report, the same authors demonstrated fabrication of porous AAOs with laterally perforated pores (i.e., pores with nanoholes along horizontal directions) by a combination of CA and wet-chemical etching of the resulting porous AAOs [171].

As mentioned in Sect. 4.4.3.3, porous AAOs with periodically modulated pore diameters may exhibit colors, when their modulation period is the order of the visible light wavelength. In other words, porous AAO with appropriately modulated pores can behave like a distributed Bragg reflector (DBR) that consists of periodically stacked layers of different refractive indices, reflecting a range of wavelengths propagating in the direction normal to the stacked layers. The range of wavelengths that are reflected is called the "photonic stopband". The central wavelength of stopband depends on the refractive index and the layer thickness. On the other

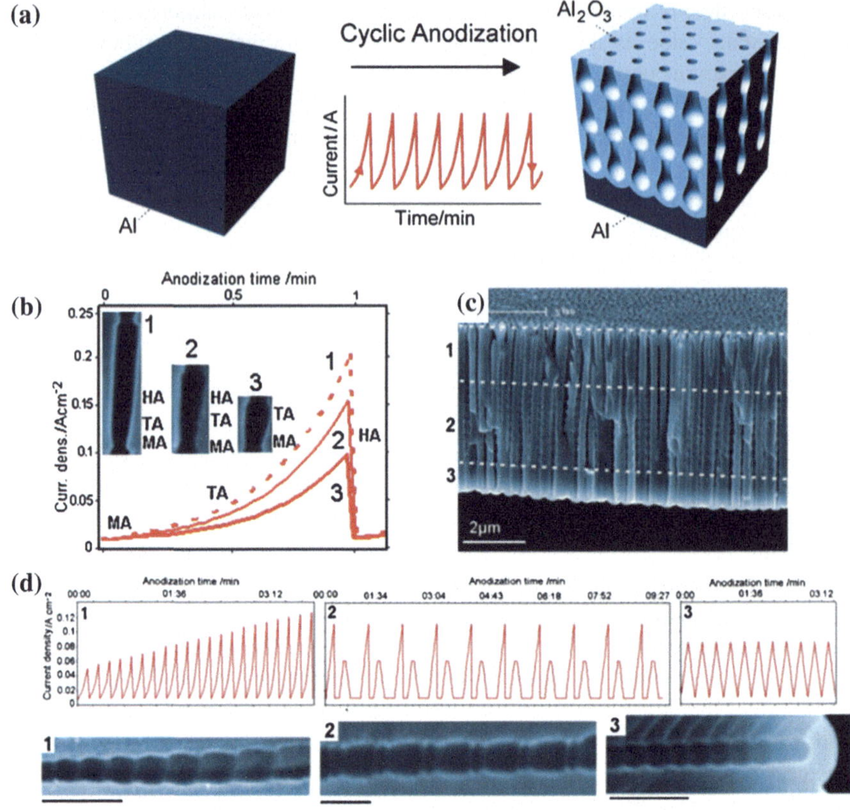

Fig. 4.22 **a** Schematics of the galvanostatic cyclic anodization (CA). **b** Influence of current amplitude on the pore shape and the length of modulated pore segments. Anodization modes (i.e., MA, TA, HA) associated with corresponding anodization currents are marked on the pore structures and graphs. **c** SEM image of porous AAO with multilayered pore architectures and different pore modulation patterns. **d** Current profiles versus pore geometries of the sample shown in (**c**). Scale bars = 500 nm. Reprinted with permission from [170]. Copyright 2009 Wiley-VCH Verlag GmbH & Co. KGaA, Weinheim

hand, its bandwidth depends on the refractive index contrast, the relative layer thickness within one period, and the total number of layers. Due to its outstanding optical characteristics, DBRs have been important components in various opto-electronic devices (e.g., vertical cavity surface-emitting laser, reflector mirrors, electroabsorptive reflection modulators), and thus cheap and robust ways of realization of DBR structures have been intensively explored recently [172–174].

Optical properties of structurally engineered porous AAOs have been intensively investigated by Fei group [175–180]. They first reported that photonic crystals made of “air pores” in AAOs can be fabricated by applying periodically varying

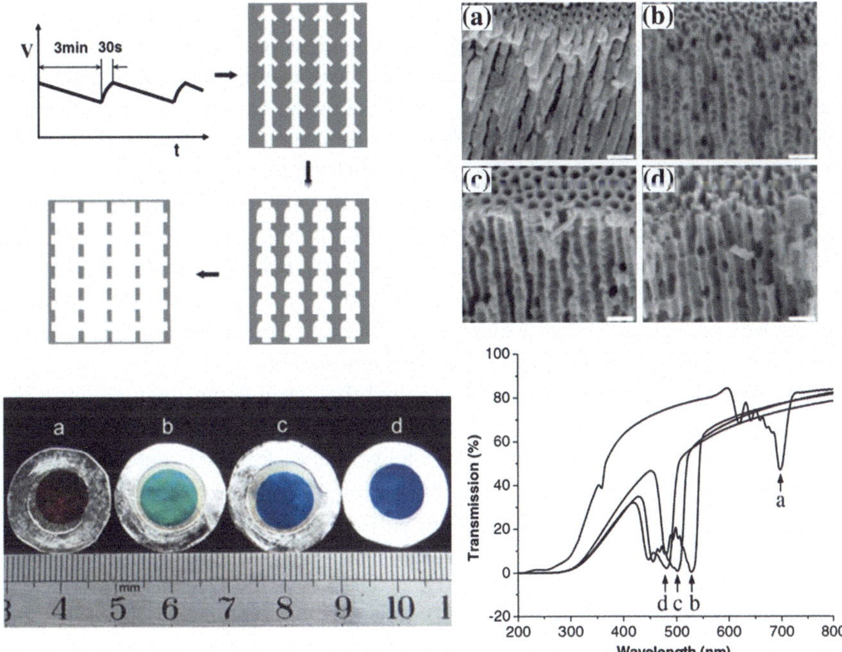

Fig. 4.23 *Top-left* Schematic procedure for the fabrication of a photonic crystal made of air pores in AAO by a combination of anodization of aluminum with periodically changing potential and wet-chemical etching. *Top-right* SEM images of porous AAO (**a**) before and after chemical etching for **b** 15 min, **c** 18 min, and **d** 20 min. Scale bars = 200 nm. *Bottom-left* A photograph and *Bottom-right* transmission spectra of the corresponding samples. Adapted with permission from [175]. Copyright 2007 IOP Publishing

potential during anodizing and subsequently performing wet-chemical etching of the resulting AAOs (Fig. 4.23) [175]. For anodization of aluminum, anodizing potential was linearly decreased from 53 to 23 V in 3 min, and then increased from 23 to 53 V in 30 s. Straight main channels were formed during the high-potential duration, and each main channel branched into several small channels during the low-voltage duration. By cycling this process, a periodic structure with main and branched channels could be formed. The stopband position in the transmission spectra could be controlled in the range from 450 to 525 nm by adjusting the time for wet-chemical etching. Later, the same group of authors put forward their approach in order to fabricate AAO-based DBR structures and demonstrated that the first Bragg condition for the main inhibition of incident light perpendicular to the surface of AAO can be modulated from 727 to 1200 nm by modifying waveform of anodizing potential [176]. With a separate report, they also showed that the transmission spectra of AAO-based DBRs can be modulated further to cover almost any wavelength range of the visible light by simply modifying the anodizing temperature [177]. The effect of incidence angle on the optical properties of AAO-based DBRs was studied by Xing et al. who observed that when the incidence angle increases progressively, a blue shift of the transmission peak occurs with gradual decrease of the peak intensity [181]. Fei group has recently demonstrated that structurally engineered porous AAOs can be used as gas sensors [178, 179]. Porous AAOs were prepared by using so-called "voltage compensation mode", which was claimed to effectively compensate the unevenness of the pore size caused by ion concentration decrease at the pore bottoms and the pore wall dissolution by the electrolyte [178, 180]. Porous AAO behaved as photonic crystal with an ultra-narrow photonic bandgap (PBG), of which the full width at the half maximum is only 30 nm. Based on shift of PBG upon exposure to ethanol vapor, porous AAO-based photonic crystal gas sensor was fabricated.

More recently, Ferré-Borrull group reported that an in-depth modulation of the pore geometry and the refractive index of AAO-based DBRs can be achieved by potentiostatic CA of aluminum with deliberately designed potential cycles (Fig. 4.24) [182]. For the modulation of pores, 150 potential cycles were applied, in which each cycle consisted of three phases: (i) a linear increasing potential ramp from 20 to 50 V at a rate of 0.5 Vs^{-1}, (ii) an interval of constant potential at 50 V lasting for time to flow a given charge (Q_o = 0, 0.5, 1 and 4 C), and (iii) a linearly decreasing potential ramp from 50 to 20 V at 0.1 Vs^{-1} (Fig. 4.24a–d). After the anodization process, wet-chemical etching of the resulting porous AAOs was carried out using 5 wt% H_3PO_4 (35 °C) in order to increase pore diameters for different period of times (0, 9, 18, and 27 min). Transmittance spectra of the AAO-based DBRs revealed that by controlling the amount of charge involved in each cycle it is possible to modulate the photonic stopband position from the UV to the near-infrared range. Further, the pore-widening causes blue shift of the stopbands and modifies their width and depth (Fig. 4.24e–h). The study well demonstrate the ability to control the photonic stopband by modulating the refractive index contrast between the layers through a combination of CA and controlled pore-widening.

Excellent examples of work on real-time and label-free biosensing applications of structurally engineered porous AAOs have been recently reported by Kumeria and coworkers [183, 184]. They fabricated porous AAO rugate filters (AAO-RFs) by anodizing aluminum in 0.3 M $H_2C_2O_4$ under different pseudosinusoidal potentiostatic conditions controlled by total amount of charge (Q). AAO-RFs were combined with reflection spectroscopy (RfS) in order to develop a biosensing system. The biosensor systems were evaluated and optimized not only by measuring shifts of the characteristic reflection peak with the effective refractive index change, which was achieved by detecting different levels of analytes, but also by performing theoretical modeling based on the Looyenga-Landau-Lifshitz model. For D-glucose sensing, the AAO-RFs biosensing system showed a large linear range from 0.01 to 1.00 M, a low detection limit of 0.01 M, a sensitivity of 4.93 nm M^{-1}, and a linearity of 0.998 [183]. For ionic mercury sensing, it showed a linear

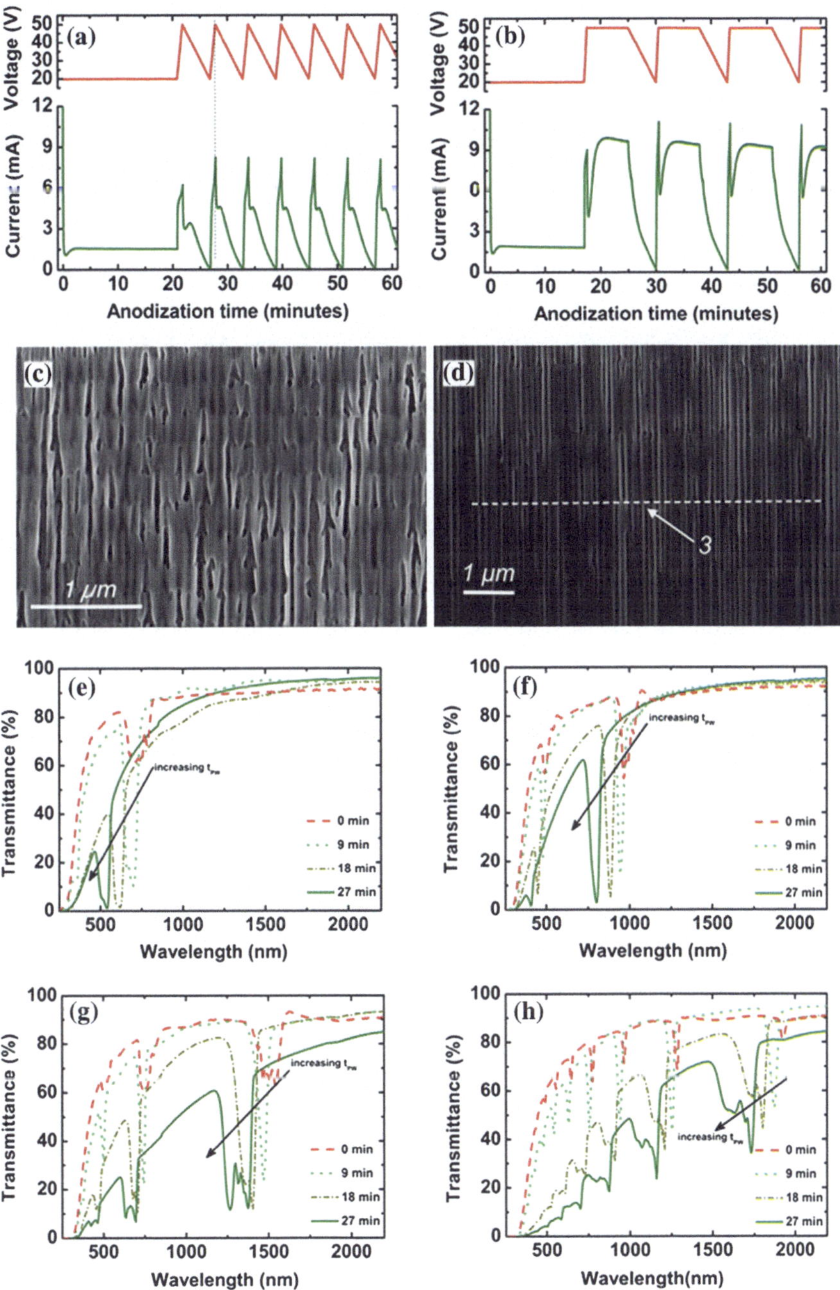
(a)
Voltage (V)
Current (mA)
Anodization time (minutes)
(b)
(c)
1 μm
(d)
3
1 μm
(e)
Transmittance (%)
Wavelength (nm)
increasing t_{PW}
0 min
9 min
18 min
27 min
(f)
(g)
(h)
Wavelength(nm)

◀ **Fig. 4.24** **a**, **b** Applied voltage profiles (*red trace*) and measured current transients (*green trace*) for the AAO samples with **a** $Q_o = 0$ C and $Q_o = 4$ C., where Q_o is the charge during the interval of constant potential at 50 V. **c**, **d** Cross-section SEM images of the respective resulting samples, respectively. The label 2 indicates conical shape of the pores. The label 3 indicates the interface between two cycles. **e–h** Transmittance spectra of the AAO-based DBRs for different Q_0 and pore-widening times: **e** $Q_o = 0$ C, **f** $Q_o = 0.5$ C, **g** $Q_o = 1$ C, and **h** $Q_o = 4$ C. The label 1 indicates the beginning of two branchings. Adapted with permission from [182]. Copyright 2013 The American Chemical Society

working range from 1 to 100 μM of Hg^{2+}, low detection limit of 1 μM, and sensitivity of 0.072 nm μM^{-1}, good chemical selectivity over different metal ions (e.g., Co^{2+}, Mg^{2+}, Ni^{2+}, Cu^{2+}, Pb^{2+}, Fe^{3+}, Ca^{2+}, Cr^{6+}, and Ag^{+}) [184].

4.5 Concluding Remarks

The anodization of aluminum and the resulting self-ordered porous anodic aluminum oxide (AAO) are currently the subject of intensive research in nanotechnology. In this chapter, the most recent advances in structural engineering of porous AAO and applications of porous AAO with tailor-made internal pore geometries have been reviewed. The structure of porous AAO can be characterized by hexagonal arrays of mutually parallel cylindrical pores. Anodization of aluminum under self-ordering regimes yields self-ordered porous AAO, whose pore depth, size, spacing, and density can be conveniently tuned by properly controlling anodizing conditions (i.e., anodizing time, electrolyte, and potential). The porous AAO possesses many useful properties, including excellent thermal stability, chemical inertness, bio-compatibility, and mechanical hardness. In addition, the surface chemistry of pore wall oxide can be conveniently engineered through various methods, including self-assembly, layer-by-layer (LbL) deposition, atomic layer deposition (ALD), chemical-vapor deposition (CVD), sol-gel deposition, and electro/electroless deposition of metals. Over the last two decades, these features have been intensively exploited for the templated syntheses of a variety of low-dimensional functional nanostructures, and also for the development of AAO-based advanced devices, which are difficult or impossible to be fabricated by existing synthetic methods or state-of-the-art lithographic techniques.

Various methods to design and engineer the pore structure of AAO have been developed. These include micromachining of porous AAO, anodization of pre-textured aluminum, multistep anodization with controlled wet-chemical oxide etching after each anodizing step, programmed potential reduction, pulse anodization (PA), and cyclic anodization (CA). These approaches have allowed the researchers to fabricate porous AAOs with a variety of structures, such as porous AAOs with high-aspect-ratio microstructures, sharp featured non-circular pores, non-hexagonal arrangement of pores, symmetrically or asymmetrically modulated pores, funnel-like pores, hierarchical three-dimensional architectures, etc.

Structurally engineered porous AAOs have provided a new degree of freedom in AAO-based syntheses of functional nanostructures. In addition, they have been utilized as platforms for investigating the adsorption and separation behaviors of diverse particles, ions, or biologically important molecules. Recent studies have shown that the structural engineering of porous AAO can result in the generation of optically active structures by imparting tailored optical properties, such as selective transmission, reflection, enhancement, confinement, or guide of light. The demonstrated optically active structures include AAO-based photonic crystals, distributed Bragg reflectors, waveguides, rugate filters, etc. Structurally engineered porous AAOs with custom-designed optical properties have been successfully utilized as platforms for developing various label-free, real-time optical sensing systems for chemical, biological, industrial and environmental analyses. Optical detections of analytes in these sensor systems are based on the change of the effective refractive index of porous AAOs. Analyte detections are not limited to optical method, but can be based on various electrochemical techniques. Apart from precise structural engineering, high selectivity and specificity towards target analytes can be achieved by deliberately engineering the surface chemistry of porous AAOs. As discussed, micromachined porous AAOs can be used as excellent platforms for MEMS devices. Therefore, continuous developments in engineering of structures and surface properties of porous AAOs may lead to a variety of smart and rugged AAO-based sensing devices with multi-performances in near future.

Many electrochemical aspects of aluminum anodization have been disclosed, especially during the past few years. However, the mechanism responsible for the self-organized formation of pores during anodization is still far from being understood and requires further extensive studies. Investigations to this direction are significant, because they may provide viable information in exploring new design rules for noble porous architectures, which are not accessible with the currently available anodization techniques. Future development of innovative anodization techniques, combined with various surface functionalization strategies, will further expand the application field of porous AAOs.

Acknowledgments The financial support from the Hankuk University of Foreign Studies Research Fund of 20151176001 is greatly acknowledged.

References

1. J.W. Diggle, T.C. Downie, C.W. Goulding, Anodic oxide films on aluminum. Chem. Rev. **69**, 365–405 (1969)
2. P.G. Sheasby, R. Pinner, *The Surface Treatment and Finishing of Aluminum and Its Alloys*, 6th edn. (Finishing Publications Ltd. & ASM International: Materials Park, Ohio, USA & Stevenage, UK, 2001)
3. H. Masuda, K. Fukuda, Ordered metal nanohole arrays made by a two-step replication of honeycomb structures of anodic alumina. Science **268**, 1466–1468 (1995)

4. W. Lee, S.J. Park, Porous anodic aluminum oxide: anodization and templated synthesis of functional nanostructures. Chem. Rev. **114**, 7487–7556 (2014)
5. W. Lee, H. Han, A. Lotnyk, M.A. Schubert, S. Senz, M. Alexe, D. Hesse, S. Baik, U. Gösele, Individually addressable epitaxial ferroelectric nanocapacitor arrays with near Tb in^{-2} density. Nat. Nanotechnol. **3**, 402–407 (2008)
6. C.J. Ingham, J. ter Maat, W.M. de Vos, Where bio meets nano: The many uses for nanoporous aluminum oxide in biotechnology. Biotechnol. Adv. **30**, 1089–1099 (2012)
7. S.B. Lee, D.T. Mitchell, L. Trofin, T.K. Nevanen, H. Söderlund, C.R. Martin, Antibody-based bio-nanotube membranes for enantiomeric drug separations. Science **296**, 2198–2200 (2002)
8. P. Banerjee, I. Perez, L. Henn-Lecordier, S.B. Lee, G.W. Rubloff, Nanotubular metal-insulator-metal capacitor arrays for energy storage. Nat. Nanotechnol. **4**, 292–296 (2009)
9. A.M.M. Jani, D. Losic, N.H. Voelcker, Nanoporous anodic aluminum oxide: advances in surface engineering and emerging applications. Prog. Mater Sci. **58**, 636–704 (2013)
10. A. Santos, T. Kumeria, D. Losic, Nanoporous anodic alumina: a versatile platform for optical biosensors. Materials **7**, 4297–4320 (2014)
11. J. Ferré-Borrull, J. Pallarès, G. Macías, L.F. Marsal, Nanostructural engineering of nanoporous anodic alumina for biosensing applications. Materials **7**, 5225–5253 (2014)
12. T. Kumeria, A. Santos, D. Losic, Nanoporous anodic alumina platforms: engineered surface chemistry an structure for optical sensing applications. Sensors **14**, 11878–11918 (2014)
13. A. Santos, T. Kumeria, D. Losic, Nanoporous anodic aluminum oxide for chemical sensing and biosensors. TrAC Trends Anal. Chem. **44**, 25–38 (2013)
14. A. Güntherschulz, H. Betz, Neue Untersuchungen über die elecktrolytishe Ventilwirkung. Z. Phys. **68**, 145–161 (1932)
15. A. Güntherschulz, H. Betz, Die Bewegung der Ionengitter von Isolatoren bei extremen electrischen Feldstärken. Z. Phys. **92**, 367–374 (1934)
16. J.A. Davies, J.P.S. Pringle, R.L. Graham, F. Brown, A radiotracer study of anodic oxidation. J. Electrochem. Soc. **109**, 999–1001 (1962)
17. J.A. Davies, B. Domeij, J.P.S. Pringle, F. Brown, The migration of metal and oxygen during anodic film formation. J. Electrochem. Soc. **112**, 675–680 (1965)
18. F. Brown, W.D. Mackintosh, The use of Rutherford backscattering to study the behavior of ion-implanted atoms during anodic oxidation of Aluminum: Ar, Kr, Xe, K. Rb, Cs, Cl, Br, and I. J. Electrochem. Soc. **120**, 1096–1102 (1973)
19. K. Shimizu, G.E. Thompson, G.C. Wood, Y. Xu, Direct observations of ion-implanted Xenon marker layers in anodic barrier films on aluminum. Thin Solid Films **88**, 255–262 (1982)
20. G.E. Thompson, Porous anodic alumina: fabrication, characterization and applications. Thin Solid Films **297**, 192–201 (1997)
21. C. Cherki, J. Siejka, Study by nuclear microanalysis and O^{18} tracer techniques of the oxygen transport processes and the growth laws for porous anodic oxide layers on aluminum. J. Electrochem. Soc. **120**, 784–791 (1973)
22. J. Siejka, C. Ortega, An O^{18} study of field-assisted pore formation in compact anodic oxide films on aluminum. J. Electrochem. Soc. **124**, 883–891 (1977)
23. P. Skeldon, G.E. Thompson, S.J. Garcia-Vergara, L. Iglesias-Bubianes, C.E. Blanco-Pinzon, A tracer study of porous anodic alumina. Electrochem. Solid-State Lett. **9**, B47–B51 (2006)
24. F. Keller, M.S. Hunter, D.L. Robinson, Structural features of oxide coatings on aluminum. J. Electrochem. Soc. **100**, 411–419 (1953)
25. M.S. Hunter, P. Fowle, Factors affecting the formation of anodic oxide coatings. J. Electrochem. Soc. **101**, 514–519 (1954)
26. T.P. Hoar, J. Yahalom, The initiation of pores in anodic oxide films formed on aluminum in acid solutions. J. Electrochem. Soc. **110**, 614–621 (1963)
27. F. Li, L. Zhang, R.M. Metzger, On the growth of highly ordered pores in anodized aluminum oxide. Chem. Mater. **10**, 2470–2480 (1998)

28. R.B. Mason, Factors affecting the formation of anodic oxide coatings in sulfuric acid electrolytes. J. Electrochem. Soc. **102**, 671–675 (1955)
29. T.P. Hoar, N.F. Mott, A mechanism for the formation of porous anodic oxide films on aluminum. J. Phys. Chem. Solids **9**, 97–99 (1959)
30. J.P. O'Sullivan, G.C. Wood, The morphology and mechanism of formation of porous anodic films on aluminum. Proc. R. Soc. London, Ser. A **317**, 511–543 (1970)
31. V.P. Parkhutik, V.I. Shershulsky, Theoretical modelling of porous oxide growth on aluminum. J. Phys. D Appl. Phys. **25**, 1258–1263 (1992)
32. S.K. Thamida, H.-C. Chang, Nanoscale pore formation dynamics during aluminum anodization. Chaos **12**, 240–251 (2002)
33. G.K. Singh, A.A. Golovin, I.S. Aranson, Formation of self-organized nanoscale porous structures in anodic aluminum oxide. Phys. Rev. B **73**, 205422 (2006)
34. J. Oh, C.V. Thompson, The role of electric field in pore formation during aluminum anodization. Electrochim. Acta **56**, 4044–4051 (2011)
35. A. Baron-Wiecheć, M.G. Burke, T. Hashimoto, H. Liu, P. Skeldon, G.E. Thompson, H. Habazaki, J.-J. Ganem, I.C. Vickridge, Tracer study of pore initiation in anodic alumina formed in phosphoric acid. Electrochim. Acta **113**, 302–312 (2013)
36. K.R. Hebert, S.P. Albu, I. Paramasivam, P. Schmuki, Morphological instability leading to formation of porous anodic oxide films. Nat. Mater. **11**, 162–166 (2012)
37. S.J. Garcia-Vergara, P. Skeldon, G.E. Thompson, H. Habazaki, A flow model of porous anodic film growth on aluminum. Electrochim. Acta **52**, 681–687 (2006)
38. S.J. Garcia-Vergara, D.L. Clere, T. Hashimoto, H. Habazaki, P. Skeldon, G.E. Thompson, Optimized observation of tungsten tracers for investigation of formation of porous anodic alumina. Electrochim. Acta **54**, 6403–6411 (2009)
39. S.J. Garcia-Vergara, P. Skeldon, G.E. Thompson, H. Habazaki, Tracer studies of anodic films formed on aluminum in malonic and oxalic acids. Appl. Surf. Sci. **254**, 1534–1542 (2007)
40. S.J. Garcia-Vergara, P. Skeldon, G.E. Thompson, H. Habazaki, A tracer investigation of chromic acid anodizing of aluminium. Surf. Interface Anal. **39**, 860–864 (2007)
41. J.E. Houser, K.R. Hebert, Stress-driven transport in ordered porous anodic films. Phys. Status Solidi A **205**, 2396–2399 (2008)
42. J.E. Houser, K.R. Hebert, Modeling the potential distribution in porous anodic alumina films during steady-state growth. J. Electrochem. Soc. **153**, B566–B573 (2006)
43. J.E. Houser, K.R. Hebert, The role of viscous flow of oxide in the growth of self-ordered porous anodic alumina films. Nat. Mater. **8**, 415–420 (2009)
44. H. Masuda, M. Satoh, Fabrication of gold nanodot array using anodic porous alumina as an evaporation mask. Jpn. J. Appl. Phys. **35**, L126–L129 (1996)
45. H. Masuda, E. Hasegawa, S. Ono, Self-ordering of cell arrangement of anodic porous alumina formed in sulfuric acid solution. J. Electrochem. Soc. **144**, L127–L130 (1997)
46. A.P. Li, F. Müller, A. Birner, K. Nielsch, U. Gösele, Hexagonal pore arrays with a 50-420 nm interpore distance formed by self-organization in anodic alumina. J. Appl. Phys. **84**, 6023 6026 (1998)
47. O. Nishinaga, T. Kikuchi, S. Natsui, R.O. Suzuki, Rapid fabrication of self-ordered porous alumina with 10-/sub-10-nm-scale nanostructures by selenic acid anodizing. Sci. Rep. **3**, 2748 (2013). 10.1038/srep02748
48. H. Masuda, K. Yada, A. Osaka, Self-ordering of cell configuration of anodic porous alumina with large-size pores in phosphoric acid solution. Jpn. J. Appl. Phys. **37**, L1340–L1342 (1998)
49. K. Nielsch, J. Choy, K. Schwirn, R.B. Wehrspohn, U. Gösele, Self-ordering regimes of porous alumina: the 10% porosity rule. Nano Lett. **2**, 677–680 (2002)
50. S. Shingubara, K. Morimoto, H. Sakaue, T. Takahagi, Self-organization of a porous alumina nanohole array using a sulfuric/oxalic acid mixture as electrolyte. Electrochem. Solid-State Lett. **7**, E15–E17 (2004)
51. C. Sun, J. Luo, L. Wu, J. Zhang, Self-ordered anodic alumina with continuously tunable pore intervals from 410 to 530 nm. ACS Appl. Mater. Interfaces **2**, 1299–1302 (2010)

52. J.M. Kape, The use of malonic acid as an anodising electrolyte. Metallurgia **60**, 181–191 (1959)
53. J.M. Kape, Unusual anodizing processes and their practical significance. Electroplat. Metal Finish. **14**, 407–415 (1961)
54. J.M. Kape, Anodizing in aqueous solutions of organic carboxylic acids. Trans. Inst. Met. Finish. **45**, 34–42 (1967)
55. S. Ono, M. Saito, M. Ishiguro, H. Asoh, Controlling factor of self-ordering of anodic porous alumina. J. Electrochem. Soc. **151**, B473–B478 (2004)
56. S. Ono, M. Saito, H. Asoh, Self-ordering of anodic porous alumina formed in organic acid electrolytes. Electrochim. Acta **51**, 827–833 (2005)
57. T. Kikuch, O. Nishinaga, S. Natsui, R.O. Suzuki, Self-ordering behavior of anodic porous alumina via selenic acid anodizing. Electrochim. Acta **137**, 728–735 (2014)
58. I. Vrublevsky, V. Parkoun, J. Schreckenbach, G. Marx, Study of porous oxide film growth on aluminum in oxalic acid using a re-anodizing technique. Appl. Surf. Sci. **227**, 282–292 (2004)
59. G.D. Sulka, K.G. Parkoła, Temperature influence on well-ordered nanopore structures grown by anodizing of aluminum in sulphuric acid. Electrochim. Acta **52**, 1880–1888 (2007)
60. W. Chen, J.-S. Wu, X.-H. Xia, Porous anodic alumina with continuously manipulated pore/cell size. ACS Nano **2**, 959–965 (2008)
61. J. Martín, C.V. Manzano, O. Caballero-Calero, M. Martín-González, High-aspect-ratio and highly ordered 15-nm porous alumina templates. ACS Appl. Mater. Interf. **5**, 72–79 (2013)
62. S. Ikonopisov, A. Girginov, M. Machkova, Post-breakdown anodization of aluminum. Electrochim. Acta **22**, 1283–1286 (1977)
63. J.M. Albella, I. Montero, J.M. Martínez-Duart, Electron injection and avalanche during the anodic oxidation of tantalum. J. Electrochem. Soc. **131**, 1101–1104 (1984)
64. S.Z. Chu, K. Wada, S. Inoue, M. Isogai, Y. Katsuta, A. Yasumori, Large-scale fabrication of ordered nanoporous alumina films with arbitrary pore intervals by critical-potential anodization. J. Electrochem. Soc. **153**, B384–B391 (2006)
65. S. Ono, M. Saito, H. Asoh, Self-ordering of anodic porous alumina induced by local current concentration: Burning. Electrochem. Solid-Sate Lett. **7**, B21–B24 (2004)
66. G.C. Tu, I.T. Chen, K. Shimizu, The temperature rise and burning for high rate anodizing of aluminum in oxalic acid. J. Jpn. Inst. Light Met. **40**, 382–389 (1990)
67. P. Csokán, Beiträge zur Kenntnis der anodischen Oxydation von Aluminium verdunnter, kalter Schwefelsaure. Metalloberfläche **15**, B49–B53 (1961)
68. V.P. Csokán, M. Holló, Beitrag zur Frage des Bildungsmechanismus von anodisch erzeugten Hartoxydschichten. Werkst. Korros. **12**, 288–295 (1961)
69. P. Csokán, C.C. Sc., Hard anodizing. Electroplat. Metal Finishing **15**, 75–82 (1962)
70. P. Csokán, Some observations on the growth mechanism of hard anodic oxide coatings on aluminium. Trans. Inst. Met. Finish. **41**, 51–56 (1964)
71. S. John, V. Balasubramanian, B.A. Shenoi, Hard anodizing aluminium and its alloys—AC in sulphuric acid—sodium sulphate bath. Met. Finish. **82**, 33–39 (1984)
72. B. Olbertz, Hartanodisieren er¨offnet aluminum vielf˙altige technische Anwendungs-möglichkeiten. Aluminium **3**, 268–270 (1988)
73. J.G. Hecker, Aluminum hard coats. Prod. Finish. **53**, 88–92 (1988)
74. A. Rajendra, B.J. Parmar, A.K. Sharma, H. Bhojraj, M.M. Nayak, K. Rajanna, Hard anodisation of aluminium and its application to sensorics. Surf. Eng. **21**, 193–197 (2005)
75. S.-Z. Chu, K. Wada, S. Inoue, M. Isogai, A. Yasumori, Fabrication of ideally ordered nanoporous alumina films and integrated alumina nanotubue arrays by high-field anodization. Adv. Mater. **17**, 2115–2119 (2005)
76. K. Schwirn, *Harte anodisation von aluminium mit Verdünnter Schwefeläure* (Martin-Luther-Universität Halle-Wittenberg, Halle (Saale), 2008)
77. K. Schwirn, W. Lee, R. Hillebrand, M. Steinhart, K. Nielsch, U. Gösele, Self-ordered anodic aluminum oxide formed by H_2SO_4 hard anodization. ACS Nano **2**, 302–310 (2008)

78. W. Lee, R. Ji, U. Gösele, K. Nielsch, Fast fabrication of long-range ordered porous alumina membranes by hard anodization. Nat. Mater. **5**, 741–747 (2006)
79. W. Lee, K. Nielsch, U. Gösele, Self-ordering behavior of nanoporous anodic aluminum oxide (AAO) in malonic acid anodization. Nanotechnology **18**, 475713 (2007)
80. Y. Li, M. Zheng, L. Ma, W. Shen, Fabrication of highly ordered nanoporous alumina films by stable high-field anodization. Nanotechnology **17**, 5101–5105 (2006)
81. Y.B. Li, M.J. Zheng, L. Ma, High-speed growth and photoluminescence of porous anodic alumina films with controllable interpore distances over a large range. Appl. Phys. Lett. **91**, 073109 (2007)
82. M.A. Kashi, A. Ramazani, M. Noormohammadi, M. Zarei, P. Marashi, Optimum self-ordered nanopore arrays with 130–270 nm interpore distances formed by hard anodization in sulfuric/oxalic acid mixtures. J. Phys. D Appl. Phys. **40**, 7032–7040 (2007)
83. Y. Li, Z.Y. Ling, S.S. Chen, J.C. Wang, Fabrication of novel porous anodic alumina membranes by two-step hard anodization. Nanotechnology **19**, 225604 (2008)
84. M.S. Hunter, P. Fowle, Determination of barrier layer thickness of anodic oxide coatings. J. Electrochem. Soc. **101**, 481–485 (1954)
85. W. Lee, J.-C. Kim, U. Gösele, Spontaneous current oscillations during hard anodization of aluminum under potentiostatic conditions. Adv. Funct. Mater. **20**, 21–27 (2010)
86. S.-S. Tan, M.L. Reed, H. Han, R. Boudreau, High aspect ratio microstructures on porous anodic aluminum oxide. *Proceedings of the 8th International Workshop on Microelectro Mechanical System (MEMS-95)*, Amsterdam, 1995; pp. 267–272
87. A.-P. Li, F. Müller, A. Birner, K. Nielsch, U. Gösele, Fabrication and microstructuring of hexagonally ordered two-dimensional nanopore arrays in anodic alumina. Adv. Mater. **11**, 483–487 (1999)
88. D. Routkevitch, A.N. Govyadinov, P.P. Mardilovich, High aspect ratio, high resolution ceramic MEMS. *Proceedings of ASME International* Mechanical Engineering, vol. 2, ASME Orlando, FL, 2000, pp. 39–44
89. D. Routkevitch, O. Polyakov, D. Deininger, C. Kostelecky, Nanostructured gas microsensor platform. In NSTI Nanotech, Anaheim **2**, 266–268 (2005)
90. H. Masuda, H. Yamada, M. Satoh, H. Asoh, M. Nakao, T. Tamamura, Highly ordered nanochannel-array architecture in anodic alumina. Appl. Phys. Lett. **71**, 2770–2772 (1997)
91. H. Masuda, H. Asoh, M. Watanabe, K. Nishio, M. Nakao, T. Tamamura, Square and triangular nanohole array architectures in anodic alumina. Adv. Mater. **13**, 189–192 (2001)
92. I. Mikulskas, S. Juodkazis, R. Tomašiūnas, J.G. Dumas, Aluminum oxide photonic crystals grown by a new hybrid method. Adv. Mater. **13**, 1574–1577 (2001)
93. J. Choi, K. Nielsch, M. Reiche, R.B. Wehrspohn, U. Gösele, Fabrication of monodomain alumina pore arrays with an interpore distance smaller than the lattice constant of the imprint stamp. J. Vac. Sci. Technol., B **21**, 763–766 (2003)
94. W. Lee, R. Ji, C.A. Ross, U. Gösele, K. Nielsch, Wafer-scale Ni imprint stamps for porous alumina membranes based on interference lithography. Small **2**, 978–982 (2006)
95. S. Fournier-Bidoz, V. Kitaev, D. Routkevitch, I. Manners, G.A. Ozin, Highly ordered nanosphere imprinted nanochannel alumina (NINA). Adv. Mater. **16**, 2193–2196 (2004)
96. H. Masuda, Y. Matsui, M. Yotsuya, F. Matsumoto, K. Nishio, Fabrication of highly ordered anodic porous alumina using self-organized polystyrene particle array. Chem. Lett. **33**, 584–585 (2004)
97. C.Y. Liu, A. Datta, Y.L. Wang, Ordered anodic alumina nanochannels on focused-ion-beam-prepatterned aluminum surfaces. Appl. Phys. Lett. **78**, 120–122 (2001)
98. B. Chen, K. Lu, Z. Tian, Gradient and alternating diameter nanopore templates by focused ion beam guided anodization. Electrochim. Acta **56**, 435–440 (2010)
99. B. Chen, K. Lu, Z. Tian, Novel patterns by focused ion beam guided anodization. Langmuir **27**, 800–808 (2011)
100. R. Krishnan, C.V. Thompson, Monodomain high-aspect-ratio 2D and 3D ordered porous alumina structures with independently controlled pore spacing and diameter. Adv. Mater. **18**, 988–992 (2007)

101. Z. Sun, H.K. Kim, Growth of ordered, single-domain, alumina nanopore arrays with holographically patterned aluminum films. Appl. Phys. Lett. **81**, 3458–3460 (2010)
102. B. Kim, S. Park, T.J. McCarthy, T.P. Russell, Fabrication of ordered anodic aluminum oxide using a solvent-induced array of block-copolymer micelles. Small **3**, 1869–1872 (2007)
103. T.S. Kustandi, W.W. Loh, H. Gao, H.Y. Low, Wafer-scale near-perfect ordered porous alumina on substrates by step and flash imprint lithography. ACS Nano **4**, 2561–2568 (2010)
104. K.-L. Lai, M.-H. Hon, I.-C. Leu, Fabrication of ordered nanoporous anodic alumina prepatterned by mold-assisted chemical etching. Nanoscale Res. Lett. **6**, 157 (2011)
105. H. Masuda, M. Yotsuya, M. Asano, K. Nishio, M. Nakao, A. Yokoo, T. Tamamura, Self-repair of ordered pattern of nanometer dimensions based on self-compensation properties of anodic porous alumina. Appl. Phys. Lett. **78**, 826–828 (2001)
106. J.T. Smith, Q. Huang, A.D. Franklin, D.B. Janes, T.D. Sands, Highly ordered diamond and hybrid triangle-diamond patterns in porous anodic alumina thin films. Appl. Phys. Lett. **93**, 043108 (2008)
107. G.E. Thompson, Y. Xu, P. Skeldon, K. Shimizu, S.H. Han, G.C. Wood, Anodic oxidation of aluminum. Philos. Mag. B **55**, 651–667 (1987)
108. G.E. Thompson, R.C. Furneaux, G.C. Wood, Electron microscopy of ion beam thinned porous anodic films formed on aluminium. Corros. Sci. **18**, 481–498 (1978)
109. I. Vrublevsky, V. Parkoun, V. Sokol, J. Schreckenbach, Analysis of chemical dissolution of the barrier layer of porous oxide on aluminum thin films using a re-anodizing technique. Appl. Surf. Sci. **252**, 227–233 (2005)
110. S. Ono, H. Ichinose, N. Masuko, The high resolution observation of porous anodic films formed on alumium in phosphoric acid solution. Corros. Sci. **33**, 841–850 (1992)
111. G.C. Wood, P. Skeldon, G.E. Thompson, K. Shimizu, A model for the incorporation of electrolyte species into anodic alumina. J. Electrochem. Soc. **143**, 74–83 (1996)
112. H. Han, S.-J. Park, J.S. Jang, H. Ryu, K.J. Kim, S. Baik, W. Lee, In-situ determination of the pore opening point during wet-chemical etching of the barrier layer of porous anodic aluminum oxide (AAO): non-uniform impurity distribution in anodic oxide. ACS Appl. Mater. Interfaces **5**, 3441–3448 (2013)
113. Y. Yamamoto, N. Baba, S. Tajima, Coloured materials and photoluminescence centres in anodic film on aluminium. Nature **289**, 572–584 (1981)
114. W.L. Xu, M.J. Zheng, S. Wu, W.Z. Sheng, Effects of high-temperature annealing on structural and optical properties of highly ordered porous alumina membranes. Appl. Phys. Lett. **85**, 4364–4366 (2004)
115. Y. Du, W.L. Cai, C.M. Mo, J. Chen, L.D. Zhang, X.G. Zhu, Preparation and photoluminescence of alumina membranes with ordered pore arrays. Appl. Phys. Lett. **74**, 2951–2953 (1999)
116. G.H. Li, Y. Zhang, Y.C. Wu, L.D. Zhang, Wavelength dependent photoluminescence of anodic alumina membranes. J. Phys.: Condens. Matter **15**, 8663–8671 (2003)
117. A. Santos, M. Alba, M.M. Rahman, P. Formentín, J. Ferré-Borrull, J. Pallarès, L.F. Marsal, Structural tuning of photoluminescence in nanoporous anodic alumina by hard anodization in oxalic and malonic acids. Nanoscale Res. Lett. **7**, 228 (2012)
118. S. Ono, N. Masuko, The duplex structure of cell walls of porous anodic films formed on aluminum. Corros. Sci. **33**, 503–507 (1992)
119. G.E. Thompson, R.C. Furneaux, G.C. Wood, STEM/EDAS analysis of the cell walls in porous anodic films formed on aluminum. J. Electrochem. Soc. **125**, 1480–1482 (1978)
120. M.C. Thornton, R.C. Furneaux, Transmission electron microscopy and digital X-ray mapping of ion-beam-thinned porous anodic films formed on aluminium. J. Mater. Sci. Lett. **10**, 622–624 (1991)
121. K. Huang, L. Pu, Y. Shi, P. Han, R. Zhang, Y.D. Zheng, Photoluminescence oscillations in porous alumina films. Appl. Phys. Lett. **89**, 201118 (2006)
122. D.A.G. Bruggeman, Derechnung verschiedener physikalischer Konstanten von heterogenen Substanzen: I. Dielektrizitätskonstanten und Leitfähigkeiten der Mischkörper aus isotropen Substanzen. Ann. Phys. **24**, 636–664 (1935)

123. A. Santos, V.S. Balderrama, M. Alba, P. Formentín, J. Ferré-Borrull, J. Pallarès, L.F. Marsal, Nanoporous anodic alumina barcodes: toward smart optical biosensors. Adv. Mater. **24**, 1050–1054 (2012)
124. A. Santos, T. Kumeria, D. Losic, Optically optimized photoluminescent and interferometric biosensors based on nanoporous anodic alumina: a comparison. Anal. Chem. **85**, 7904–7911 (2013)
125. F. Casanova, C.E. Chiang, C.-P. Li, I.V. Roshchin, A.M. Ruminski, M.J. Sailor, I.K. Schuller, Effect of surface interactions on the hysteresis of capillary condensation in nanopores. Europhys. Lett. **81**, 26003 (2008)
126. F. Casanova, C.E. Chiang, C.-P. Li, I.V. Roshchin, A.M. Ruminski, M.J. Sailor, I.K. Schuller, Gas adsorption and capillary condensation in nanoporous alumina films. Nanotechnology ***19***, 315709 (2008)
127. S.D. Alvarez, C.-P. Li, C.E. Chiang, I.K. Schuller, M.J. Sailor, A label-free porous alumina interferometric immunosensor. ACS Nano **3**, 3301–3307 (2009)
128. T. Kumeria, D. Losic, Reflective interferometric gas sensing using nanoporous anodic aluminium oxide (AAO). Phys. Status Solidi RRL **5**, 406–408 (2011)
129. T. Kumeria, L. Parkinson, D. Losic, A nanoporous interferometric micro-sensor for biomedical detection of volatile sulphur compounds. Nanoscale Res. Lett. **6**, 634 (2011)
130. T. Kumeria, D. Losic, Controlling interferometric properties of nanoporous anodic aluminum oxide. Nanoscale Res. Lett. **7**, 88 (2012)
131. T. Kumeria, A. Santos, D. Losic, Ultrasensitive nanoporous interferometric sensor for label-free dection of gold(III) ions. ACS Appl. Mater. Interfaces **5**, 11783–11790 (2013)
132. B. He, S.J. Son, S.B. Lee, Shape-coded silica nanotubes for biosensing. Langmuir **22**, 8263–8265 (2006)
133. A. Santos, P. Formentín, J. Pallarès, J. Ferré-Borrull, L.F. Marsal, Structural engineering of nanoporous anodic alumina funnels with high aspect ratio. J. Electroanal. Chem. **655**, 73–78 (2011)
134. B. He, S.J. Son, S.B. Lee, Suspension array with shape-coded silica nanotubes for multiplexed immunoassays. Anal. Chem. **79**, 5257–5263 (2007)
135. J. Li, C. Li, C. Chen, Q. Hao, Z. Wang, J. Zhu, X. Gao, Facile method for modulating the profiles and periods of self-ordered three-dimensional alumina taper-nanopores. ACS Appl. Mater. Interfaces **4**, 5678–5683 (2012)
136. A. Santos, T. Kumeria, Y. Wang, D. Losic, In situ monitored engineering of inverted nanoporous anodic alumina funnels: on the precise generation of 3D optical nanostructures. Nanoscale **6**, 9991–9999 (2014)
137. T. Nagaura, F. Takeuchi, Y. Yamauchi, K. Wada, S. Inoue, Fabrication of ordered Ni nanocones using a porous anodic alumina template. Electrochem. Commun. **10**, 681–685 (2008)
138. T. Nagaura, F. Takeuchi, S. Inoue, Fabrication and structural control of anodic alumina films with inverted cone porous structure using multi-step anodizing. Electrochim. Acta **53**, 2109–2114 (2008)
139. T. Yanagishita, T. Kondo, K. Nishio, H. Masuda, Optimization of antireflection structures of polymer based on nanoimprinting using anodic porous alumina. J. Vac. Sci. Technol., B **26**, 1856–1859 (2008)
140. G. Macias, L.P. Hernández-Eguía, J. Ferré-Borrull, J. Pallares, L.F. Marsal, Gold-coated ordered nanoporous anodic alumina bilayers for future label-free interferometric biosensors. ACS Appl. Mater. Interfaces **5**, 8093–8098 (2013)
141. H. Takahashi, M. Nagayama, H. Akahori, A. Kitahara, Electron-microscopy of porous anodic oxide films on alumium by ultra-thin sectioning technique. Part 1. The structural change of the film during the current recovery period. J. Electron Microscopy **22**, 149–157 (1973)
142. J.W. Diggle, T.C. Downie, C.W. Goulding, Processes involved in reattainment of steady-state conditions for the anodizng of aluminum following formation voltage changes. J. Electrochem. Soc. **116**, 737–740 (1969)

143. W. Lee, J.-C. Kim, Highly ordered porous alumina with tailor-made pore structures fabricated by pulse anodization. Nanotechnology **21**, 485304 (2010)
144. W. Lee, K. Schwirn, M. Steinhart, E. Pippel, R. Scholz, U. Gösele, Structural engineering of nanoporous anodic aluminum oxide by pulse anodization of aluminum. Nat. Nanotechnol. **3**, 234–239 (2008)
145. R.C. Furneaux, W.R. Rigby, A.P. Davidson, The formation of controlled porosity membranes from anodically oxidized aluminium. Nature **337**, 147–149 (1989)
146. J.M. Montero-Moreno, M. Belenguer, M. Sarret, C.M. Müller, Production of alumina templates suitable for electrodeposition of nanostructures using stepped techniques. Electrochim. Acta **54**, 2529–2535 (2009)
147. K. Nielsch, F. Müller, A.-P. Li, U. Gösele, Uniform nickel deposition into ordered alumina pores by pulsed electrodeposition. Adv. Mater. **12**, 582–586 (2000)
148. G. Sauer, G. Brehm, S. Chneider, K. Nielsch, R.B. Wehrspohn, J. Choi, H. Hofmeister, U. Gösele, Highly ordered monocrystalline silver nanowire arrays. J. Appl. Phys. **91**, 3243–3247 (2002)
149. W. Cheng, M. Steinhart, U. Gösele, R.B. Wehrspohn, Tree-like alumina nanopores generated in a non-steady-state anodization. J. Mater. Chem. **17**, 3493–3495 (2007)
150. J. Choi, G. Sauer, K. Nielsch, R.B. Wehrspohn, U. Gösele, Hexagonally arranged monodisperse silver nanowires with adjustable diameter and high aspect ratio. Chem. Mater. **15**, 776–779 (2003)
151. W. Lee, R. Scholz, K. Nielsch, U. Gösele, A template-based electrochemical method for the synthesis of multisegmented metallic nanotubes. Angew. Chem. Int. Ed. **44**, 6050–6054 (2005)
152. G. Jeon, M. Jee, S.Y. Yang, B.-Y. Lee, S.K. Jang, J.K. Kim, Hierarchically self-organized monolithic nanoporous membrane for excellent virus enrichment. ACS Appl. Mater. Interfaces **6**, 1200–1206 (2014)
153. J. Li, C. Papadopoulos, J. Xu, Growing Y-junction carbon nanotubes. Nature **402**, 253–254 (1999)
154. C. Papadopoulos, A. Rakitin, J. Li, A.S. Vedeneev, J.M. Xu, Electronic transport in Y-junction carbon nanotubes. Phys. Rev. Lett. **85**, 3476–3479 (2000)
155. Y.C. Sui, J.A. González-León, A. Bermúdez, J.M. Saniger, Synthesis of multi branched carbon nanotubes in porous anodic aluminum oxide template. Carbon **39**, 1709–1715 (2001)
156. T. Gao, G. Meng, J. Zhang, S. Sun, L. Zhang, Template synthesis of Y-junction metal nanowires. Appl. Phys. A **74**, 403–406 (2002)
157. G. Meng, Y.J. Jung, A. Cao, R. Vajtai, P.M. Ajayan, Controlled fabrication of hierarchically branched nanopores, nanotubes, and nanowires. Proc. Natl. Acad. Sci. U.S.A. **102**, 7074–7078 (2005)
158. C. Shuoshuo, L. Zhiyuan, H. Xing, L. Yi, Controlled growth of branched channels by a factor of $1/\sqrt{n}$ anodizing voltage? J. Mater. Chem. **19**, 5717–5719 (2009)
159. A.Y.Y. Ho, H. Gao, Y.C. Lam, I. Rodríguez, Controlled fabrication of multitiered three-dimensional nanostructures in porous alumina. Adv. Funct. Mater. **18**, 2057–2063 (2008)
160. K. Pitzschel, J.M.M. Moreno, J. Escrig, O. Albrecht, K. Nielsch, J. Bachmann, Controlled introduction of diameter modulations in arrayed magnetic iron oxide nanotubes. ACS Nano **3**, 3463–3469 (2009)
161. D.J. Arrowsmith, A.W. Clifford, D.A. Moth, Fracture of anodic oxide formed on aluminum in sulphuric acid. J. Mater. Sci. Lett. **5**, 921–922 (1986)
162. K. Wada, T. Shimohira, M. Amada, N. Baba, Microstructure of porous anodic oxide films on aluminium. J. Mater. Sci. **21**, 3810–3816 (1986)
163. W. Lee, R. Scholz, U. Gösele, A continuous process for structurall well-defined Al_2O_3 nanotubes based on pulse anodization of aluminum. Nano Lett. **8**, 2155–2160 (2008)
164. K.E. La Flamme, G. Mor, D. Gong, T. La Tempa, V.A. Fusaro, C.A. Grimes, T.A. Desai, Nanoporous alumina capsules for cellular macroencapsulation: transport and biocompatibility. Diabetes Technol. Ther. **7**, 684–694 (2005)

165. E.E.L. Swan, K.C. Popat, C.A. Grimes, T.A. Desai, Fabrication and evaluation of nanoporous alumina membranes for osteoblast culture. J. Biomed. Mater. Res., Part A **72A**, 288–295 (2005)
166. B. Jongsomjit, J. Panpranot, J.G. Goodwin Jr, Co-support compound formation in alumina-supported cobalt catalysts. J. Catal. **204**, 98–109 (2001)
167. E.P. Gusev, M. Copel, E. Cartier, I.J.R. Baumvol, C. Krug, M.A. Gribelyuk, High-resolution depth profiling in ultrathin Al_2O_3 films on Si. Appl. Phys. Lett. **76**, 176–178 (2000)
168. Y. Wang, A. Santos, G. Kaur, A. Evdokiou, D.L. Losic Jr, Structurally engineered anodic alumina nanotubes as nano-carriers for delivery of anticancer therapeutics. Biomaterials **35**, 5517–5526 (2014)
169. A. Santos, L. Vojkuvka, M. Alba, V.S. Balderrama, J. Ferré-Borrull, J. Pallarès, L.F. Marsal, Understanding and morphology control of pore modulations in nanoporous anodic alumina by discontinuous anodization. Phys. Status Solidi A **209**, 2045–2048 (2012)
170. D. Losic, M. Lillo, D. Losic Jr, Porous alumina with shaped pore geometries and complex pore architectures fabricated by cyclic anodization. Small **5**, 1392–1397 (2009)
171. D. Losic, D. Losic Jr, Preparation of porous anodic alumina with periodically perforated pores. Langmuir **25**, 5426–5431 (2009)
172. E.A. Avrutin, V.B. Gorfinkel, S. Luryi, K.A. Shore, Control of surface-emitting laser diodes by modulating the distributed Bragg mirror reflectivity: small-signal analysis. Appl. Phys. Lett. **63**, 2460–2462 (1993)
173. J. Yoon, W. Lee, E.L. Thomas, Optically pumped surface-emitting lasing using self-assembled block-copolymer-distributed Bragg reflectors. Nano Lett. **6**, 211–2214 (2006)
174. L. Chen, E. Towe, Nanowire lasers with distributed-Bragg-reflector mirrors. Appl. Phys. Lett. **89**, 053125 (2006)
175. B. Wang, G.T. Fei, M. Wang, M.G. Kong, L.D. Zhang, Preparation of photonic crystals made of air pores in anodic alumina. Nanotechnology **18**, 365601 (2007)
176. W.J. Zheng, G.T. Fei, B. Wang, Z. Jin, L.D. Zhang, Distributed Bragg reflector made of anodic alumina membrane. Mater. Lett. **63**, 706–708 (2009)
177. W.J. Zheng, G.T. Fei, B. Wang, L.D. Zhang, Modulation of transmission spectra of anodized alumina membrane distributed bragg reflector by controlling anodization temperature. Nanoscale Res. Lett. **4**, 665–667 (2009)
178. G.L. Shang, G.T. Fei, Y. Zhang, P. Yan, S.H. Xu, L.D. Zhang, Preparation of narrow photonic bandgaps located in the near infrared region and their applications in ethanol gas sensing. J. Mater. Chem. C **1**, 5285–5291 (2013)
179. G.L. Shang, G.T. Fei, Y. Zhang, P. Yan, S.H. Xu, H.M. Ouyang, L.D. Zhang, Fano resonance in anodic aluminum oxide based photonic crystals. Sci. Rep. **4**, 3601 (2014)
180. G.L. Shang, G.T. Fei, S.H. Xu, P. Yan, L.D. Zhang, Preparation of the very uniform pore diameter of anodic alumina oxidation by voltage compensation mode. Mater. Lett. **110**, 156–159 (2013)
181. H. Xing, L. Zhi-Yuan, C. Shuo-Shuo, H. Xin-Hua, Influence of light scattering on transmission spectra of photonic crystals of anodized alumina. Chin. Phys. Lett. **25**, 3284–3287 (2008)
182. M.M. Rahman, L.F. Marsal, J. Pallarès, J. Ferré-Borrull, Tuning the photonic stop bands of nanoporous anodic alumina-based distributed bragg reflectors by pore widening. ACS Appl. Mater. Interfaces **5**, 13375–13381 (2013)
183. T. Kumeria, M.M. Rahman, A. Santos, J. Ferré-Borrull, L.F. Marsal, D. Losic, Structural and optical nanoengineering of nanoporous anodic alumina rugate filters for real-time and label-free biosensing applications. Anal. Chem. **86**, 1837–1844 (2014)
184. T. Kumeria, M.M. Rahman, A. Santos, J. Ferré-Borrull, L.F. Marsal, D. Losic, Nanoporous anodic alumina rugate filters for sensing of ionic mercury: toward environmental point-of analysis systems. ACS Appl. Mater. Interfaces **6**, 12971–12978 (2014)

Chapter 5
Soft and Hard Surface Manipulation of Nanoporous Anodic Aluminum Oxide (AAO)

Abdul Mutalib Md Jani, Hanani Yazid, Anisah Shafiqah Habiballah, Abdul Hadi Mahmud and Dusan Losic

Abstract Nanoporous materials with straight channels have attracted considerable interest due to their unique physical properties and many potential applications such as separation, sensing, biomedical and electronics. For the last few decades, nanoporous alumina or anodic aluminium oxide (AAO) membrane is gaining attention due to its broad applicability in various applications. The unique properties of AAO membrane coupled with tunable surface modification and properties is playing an increasingly important platform in a diverse range of applications such as separation, energy storage, drug delivery and template synthesis, as well as biosensing, tissue engineering and catalytic studies. This chapter aims to introduce the recent advances and challenges for surface manipulation of AAO following the 'soft' and 'hard' modification strategies. The functions of these modified nanostructures materials and latest important applications are evaluated with respect to improved performance and possible implications of those strategies for the future trends of surface engineering are discussed.

5.1 Introduction

AAO surfaces are insulating and suffer from chemical instability in the acidic environment [1]. This limitation can be overcome by changing the surface properties and by adding new surface functionalities. The rich content of hydroxyl groups on the AAO surface allows them to be easily modified via modification with organic molecules with the desired functionality. The motivation for surface

A.M.M. Jani (✉) · H. Yazid · A.S. Habiballah · A.H. Mahmud
Chemistry Department, Faculty of Applied Sciences,
Universiti Teknologi MARA, 02600, Arau, Perlis, Malaysia
e-mail: abdmutalib@perlis.uitm.edu.my

D. Losic
School of Chemical Engineering, University of Adelaide,
North Engineering Building, Adelaide, SA 5000, Australia

D. Losic and A. Santos (eds.), *Nanoporous Alumina*,
Springer Series in Materials Science 219, DOI 10.1007/978-3-319-20334-8_5

manipulation of AAO membrane is twofold. First, soft technique approaches include self-assembly processes (thiol, silanes), polymer modifications (polymer grafting and layer-by-layer). Subsequent modifications of the thus introduced functionality with biomolecules or nanoparticles can be carried out. Second, hard surface modification techniques relevant to AAO include thermal vapour chemical and electrochemical deposition, electroless, chemical vapour deposition (CVD) and atomic layer deposition (ALD). Lesser consideration is given to sol-gel derived and template synthesis on AAO since they lie beyond the scope of this chapter and have been previously reviewed elsewhere [2, 3]. With surface manipulation in mind, this chapter will present the fundamentals, important elements and current work on the surface manipulation of AAO using different strategies mentioned above. Such modified AAO offers an attractive feature and open up the new application to which unmodified membranes cannot be applied.

5.1.1 Soft Techniques

5.1.1.1 Self Assembly Monolayers

Self-assembly monolayer (SAMs) is an organic formation that formed spontaneously by adsorption, rearrangement and chemical reaction of molecules species from the liquid or gas phase onto solid surfaces. SAM's preparation is straightforward, and they provide a convenient, uniform, flexible and simple system which displays high chemical and thermal stability. The early work sought to chemically modify the surface using organic acids, thiolates, organosilanes, alkanethiolates, alcohols, amines and carboxylic acids assemblies on AAO-gold-coated surface with the goal of introducing functional groups for specific orientation, functionalities and selectivity. Considerable work has been done involving SAMs focusing on octadecyltrichlorosilane (OTS) that resulted in a rapidly growing body of research [4]. Formation of SAMs from organic acids i.e. carboxylic acid, α-hydroxy-carboxylic acid, alkyne, alkene, phosphonic acid, on AAO possessed a good thermal stability because of their ability to spontaneously adsorb via physical adsorption or chemisorption [5]. In general, unique approaches that combine gold-coating and SAMs on AAO surfaces may play an active role, used solely as immobilization matrix for the biological recognition entity.

For instance, Wu et al. [6] showed that a biosensor to detect the A-beta (1–42) peptide in predicting Alzheimer disease could be fabricated in which SAM (11-mercaptoundecanoic acid) modified gold coated AAO was used as sensing platform. In this configuration, before SAM was done, a thin gold film was sputtered onto the AAO substrate to serve as an electrode for metal deposition. Subsequently, the 11-mercaptoundecanoic acid (MUA) modified surface was further functionalised with EDC/NHS. The uniform distribution of gold nanoparticles on the surface of AAO has increased the number of MUA molecules hence increasing the binding of EDC/NHS and monoclonal antibody IgG molecules on the AAO membrane that lead to lower limit of detection. Such phenomena have also been observed when densely packed SAMs (MUA-gold-AAO) were attached with another type of bio-receptors. For example, Kumeria et al. [7] investigated the mechanism of antibody-antigen interactions at SAMs (MUA-gold-AAO) electrode using reflectometric interference spectroscopy (RIfS). The mechanism of the system is shown in Fig. 5.1. This AAO biosensor specifically binds to circulating tumor cell, and offered direct and label-free detection of tumor cell with the limit of detection as low as (1000–100,000 cells/mL), hence allowing analysis to be done with little amount of sample (<50 µL). Recently, Tung et al. [8] have described MUA-gold-AAO based biosensor for the impedimetric detection of dengue virus. In this event, IgG-like sensing probes was found to conform the binding event between the dengue virus and the bioreceptor (CLEC5A). Recently, the use of gold-nanoparticle-labeled antibodies to enhance the sensitivity of the localized surface plasmon resonance (LSPR) biosensor was demonstrated by Yeom et al. [9]. The system incorporated MUA-gold-AAO as the immobilising matrix for the antibody (C-reactive protein-CRP), detection and the immunosensor showed high sensitivity over a wide concentration range and enabled a selective immunoassay to be performed. See [10] for more detailed discussion on preparative issues of AAO for chemo and biosensing applications. As the chemistry of AAO surfaces continues to advance, it is expected that additional methods for the preparation of stable monolayers will be realized, contributing further to the technological development of AAO and to a complete understanding of its fundamental and properties.

5.1.1.2 Thiol Surface Chemistry

Most of common method attaching thiol group on AAO membranes are prepared on gold-coated surfaces. This is due to high affinity of the gold surface towards thiol group that binds strongly on its surface without any further side reactions. Generally, thiol functionalization on gold coated AAO membranes is achieved via

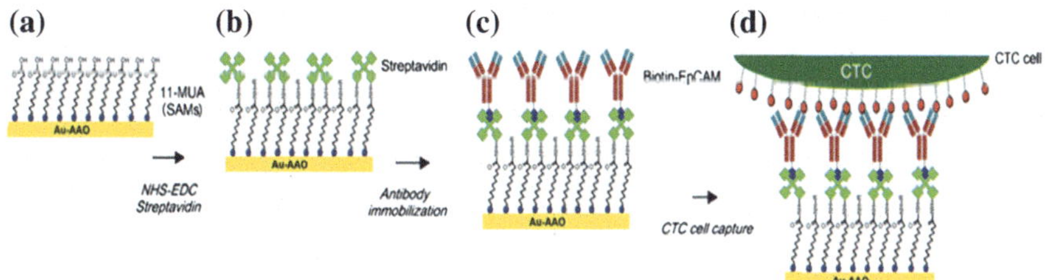

Fig. 5.1 Schematic representation of the surface functionalization steps of gold modified AAO used for RIfS biosensing. The system begins with **a** the formation of self-assembled monolayers (SAMs) of carboxyl-containing thiol followed by **b** covalent attachment of streptavidin on activated SAMs after activation with coupling (NHS/EDC) agents and finally **c** immobilization of biotinylated Anti-EpCAM antibodies; **d** scheme of binding of CTC cell on anti-EpCAM antibodies. Adapted with permission from [7]

immersion of freshly prepared membrane into dilute alkylthiol in an ethanoic solution for 12–24 h at room temperature. A series of experiments were carried out by Le et al. [11] for separation studies using five different kinds of thiol-modified gold coated AAO membranes. These studies suggested that highly polar carboxylic group increases the hydrophilicity while non-polar alkyl group increases the hydrophobicity of the golad-coated AAO membrane. Likely, the hydrophobicity increased with longer alkyl chain thiol modified gold coated AAO membranes. Smuleac et al. [12] demonstrated that the interaction of 3-glycidoxypropyl trimethoxysilane modified AAO surface with certain ligands (e.g., polyglutamic acid) could cause a conformational transition of the chain to create a polythiol containing 240 repeat units on AAO surface. Such phenomena have also been observed with an ingenious connection to create a polythiol containing 240 repeat units on AAO surface. Building up the concept of gold coated AAO surfaces, Steinem group [13–15] showed that the suspended lipid bilayers could be done on alkanethiols coated AAO supports with the ability to mimic biological membrane and monitor ion channel activities (Fig. 5.2). Their fabrication approach consists of gold coating on AAO followed by formation of alkanethiols with a negatively charged head group. Fusion of lipid vesicle featuring positively charged lipid head groups leads to a suspended bilayer that prevents vesicle fusion inside the pore. In a similar configuration, purple membranes can be adsorbed on freestanding lipid bilayers, termed nano-black lipid membranes (nano-BLMs) for photocurrents measurements [16]. A recent review of suspended bilayers on alkanethiols coated AAO surface has been described in [17].

5.1.1.3 Silanization

Silanization has been demonstrated an effective and flexible method for changing the wetting and adsorption properties of AAO membranes. The versatility of this method resides in its straightforwardness, fast and stable as it can be carried out at

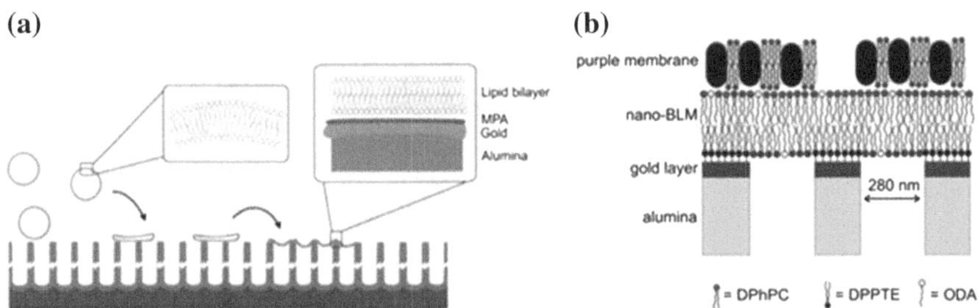

Fig. 5.2 **a** Schematic representation of pore suspending bilayers made by fusion of large unilamellar vesicles on 3-mercaptopropionic acid monolayer functionalized Au-coated AAO substrates. **b** A schematic illustration of adsorbed purple membranes on nano-BLMs. Adapted with permission from [15, 16]

moderate temperatures, and no particular conditions or expensive equipment are required. This characteristic result in the essential ability of this technique to be applicable to any materials and a wide range of application. The uniformity and stability of this technique are extremely sensitive and depends on the details of the organosilane chemistry and the functionalization conditions. For the attachment of silanes on the surface, the principle involves activation of silane molecules to promote condensation reaction between the silanol group of silane and hydroxysil groups of the surface (in this case the AAO). This reaction will form a stable bond on the surface and releases free alcohols as a side product [18]. Wide varieties of substituted organosilanes are commercially available, and the presences of these silanes significantly change the local environment and properties of the AAO membranes. The main properties of surface modification include hydrophobicity, hydrophilicity, absorption, orientation, charge conduction.

For instance, AAO membranes were rendered hydrophobic by functionalization with octadecyltrimethoxysilane (ODS) has been reported by few studies. ODS modification was accomplished by immersing the membrane into a solution containing ODS with 15 mL of absolute alcohol and 50 mM of sodium acetate buffer [4]. Likewise, a different chain of alkyltricholrosilanes with tail length from C1–C8 can be functionalized on the AAO nanopores [19]. Modification of membrane consists of silane with fluoro tails, i.e., perfluorodecydimethylchlorosilane has changed the wettability of the membrane from 20 to 160° [20, 21]. Other approach employed PEG-silane that is formed by reacting PEG-silicon tetrachloride in the presence of triethylamine as a catalyst. This combination was then reacted on the hydroxylated surfaces of AAO to form a network of Si–O–Si bonds resulting immobilization of PEG on the surface. This study investigated the biocompatibility aspects of PEG coated AAO surfaces for possible use to prevent membrane fouling and immunoisolation [22, 23]. In addition, Steinle et al. [24] derivatized AAO membranes with a silane that terminated with carboxylic acids. This was achieved by incubating the APTES terminated AAO membrane with diacids chloride and diisopropylethylamine. In more advanced work, an interesting development of multi-layered silanes with different types of functionalities and wettability was demonstrated by Jani et al. [25] on AAO nanochannels. This process was based on combination series of anodisations and silanization during fabrication and allowed to further tune the transport properties on the nanochannels (Fig. 5.3) [26].

Silane modifier typically acts as the coupling agent or chemical linker to help attach other molecules or compounds in preparing customized product (Fig. 5.4). Indeed many works have been successfully attached enzyme [27], antibody [28], urease [29], DNA molecules [30], biotin [31], lipid bilayers [32, 33], polymers, cells [34] and nanoparticles [26, 35] to AAO surfaces via silanization. For example, Zhang and co-workers utilized an elegant approach to biosensor development that involved covalent attachment of glucose oxidase (GOD) to the 3-glycidoxypropyl-trimethoxysilane (GPTMS) modified AAO nanochannels. From this, a novel enzyme reactor system that enables a controlled catalysing rate was developed [27]. Shi et al. [36] also utilized GPMTS modified AAO surfaces as the separation and purification sites for haemoglobin from the red cell. This was achieved by reacting

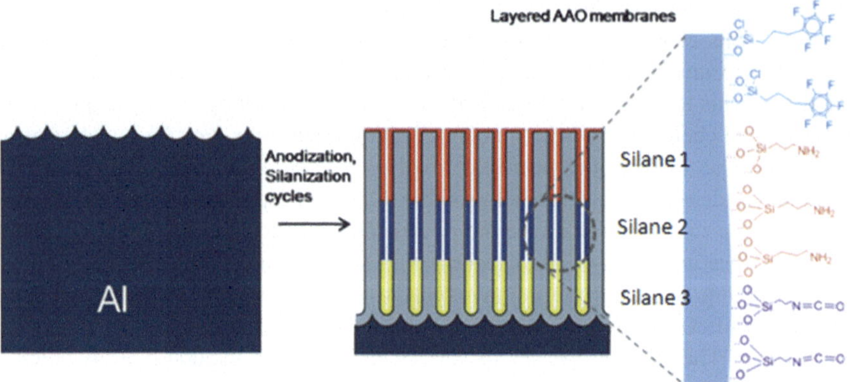

Fig. 5.3 Schematic illustration of anodization and silanization cycles to produce an AAO membrane with multiple silane layers. Adapted with permission from [26]

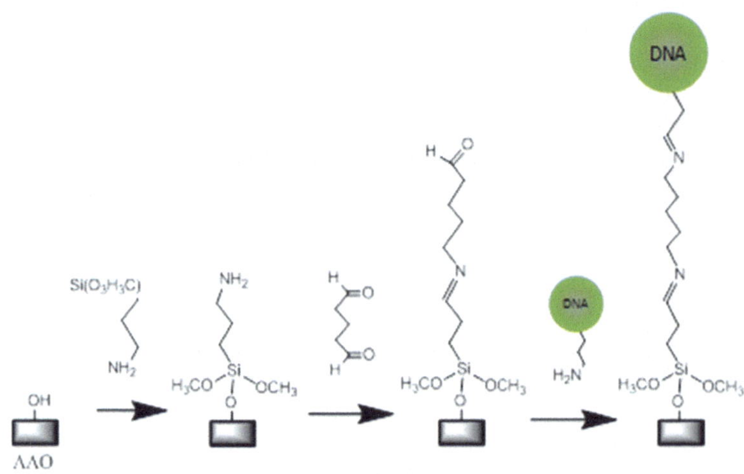

Fig. 5.4 Silanization of hydroxylated AAO surface with isocyanatopropyl triethoxysilane and subsequent immobilization of amino-terminated DNA. Adapted with permission from [45]

GPMTS with chitosan, and the chitosan–AAO composite membrane was successfully used for affinity protein separation specifically for haemoglobin from the haemoglobin-phosphate solution and hemolysate. An alternative approach to covalent immobilization involves antibodies recognition sites. This approach introduced by Joung and co-workers utilized hyaluronic acid that enable ring opening reaction of GPMTS modified AAO surface to be carried out. The hyaluronic acid then functioned as coupling sites with sulfo-NHS esters for antibodies immobilization purpose. This novel label-free immunosensor system demonstrated a remarkable capacity for detecting pathogenic E. coli bacteria [28]. Evidence for incorporating DNA molecules on silane functionalised AAO surface has been demonstrated by Valssiouk et al. [30, 37]. These have resulted in covalent attachment of DNA oligomer that helps hybridization process inside AAO nanopores [38, 39]. Silane coated surfaces also have been used to suspend lipid bilayers in AAO membranes. The suspended bilayers termed polymer-cushioned lipid bilayers were first achieved by activated the surfaces with aminopropyl-dimethylethoxysilane (ADMS) and the second step consists of grafting the N-hydroxy-succinimidyl carbonate (NHS-PEG) to amino coated surfaces. It was suggested that grafted PEG chains triggered the vesicle fusion and maintained the fluidity of the bilayers [33]. A recent example of silane-glutaraldehyde AAO surfaces is the covalent immobilization of glucose-6-phosphate dehydrogenase (G6PD) was demonstrated. In a similar approach, silanized membranes were incubated with N,N-dimethyl-formamide (DMF) containing 25 mg of N-succinimidyl-3-meleimidoproprionate. The maleimide-grafted membranes were then immobilized with human CYP2E1 by cross-flow filtration at different times [40]. Similarly, a detailed investigation of cellular adhesive peptide (RGDC) dependence of osteoblasts (bone forming cells) on maleimide-grafted AAO membrane was subsequently demonstrated by Leary Swan et al. [34]. APTES coated AAO surface also has been used as the surface attached initiator for grafting polymer brushes on membrane surfaces. For example grafting of poly(γ-benzyl-Lglutame) PBLG [41] and PNIPAM (poly(N-isopropylacrylamide) [42] via surface initiated polymerization. This work determines that, the density of the polymers can be controlled by controlling the density of initiator. This work was later refined by Bruening and co-workers by a method in which the initial step of activating the membrane surface with silane can be excluded by employing synthesized trichlorosilanes initiator (11-[42]undecyltricholorsilane) to grafted PHEMA (poly(2-hydroxyethyl methacrylates) brushes on AAO surfaces [43]. Silanized AAO surfaces also have been used to as a template for fabricating various nanotubes. For example, Sehayek et al. [44] used APTES functionalized AAO membrane to form solid, porous, nanoparticle-based nanotubes. The nanotubes are prepared by passing a citrate-stabilized metal (Au, Ag) colloid solution through the pores of an aminosilane-modified AAO membrane. As a result, mechanically stable and electrically conducting gold or silver naotubes were fabricated.

An interesting derivative of silanisation work on AAO surfaces is a process in which silanization was carried out on gold-coated AAO surface. This work involved depositing a thin layer of gold on AAO using either electrodeposition, electroless or sputtering technique. The ability of depositing tethered lipid bilayer on silanized gold coated surface was investigated by Largueze et al. [46]. In this approach, the bottom part of AAO membrane was coated with gold and prior to lipid deposition the gold layer was functionalized with silanes, undecanethiol and the last step being a PEG-triggered fusion of the surface-attached liposomes. By means of CV, the alterations of lipid membranes can be observed by inducing it with detergent.

5.1.1.4 Polymer Functionalization

Polymeric coatings on AAO membranes offer a high potential in respective areas of applications and have been examined by several research groups. This surface modification technique allows a fine control of surface chemistry, functionality, density and thickness of the coating on the AAO membrane. Polymer layers can be either attached to the top of the pores or inside the pores. Polymer-modified AAO membranes have shown improved binding capacity, selectivity, biocompatibility, stability and lubricant properties in comparison to non-modified AAO membranes. The most common approaches for a polymer functionalization on AAO are polymer grafting, plasma polymerization and polyelectrolytes multilayers.

The modification of AAO membranes with polymer brushes rely on the principle of grafting approaches that can be achieved via non-covalent and covalent interactions between the polymer chains and the membrane. Qi et al. [47] investigated the use of poly(2-methoxy-5-(2′-ethyl-hexyloxy)-p-phenylenevinylene (MEHPPV) or poly(2,3,-diphenyl)phenylenevinylene) DP-PPV polymer brushes into the pores of AAO membranes via physisorption. However, the adsorbed polymer brushes appeared to be suffering from poor stability. Covalent grafting generally occurs through either grafting to or grafting form approaches via atom transfer radical polymerization (ATRP), reversible addition-fragmentation chain transfer polymerization (RAFT), plasma-induced graft polymerization, ring-opening metathesis polymerization (ROMP), and the self-polymerization of dopamine (DOP-SP) methods [48, 49]. The resulting polymer brushes contain multiple binding sites that give rise to high binding capacities. Moreover, the polymers can be asymmetrically modified within the nanochannels of the AAO that exhibit specific biological recognition and provide high selectivity. Nagale et al. [50] reported the grafting of amino-terminated poly(tert-butl acrylate PTBA) to the top layer of AAO surfaces functionalized with of carboxylic acid terminated thiol. In this work, the pores channel is free from a polymer, thereby providing high-flux composite membranes. Sun et al. [51] have developed the ATRP approach to modify the AAO membranes utilizing poly (2-hydroxyethyl methacrylates) (PHEMA) brushes with nitrilotriacetate-Cu^{2+} (NTA–Cu^{2+}) complexes that yielded membranes with bovine serum albumin (BSA)-binding capacity. Static binding capacities of 150 mg BSA/ml and saturation of the membranes with BSA or myoglobin in less than 15 min were determined. Later on, the same group synthesizing different brushes of (PHEMA)-NTA–Ni^{2+} on AAO pores that allow to bind oligohistidine-tagged ubiquitin for protein purification [48]. Other general attributes related to polymer brushes grafted AAO membranes have been described and illustrated by Bruneing et al. [43] in details.

Thermo-responsive polymer brushes of PNIPAM are widely acknowledged and received considerable attention in the recent years. The switchable properties such as hydrophilic/hydrophobic switching at the lower critical solution temperature (LCST) make this polymer attractive for a variety of applications. Generally, ATRP technology has been used to synthesis PNIPAM grafted on AAO membranes. This was first demonstrated by Fu et al. [52] by utilizing one step approach using an

AAO surface functionalized with the ATRP initiator 1-(trichlorosilyl)-2-[m/p (chloromethyl)phenyl]ethane followed by PNIPAM grafting. This result suggests that the length of the polymer in the porous network can be used to control the size of surface pores, surface roughness and the interfacial energy. Changes in initiators have been described as potentially useful in ATRP applications [42, 53]. For example, Li et al. [42] employed 2-boromoisobutyryl bromide (BIBB) on the membrane surface to graft PNIPAM using ATRP method. Likely, Wang et al. [53] demonstrated the preparation of molecular imprinted polymers (MIPs) using surface-initiated ATRP on an AAO membrane. In one recent example, Ma et al. [54] demonstrated by combining ATRP, ROMP and DOP-SP methods, various asymmetric polymer brushes-stabilized Au-Pd in the AAO channels can be formed. The device consisted of poly(3-sulfopropyl methacrylate potassium salt) (PSPMA) on one side and poly(2-(methacryloyloxy)ethylmethylammonium chloride) (PMETAC) on the other side of the membrane fabricated in the reaction cell (Fig. 5.5). This effect was then exploited so as to produce an asymmetrical catalytic array of Au-PMETAC@PSPMA-Pd by depositing Au and Pd on both sides of the AAO membranes, respectively. This new sandwich membrane effectively demonstrates excellent flow-through catalysis.

Polyelectrolytes (PEs) are polymers with a variety interesting properties and applications. Since 1997, a large academic interest in the field of layer-by-layer (LBL) arose from the discovery by Decher [55] that alternate depositing of oppositely charged polyelectrolytes (PEs) polymers forming PE multilayers (PEMs). The LBL assembly is a simple, yet versatile and inexpensive technique. Moreover, the film thickness can be easily controlled at the nanometer scale, and the resulting layers can be further functionalized with biomolecules or nanoparticles [43, 56].

AAO membrane consisting PEMs was first demonstrated by Balachandra et al. [57] utilizing polyacrylic acid (PAA) partially complexed with Cu^{2+} ions onto a UV/ozone treated AAO membrane followed by immersion in a solution of polyallylamine hydrochloride (PAH). The above procedure was repeated until the desired numbers of bilayers were deposited and conformal and smooth films were formed on the AAO. Subsequent work show AAO pores were not clogged after deposition of 5–7 PEM bilayers thus enabling this system to be employed molecular size selective transport. The number of PEMs layers combined with different compositions of PEMs significantly influenced the flux, transport, rejection rate and selectivity of solutes through these functionalized membranes [43]. For example, Hong et al. [58] showed that PEMs consisting 4.5 bilayers of polystyrene sulfonate/poly(diallyldimethylammonium chloride) (PSS/PDADMAC) on AAO pore gave Fluoride ions rejections more than 70 % with a selectivity factor more than 3 compared to other monovalent anions. They additionally showed that the similar PEMs on AAO revealed 98 % rejection of phosphate ions as compared to the commercial NF90 membrane [59]. More recently, Ouyang et al. [60] investigate the effect of PEMs with five bilayers of PSS/PAH deposited on AAO showing effective rejection of Mg^{2+} and Ca^{2+} cations by 95 %. The selectivity of cations rejection could be enhanced by increasing the charge of the terminated PAH layer by means of increasing ionic strength in the PAH deposition solution. Hong et al.

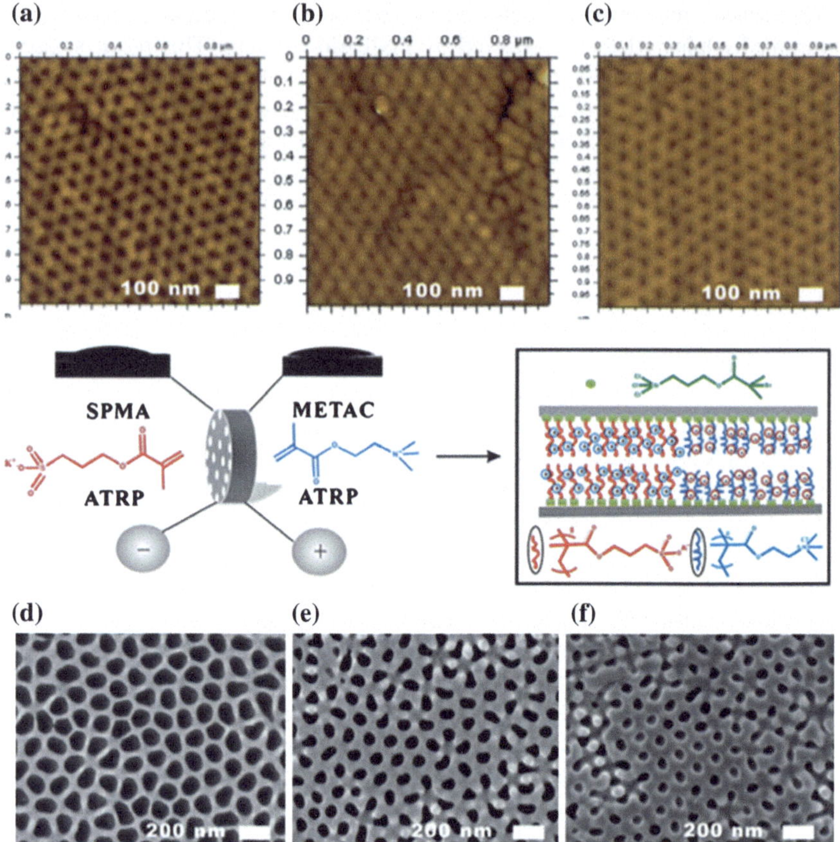

Fig. 5.5 AFM images of AAO membrane (pore size of 30 nm) before (**a**) and after modification with polymer brushes of PMETAC (**b**) and PSPMA (**c**). Schematic illustration of asymmetrical modification with double hydrophilic polymer brushes within AAO channels (*middle*). **d**–**f** FESEM images of AAO membrane (pore size of 130 nm) before (**d**) and after modification with PMETAC (**e**) and PSPMA (**f**). Adapted with permission from [54]

[61] later showed, seven bilayers of PSS/PAH on AAO yielded a high selectivity of glycine over L-glutamine, whereas 4.5 bilayers of PSS/PAH coated AAO membrane showed a sucrose rejection of about 99 % and increasing the flux of the solution. Both studies demonstrated that the performance of selectivity and rejection are comparable with commercially available membranes.

As noted earlier, biomolecules or nanoparticles could be deposited onto PEMs layer. Dai et al. [62] evaluated the performance of PEMs of PAA/PAH followed by immobilization of antibodies on the carboxylic groups of the PAA layers following the carbodiimide coupling approach. The resulting coating on AAO resisted non-specific protein adsorption. Charged polyelectrolytes on AAO membranes also allowed the immobilization of citrate-stabilized gold nanoparticles on PEMs under retention of the nanoparticles' catalytic activity as demonstrated by Dotzauer et al. [63, 64]. For a detailed discussion on the synthesis and properties of LbL films, the reader is referred to an excellent review written by Joseph et al. [65].

5.1.2 Hard Modifications

Metal coating by using electrochemical deposition, electroless deposition, thermal vapour deposition, plasma polymerization, atomic layer deposition (ALD) and chemical vapour deposition (CVD) are the examples of hard techniques used to improve the physical, chemical and structural engineering properties of AAO. These innovated modification approaches are utilized to deposit a wide range of materials from metals, metal oxides, semiconductors, ceramics, nitrides, as well as carbon nanotubes onto the AAO membrane [66–69]. It is also highlighted that these developed methods are suitable for the synthesis of diverse nanostructured materials such as nanowires, nanotubes or nanorods with the assistance of that particular AAO as a template. As such, these methods provide a broad scope to engineer AAO-progress in technologies for specific applications by controlling and designing structural growth of AAO with different sizes, arrangements, structures, geometries and pore architectures [45].

5.1.2.1 Electrochemical Deposition

Electrochemical deposition is used extensively to deposit metal or alloy on AAO membrane, mostly owing to its high material transfer efficiency, cost effective and controllable film thickness. In comparison to thermal vapour deposition, ALD or CVD, the electrodeposition route is easy, as well as less skill dependent, and does not require sophisticated instrumentation [70]. The experimental set-up is quite similar to that of anodization, while the AAO serves as the cathode. In most cases, a thin metal layer (commonly Au) is first evaporated on one side of the nanoporous AAO membrane to serve as a working electrode. In particular with regards to electrodeposition method, it is emerged as an accepted versatile approach for AAO template synthesis-pore filling with conducting metals to obtain continuous arrays of nanowires, nanotubes, nanorods or nanoparticles with large aspect ratios [71–75]. See [76] for a more detailed discussion including bottom-up and top down approaches on metal deposition on AAO surface.

For metal nanoparticles deposited on the AAO membrane; Platt et al. [77] used the electrodeposition of Pd nanoparticles at the liquid-liquid interface using AAO template. The resulting nanoparticles were formed at the mouth of the AAO pores, and the locus of their formation is being dictated by the position of the organic-water interface. However, many issues concerning the mechanism of this interfacial deposition process remain to be clarified, and such studies are in progress. Another approach by Lee et al. [78] has developed an alternating current (AC) technique for dielectrophoretic assembly of Au nanoparticles in the AAO template. This simple preparation mode showed a selectively patterned of Au nanoparticles is successfully fabricated in the pores of AAO, which is expected to lead for exciting sensor applications in various fields. Later it was shown by Masuda group [79] that developed a method to fabricate layered three dimensional (3-D) Au nanoparticles inside an AAO matrix for use as a surface-enhanced Raman scattering (SERS) substrate. The electrodeposition of Au nanoparticles inside the AAO channel is demonstrated by alternating current (AC) electrolysis of AAO. Then, the number of layers of the Au nanoparticle arrays is determined by the number of repetitions of this process, and the gap size between Au nanoparticles is determined by the duration of anodization after the electrochemical anodization of Au.

One of the most effective approaches of electrodeposition is to fabricate nanowires by embedding metal in the AAO matrix. Whitney [80] recently employed electrodeposition approach prepared by nuclear track etching to fabricate Ni and Co nanowires into the AAO template. In addition, high uniformity of metal/alloy deposition has been demonstrated by the group of GÖsele using current-controlled deposition sequences [81–83]. A few years later, this group has successfully prepared segmented Ag–Au alloy nanowires inside the AAO membrane via the electrodeposition approach [84]. The Ag component could then be selectively removed using nitric acid. Other critical preparative issue during electrodeposition, which is an unstable electrical current that causes uniform filling of the pores cannot achieve smoothly. Therefore, it is of paramount interest by Kim et al. [85] to fabricate free-standing Pd nanowires via pulsed electrodeposition (PED), in which the electrodeposition of Pd is carried out galvanostatically with a pulse mode. In this work, PED method is more reliable for deposition into high aspect ratio materials and can compensate for the slow diffusion-driven transport in the narrow pores. The prepared Pd nanowire arrays are standing freely on a Ti-coated Si wafer after removing the AAO template. The most recent work was demonstrated by Pecko et al. [86], they synthesized Fe-Pd nanowires by applying both potentiostat and PED regimes. When using potentiostat deposition, only fragmented nanowires are obtained, and the use of PED is shown to be effective for producing solid nanowires. In particular, another work by Wang et al. [87] used supercritical electrodeposition method, an electroplating reaction with the emulsion of carbon dioxide (CO_2), surfactant and electroplating solution to prepare highly ordered of Ni nanowires. The plating in the emulsion is similar to pulse plating, and it is also highlighted that the developed method showed a wide range of variable parameters for obtaining Ni nanowire arrays with uniform characteristics using AAO template.

On the other hand, electrodeposition technique for semiconductor nanowires such as CdS was used after pioneering work by Routkevitch and co-workers [88] in 1996. It also has been reported by Xu et al. [89–91] that single crystal compound semiconductor nanowires of CdS, CdSe and CdTe can be fabricated into the nanoporous AAO template by sequential deposition of cationic and anionic components of the semiconductor by reverse potential sweeping. Taking CdSe as an example, a small amount of Se is first deposited followed by the deposition of excess Cd and subsequently stripping of the Cd during the reverse potential sweep

[92]. The most recent work by Chen et al. [93] showed transitional metal chalcogenide compound CuS nanowires are successfully fabricated via electrodeposition and sulfurization. Cu nanowires are first synthesized and then sulfurized to form CuS nanowires. Results indicated an appropriate pH value of a deposition solution, a sufficient amount of sulphur powder and temperature during sulfurization is crucial to forming well crystallized CuS nanowires.

The AAO template method for the synthesis of one dimensional (1-D) nanostructured material such as nanorods or nanotubes is first introduced by Martin and the co-workers [71]. The method is based on the electrochemical deposition of Au within the pores of nanoporous AAO template membranes. The authors showed that the Au/AAO composites can be optically transparent in the visible and also that by changing the aspect ratio of the prepared nanocylinders; the colour of the composite membrane can be varied. In another approach, Burdick et al. [94] demonstrated the high-throughput fabrication of alternate multisegment deposition of Au with short gaps of Ag. Likewise, Hoang et al. [95] demonstrated the growth of multisegmented nanorods comprising Au and sacrificial Ag segments using an electrochemical wet etching approach. In particular, Kim et al. [69] have fabricated striped multimetal nanorods by direct current (DC) electrodepositing three different metals of Au, Ni and Cu sequentially in the pores of AAO templates. The lengths and the sequence of metal segments in a striped rod can be tailored readily by controlling the durations of electrodeposition and the order of electroplating solutions, respectively. Another study by Wang et al. [96], they reported a novel approach for the fabrication of TiO_2 nanotubes arrays by electrodeposition on the nanoporous AAO template from inorganic aqueous containing $TiCl_3$ and Na_2CO_3. The method applied in this work can be used for the preparation of other order 1-D semiconductor nanotubes as well.

Up till now, studies of the growth mechanism of nanotubes have been concerned with the nature of the material and the chemistry of the pore wall. However, the effect of a back electrode on the formation of electrodeposited nanotubes has not been reported. Herein, Atalay et al. [97] studied the electrodeposition of CoNiFe nanotubes on vitreous templates placed on highly ordered nanoporous AAO. It was found that the back electrode placed on AAO template was an important factor in controlling the shapes and properties of the nanostructure. It is remarkable that a nano-engineering strategy for fabricating SnO_2/MnO_2 core/shell nanotube arrays is innovated by Grote et al. [98]. The SnO_2 core inside the nanoporous AAO is fabricated by atomic layer deposition and subsequently a thin MnO_2 shell is coated by electrochemical deposition onto the SnO_2 core. The fabrication process allows precise controlling of the length, spacing, diameter, wall thickness and the selection of open-end or closed-end nature of the prepared nanotubes. Another advancement of nanostructured materials prepared by electrodeposition has been reviewed by Gurappa et al. [99], Bicelli et al. [100] and Hurst et al. [70].

5.1.2.2 Electroless Deposition

Electroless deposition is defined as a deposition of a metal coating either by immersion of a metal in a suitable bath containing a chemical reducing agent. Compared with electrochemical deposition, the advantage of the electroless deposition is that there is no necessity for the surface to be electronically conductive [101] and the deposition starts from the pore walls and growth inwardly [102]. Martin et al. [71] first applied this approach to AAO membranes using Au deposition. The Au nanotube/AAO composite membranes are successfully used for molecular separations showing size, charge and chemistry based selectivity. By varying the electroless deposition time, hollow tubes can be obtained with short deposition time while solid nanorods can be obtained using longer deposition time [70]. Another work by Cheng et al. [103] used the electroless deposition technique to fabricate high performance SERS-active Au nanoparticles arrays with tuneable particle gaps on AAO substrates. They reported that pH and the temperature were the main factors controlling the size, shape and aggregation of Au nanoparticles, as well as the inter-particle distance. As a matter of fact, catalytic and SERS properties of Au nanotubes inside the AAO pores are demonstrated as the results of clustered Au surface inside the nanotubes, as well as the Au nucleation process during chemical deposition. Interestingly, the SERS technique also can be used for mapping the gold layer integrated with 3-mercaptobenzoic acid (mMBA) inside AAO pore channels as investigated by Velleman et al. [104]. In this fashion, the gold layer was deposited by electroless deposition prior to the formation of SAMs of mMBA inside pore channels (Fig. 5.6). Their findings suggested that the developed SERS active surface can be applied for ultrasensitive sensing applications.

Although this method seems convenient and versatile in preparing nanotubes, one problem is that the plating reaction occurs not only on the pore walls but also on the surface of AAO membrane. If the deposition rate is too fast, the pore entrance is

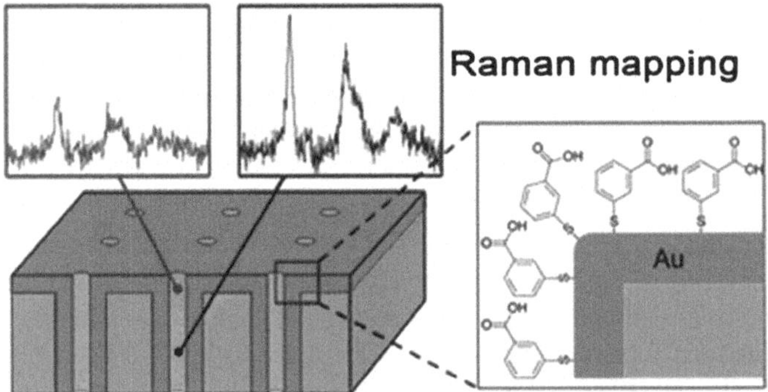

Fig. 5.6 Schematic diagram of AAO membrane coated with gold layer via electroless deposition and subsequent functionalized with 3-mercaptobenzoic acid (mMBA). Surface enhanced Raman scattering (SERS) conformed spatial distribution of mMBA inside the channels. Adapted with permission from [104]

sealed before the reactant can get through the full length of the pore. Therefore, further progress in this area has seen the developed meet of a variety of fabrication approach to design exotic patterns and arrangements of nanotubes inside the AAO membranes. Li et al. [105] showed an elegant approach for preparing Cu nanotube arrays with open-ends by electroless deposition. This is accomplished by a four steps procedure, i.e. pore-wall modification, polishing treatment, sensitization-activation and electroless deposition. Additionally, sensitization-activation is necessary to produce a conformal coating. This nanotube arrays with open-ends have potential application in preparing novel core-shell nanocable metal alloys and other interesting nanomaterials. Another work by Zhang et al. [106], they synthesized Ag nanotubes by electroless deposition in nanoporous AAO templates. The reaction used for the formation Ag nanotubes within AAO is well-known Tollens' test in sugar chemistry. Ag nanotubes with lengths over 10 μm are successfully fabricated inside the AAO membranes. Likewise, Ag nanoparticles with uniform size and smooth surface are deposited inside AAO using this technique [67, 107].

In particular, Wang et al. [101] synthesized a wide range of metals nanotube arrays in AAO templates including Co, Ni and Cu arrays via electroless deposition on APTES-functionalized AAO membranes. The inner diameter of the nanotubes can be tailored by adjusting the deposition times while the length of the nanotubes is determined by the thickness of the AAO template. Remarkably, a novel approach is obtained during fabrication of end-closed NiFeCo-B nanotube arrays via electroless deposition by Azizi et al. [102]. This work is different from previous efforts as the formation of end-closed nanotubes is successfully fabricated for the first time. A different approach has been applied by Kang et al. [108]; superhydrophilic/superhydrophobic Ni micro-arrays structure and a substrate with strong mechanical strength are fabricated by combining a simple electroless deposition and self-assembly silanization process. These Ni nanoparticles are deposited on an etched AAO membrane template by electroless deposition without activation and sensitization. Recently, boron-doping Ni@Au hybrid magnetic nanotubes are deposited rapidly on AAO membrane via electroless method [109]. This peculiar work differs from previous efforts because of the more rapid and facile synthesis of Ni@Au nanotubes. Moreover, this study is the first to explore the in vitro cytotoxicity of Ni@Au nanotubes on Molt-4-cells, which makes them potentially applicable for biomedicine. Another comprehensive review on electroless deposition method can be found by Ali et al. [110] and Stojan et al. [111].

5.1.2.3 Thermal Vapour Deposition

Thermal vapour deposition is one of the simplest and most popular approaches for fabricating various types of nanostructures with certain characteristics. The basic process of this method is sublimating source material(s) in powder form at high temperature, and a subsequent deposition of the vapour in a certain temperature region to form desired nanostructures. The aim of this modification is not only typical to improve conductivity and chemical stability of the AAO material, but also to introduce some intriguing properties such as catalytic, electrochemical, magnetic, optical and transport [66, 112]. Lei et al. [66] has reviewed the AAO coated with metal such as Ni or Co which improved magnetic properties whereas catalytic properties have been introduced by coating with Au, Pd, Pt, Ti/TiO_2. In addition, metal coating also served as the basis for a further chemical modification approach to help in binding AAO with various chemical and biological species for optical sensing and molecular separation applications. Arguably, the main disadvantage of this technique is that only the top part of the nanopores can be modified and the deposition of internal pore surface is limited. Therefore, this will provide a glimpse of opportunities to many researchers for the metal deposition approaches on the AAO to be explored.

Toh et al. [113] has successfully prepared Pt-coated AAO membranes for selective transport and separations of charged proteins by sputtering the AAO membranes with Pt. During separation experiment of charged proteins, electrical potential is applied to create a high electric field. In fact, further size selective separation is achieved by increasing the thickness of sputtered Pt layer. Another approach by Qiu et al. [114] has fabricated a highly ordered hemispherical Ag nanocap arrays template on AAO membrane by using a direct current magnetron sputtering system. The surface structure can be tuned further to optimize the enhancement factor according to optional AAO fabrication and Ag deposition parameters. The design of Ag nanocap array with uniform and highly reproducible SERS-active properties may provide a breakthrough for the fabrication of robust, cost-effective, exceptionally sensitive and large-area SERS-based sensors. Besides, Yin et al. [115] reported a facile method for the synthesis of silica nanotubes by thermal decomposition of polydimethylsiloxane (PDMS) rubbers. At elevated temperature, PDMS rubbers could be first decomposed into volatile cyclic oligomers while under inert conditions. When the oligomers are oxidized in air, vapour phase silica is formed, which nucleated and grew heterogeneously on the walls of AAO templates. The diameter and length of the silica nanotubes are mainly determined by the pore size of the AAO template while the wall thickness of the nanotubes could be controlled by the initial amount of PDMS rubbers.

For the synthesis of metal and metal oxide nanotube arrays, Whitesides et al. [116] reported the synthesis of Au and ITO nanotube arrays with a thin backing using shadow evaporation. In the synthesis, the AAO membranes are tilted with respect to the column of evaporating materials, controlling the geometry of the tubes. Meanwhile, rotation of the membrane during evaporation ensured the uniformity of the resultant materials. In another approach, Pereira et al. [117] reported on a simple process for the formation of a functionally modified AAO channel by using pulsed laser deposition (PLD) method to deposit a metal (i.e. Au, Pt) or a mixture of metal (i.e. Pt–Ru) on the AAO membrane. PLD method is a versatile and powerful tool for the growth of high purity thin films and nanostructured materials. This work has achieved an optimizing kinetic energy of the deposited metal in contact with the AAO substrate. Particularly, Wang et al. [118] combined physical vapour deposition, gating ion milling and thermal annealing techniques for the fabrication of metal nanodot arrays on AAO. This new approach is based on the

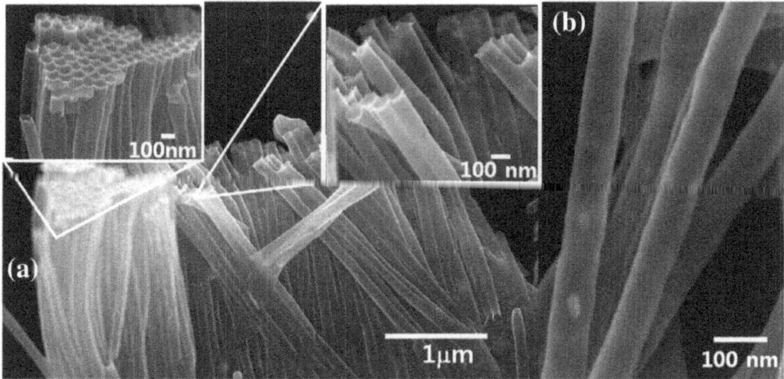

Fig. 5.7 Scanning electron microscope (SEM) images of Au-CNT-AAO prepared at 973 K. Inset images in **a** are magnified images of indicated areas, **b** is magnified SEM image of another part in the same sample. Adapted with permission from [119]

fabrication of ring and half-moon shaped nanostructured arrays and a further annealing process. Such approach will probably broaden the potential applications of template based nanostructures. More recently, Lee and co-workers [119] have demonstrated a facile and efficient method for preparation of bamboo-like Au-carbon nanotubes by thermal decomposition of sucrose in an Au-AAO template (Fig. 5.7). The crystalline Au nanoparticles attached on the bamboo-like conducting carbon nanotubes obviously improved the electrochemical response of the nanotubes.

5.1.2.4 Plasma Polymer Deposition

Plasma polymer deposition (or plasma polymerization) is defined as the formation of polymeric materials under the influence of plasma condition [120]. In the plasma polymerization process, a monomer gas is pumped into a vacuum chamber where it is polymerized by plasma to form a thin, clear coating. This technique is particularly efficient and convenient because the ultrathin polymer films of various functionalities can be deposited on any surfaces without the need of surface pre-modification [121]. It is well-established by the first report of plasma treatment on AAO membrane from Brevnov et al. [122], who prepared Janus-type membranes with a hydrophilic and hydrophobic side. This approach reported that a super-hydrophobic surface is obtained on AAO membranes using inductively coupled plasma polymerization of fluorocarbon (C_4F_8) monomer, with a water contact angle (WCA) of 150°. The other side of the membrane is left unmodified and showed a WCA less than 5°. In comparison with other methods used for the surface modification of AAO membrane, the advantages of plasma polymerization are that it is a one-step, fast and low-temperature deposition process with a sterile and solvent free technique [123]. Considering the important of the above backdrop,

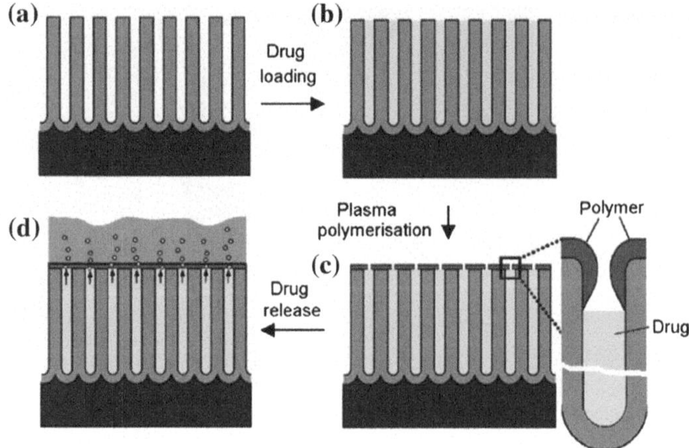

Fig. 5.8 Schematic illustration of Plasma modification of AAO membrane for controlled drug release. **a** AAO porous layer fabricated by electrochemical anodization, **b** drug loading (vancomycin) inside of pores, **c** the deposition of the plasma polymer layer (allylamine) on the top of the pores and finally **d** the release of drug from the pores into solution. Adapted with permission from [124]

hence Losic et al. [123] aim to investigate the potential of plasma polymerization of *n*-heptylamine in order to create *n*-heptylamine layers with surface amine groups on AAO membrane. The modified AAO membranes with active amino groups are a promising platform for the development of novel functional and smart membranes for advanced controlled drug release (Fig. 5.8) [124].

5.1.2.5 Chemical Vapor Deposition

Chemical vapor deposition (CVD) involves the dissociation of gaseous molecules in an activated (heat, light, plasma) environment followed by the formation of stable and conformal films on a substrate [125]. CVD has the capability of producing highly dense and pure materials, offers an excellent control over coating thickness and coverage as well as sustains fast growth rates as compared to other deposition processes [126]. The CVD method is widely used to modify AAO membrane for the fabrication of carbon nanotubes (CNTs) [127–129]. An aligned multi walled CNTs can be further applied as catalyst support [130], drug delivery [131] and field emitters [132]. Fabrication of CNTs by CVD can be catalyst-assisted or without the use of a catalyst.

Rana et al. [128] produced aligned and dense CNT forests over AAO and AAO/Si substrates with the use of ethanol based precursor. These hybrid structures can potentially be applied in the electronics industry as structures for building light emitting diodes, solar cells and super capacitors [132]. Meanwhile, Fang et al. [133] studied the fabrication of nanoporous alumina–carbon nanohybrid by plasma CVD.

The process starts from the electrochemical fabrication of free-standing AAO membranes (Fig. 5.9). The membrane is then coated with a thin gold layer followed by the growth of carbon nanowall in the presence of methane and argon gas mixture. The presence of thin layer gold is reported to govern the density and size of the carbon nanowall.

Fabrication of well-aligned, open-ended CNTs inside the pores of AAO template without using any metallic catalyst was reported by Sarno et al. [134]. An effective volume conductivity of CNT/AO composite was obtained from a few up to 10 kS/m, which is in line with the literature [135]. Recently, Altalhi et al. [136] introduced the well-organized CNT membranes with tunable molecular transport properties using a plastic bag as the carbon source by catalyst/solvent free CVD approach (Fig. 5.10). In this study, a small piece of the plastic bag was put into the pyrolysis zone while AAO membrane was placed in the deposition zone of the CVD reactor. The optimal deposition achieved were 850 °C and 30 min with further deposition time, resulted in an increased of CNTs wall thickness. The fabricated CNTs-AAO membrane demonstrates the ability to selectively tune molecular transport as a function of the interaction between molecules and the inner surface of CNTs. Later it was shown by Segura's group [137], who reported the synthesizing of gold nanoparticles and carbon nanotubes (Au-CNT) hybrid structures inside the pores of AAO by the non-catalytic decomposition of acetylene. The CNT-AAO

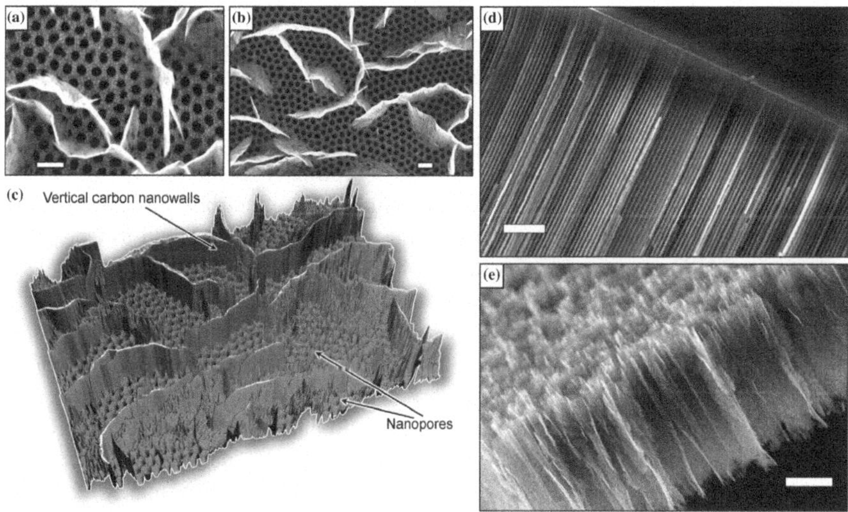

Fig. 5.9 Carbon nanowalls are grown on AAO templates at the methane flow rate of 10 sccm for 10 min. **a** and **b** High-resolution SEM images of the nanowall patterns on gold-covered and bare membranes with the pore sizes of 60 nm. **c** 3D reconstruction of the pattern shown in (**b**). **d** Side view of the AAO template, straight channels are clearly visible. **e** Side view of the carbon nanowalls with the height of about 2 lm, grown on the gold-covered membrane with the pore sizes of 60 nm. Scale bars are 200 nm (**a** and **b**) and 1000 nm (**d** and **e**). Adapted with permission from [133]

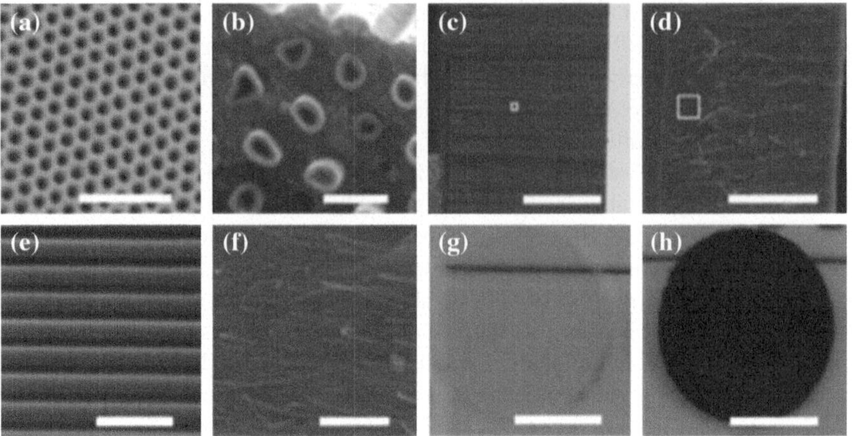

Fig. 5.10 Set of SEM images of as-produced NAAMs and CNTs–NAAMs fabricated by CVD synthesis from commercially available non-degradable plastic bags. **a** NAAM template *top view* (scale bar = 500 nm). **b** Detail of CNTs embedded in a NAAM template (scale bar = 100 nm). **c** NAAM template cross-section view (scale bar = 25 μm). **d** CNTs–NAAM cross-section view (scale bar = 25 μm). **e** and **f** Magnified views of *white squares* shown in (**c**) and (**d**), respectively (scale bars = 250 nm and 2 μm, respectively). **g** and **h** Digital photographs of a NAAM before and after CVD of CNTs, respectively (scale bar = 0.5 cm). Adapted with permission from [136]

composite membranes were impregnated with a HAuCl4/2-propanol solution by dip-coating or drop-casting. The AAO template was then removed with NaOH solution to produce an Au-CNT hybrid. This Au-CNT hybrid is supported on interdigitated microelectrodes (IME) chips to be applied as a sensor for the detection of hydrocarbon gas as acetylene.

Besides CNTs, the CVD method is also been utilized for fabrication of silicon nanowires (SiNWs). Significant improvement of the electrode capacitance per planar surface area and nanowires length unit are achieved compared to electrodes made from SiNWs grown from gold colloids [138]. While Lefeuvre et al. [139] developed highly organized SiNWs by hot wire assisted chemical vapor deposition (HWCVD) process. Tungsten hot wire is used to prevent the deposition of parasitic amorphous silicon (a-Si) that can clog the pores and thus prevent the SiNWs growth. The obtained SiNWs have a high potential for device realization, like PIN junctions, FETs or electrodes for Li-ion batteries. In the meantime, Zhao et al. [140] reported the fabrication of vertically aligned CrO_2 nanowire arrays via atmospheric-pressure CVD assisted by AAO templates. The CrO_2 nanowire arrays show remarkably enhanced coercivity compared with CrO_2 films or bulk. It was reported that the length of CrO2 nanowire was greatly influenced by the pore diameter of the AAO template used. This highly ordered nanowire arrays have important applications in ultrahigh-density perpendicular magnetic recording devices and the mass production of spintronic nanodevices. Polymeric nanotubes can also be synthesized within the AAO membranes based on initiated chemical vapor deposition (iCVD) technique. Ince et al. [141] synthesized polymeric nanotubes for biorecognition of immunoglobulin G (IgG). The imprinted polymeric nanotubes possess relatively good monodispersity, high binding capacity and significant specific recognition ability toward target molecules.

5.1.2.6 Atomic Layer Deposition (ALD)

Atomic layer deposition (ALD) is a thin film growth technique which employs a cyclic process of self-limiting chemical reactions between gaseous precursors and a substrate [142]. Suntola and Antson pioneered this technique last few decades by producing highly oriented ZnS thin film from H_2S and $ZnCl_2$ precursors [143, 144]. The ALD technique allows the deposition of various materials such as oxides, nitrides, sulfides and metals with coating with precise thickness and compositional control as well as high conformality [45, 145, 146]. This technique has been exploited for surface modification of AAO membrane [147].

The ALD of silica, titania and alumina on AAO membrane enables a controlled reduction of the AAO pore dimension, hence improve its catalytic, optical and transport properties [148–150]. For a review on specific surface chemistry employing ALD of silica on AAO that improved properties for desirable applications in molecular separation, tissue engineering, biosensing and drug delivery see [151]. Meanwhile, Comstock et al. successfully synthesized activated iridium oxide (IrOx) films by utilizing ALD to deposit a thin conformal Ir film within an AAO membrane [152]. The Ir film is then activated by potential cycling in H_2SO_4 to form an activated IrOx films. The ALD method provides a control of film porosity that results in an enhancement of cathodal charge storage capacities for the application in electrochromic devices, pH sensing, and neural stimulation.

Different ALD strategies have been applied to coat AAO nanopores using single metal oxide (Fe_2O_3, ZnO, TiO), mixtures of metal oxides ($SiO_2/Fe_2O_3/SiO_2$), nitrides (WNx) and composite materials (TiN–Al_2O_3–TiN) by using conventional gas or even liquid phase ALD. Bachmann et al. [153] have fabricated ordered magnetic Fe_2O_3 nanotube arrays by means of ALD using AAO membranes as templates for potential application in high-density data storage. Recently, Norek et al. [154] synthesized AAO/ZnO and AAO/ZnO/Ag composite by ALD. An enhancement of luminescence properties by a factor of ~ 2.5 for AAO/ZnO/Ag has been reported upon Ag deposition. Pitzschel et al. [155] used regulated AAO membrane prepared by a combination of HA and MA anodization techniques for fabrication of layered nanotubes composed of $SiO_2/Fe_2O_3/SiO_2$ using ferrocene and APTES as precursors for ALD. Later on, Banerjee et al. [156] prepared alternate layers of metal–insulator–metal (MIM) by depositing TiN–Al_2O_3–TiN multilayers inside AAO nanopores yielding nanotubular capacitors with equivalent planar capacitance up to 100 μF cm^{-2}. AAO as a template for fabrication of multiple-walled coaxial nanotubes of five nested layers using ALD was reported by Gu et al. [157]. This template-guided ALD allows remarkable control of nanotube thickness, diameter and spacing within the atomic resolution. Free-standing

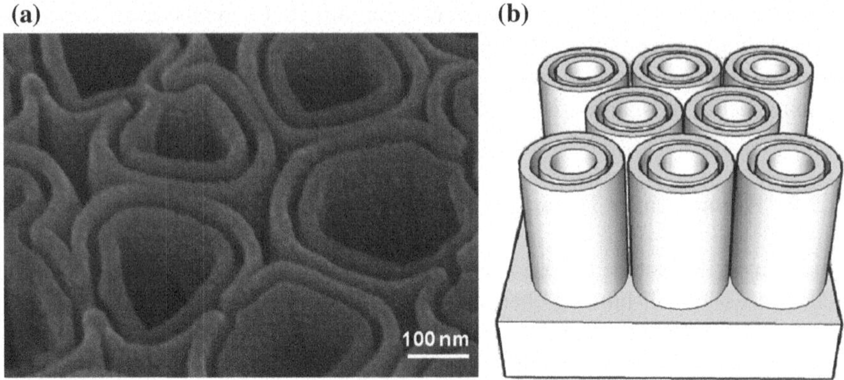

Fig. 5.11 **a** High-magnification tilted SEM *top view* of resultant coaxial HfO_2 nanotubes following release from the AAO template and removing the sacrificial spacer Al_2O_3 layer to expose both the inner and outer surfaces of the nested nanotubes. **b** Schematic model highlighting the design of arrays of free-standing coaxial nested nanotubes. Adapted with permission from [158]

double-walled HfO_2 tube-in-tube structure was successfully synthesized with 15 nm HfO_2 film and 25 nm sacrificial spacer. The dimension of the spacer gap between two active nanotubes can be engineered with the precise control afforded by ALD. This reported fabrication technique can be extended to produce nanostructures of different materials (metal oxides, semiconductors, and metals) which enable the custom design of broadband sensors and detectors. Recently, Pardon et al. [158] presented the functionalization of AAO membrane by ALD of Pt and Al_2O_3 for potential development as nanofluidic biosensors (Fig. 5.11). Conformal deposition of Al_2O_3 layer on Pt film forms a metal-insulator stack inside the AAO nanopore which allows the field effect control of the nanofluidic properties, thus opens novel possibilities for the potential development of nanofluidic biosensors.

5.2 Concluding Remarks

There are numerous ways in which AAO membrane can interact with active compounds. The strategy of manipulating AAO membranes using soft and hard technique has been described in this chapter. These approaches can be used to create a collection of AAO with different surface functionalities. With regards to soft technique, a different approach that gave impact on properties of the resulting materials, as well as applications enabled by these surface modifications has been highlighted. In addition, manipulating the AAO surface by hard technique through metal or metal oxide coating facilitates new synergetic effects which are greatly influenced by the processing conditions. This technique can be easily extended to produce more delicate inverse nanostructures and the diameter, length and degree of orientation of the resultant nanostructures can be manipulated by controlling the shape of the AAO template. Despite all this progress, the future application of functionalized AAO membranes will depend mainly on the unpredictable needs of the market, however this will be influenced by the developments in device performance and cost. In summary, this field still holds many challenges and opportunities for researchers to explore.

Acknowledgments The authors like to thank graduate students and collaborators whose has contributed greatly to this work. The support of our research by Malaysian's Ministry of Higher Education (MOHE) under Grant FRGS 600-RMI/FRGS 5/3 (10/2013) and Ministry of Science, Technology and Innovation under Grant 100-RMI/SF 16/6/2 (8/2014) are gratefully acknowledged.

References

1. Y.S. Lin, Microporous and dense inorganic membranes: current status and prospective. Sep. Purif. Technol. **25**(1–3), 39–55 (2001)
2. Y. Liu, J. Goebl, Y. Yin, Templated synthesis of nanostructured materials. Chem. Soc. Rev. **42**(7), 2610–2653 (2013)
3. S. Meoto, M.-O. Coppens, Anodic alumina-templated synthesis of mesostructured silica membranes—current status and challenges. J. Mater. Chem. A **2**(16), 5640–5654 (2014)
4. D.J. Odom, L.A. Baker, C.R. Martin, Solvent-extraction and langmuir-adsorption-based transport in chemically functionalized nanopore membranes. J. Phys. Chem. B **109**(44), 20887–20894 (2005)
5. A. Debrassi et al., Stability of (Bio)functionalized porous aluminum oxide. Langmuir **30**(5), 1311–1320 (2014)
6. C.-C. Wu et al., Electrochemical impedance spectroscopy analysis of A-beta (1–42) peptide using a nanostructured biochip. Electrochim. Acta **134**, 249–257 (2014)
7. T. Kumeria et al., Label-free reflectometric interference microchip biosensor based on nanoporous alumina for detection of circulating tumour cells. Biosens. Bioelectron. **35**(1), 167–173 (2012)
8. Y.-T. Tung et al., Nanostructured electrochemical biosensor for the detection of the weak binding between the dengue virus and the CLEC5A receptor. Nanomed. Nanotechnol. Biol. Med. **10**(6), 1335–1341 (2014)
9. S.-H. Yeom et al., Enhancement of the sensitivity of LSPR-based CRP immunosensors by Au nanoparticle antibody conjugation. Sens. Actuators B: Chem. **177**, 376–383 (2013)
10. A. Santos, T. Kumeria, D. Losic, Nanoporous anodic aluminum oxide for chemical sensing and biosensors. TrAC Trends Anal. Chem. **44**, 25–38 (2013)
11. Q.T.H. Le et al., Ultra-thin gates for the transport of phenol from supported liquid membranes to permanent surface modified membranes. J. Membr. Sci. **205**(1–2), 213–222 (2002)
12. V. Smuleac et al., Polythiol-functionalized alumina membranes for mercury capture. J. Membr. Sci. **251**(1–2), 169–178 (2005)
13. E.K. Schmitt et al., Electrically insulating pore-suspending membranes on highly ordered porous alumina obtained from vesicle spreading. Soft Matter **4**(2), 250–253 (2008)
14. A. Janshoff, C. Steinem, Transport across artificial membranes–an analytical perspective. Anal. Bioanal. Chem. **385**(3), 433–451 (2006)
15. C. Hennesthal, J. Drexler, C. Steinem, Membrane-suspended nanocompartments based on ordered pores in alumina. ChemPhysChem **3**(10), 885–889 (2002)
16. C. Horn, C. Steinem, Photocurrents generated by bacteriorhodopsin adsorbed on nano-black lipid membranes. Biophys. J. **89**(2), 1046–1054 (2005)

17. I. Mey, C. Steinem, A. Janshoff, Biomimetic functionalization of porous substrates: towards model systems for cellular membranes. J. Mater. Chem. **22**(37), 19348–19356 (2012)
18. L. Treccani et al., Functionalized ceramics for biomedical, biotechnological and environmental applications. Acta Biomater. **9**(7), 7115–7150 (2013)
19. A.Y. Ku et al., Evidence of ion transport through surface conduction in alkylsilane-functionalized nanoporous ceramic membranes. Langmuir **22**(20), 8277–8280 (2006)
20. L. Velleman et al., Structural and chemical modification of porous alumina membranes. Microporous Mesoporous Mater. **126**(1–2), 87–94 (2009)
21. Z.D. Hendren, J. Brant, M.R. Wiesner, Surface modification of nanostructured ceramic membranes for direct contact membrane distillation. J. Membr. Sci. **331**(1–2), 1–10 (2009)
22. K.C. Popat et al., Surface modification of nanoporous alumina surfaces with poly(ethylene glycol). Langmuir **20**(19), 8035–8041 (2004)
23. K.E. La Flamme et al., Biocompatibility of nanoporous alumina membranes for immunoisolation. Biomaterials **28**(16), 2638–2645 (2007)
24. E.D. Steinle et al., Ion channel mimetic micropore and nanotube membrane sensors. Anal. Chem. **74**(10), 2416–2422 (2002)
25. A.M.M. Jani et al., Nanoporous anodic aluminium oxide membranes with layered surface chemistry. Chem. Commun. **21**, 3062–3064 (2009)
26. A.M.M. Jani et al., Dressing in layers: layering surface functionalities in nanoporous aluminum oxide membranes. Angew. Chem. **122**(43), 8105–8109 (2010)
27. J. Yu, Y. Zhang, S. Liu, Enzymatic reactivity of glucose oxidase confined in nanochannels. Biosens. Bioelectron. **55**, 307–312 (2014)
28. C.-K. Joung et al., A nanoporous membrane-based impedimetric immunosensor for label-free detection of pathogenic bacteria in whole milk. Biosens. Bioelectron. **44**, 210–215 (2013)
29. Z. Yang et al., Piezoelectric urea biosensor based on immobilization of urease onto nanoporous alumina membranes. Biosens. Bioelectron. **22**(12), 3283–3287 (2007)
30. I. Vlassiouk et al., "Direct" detection and separation of DNA using nanoporous alumina filters. Langmuir **20**(23), 9913–9915 (2004)
31. P. Takmakov, I. Vlassiouk, S. Smirnov, Application of anodized aluminum in fluorescence detection of biological species. Anal. Bioanal. Chem. **385**(5), 954–958 (2006)
32. M.M.J. Abdul et al., Pore spanning lipid bilayers on silanised nanoporous alumina membranes, in *SPIE* (2008)
33. B. Demé, D. Marchal, Polymer-cushioned lipid bilayers in porous alumina. Eur. Biophys. J. **34**(2), 170–179 (2005)
34. E.E. Leary Swan, K.C. Popat, T.A. Desai, Peptide-immobilized nanoporous alumina membranes for enhanced osteoblast adhesion. Biomaterials **26**(14), 1969–1976 (2005)
35. J.R. Stephens, J.S. Beveridge, M.E. Williams, Diffusive flux of nanoparticles through chemically modified alumina membranes. Analyst **136**(18), 3797–3802 (2011)
36. W. Shi et al., Functionalized anodic aluminum oxide (AAO) membranes for affinity protein separation. J. Membr. Sci. **325**(2), 801–808 (2008)
37. I. Vlassiouk, P. Takmakov, S. Smirnov, Sensing DNA hybridization via ionic conductance through a nanoporous electrode. Langmuir **21**(11), 4776–4778 (2005)
38. W.W. Ye et al., A nanoporous membrane based impedance sensing platform for DNA sensing with gold nanoparticle amplification. Sens. Actuators B: Chem. **193**, 877–882 (2014)
39. X. Wang, S. Smirnov, Label-free DNA sensor based on surface charge modulated ionic conductance. ACS Nano **3**(4), 1004–1010 (2009)
40. S. Tanvir et al., Covalent immobilization of recombinant human cytochrome CYP2E1 and glucose-6-phosphate dehydrogenase in alumina membrane for drug screening applications. J. Membr. Sci. **329**(1–2), 85–90 (2009)
41. K.H.A. Lau, H. Duran, W. Knoll, In situ characterization of N-carboxy anhydride polymerization in nanoporous anodic alumina. J. Phys. Chem. B **113**(10), 3179–3189 (2009)
42. P.-F. Li et al., Thermo-responsive gating membranes with controllable length and density of poly(N-isopropylacrylamide) chains grafted by ATRP method. J. Membr. Sci. **337**(1–2), 310–317 (2009)

43. M.L. Bruening et al., Creation of Functional membranes using polyelectrolyte multilayers and polymer brushes. Langmuir **24**(15), 7663–7673 (2008)
44. T. Sehayek et al., Template Synthesis of nanotubes by room-temperature coalescence of metal nanoparticles. Chem. Mater. **17**(14), 3743–3748 (2005)
45. A.M.M. Jani, D.L.N.H. Voelcker, Nanoporous anodic aluminium oxide: Advances in surface engineering and emerging applications. Prog. Mater Sci. **58**, 636–704 (2013)
46. J.B. Largueze, K.E. Kirat, S. Morandat, Preparation of an electrochemical biosensor based on lipid membranes in nanoporous alumina. Colloids Surf., B **79**(1), 33–40 (2010)
47. D. Qi et al., Optical emission of conjugated polymers adsorbed to nanoporous alumina. Nano Lett. **3**(9), 1265–1268 (2003)
48. P. Jain et al., High-capacity purification of his-tagged proteins by affinity membranes containing functionalized polymer brushes. Biomacromolecules **8**(10), 3102–3107 (2007)
49. R. Barbey et al., Polymer brushes via surface-initiated controlled radical polymerization: synthesis, characterization, properties, and applications. Chem. Rev. **109**(11), 5437–5527 (2009)
50. M. Nagale, B.Y. Kim, M.L. Bruening, Ultrathin, hyperbranched poly(acrylic acid) membranes on porous alumina supports. J. Am. Chem. Soc. **122**(47), 11670–11678 (2000)
51. L. Sun et al., High-capacity, protein-binding membranes based on polymer brushes grown in porous substrates. Chem. Mater. **18**(17), 4033–4039 (2006)
52. Q. Fu et al., Reversible Control of free energy and topography of nanostructured surfaces. J. Am. Chem. Soc. **126**(29), 8904–8905 (2004)
53. H.-J. Wang et al., Template synthesized molecularly imprinted polymer nanotube membranes for chemical separations. J. Am. Chem. Soc. **128**(50), 15954–15955 (2006)
54. S. Ma et al., A general approach for construction of asymmetric modification membranes for gated flow nanochannels. J. Mater. Chem. A **2**(23), 8804–8814 (2014)
55. G. Decher, Fuzzy nanoassemblies: toward layered polymeric multicomposites. Science **277** (5330), 1232–1237 (1997)
56. R.R. Costa, J.F. Mano, Polyelectrolyte multilayered assemblies in biomedical technologies. Chem. Soc. Rev. **43**(10), 3453–3479 (2014)
57. A.M. Balachandra, J. Dai, M.L. Bruening, Enhancing the anion-transport selectivity of multilayer polyelectrolyte membranes by templating with Cu2+. Macromolecules **35**(8), 3171–3178 (2002)
58. S.U. Hong, R. Malaisamy, M.L. Bruening, Separation of fluoride from other monovalent anions using multilayer polyelectrolyte nanofiltration membranes. Langmuir **23**(4), 1716–1722 (2007)
59. S.U. Hong, L. Ouyang, M.L. Bruening, Recovery of phosphate using multilayer polyelectrolyte nanofiltration membranes. J. Membr. Sci. **327**(1–2), 2–5 (2009)
60. L. Ouyang, R. Malaisamy, M.L. Bruening, Multilayer polyelectrolyte films as nanofiltration membranes for separating monovalent and divalent cations. J. Membr. Sci. **310**(1–2), 76–84 (2008)
61. S.U. Hong, M.L. Bruening, Separation of amino acid mixtures using multilayer polyelectrolyte nanofiltration membranes. J. Membr. Sci. **280**(1–2), 1–5 (2006)
62. J. Dai, G.L. Baker, M.L. Bruening, Use of porous membranes modified with polyelectrolyte multilayers as substrates for protein arrays with low nonspecific adsorption. Anal. Chem. **78** (1), 135–140 (2005)
63. D.M. Dotzauer et al., Catalytic membranes prepared using layer-by-layer adsorption of polyelectrolyte/metal nanoparticle films in porous supports. Nano Lett. **6**(10), 2268–2272 (2006)
64. D.M. Dotzauer et al., Nanoparticle-containing membranes for the catalytic reduction of nitroaromatic compounds. Langmuir **25**(3), 1865–1871 (2009)
65. N. Joseph et al., Layer-by-layer preparation of polyelectrolyte multilayer membranes for separation. Polym. Chem. **5**(6), 1817–1831 (2014)

66. Y. Lei, W. Cai, G. Wilde, Highly ordered nanostructures with tunable size, shape and properties: a new way to surface nano-patterning using ultra-thin alumina masks. Prog. Mater. Sci. **52**, 465–639 (2007)
67. G. Wang, C. Shi., N. Zhao, X. Du, Synthesis and characterization of Ag nanoparticles assembled in ordered array pores of porous anodic alumina by chemical deposition. Mater. Lett. **61**, 3795–3797 (2007)
68. H.P. Xiang, L. Chang, S. Chao, C.H. Ming, Carbon nanotubes prepared by anodic aluminum oxide template method. Chin. Sci. Bull. **57**, 187–204 (2012)
69. M.R. Kim, D.K. Lee, D.-J. Jang, Template-based electrochemically controlled growth of segmented multimetal nanorods. J. Nanomater. **2010**, 1–8 (2010)
70. S.J. Hurst, E.K. Payne, L. Qin, C.A. Mirkin, Multisegmented one-dimensional nanorods prepared by hard-template synthetic methods. Angew. Chem. Int. Ed. **45**(17), 2672–2692 (2006)
71. C.R. Martin, Nanomaterials: a membrane-based synthetic approach. Science **5193**(6), 1961–1966 (1994)
72. M.S. Sachiko Ono, Hidetaka Asoh, Self-ordering of anodic porous alumina formed in organic acid electrolytes. Electrochim. Acta **51**, 827–833 (2005)
73. C.-H. Peng, T.-Y. Wu, C.-C. Hwang, A preliminary study on the synthesis and characterization of multilayered Ag/Co magnetic nanowires fabricated via the electrodeposition method. Sci. World J. **2013**, 1–6 (2013)
74. C.-K. Chen, D.-S. Chan, C.-C. Lee, S.-H. Chen, Fabrication of orderly copper particle arrays on a multi-electrolyte-step anodic aluminum oxide template. J. Nanomater. **2013**, 1–8 (2013)
75. A. Santos, L. Vojkuvka, J. Pallare´s, J. Ferre´-Borrull, L.F. Marsal, Cobalt and nickel nanopillars on aluminium substrates by direct current electrodeposition process. Nanoscale Res. Lett. **4**, 1021–2028 (2009)
76. L.F. Dumee et al., The fabrication and surface functionalization of porous metal frameworks —a review. J. Mater. Chem. A **1**(48), 15185–15206 (2013)
77. M. Platt, R.A.W. Dryfe, P.L. E.P.L. Roberts, Electrodeposition of palladium nanoparticles at the liquid/liquid interface using porous alumina templates. Electrochim. Acta **48**, 3037–3046 (2003)
78. H.J. Lee, T. Yasukawa, M. Suzuki, S.H. Lee, T. Yao, Y. Taki, A. Tanaka, M. Kameyama, H. Shiku, T. Matsue, Simple and rapid preparation of vertically aligned gold nanoparticle arrays and fused nanorods in pores of alumina membrane based on positive dielectrophoresis. Sens. Actuators B, **136**, 320–325 (2009)
79. T. Kondo, K. Nishio, H. Masuda, Surface enhanced Raman scattering in multilayered Au nanoparticles in anodic porous alumina. Appl. Phys. Exp. **9**, 0320001 (2009)
80. T.M. Whitney, J.S. Jiang, P.C. Searson, C.L. Chien, Fabrication and magnetic properties of arrays of metallic nanowires. Science **261**, 1316–1319 (1993)
81. K. Nielscha, R.B. Wehrspohn, J. Barthela, J. Kirschnera, S.F. Fischerb, H. Kronmuller, T. Schweinbock, D. Weissc, U. Gosele, High density hexagonal nickel nanowire array. J. Magn. Magn. Mater **429**, 234–240 (2002)
82. K. Nielsch, F. Müller, A.P. Li, U. Gösele, Uniform nickel deposition into ordered alumina pores by pulsed electrodeposition. Adv. Mater. **12**(8), 582–586 (2000)
83. J. Choi, G. Sauer, K. Nielsch, R.B. Wehrspohn, U. Gosele, Silver infiltration into monodomain porous alumina with adjustable pore diameter and with high aspect ratio. Chem. Mater. **15**, 776–779 (2003)
84. L. Liu et al., Fabrication and characterization of a flow-through nanoporous gold nanowire/AAO composite membrane. Nanotechnology **33**, 335604 (2008)
85. K. Kim, M. Kim, S.M. Cho, Pulsed electrodeposition of palladium nanowire arrays using AAO template. Mater. Chem. Phys. **96**, 278–282
86. D. Pecko, K.Z. Rožman, N. Kostevšek, M.S. Arshad, B. Markoli, Z. Samardzija, S. Kobe, Electrodeposited hard-magnetic Fe50Pd50 nanowires from an ammonium-citrate-based bath. J. Alloy. Compd. **605**, 71–79 (2014)

87. J.K. Wang, J.M. Char, P.J. Lien, Optimization study on supercritical electrodeposition of nickel nanowire arrays using AAO template. ISRN Chem. Eng. **2012**, 1–9 (2012)
88. D. Routkevitch, T. Bigioni., M. Moskovits, J.M. Xu, Electrochemical fabrication of CdS nanowire arrays in porous anodic aluminium oxide templates. J. Phys. Chem. **100**, 14037–14047 (1996)
89. D.S. Xu, Y.J. Xu, D.P. Chen, X. Shi, G.L. Guo, L.L. Gui, Y.Q. Tang, Preparation of CdS single-crystal nanowires by electrochemically induced deposition. Adv. Mater. **12**, 520–522 (2000)
90. D.S. Xu, Y.J. Xu, D.P. Chen, G.L. Guo, L.L. Gui, Y.Q. Tang, Preparation and characterization of CdS nanowire arrays by dc electrodeposit in porous anodic aluminium ocide templates. Chem. Phys. Lett. **325**, 340–344 (2000)
91. D.S. Xu, D.P. Chen, Y.J. Xu, X. Shi, G.L. Guo, L.L. Gui, Y.Q. Tang, Preparation of II-VI group semiconductor nanowire arrays by dc electrochemical deposition in porous aluminium oxide templates. Pure Appl. Chem. **72**, 127–135 (2000)
92. J.D. Klein, R.D. Herrick, D. Palmer, M.J. Sailor, C.J. Brumlik, C.R. Martin, Electrochemical fabrication of cadmium chalcogenide microdiode arrays. Chem. Mater. Res. **5**, 902 (1993)
93. H. Chen, Y.-M. Yeh, Y.T. Chen, Y.L. Jiang, Influence of growth conditions on hair-like CuS nanowires fabricated by electro-deposition and sulfurization. Ceram. Int. **40**, 9757–9761 (2014)
94. J. Burdick, E. Alonas, H.-C. Huang, K. Rege, J. Wang, High-throughput templated multisegment synthesis of gold nanowires and nanorods. Nanotechnology **20**, 065306 (2009)
95. N. Van Hoang, S. Kumar, G.H. Kim, Growth of segmented gold nanorods with nanogaps by the electrochemical wet etching technique for single-electron transistor applications. Nanotechnology **20**(12), 125607 (2009)
96. H. Wang, Y. Song, W. Liu, S. Yao, W. Zhang, Template synthesis and characterization of TiO2 nanotube arrays by the electrodeposition method. Mater. Lett. **93**, 319–321 (2013)
97. F.E. Atalay, H. Kaya, V. Yagmur, S. Tari, S. Atalay, D. Avsar, The effect of back electrode on the formation of electrodeposited CoNiFe magnetic nanotubes and nanowires. Appl. Surf. Sci. **256**, 2414–2418 (2010)
98. F. Grote, L. Wen, Y. Le, Nano-engineering of three-dimensional core/shell nanotube arrays for high performance supercapacitors. J. Power Resour. **256**, 37–42 (2014)
99. L. Binder, I. Gurrappa, Electrodeposition of nanostructured coatings and their characterization—a review. Sci. Technol. Adv. Mater. **9**, 043001 (2008)
100. L.P. Bicelli, B. Bozzoni, C. Mele, L. D'Urzo, A review of nanostructural aspects of metal electrodeposition. Int. J. Electrochem. Sci. **3**, 356–408 (2008)
101. W. Wang, N. Li, X. Li, W. Geng, S. Qiu, Synthesis of metallic nanotube arrays in porous anodic aluminum oxide template through electroless deposition. Mater. Res. Bull. **41**, 1417–1423 (2006)
102. A. Azizi, M. Mohammadi., S.K. Sadrnezhaad, End-closed NiCoFe-B nanotube arrays by electroless method. Mater. Lett. **65**, 289–292 (2011)
103. M. Cheng, Au nanoparticle arrays with tunable particle gaps by template-assisted electroless deposition for high performance surface-enhanced Raman scattering. Nanotechnology **21**(1), 015604 (2010)
104. L. Velleman et al., Raman spectroscopy probing of self-assembled monolayers inside the pores of gold nanotube membranes. Phys. Chem. Chem. Phys. **13**(43), 19587–19593 (2011)
105. N. Li, X. Li, X. Yin, W. Wang, S. Qiu, Electroless deposition of open-end Cu nanotube arrays. Solid State Commun. **132**, 841–844 (2004)
106. S.H. Zhang, Synthesis of silver nanotubes by electroless deposition in porous anodic aluminium oxide templates. Chem. Phys. Lett. **9**, 1106–1107 (2004)
107. Y. Piao, H. Lim, J.Y. Chang, W.-Y. Lee, H. Kim, Nanostructured materials prepared by use of ordered porous alumina membranes. Electrochim. Acta **50**, 2997–3013 (2005)
108. C. Kang, H. Lu, S. Yuan, D. Hong, K. Yan, B. Liang, Superhydrophilicity/superhydrophobicity of nickel micro-arrays fabricated by electroless deposition on an etched porous aluminum template. Chem. Eng. J. **203**, 1–8 (2012)

109. X. Li, M. Wang, Y. Ye, K. Wu, Boron-doping Ni@Au nanotubes: facile synthesis, magnetic property, and in vitro cytotoxicity on Molt-4 cells. Mater. Lett. **108**, 222–224 (2013)
110. I.R.A. Christie, H.O. Ali, A review of electroless gold deposition processes. Circuit World **11** (4), 10–16 (1985)
111. S. Stojan, P.L.C. Djokić, Electroless deposition: theory and applications. Mod. Aspects Electrochem. **48**, 251–289 (2010)
112. P.S. Cheow, Transport and separation of proteins across platinum-coated nanoporous alumina membranes. Electrochim. Acta **53**(14), 4669–4673 (2008)
113. B.T.T. Nguyen, E.Z.C. Ting, C.S. Toh, Development of a biomimetic nanoporous membrane for the selective transport of charged proteins. Bioinspirat Biomim. **3**(3), 035008 (2008)
114. T. Qiu, W. Zang, X. Lang, Y. Zhou, T. Cui, P.K. Chu, Controlled assembly of highly Raman-enhancing silver nanocap arrays templated by porous anodic alumina membranes. Small **20**, 2333–2337 (2009)
115. Y. Hu, J. Ge, Y. Yin, PDMS rubber as a single-source precursor for templated growth of silica nanotubes. Chem. Commun. **8**, 914–916 (2009)
116. M.D. Dickey, E.A. Weiss., E.J. Smythe, R.C. Chiechi, F. Capasso, G.M. Whitesides, Fabrication of arrays of metal and metal oxide nanotubes by shadow evaporation. ACS Nano. **2**(4), 800–808 (2008)
117. A. Pereira, F. Laplante, M. Chaker, D. Guay, Functionally modified macroporous membrane prepared by using pulsed laser deposition. Adv. Funct. Mater. **17**(3), 443–450 (2007)
118. S. Wang, G.J. Yu, J.L. Gong, D.Z. Zhu, H.H. Xia, Large-area uniform nanodot arrays embedded in porous anodic alumina. Nanotechnology **18**(1), 015303 (2007)
119. M. Lee, S.C. Hong, D. Kim, Formation of bamboo-like conducting carbon nanotubes decorated with Au nanoparticles by the thermal decomposition of sucrose in an AAO template. Carbon **50**, 2465–2471 (2012)
120. K.S. Siow, L. Britcher, S. Kumar, H.J. Griesser, Plasma methods for the generation of chemically reactive surfaces for biomolecule immobilization and cell colonization—a review. Plasma Process Polym. **3**(6–7), 392–418 (2006)
121. K. Vasilev, A. Michelmore, H.J. Griesserb, R.D. Short, Substrate influence on the initial growth phase of plasma-deposited polymer films. Chem. Commun. **2009**, 3600–3602 (2009)
122. D.A. Brevnov, M.J. Brooks, G.P. López, P.B. Atanassov, Fabrication of anisotropic super hydrophobic/hydrophilic nanoporous membranes by plasma polymerization of C4F8 on anodic aluminum oxide. J. Electrochem. Soc. **151**(8), 484–489 (2004)
123. D. Losic, M.A. Cole, B. Dollmann, K. Vasilev, H.J. Griesser, Surface modification of nanoporous alumina membranes by plasma polymerization. Nanotechnology **19**(24), 245704 (2008)
124. S. Simovic, D. Losic, K. Vasilev, Controlled drug release from porous materials by plasma polymer deposition. Chem. Commun. **46**(8), 1317–1319 (2010)
125. K. Choy, Chemical vapour deposition of coatings. Prog. Mater Sci. **48**(2), 57–170 (2003)
126. J.-H. Park, T. Sudarshan, *Chemical Vapor Deposition,* 2nd ed., vol. 2. (ASM International, Ohio, 2001)
127. A. Popp, J. Engstler, J. Schneider, Porous carbon nanotube-reinforced metals and ceramics via a double templating approach. Carbon **47**(14), 3208–3214 (2009)
128. K. Rana, G. Kucukayan-Dogu, E. Bengu, Growth of vertically aligned carbon nanotubes over self-ordered nano-porous alumina films and their surface properties. Appl. Surf. Sci. **258** (18), 7112–7117 (2012)
129. S. Park et al., Carbon nanosyringe array as a platform for intracellular delivery. Nano Lett. **9** (4), 1325–1329 (2009)
130. S. Sigurdson et al., Effect of anodic alumina pore diameter variation on template-initiated synthesis of carbon nanotube catalyst supports. J. Mol. Catal. A: Chem. **306**(1), 23–32 (2009)
131. J.L. Perry, C.R. Martin, J.D. Stewart, Drug delivery strategies by using template synthesized nanotubes. Chem. Eur. J. **17**(23), 6296–6302 (2011)

132. C.-S. Li et al., Application of highly ordered carbon nanotubes templates to field-emission organic light-emitting diodes. J. Cry. Growth **311**(3), 615–618 (2009)
133. J. Fang et al., Multipurpose nanoporous alumina-carbon nanowall bi-dimensional nano-hybrid platform via catalyzed and catalyst-free plasma CVD. Carbon **78**, 627–632 (2014)
134. M. Sarno et al., Electrical conductivity of carbon nanotubes grown inside a mesoporous anodic aluminium oxide membrane. Carbon **55**, 10–22 (2013)
135. D. Mattia et al., Effect of graphitization on the wettability and electrical conductivity of CVD-carbon nanotubes and films. J. Phys. Chem. B **110**(20), 9850–9855 (2006)
136. T. Altalhi et al., Synthesis of well-organised carbon nanotube membranes from non-degradable plastic bags with tuneable molecular transport: towards nanotechnological recycling. Carbon **63**, 423–433 (2013)
137. A. Tello et al., The synthesis of hybrid nanostructures of gold nanoparticles and carbon nanotubes and their transformation to solid carbon nanorods. Carbon **46**(6), 884–889 (2008)
138. F. Thissandier et al., Ultra-dense and highly doped SiNWs for micro-supercapacitors electrodes. Electrochim. Acta **117**, 159–163 (2014)
139. E. Lefeuvre et al., Optimization of organized silicon nanowires growth inside porous anodic alumina template using hot wire chemical vapor deposition process. Thin Solid Films. **519** (14), 4603–4608 (2011)
140. Q. Zhao et al., Synthesis of dense, single-crystalline CrO_2 nanowire arrays using AAO template-assisted chemical vapor deposition. Nanotechnology **22**(12), 125603 (2011)
141. G.O. Ince et al., One-dimensional surface-imprinted polymeric nanotubes for specific biorecognition by initiated chemical vapor deposition (iCVD). ACS Appl. Mater. Inter. **5** (14), 6447–6452 (2013)
142. S.M. George, Atomic layer deposition: an overview. Chem. Rev. **110**(1), 111–131 (2009)
143. J. Antson, T. Suntola, *Method for producing compound thin films*. Asm America Inc., 1997, p. 430
144. S. Skoog, J. Elam, R. Narayan, Atomic layer deposition: medical and biological applications. Inter. Mater. Rev. **58**(2), 113–129 (2013)
145. M. Knez, K. Nielsch, L. Niinista, Synthesis and surface engineering of complex nanostructures by atomic layer deposition. Adv. Mater. **19**(21), 3425–3438 (2007)
146. D.J. Comstock et al., Tuning the composition and nanostructure of Pt/Ir films via anodized aluminum oxide templated atomic layer deposition. Adv. Funct. Mater. **20**(18), 3099–3105 (2010)
147. K. Grigoras, V.-M. Airaksinen, S. Franssila, Coating of nanoporous membranes: atomic layer deposition versus sputtering. J. Nanosci. Nanotechnol. **9**(6), 3763–3770 (2009)
148. A. Ott et al., Atomic layer controlled deposition of Al_2O_3 films using binary reaction sequence chemistry. Appl. Surf. Sci. **107**, 128–136 (1996)
149. G. Xiong et al., Effect of atomic layer deposition coatings on the surface structure of anodic aluminum oxide membranes. J. Phys. Chem. B **109**(29), 14059–14063 (2005)
150. V. Romero et al., Changes in morphology and ionic transport induced by ALD SiO_2 coating of nanoporous alumina membranes. ACS Appl. Mater. Inter. **5**(9), 3556–3564 (2013)
151. L. Velleman et al., Structural and chemical modification of porous alumina membranes. Micropor. Mesopor. Mat. **126**(1), 87–94 (2009)
152. D.J. Comstock et al., Synthesis of nanoporous activated iridium oxide films by anodized aluminum oxide templated atomic layer deposition. Electrochem. Comm. **12**(11), 1543–1546 (2010)
153. J. Bachmann et al., Ordered iron oxide nanotube arrays of controlled geometry and tunable magnetism by atomic layer deposition. J. Am. Chem. Soc. **129**(31), 9554–9555 (2007)
154. M. Norek et al., Plasmonic enhancement of blue emission from ZnO nanorods grown on the anodic aluminum oxide (AAO) template. Appl. Phys. A **111**(1), 265–271 (2013)
155. K. Pitzschel et al., Controlled introduction of diameter modulations in arrayed magnetic iron oxide nanotubes. ACS Nano **3**(11), 3463–3468 (2009)

156. P. Banerjee et al., Nanotubular metal-insulator-metal capacitor arrays for energy storage. Nat. Nano. **4**(5), 292–296 (2009)
157. D. Gu et al., Synthesis of nested coaxial multiple-walled nanotubes by atomic layer deposition. ACS Nano **4**(2), 753–758 (2010)
158. G. Pardon et al., Pt-Al_2O_3 dual layer atomic layer deposition coating in high aspect ratio nanopores. Nanotechnology **24**, 1–11 (2013)

Chapter 6
Optical Properties of Nanoporous Anodic Alumina and Derived Applications

Josep Ferré-Borrull, Elisabet Xifré-Pérez, Josep Pallarès and Lluis F. Marsal

Abstract Inexpensive formation of periodically ordered structures with periodicities and feature sizes lower than 100 nm has triggered a vast amount or research in recent years. Of particular interest in nanotechnology is the nanoporous alumina, which can be produced with a self-organized arrangement of pores in the adequate anodization conditions. Most of the interest of this material is based in its outstanding physical and chemical and properties, and more specifically in its optical properties. The interaction of light with the nanostructured porous alumina gives rise to a wealth of optical properties that have their interest both in research level and also in application. In this work we aim at giving an extensive review of the published research on the optical properties of nanoporous alumina. The review will account for the different studied optical properties of this material such as the existence of a photonic stop band originated from its quasi-random nanostructure, the interferometric and light guiding properties that can be applied to biosensing, the structure nanoengineering to achieve more complex photonic behaviour such as distributed-Bragg reflectors or rugate filters, and the photoluminescence properties. In a second part, a summary of the different applications proposed on the basis of these properties, such as in biotechnology or energy will be given.

6.1 Introduction

Among the different features and properties of porous anodic alumina (p-AAO), its optical properties are of special interest and have been the subject of very intensive research. The interaction of light with p-AAO can be analyzed from different points of view: the material interaction, the porous structure interaction or the relation of geometrical features of the nanostructure (periodicity or quasi-periodicity, distance

J. Ferré-Borrull · E. Xifré-Pérez · J. Pallarès · L.F. Marsal (✉)
Departament d'Enginyeria Electrònica, Universitat Rovira i Virgili,
Elèctrica i Automàtica. Av. Paisos Catalans, 26, 43007 Tarragona, Spain
e-mail: lluis.marsal@urv.cat

D. Losic and A. Santos (eds.), *Nanoporous Alumina*,
Springer Series in Materials Science 219, DOI 10.1007/978-3-319-20334-8_6

between pores) with the light wavelength. In this chapter we offer a review of these properties in the view of these different types of interaction. We also aim at providing examples of the applications of p-AAO that are based on these optical properties. This chapter is organized therefore in two main sections, the first one devoted to the optical properties, classifying them in the intrinsic material properties (such as refractive index and photoluminescence), the properties as a porous material (focusing on the effective refractive index) and the structural-photonic properties (such as photonic stop bands in 1-D or 2-D). The subsequent section gives an overview of the different optical properties-based applications of p-AAO organized in different aspects such as sensing by reflection interference spectroscopy or optical waveguide spectroscopy, improving photovoltaic conversion efficiency or light extraction efficiency in light emitting diodes (LED) by nanopatterning the devices surface or creating unique barcode labeling nanostructures.

6.2 Interaction of Light with Porous Anodic Aluminum Oxide

6.2.1 Anodic Aluminum Oxide: The Host Material

The aluminum oxide is the solid matrix of the p-AAO, thus, the optical properties of this matrix influence directly any other derived optical property, both because of the porous nature of the material and from its geometrical characteristics. Nevertheless, few investigations have been carried out to study these intrinsic optical properties, although it is commonly accepted that the obtained aluminum oxide has optical properties much different than those of the pure bulk crystal material, since natural crystalline aluminum oxide is transparent and non-luminescent [1–3]. Instead, p-AAO is clearly luminescent and absorbing in the ultra-violet (UV) and blue region of the visible spectrum. In fact, the main studied feature of the anodic aluminum oxide are the photoluminescence and its associated absorption [4–8]. More specific studies on these host material optical properties are scarce. For instance, in [9] transmittance and reflectance measurements in the infra-red of p-AAO films produced with phosphoric acid are used to obtain the optical properties, although in such range the absorption is smaller. Yakovlev et al. [10] also study the composition and optical constants of the porous alumina from infrared reflectance measurements.

Instead, a vast amount of studies is devoted to the origin of the photoluminescence (PL) in p-AAO. There is a general consensus in that this luminescence is produced by the existence of singly ionized oxygen vacancies (F^+ centers) [11–17], and in some cases doubly ionized oxygen vacancies (F^{++} centers) [16], produced in the course of the formation of aluminum oxide barrier material. Furthermore, the most luminescent p-AAO is found among those produced with oxalic acid. This happens because oxalic ions diffused into the material during anodization show strong emission [16, 18–20]. All these studies agree in the fact that PL bands appear at 390 nm if they are originated by F^+ centers [21], while the band for oxalic impurities is in the range between 350 and 400 nm. The different acid electrolytes used in the fabrication of the p-AAO influence the PL spectra. Oxalic acid electrolyte results in bright and wide fluorescence peaks [16, 18], while for the rest of electrolytes fluorescence is weaker and corresponds to the F^+ centers [5, 11, 12].

Luminescence in p-AAO can be tuned up to some extent in different ways. One possibility is to change the kind of acid in the electrolyte. For instance, by using chromic acid and varying the anodization process between 20 and 50 V, the PL maximum can be tuned between 410 and 450 nm [22] due to the different amount of incorporated chromate ions and of produced oxygen vacancies with the anodization voltage and also because the interaction between the two kinds of PL sources. However, as it is well-known [23], self-arrangement of the pores in an hexagonal lattice only happens in a relatively narrow range of anodization voltages, thus to achieve the mentioned PL modulation may result in p-AAO with a limited pore ordering. This effect, although in a smaller extent has also been observed for oxalic-produced p-AAO [14].

A detailed study of the influence of the use of different acid electrolytes and applying different post-processing steps to p-AAO films is given by Santos et al. [24]. In this work, PL spectra of samples obtained by the hard anodization procedure under oxalic and malonic acids are reported and their dependence with respect to pore widening treatments after sample production are studied. Figure 6.1 summarizes the obtained results. Figure 6.1a shows the PL spectrum peak intensity as a function of the voltage applied during anodization, while Fig. 6.1b shows the wavelength of such peak. The behavior of the peak intensity is slightly increasing for the samples produced with oxalic acid, while for malonic the intensity clearly decreases with the anodization voltage. This difference in behavior is related to the amount of incorporated ionic species from the electrolyte to the aluminum oxide and to their luminescence efficiency. It is also remarkable that the peak wavelength is independent of the applied anodization voltage, what indicates that the cause of the observed PL is closely related to the impurities. In Fig. 6.1c, d, the same magnitudes, PL spectrum peak intensity and central wavelength are depicted as a function of the diameter of the pore (or equivalently as a function of the pore widening time applied after the anodization). The peak intensity increases with increasing pore widening time, what indicates that the outer oxide layers are absorbing part of the PL emitted light. The sudden decrease of the PL for the case of the malonic acid indicates that the oxide pore walls are almost consumed by the etching procedure. These results lead to the conclusion that p-AAO pore walls are composed of layers with changing composition as a function of the distance from the pore axis. In this work, it is concluded that there exist three layers in the oxide, as depicted in Fig. 6.2. in agreement with other works such as [25, 26].

Photoluminescence is also influenced by other process parameters, in particular the post-fabrication annealing in order to obtain a more crystalline aluminum oxide [13]. In this work, by jointly analyzing X-ray spectra, micro-Raman, FTIR transmission and PL for samples annealed at temperatures up to 1100 °C it is shown that

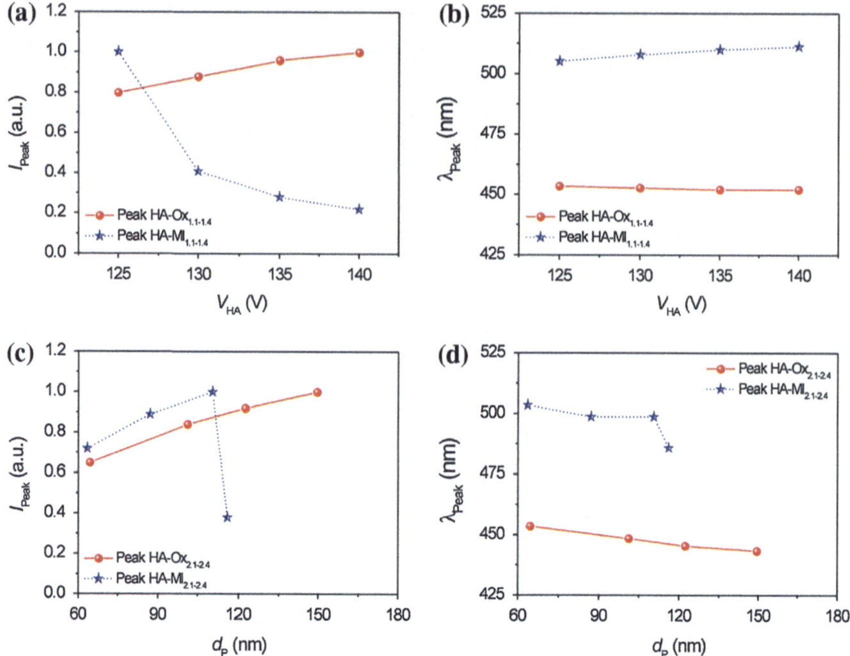

Fig. 6.1 Dependence of photoluminescent properties of p-AAO obtained with oxalic (Ox) and malonic (Ml) electrolytes under different anodization conditions. **a** Peak PL spectrum intensity as a function of the anodization voltage, **b** peak wavelength as a function of the anodization voltage, **c** peak PL spectrum intensity as a function of pore diameter (or equivalently pore widening time), **d** peak wavelength as a function of the pore diameter. Adapted with permission from [24]

the increase in crystallinity quenches significantly the PL and red-shifts its maximum. Furthermore, combining p-AAO with other luminescent materials can contribute to enhance the emission efficiency, as reported by Shi et al. [15], where the interaction of the F^+ centers of the AAO and of the ZnO nanoparticles grown within the p-AAO porous matrix contribute to this PL enhancement.

6.2.2 *Porous Anodic Aluminum Oxide as an Effective Medium*

Most of the works related with the optical properties of porous anodic aluminum oxide consider the material as an effective medium to model its interaction with light propagating in the porous matrix [27–32]. The nanopore size and interpore distance are usually much smaller than the wavelength of visible light, thus, the different effective-medium approximations (EMA) can be considered (Bruggeman, Lorentz-Lorenz [33–37]). Applications that are based in this effective medium

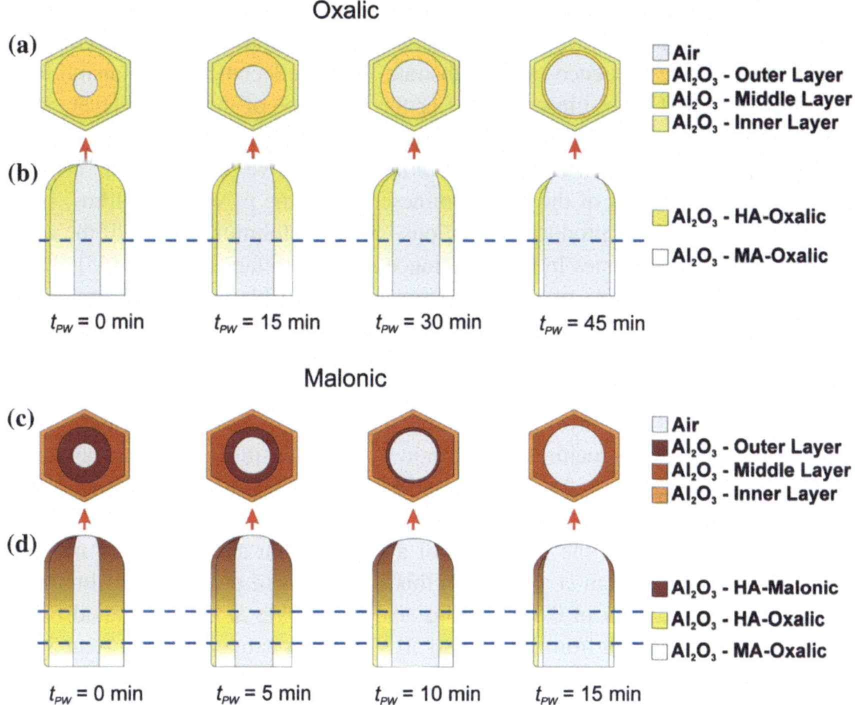

Fig. 6.2 The study of the PL spectra and their evolution as a function of different anodization parameters permits to reach the conclusion that the oxide in the p-AAO is composed of three distinct layers with different composition and optical properties. This structure is observed both for p-AAO obtained with oxalic and with malonic acid electrolytes. Adapted with permission from [24]

approximation are of very different kind. For instance, if the p-AOO thin film on the substrate is thin enough to sustain multiple-reflection interferences a Fabry–Pérot (F–P) cavity is formed. These interferences are used to obtain information about the properties of the material by means of analyzing their reflectance or transmittance spectra. The spectra show a series of maxima and minima that follow the well-known relation:

$$K\lambda = 2n_{eff}d\cos(\theta),$$

where n_{eff} is the refractive index of the effective medium composed of the oxide host material and the pores (with everything that might be filling them or covering their walls), d is the thickness of the p-AAO film and θ is the angle of propagation of the wave inside the film. An analysis of the wavelengths at which these maxima and minima occur [38, 39] leads to the extraction of the value of 2 n_{eff} d, known usually as the effective optical thickness (EOT). A more accurate analysis [40, 41] in which the actual transmittance or reflectance values of the maxima and minima and in which information about the substrate is taken into account would permit even to know separately the n_{eff} and d. Examples of the application of such concepts are the study of the influence of the substrate in the reflectance and transmittance properties of p-AAO thin films [42, 43] or the influence of the barrier layer, what leads to obtain information about its dissolution rate [44].

The optical properties of the p-AAO as an effective medium can also be investigated on the basis of the photoluminescence of the p-AOO thin films, since the F–P cavity is able to produce oscillations in the photoluminescence of the same nature as the observed ones in the reflectance or transmittance spectra [45]. Thus, by analyzing the positions of the oscillations it is possible also to determine the refractive index and even thickness of the p-AAO films [46], or to examine the influence of the use of Si substrates [47]. The control of the interferences in the p-AOO layers leads to interesting applications as for instance the development of a color tuning method by filling the porous alumina membranes with metal-dielectric-metal structures by atomic layer deposition [48], or obtaining materials with super-low refractive index (less than 1) by electrodepositing a thin layer of Ag within the p-AAO pores [49].

A systematic study of the properties of a p-AAO film as an effective medium was performed by Rahman et al. [50]. In this study, sets of p-AAO thin films were produced by anodization in 0.3 M oxalic acid electrolyte at different anodization potentials (V_{anod}) ranging from 20 V up to 50 V and at a temperature between 5 and 7 °C. This range of potentials was chosen in order to maintain the self-arrangement of the pores in the mentioned anodization conditions. Furthermore, for each anodization potential different anodization times (t_{anod}) were applied in the range between 10 and 25 min. Finally, a pore widening (a wet etching process carried out after anodization in order to modulate the pore diameter) was applied for different times (t_{PW}) in order to study the pore widening rate as a function of the applied anodization voltage. The study was based on the measurement of the optical properties of the p-AAO films after the different processes by means of ellipsometry [51–53] using a polarimeter. Polarimetry, and particularly ellipsometry, are techniques generally used in the investigation of thin films and their optical properties, but also many other kinds of samples such as biological specimens [54, 55]. The techniques are based on the measurement of the change in the polarization state of an incident beam upon reflection or transmission. Figure 6.3 shows an example of the analysis of ellipsometric spectra for the sample obtained with $V_{Anod} = 20$ V and $t_{Anod} = 10$ min and with no pore widening, where each graph shows the ellipsometric angles (Δ and Ψ) for different angles of incidence. The analysis is performed by fitting theoretical ellipsometric spectra calculated on the basis of a model for the sample to the experimentally measured spectra. The best fit is obtained considering simultaneously the spectra for all angles of incidence [56, 57]. The diagram in Fig. 6.4 depicts the considered model. The left drawing depicts the ideal cross section of the p-AOO thin film while the optical model considered to obtain the fitting is on the right. The porous layer is modeled as a mixture of materials with a Bruggeman effective medium approximation, where the fitting parameters are the porosity (P) and the thickness. An intermediate layer in the thin film needs to be

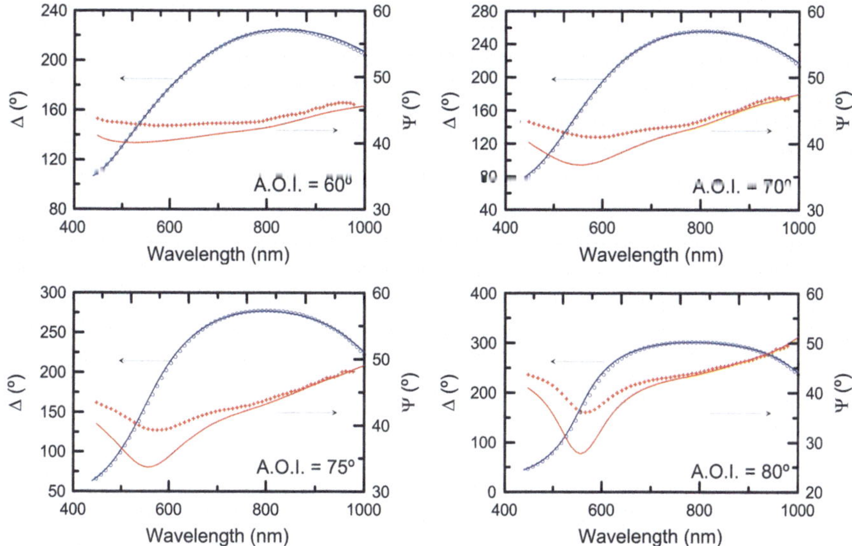

Fig. 6.3 Ellipsometric characterization of p-AAO thin films. The plots show the ellipsometry spectra of a p-AAO sample obtained with oxalic acid electrolyte (0.3 M) at an anodization voltage V_{Anod} = 20 V and for an anodization time t_{Anod} = 10 min and with no pore widening. Symbols represent the experimental measurements while the lines correspond to the best fit obtained on the basis of a theoretical model of the optical properties of the porous material. Adapted with permission from [50]

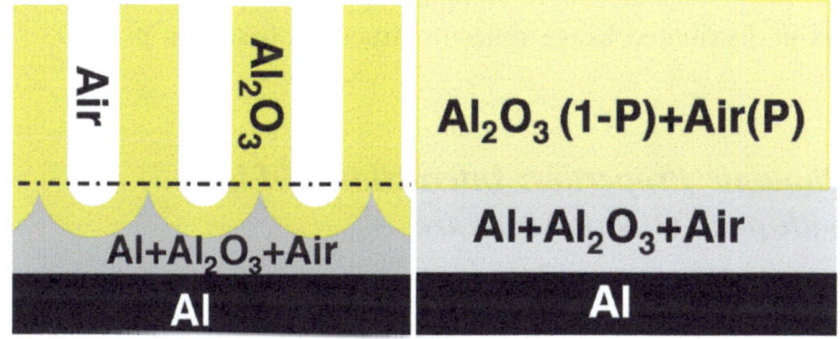

Fig. 6.4 Optical model of p-AAO used to analyze the ellipsometric spectra in Fig. 6.3. The porous material is modeled as an effective medium consisting of a mixture of aluminum oxide and air, where P is the porosity. An additional thin layer between the aluminum substrate and the porous material is also considered to take into account for the barrier layer and the textured aluminum surface. This additional thin layer is also modeled as an effective medium mixture of aluminum, aluminum oxide and air. Adapted with permission from [50]

considered in the model to take the barrier layer into account and the rough aluminum surface at the oxide-metal interface.

Figure 6.5 is the summary of the obtained results: the thicknesses and porosities for all the investigated samples obtained from the analysis of the ellipsometric

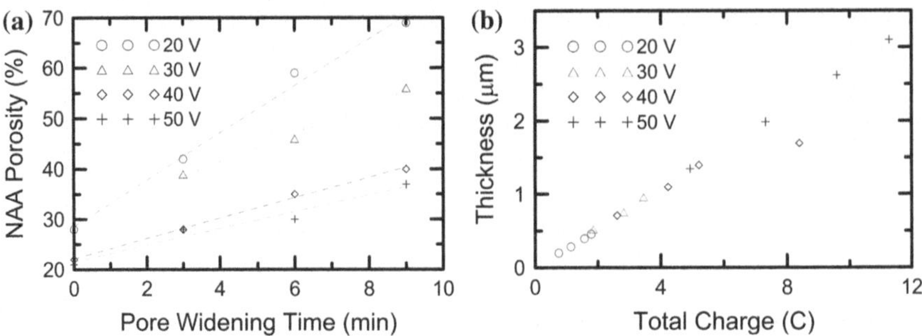

Fig. 6.5 Results of the study on the dependence of the pore widening rate on the applied anodization voltage. **a** Porosity as a function of the pore widening time for the different applied anodization voltages, **b** thickness as a function of the charge spent in the anodization. Adapted with permission from [50]

spectra. The top graph depicts the porosity *P* as a function of the pore widening time (t_{PW}). It can be seen that the rate of porosity increase is bigger for the smaller applied anodization voltage. This is translated in terms of optical properties as an increase in refractive index difference for increasing pore widening time for different applied anodization voltage. Another remarkable conclusion of this study is that the thickness of the p-AOO film is proportional to the total charge employed in the electrochemical etching process, with a rate of 250 nm/C and independently of the applied anodization voltage. This independence permits to use the charge to control the thickness of the produced p-AAO films. These two last points are fundamental to obtain other more complex optical properties such as high reflectance bands in distributed Bragg reflector structures based on p-AAO.

6.2.3 Photonic Properties: Interaction of Light with p-AAO Nanostructure

In the previous section, we have discussed how light interacts with the porous anodic aluminum oxide two-dimensional nanostructure in the cases in that the wavelength is much bigger than the interpore distance or the pore diameter. However, under the appropriate conditions (such as mild anodization in phosphoric acid electrolytes or hard anodization in oxalic acid electrolytes [58]) the interpore distances can reach values near the wavelength of visible or UV light and give rise to structure-related photonic properties. Furthermore, multilayer porous structures can be achieved by many several methods [59–63]. Photonic properties related to one-dimensional nanostructures such as the distributed Bragg reflectors (DBR) arise when such multilayers have thicknesses of the order of magnitude of the wavelength.

Research devoted to porous anodic aluminum oxide experienced an exponential increase since the publication of the renowned paper from Masuda and Fukuda [64].

The main factors for such outbreak are that the work reveals a new material for nanotechnology and most specially because of the self-ordering of the pores in a triangular/hexagonal arrangement. This arrangement was reminiscent of the ordered nanostructures of photonic crystals [65, 66] and thus a great effort has been devoted to the production of photonic crystals based on p-AAO. Although an accurate control of the fabrication conditions can lead to a good ordering of the pores [67–69], the development of more sophisticated technologies involving prepatterning with different methods permits perfect ordering over a long range. The preferred methods for such prepatterning are the nanoimprint lithography [70] and the nanosphere lithography (also known as colloidal lithography) [71]. Although nanosphere lithography is very promising as it takes advantage of the self-arrangement properties of nanospheres and their size tunability, nanoimprint lithography is more versatile in the kind of structures that can be fabricated, making possible the achievement of hexagonal or square lattices and pore shapes different than the usual circular section [69]. Another method for aluminum substrate prepatterning consists of using commercial optical 1-D gratings and performing the nanoimprinting in two steps with a 60° rotation of the grating between them [72]. With this, a rhombic pattern is formed on the aluminum surface. By adjusting adequately the anodization conditions, pores nucleate at the intersections of the imprinted lines and also at the center of the rhombi thank to the self-repairing mechanism described by Masuda et al. [73]. Finally, laser interference lithography is also specially adequate for prepatterning aluminum in order to produce highly ordered p-AAO, as the interference pattern is highly periodic and naturally extending on a long range [74, 75].

The photonic crystals based on p-AAO are especially adequate for the use in different nanophotonic applications. Lasing has been demonstrated by Masuda et al. [76] using a monodomain p-AAO covered with a layer of dendrimer-encapsulated fluorescent dye (pyrromethene 597). The emission band of the fluorophore and the lattice constant of the p-AAO need to be properly adjusted in order to tune the stop band edge of the photonic crystal structure at the maximum of luminescence emission. By pumping this structure, lasing can be observed and tuning can be achieved by changing the p-AOO geometry with a pore widening treatment. These applications can be even improved when the p-AAO is used as a template to pattern other materials such as ordered polymer nanopillar arrays [77, 78]. This nanopatterning technique has been demonstrated useful for improving the light extraction in GaN-based LEDs where the highly ordered p-AAO was used as a selective dry etching mask on the entire surfaces of the devices, achieving a light output improvement of as much as 94 % [79–81]. The photonic crystal properties of p-AAO have even been used to investigate the optical properties and composition of the material itself, as in the work of Choi et al. [25], where the combination of modeling and measurement of transmittance through the 2-D nanostructure of perfect p-AAO-based photonic crystals makes possible to demonstrate that the oxide layers at the pore walls are actually composed of two layers: the outermost with the highest concentration of incorporated electrolyte anions and innermost with a more pure barrier-like aluminum oxide.

Other authors have investigated the properties of photonic crystals arising from the combination of the triangular structure of p-AAO with metals and conclude with the existence of photonic band gaps in the visible region of the spectrum for the TM polarization when the pore radius is 200 nm [82]. These photonic properties have been demonstrated experimentally by means of polarimetry measurements [83].

Photonic properties of p-AAO can also be observed in the case of naturally self-assembled pores, this is: without prepatterning. In such case, the self arrangement takes place in domains, with domains of lateral sizes that can range between tens and hundreds of pores [23, 84]. Each well-ordered domain is surrounded by neighboring randomly oriented domains separated by lattice dislocations. Although such structure is essentially random, it preserves some information about lattice ordering and especially about lattice constants. It can be demonstrated by numerical modeling [85] that such quasi-random structure possesses a photonic stop band. To demonstrate this, the Finite-Difference Time-Domain (FDTD) method was applied to simulate propagation of light through the material in a direction perpendicular to the p-AAO pores. Figure 6.6a depicts schematically the computational domain used in the simulations, which consist in calculating the transmittance through a given length of material. Since the structure is essentially random, simulations were carried out for groups of different p-AAO distributions obtained from images of real samples (Such as the one in Fig. 6.6b) and the transmittance for all the distributions were averaged. The result is summarized in Fig. 6.6c, where it is shown that there exists a range of frequencies with a transmittance several orders of magnitude below that of other frequencies: a stop band. More systematic studies consisting of modifying numerically characteristic dimensions (the pore diameter or the interpore distance) confirm that this stop band

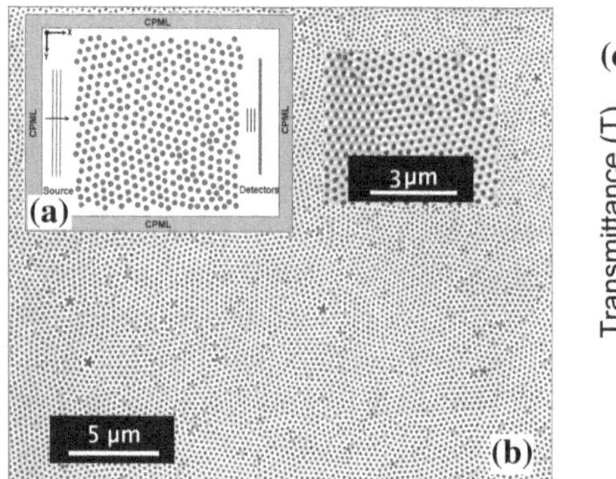

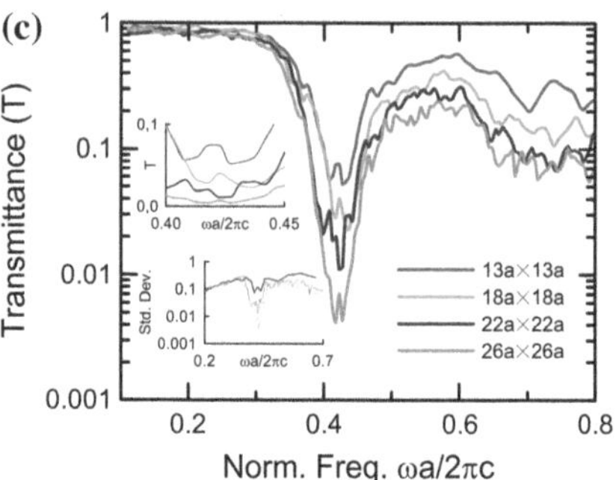

Fig. 6.6 Demonstration of the existence of a photonic stop band in self-ordered p-AAO. **a** Computational domain used in FDTD studies of transmission properties of p-AAO, **b** example of real sample used to obtain the structures considered in the simulations, **c** transmittance through a simulated p-AAO structure of the size specified in the legend, where 'a' is the average interpore distance. Reprinted from [85], Copyright (2012), with permission from Elsevier

appears at the same range as that of a perfect triangular structure of pores in an aluminum oxide matrix with the same characteristic dimensions [86].

Another widespread approach to obtain photonic properties from p-AAO structures is the application of a periodicity to one of the anodization parameters (usually anodization voltage [87], although others as current [63] or temperature [88] have been demonstrated). This periodicity produces a modulation with the depth of the structure porosity. This, in consequence, produces a variation in the refractive index with the depth, constituting a 1-D photonic crystal. These structures are also known as distributed-Bragg reflectors, as the photonic band gap results in a high reflection band. Early reports on the application of this technique can be found in [61, 89], where the anodization voltage is varied periodically with a carefully controlled profile that combines sinusoidal variations with linear variations. Such profile is based on the knowledge that a reduction of the anodization voltage during anodization in a factor $1/\sqrt{2}$ causes pores to divide in two new pores [90]. With a periodic profile it is possible to promote pore bifurcation and coalescence in a periodic manner. With this, the existence of a photonic band gap (PBG) is demonstrated and with a subsequent chemical etching (a pore widening) the position of such band gap can be adjusted. In a further work, the same group demonstrated a different approach to tune the position of the stop band by controlling the anodization temperature [91] and that the band gap can be tuned within all the range of the visible spectrum. A combination of the mentioned processes in two different variations of the periodicities gives rise to p-AAO-based photonic crystal heterosturctures [92]. Different combinations of band structures (overlapped, staggered and split PBGs) are introduced, what offers a great versatility in the adjustment of the band gap. Another technique to obtain DBR structures is pulsed anodization [93], where a periodic change between mild anodization and hard anodization is applied (in this case using a sulphuric acid electrolyte and cycling between 25 and 35 V). These structures are the basis of several applications, mainly in chemical sensing or bio-sensing [28, 30, 94]. More elaborated sensors take advantage of the Fano resonances that occur when the photonic states of the 1-D DBR interact with the scattering modes of the porous nanostructure [95].

In the fabrication of these 1-D photonic crystal structures based on p-AAO is of special importance the application of the subsequent pore widening process. This is demonstrated in [62] where a systematic study of the effect of pore widening on DBR structures is carried out. On the basis of the previous demonstration by the same authors that the pore widening rate in single-layer p-AAO samples depends on the applied anodization voltage [50], the material produced with a periodic anodization voltage preserves the information about that variation in its depth, and such information is developed when a subsequent pore widening is applied. Figures 6.7 and 6.8 summarize the results. Figure 6.7a shows the applied anodization voltage for one of the samples and the recorded anodization current for the first 60 min of the process. Figure 6.7b shows the recorded current corresponding to one single period (the third one in Fig. 6.7a). The voltage profile within one period is especially simple, with three phases: (i) a linearly increasing voltage ramp, (ii) a phase at a constant voltage and (iii) a decreasing linear voltage ramp. The increasing and

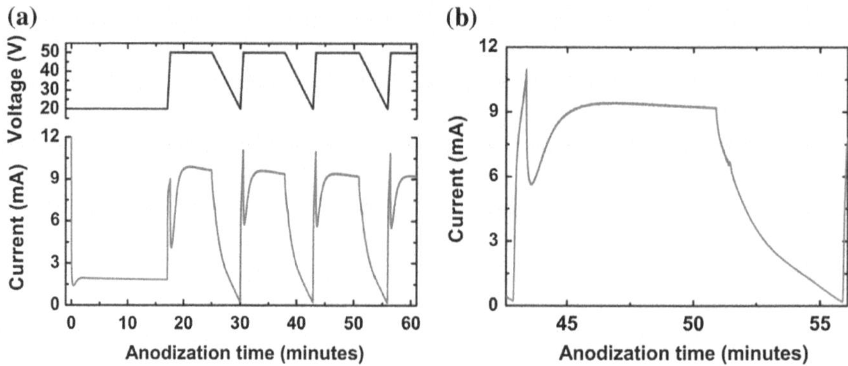

Fig. 6.7 Cyclic anodization procedure applied to obtain p-AAO DBR structures. **a** First three cycles of applied anodization voltage and recorded anodization current, **b** close-up of the third cycle of the recorded anodization current. Reprinted with permission from [62]. Copyright (2013) American Chemical Society

decreasing ramps constitute transition zones in the growing pores, thus they are chosen as the fastest possible in order to maintain the anodization while obtaining a minimum thickness corresponding to the increasing and decreasing voltage phases. The constant voltage phase, instead, can be used to control the thickness of each cycle in the DBR structure by controlling the amount of charge employed in the phase, Q_0. Given the previous result that the pore growth rate is linearly dependent of the anodization charge, regardless of the applied anodization voltage [50] permits the tuning of the photonic stop band central wavelength. From the corresponding current transient it can be concluded that the dynamics of the pore development show two trends within each phase: a first part in which the dynamics of the growing pores show a resistance to change its state of growth and a second trend with a stable growth state.

Figures 6.8a, b show the transmittance spectra for two of the samples corresponding to $Q_0 = 0.5$ C and $Q_0 = 4$ C, for the as-produced samples and for the same samples after different times of pore widening (t_{PW}). Concerning the as-produced samples, the spectra show stop bands (wavelength ranges with reduced transmittance). However, the spectral shape of such stop bands is irregular with a moderate minimum in the transmittance and even with a local maximum. This is a consequence of the small refractive index contrast between the different phases within each cycle in the microstructure. Instead, after pore widening, the stop bands become well-defined with a minimum transmittance close to zero and much lower than the transmittance out of the stop band range. This leads to the conclusion that the pore widening treatment after the p-AAO-based DBR is crucial in order to obtain photonic crystal properties and to exploit their possible applications. An analysis of the photonic stop band wavelengths on the basis of a three-layer model for each period, permits to estimate the variations of the effective refractive index with the depth. As a summary of such study, Fig. 6.8c shows the dependence of the refractive indexes of the different layers within a cycle and with t_{PW}.

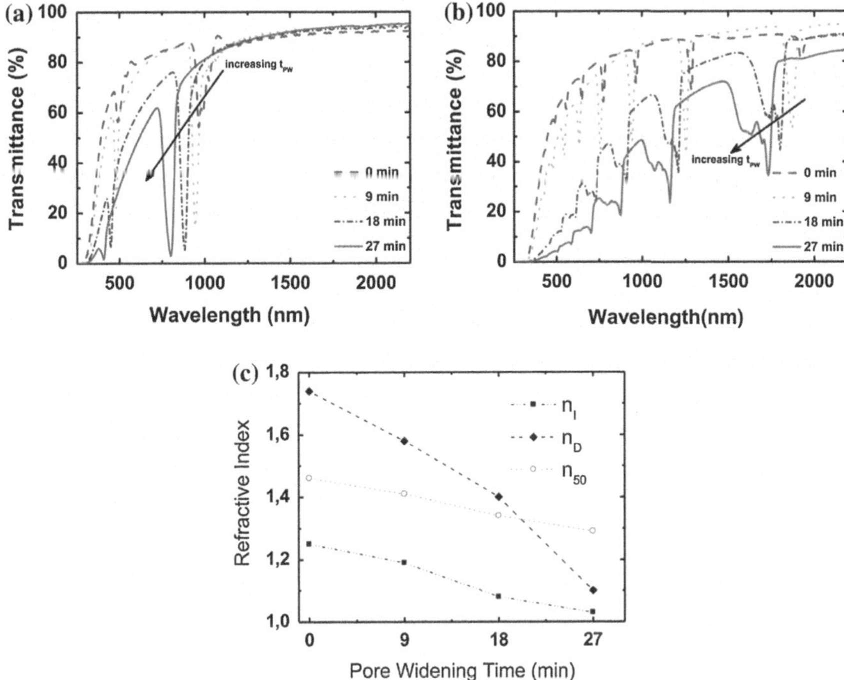

Fig. 6.8 Transmittance spectra of two of the samples obtained with the cyclic anodization process of Fig. 6.7 after different times of pore widening and values of the refractive indexes of the layers composing the DBR estimated from the spectra. **a** Transmittance spectra of the sample with $Q_0 = 0.5$ C for different pore widening times, **b** transmittance spectra corresponding to the sample with $Q_0 = 4$ C, **c** refractive index of the different layers corresponding to the phase of increasing anodization voltage (n_I), the phase of decreasing voltage (n_D) and the phase at constant voltage, $V_{anod} = 50$ V (n_{50}). Reprinted with permission from [62]. Copyright (2013) American Chemical Society

6.3 Applications Based on the Optical Properties of Porous Anodic Aluminum Oxide

The optical properties of porous Anodic Aluminum Oxide are on the basis of many of the applications of this nanomaterial. Optical methods are ideal for readout strategies in many situations, such as in sensing, as they offer a direct, non-invasive and non-destructive transduction mechanism from the change in structural or surface properties at the nanoscale to a macroscopic easily measurable magnitude (reflectance, resonant wavelength, etc.). This, together with the structural properties of p-AAO (high surface-to-volume ratio, great versatility in the geometry engineering, etc.) justify the wide range of applications of p-AAO, and in particular those based on its optical properties.

The most preferred application found in literature of p-AAO is sensing. This happens because of two main factors: (i) its porous structure is adequate for integrating the necessary fluidic circuits with the optical setup and (ii) its great surface-to-volume ratio permits a great amplification factor of the changes in the surface, even for small amounts of the species to be detected. Nevertheless, optical properties of p-AAO are also the basis of other applications not directly related to sensing, as for instance the extraction of light from LEDs [79, 96] (where the p-AAO nanostructure serves as a texturizing template for the LED surface), the generation of laser emission [76] (where the self-arrangement of pores contributes to create cavity modes to amplify laser light emitted by a gain material infiltrated within the pores), to texturize surfaces and interfaces of photovoltaic cells to improve light absorption or charge collection in the energy conversion process [97], or even to create unique photonic barcodes for labeling applications [98]. Doing an extensive review of all the applications based on the optical properties would require a much longer extent than this chapter. Consequently, in this section we aim at offering a classification of the different existing applications and an overview of the most relevant works on each application with the aim of putting into evidence the different relation between them and with the different optical properties discussed in the previous section.

We have divided the section into six major uses of p-AAO. The first five are (i) use as waveguides, (ii) use as templates to obtain surface-enhanced Raman scattering (SERS) substrates, (iii) use as reflectometric probes in sensing, (iv) applications of the photoluminescent properties and (v) photovoltaic applications. Finally, as the range of applications is continuously growing, latest related concepts have been discussed in a sixth group.

6.3.1 Waveguides Based on Porous Anodic Aluminum Oxide

The good optical properties of p-AAO and specially the possibility of finely tuning the thickness of the material as a film lead to the natural application as a waveguide. The use of p-AAO as an optical waveguide can be found as early after the Masuda and Fukuda proposal as 1994 [99]. In this work, the authors fabricated channel waveguides by filling selectively the alumina pores with photocurable resins of suitable refractive index, obtaining even y-junction waveguides. Later, a more complete study of the application of p-AAO waveguides to sensing can be found in the work by Lau et al. [31]. In this paper, the p-AAO films are used as planar optical waveguides investigated by optical waveguide spectroscopy (OWS). This technique consists in coupling light through a prism onto the p-AAO film and recording the reflectance of light as a function of the angle of incidence. The different waveguide modes appear as sharp valleys at different angles of incidence. With this approach in combination with a microfluidic system the authors are able to measure the adsorption and desorption of Bovine Serum Albumin (BSA) with subangstrom sensitivity in the effective thickness of the adsorbed protein. Furthermore, they are able to monitor the process of pore widening or the exchange of the pore-filling medium between phosphate buffer solution and ethanol. Yamaguchi et al. [100] use a similar approach to investigate the sensitivity of OWS by monitoring the complex formation of bathophenanthroline (Bphen) with Fe(II). By comparing their experimental results with modeling they could conclude that such sensors are specially sensitive to the change in the imaginary part of the effective refractive index and that such a system can provide higher resolution tan conventional equivalent surface plasmon resonance sensors.

The combination of the waveguide properties of p-AAO films with the possibility of infiltrating the pores with different active materials opens up further possibilities. For instance, Trivinho-Strixino et al. [101] suggested the use of photoluminescent polymers deposited over the p-AAO film to demonstrate that the emission from the polymeric and from the p-AAO interact and give rise to different waveguide modes that can be measured directly from the edge of the film. A demonstration of the capability of such a system is given in the detection of a common pesticide, chlorpyriphos. An alternative approach is using the p-AAO film as a template to fabricate free-standing polycyanurate arrays that constitute themselves a fluidic-compatible optical waveguide that can be interrogated by OWS [102]. A further step in this strategy consists in the infiltration of the pores with active species that can remain within the p-AAO structure and leave room for the fluidic infiltration of the analytes [103]. In this work, several layers of polyelectrolyte are applied in a layer-by-layer method to demonstrate their use as protective coating (even against AAO etchants) or as linkers for single-probe DNA for application in DNA–DNA hybridization assays.

A more systematic study on the possibilities of applying p-AAO as optical waveguides in biosensing, and particularly in protein adsorption, can be found in [104]. By means of finite-element simulations and OWS the authors measure a range of p-AAO waveguides with pore radii between 10 and 40 nm and depths from 0.8 to 9.6 μm. The combination of the two techniques permits to interpret the adsorption kinetics, concluding that their initial slope is inversely proportional to the pore depth and linearly proportional to the protein concentration. Furthermore, they prove that adsorption kinetics allows for an accurate determination of protein concentration. The waveguide modes concentrate most of electromagnetic field energy within the p-AAO film, however, the modes have also an evanescent tail that is highly suitable for interaction with metallic nanoparticles deposited on the film, generating localized surface plasmon resonance (LSPR) in the nanoparticles. This, together with the high surface-to-volume ratio of the film permits an enhancement of the SERS signal that can be obtained from the nanoparticles. This is demonstrated in [105] by means of an application to immune recognition with a concentration as low as 0.1 ng/ml.

6.3.2 Porous Anodic Aluminum Oxide for Surface-Enhanced Raman Spectroscopy Applications

A complementary approach to biosensing based on evanescent waves in waveguides is the use of the nanostructure of p-AAO to shape metallic nanoparticles and obtain SERS substrates. A first example is found in the work by Du et al. [106] where silver nanowire arrays with different diameters and aspect ratios are used to enhance the SERS signal of 4-mercaptopyridine (4-mpy). The authors were able to determine optimal nanorod diameters and validate the results with electromagnetic models. More recently, p-AAO has been used to form hexagonally ordered Au nanoparticles and nanoshells [107]. The hexagonal arrangement contributes to the amplification of SERS signal and demonstrates the capability to detect Hg^{2+} ions down to a concentration of 1 ppb in water and distinguish them from other ionic species such as Pb^{2+} or Cd^{2+}.

An alternative approach to apply the optical properties gold-capped p-AAO structures consists of measuring the changes in the localized LSPR wavelength as a response of a change in the structure environment. This approach can be found in the works from Kim et al. [108, 109] and of Yeom et al. [110]. In these works, p-AAO with sputtered gold nanoparticles at the top wall of the porous structures is functionalized to selectively detect C-protein antibody/antigen or aptamer-protein interactions by means of a reflectometric setup.

6.3.3 Reflection Interference Spectroscopy

Most of the applications based in evaluating the change in the effective refractive index of the porous Anodic Aluminum Oxide upon attachment of different chemical or biological species within the inner surface of the pores rely on the measurement of the changes in the reflectance spectra of the structure. This technique is generally called Reflection Interference Spectroscopy (RIfS) [28] although it can also be referred to as the Reflection Interference Fourier Transform Spectroscopy (RIFTS) [29], when an additional Fourier transform of the spectrum is applied to interpret the spectral shifts. The technique was previously introduced for other materials such as porous silicon [111–114] and was naturally adapted to porous alumina. Porous alumina offers different advantages with respect to porous silicon, for instance the well-defined geometry of the pores, the possibilities of engineering such geometry and its optical properties [87] and its chemical stability over a great range of pH [29]. In RIfS the spectra are usually monitored in real-time while the sensing reactions are taking place by the use of a compact fiber CCD-spectrometer [115]. The fiber probe end is composed of six illuminating fibers distributed around a central fiber connected as a bundle to a source with a wide spectrum (usually in the range between 200 and 1100 nm, obtained from deuterium and tungsten lamps). On the other hand, the central fiber is connected to the spectrometer itself. A converging lens produces an image of the illuminating fibers onto the p-AAO structure in a flow cell with a transparent window. Finally the same lens focuses the reflected light onto the central fiber.

The applications of RIfS can be divided onto two main groups: (i) those using the p-AAO thin film as a Fabry–Pérot (FP) interferometer and (ii) those that use an engineered structure of p-AAO in the form of a distributed-Bragg reflector. The first application of p-AAO as a FP interferometer in the detection of specific biomolecules is reported by Pan et al. [116], where they demonstrated detection of complementary DNA down to a concentration of 2 nm/cm^2 within p-AAO pores by registering the shift in the reflectance spectra. After this, many examples have been proposed, for instance Lee et al. [117] evaluated the sensitivity of a p-AAO-based sensor functionalized with Prolinker A for detecting β-galactoxidase and estimated a 200-fold enhancement with respect to an equivalent porous Silicon-based sensor. A similar approach is used in [118, 119] where sulphurous gases (H_2S) are detected on human breath down to a level of 0.5 % by adsorption of the gaseous molecules onto gold-coated p-AAO.

An important improvement in the RIfS technique is the analysis of the shifts in the reflectance spectra by fast Fourier transform, which yields direct information on the effective optical thickness of the p-AAO layer with the different substances filling the pores or covering the inner pore walls. A remarkable application is as immunosensor [29] using the immunoglobulin (IgG) as model. In this work it is shown that the p-AAO displays a greater affinity for Protein A than for IgG, so the sensor surface is first modified with Protein A and subsequently with IgG. Then, selectivity is confirmed by showing significant system response to IgG derived from rabbit but none to IgG derived from chicken. A wealth of applications have been proposed since then, for instance to detect gold ions with a limit of detection of 0.1 μM in water by exploiting the affinity of the metal ions for the surface functionalization agent MPTES [120]. The same technique is also applicable to more sophisticated substances and even to tumor cells as for instance in [121] where by the functionalization with biotinylated anti-EpCAM antibody that specifically binds to such cancer cells are able to detect down to 1000 cells/ml in blood serum with a sample volume as small as 50 μl. Besides to sensing the presence of biological molecules in aqueous media, the method can be used to monitor the release of drugs loaded onto p-AAO matrices in a flowing medium. This method provides more reliable and relevant information on drug release kinetics than usual in vitro assays. In this work it is demonstrated that the release mechanism in p-AAO loaded indomethacin is diffusion driven, and a quantitative relation between the release rate and the flow rate is obtained. Other authors propose to obtain the F–P p-AAO structure onto a transparent substrate [122]. In this way, sensitivity might be increased by simultaneous measuring reflection and transmission. An alternative transduction mechanism is by using polarimetry, as it is demonstrated in [123]. In this work an immunoassay for detection of IgG with a limit of detection (LOD) of 225 pM is achieved. Finally, there exist also studies on the optimization of the Fabry–Pérot structures to obtain a maximum sensitivity, such as in [124, 125].

As stated above, a second possible approach to measuring changes in effective optical thickness of p-AAO thin films upon attachment of biological molecules to the pore walls is to incorporate part of the necessary processing to estimate the change into the very p-AAO structure. This can be accomplished by structuring the p-AAO thin film as a DBR. As stated in Sect. 6.2.3, this is a structure with a periodic variation of the refractive index in the direction perpendicular to the film surface that possesses a photonic stop band, in which incident light cannot propagate and is reflected. Since the position of such photonic stop band depends strongly on the effective refractive indices of the different layers composing the nanostructure, it is also sensitive to the medium infiltrated into the pores and to the species adsorbed onto the pores surfaces. A first proof-of-concept to using p-AAO Bragg stacks as chemical sensors is found in [126], although in this work, instead of using photonic stop band shift as transduction measurement, they rely on the change in intensity of the transmittance dip upon infiltration with substances such as alcohols or alkanes. Li De Zhang et al. have a vast experience in the preparation of p-AAO Bragg stacks and some examples of their application to biosensing [127] or to gas sensing [128]. Sensitivity and limit-of-detection evaluation of this biosensing systems for glucose concentration measurement can be found in [115]. In this work, the authors also study the dynamics of the change in the stop band central wavelength with time and validate the results with a Looyenga-Laundau-Lifshitf isotherm model. This result is of special interest as it points out that the binding dynamics can be monitored with this technique and that such binding dynamics provide valuable information for the sensing. The same approach has been also demonstrated to the detection of Hg^{2+} ions in water wit a limit of detection of 1 μM. Finally, a comparison between the two RIfS methods presented in this section can be found in [129] and more detailed reviews on these applications in [130, 131].

It is also worth mentioning an intermediate RIfS approach between F–P cavities or DBR stop bands. The very nature of the porous structure of p-AAO permits to provide the structure with other interesting properties in addition to the optical properties. For instance, by obtaining a double layer of p-AAO where the straight pores show a diameter reduction at a given depth, permits to provide size exclusion to the biosensor. This is demonstrated in [30] where a double layer is produced by stopping the p-AAO growth in the second step, applying a pore widening and then resuming the second step again. With this, the structure depicted in Fig. 6.9 is obtained. Such structure can be understood as three coupled F–P cavities, two of them corresponding to each of the actual layers of the film (L_1 and L_2), and a third one corresponding to the union of the two layers. The reflectance spectrum of this structure is complex, but the Fourier transform reveals the existence of three peaks corresponding to each of these F–P cavities, as depicted in Fig. 6.9b. The pore diameters can be carefully designed to permit the infiltration of the species to be recognized (in this case the protein BSA) in the upper section of the pores and avoid infiltration in the lower. With this strategy, the lower layer can be used as a reference in the RIfS procedure to compensate for changes in the solvent or drifts in the system. In addition to this, if a thin gold film is deposited on top of the p-AAO bilayer, an increase in the refractive index contrast between the layers and the

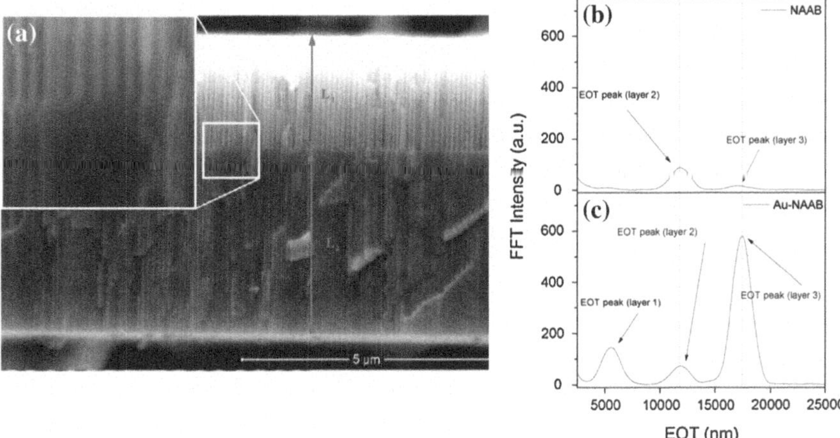

Fig. 6.9 **a** Porous anodic aluminum oxide bilayer used for immunosensing, the bilayer can be seen as a system of three F–P cavities, two corresponding to each of the actual layers in the structure (L_1 and L_2) and the third one corresponding to the union of the two layers (L_3). **b** Fourier transform of the spectra of the bilayer without (*upper graph*) and with (*lower graph*) a thin gold layer on top of the p-AAO. The gold layer increases the visibility of the peaks in the Fourier transform corresponding to the L_1 and L_3 cavities, contributing to an increase in sensitivity of the structure. Reprinted with permission from [30]. Copyright (2013) American Chemical Society

incident medium is produced, which results in an increase in the visibility of the peaks in the spectrum Fourier transform, as depicted in Fig. 6.9c. This increase in visibility permits to reduce the noise in the measurement of the shifts or the height of the peaks upon adsorption of the protein and to achieve lower LODs.

6.3.4 *Photoluminescence-Based Applications of Porous Anodic Aluminum Oxide*

Photoluminescence emission is an outstanding property of p-AAO widely used for developing interesting optical devices/applications. Porous-AAO presents a distinctive PL fingerprint in the UV-visible region. Furthermore, this PL emission can be modulated thank to the F–P effect to obtain oscillations in the PL spectrum [24, 45, 98]. Noteworthy, the p-AAO PL fingerprint can be tuned [24] as the number, intensity and position of these oscillations rely on the geometric structure of p-AAO, i.e. the pore length and its diameter. This interesting characteristic offers the possibility of designing versatile p-AAO structures optically optimized that provide intense and well-resolved PL spectra.

PL intensity for p-AAO is also closely dependent on the acid electrolyte used. Although PL emission is observed for p-AAO fabricated with any of the usual electrolyte acids, it has been demonstrated that the PL intensity of p-AAO fabricated under mild anodization conditions in oxalic is rather higher than that of p-AAO produced in sulphuric, phosphoric and malonic acids for mild or hard anodization conditions [16, 132].

Most of the works related to the application of p-AAO PL are focused on the development of optical biosensors. PL biosensors take advantage not only of the exclusive PL fingerprint of p-AAO but also of other interesting properties of p-AAO such as its thermal stability, environment-resisting and biodegradability. Biosensors based on p-AAO PL are characterized by their high degree of resolution, sensibility and biocompatibility [100, 133]. Moreover, their performed measurements are reliable and reproducible as the PL spectrum of p-AAO remains stable over time [98].

A selective and highly sensitive enzymatic PL biosensor is presented in [134]. It highlights for its accurate detection and quantification of the analyzed sample, trypsin. The sensing technique for this biosensor is based on the change in the wavelength of the oscillation maxima in the PL spectrum, which is the same as in RIfS. The presence of trypsin inside the p-AAO pores implies a modification of the effective refractive index and thus a change in the effective optical thickness (EOT). This sensing principle, highly sensitive to the presence of small quantities of analyte, allows the qualitative detection of the immobilized enzymes by means of the increment in the effective optical thickness (ΔEOT) [134].

In this work, trypsin immobilization inside the p-AAO pores is performed in a controlled manner by functionalizing and activating the p-AAO structure. The surface functionalization of p-AAO with different functional substances is widely used. It consists on the surface modification of p-AAO by introducing organic compounds that act as chemical selective linkers for the substance to be detected. Accordingly, with the functionalization, the selectivity and efficiency of the biosensor are highly increased. The high sensitivity of the biosensor presented in [134] is so especially remarkable that the sensor itself is used for monitoring each stage of the process (functionalization, activation and final immobilization of trypsin) as each stage implies a modification of the EOT.

It is interesting to compare the close similarities between the sensing principle of this PL biosensor [134] and the sensing principle of RIfS sensors [29, 125]. Both sensing principles are based on the interaction between the light and the analyzed sample: both study the shift in the F–P oscillations due to the presence of the target analyte inside the pores. However, whereas RIfS sensors analyze the reflectance spectrum, PL sensors analyze the spectrum of an inherent property of the material itself. Both PL and RIfS principles have revealed to be interesting and successful optical techniques for sensors development. A comparative study about the sensing performance of PL and RIfS p-AAO based biosensors is reported in [32] for several analytes. In this work, the sensitivity and limit of detection of PL and RIfS p-AAO sensors are compared for glucose and L-cysteine sensing under nonspecific and specific adsorption conditions, respectively. It is noteworthy that the sensors studied in this work were structurally optimized for the corresponding analyte detection.

This study concludes that the sensitivity of the sensors is highly dependent on the properties of the analyte molecules and the adsorption conditions through the nanoporous matrix. The size and the optical properties of the analyte molecules together with their physical and optical interaction with the nanoporous matrix surface highly determine the sensitivity. The obtained results demonstrate that in general, PL sensors are more sensitive than RIfS sensors although the difference between them is closely analyte dependent. Whereas PL biosensors are much more sensitive for glucose detection than RIfS sensors, both sensing platforms showed comparable performance for L-cysteine detection, although the sensitivity of the PL sensor was slightly better. The reason that explains this completely different sensitivity for glucose and L-cystein is that glucose is much more photoluminescent than L-cysteine and therefore the sensitivity of PL sensor is higher to the glucose immobilization.

Among the interesting advantages that porous alumina offers it is worthy to remark its compatibility to be integrated with microdevices and systems. Taking advantage of this attractive property, Li et al. [135] have developed micropatterns of p-AAO on indium tin oxide (ITO) glass substrate that can enhance the fluorescence signals of the fluorophores and labeled biomolecules on the substrate up to 2 or 3 orders of magnitude compared to the glass substrate. This would significantly reduce the consumption of the biosamples for fluorescence-based sensing, imaging, and analysis, and would allow the detection of biomolecules of ultralow concentration [135].

The p-AAO micropatterns presented in this study are fabricated rapidly and very cost-effectively by combining a lift-off process and a one-step anodization. The fluorescent images of several fluorescent dyes and labeled biomolecules are observed to be significantly enhanced when they are adsorbed on the micropatterned p-AAO as compared to those adsorbed directly on the glass substrate. However, the geometry of the p-AAO substrate, that is the nanopore size and the interspacing, will significantly affect the enhancement of one fluorescent dye or the other. This means that the geometric properties, and therefore the optical properties, of p-AAO should be tuned to optimize the enhancement of a specific fluorescent dye. Advantageously, the geometry of p-AAO can be easily tuned during its formation and its fabrication process is compatible with standard lithography-based microfabrication, which means that p-AAO micropatterns can be easily scalable and their shapes can be readily varied or modified. These features suggest the simplicity for integrating p-AAO micropatterns into microdevices or microfluidic devices for fluorescence-based bioassay applications. Another advantage of using p-AAO for developing a fluorescence enhancer is the very good stability and repeatability that p-AAO offers being possible to reuse the same p-AAO substrate many times. Besides, differently to other fluorescence enhancers [135], a thin layer (i.e., several nanometers) of dielectric material is not required as a spacer to separate the p-AAO surface and the fluorophores.

PL emission of p-AAO can also be used as an active photonic mark for labelling. A barcode system based on the PL spectrum of p-AAO in the UV-visible range is proposed in [98]. In this system, a unique barcode is related with every p-AAO geometry (i.e. the pore length and its diameter) by means of its PL spectrum, which can be structurally tuned at will. Each bar position corresponds to the wavelength of each oscillation in the PL spectrum and the higher the oscillation intensity the wider the bar. This barcode is an interesting tool for developing a quick and simple identification method for p-AAO substrates where even close structure p-AAO geometries can be fast and easily differentiated by their unique PL barcode.

This barcode system has already been successfully tested for detecting infiltrated quantities of analytes inside the p-AAO pores, more precisely oxazine dye and glucose [98]. The results obtained from these experiments have demonstrated that this system can give quick information about the presence of the infiltrated analyte and is a promising method for developing optical biosensors.

6.3.5 Porous Anodic Aluminum Oxide in Photon-Energy Conversion

The geometrical properties of p-AAO and the optical properties based on its nanostructured geometry can be found also on the basis of photovoltaic (PV) energy applications. The uses of p-AAO for such purposes can be classified into two main categories: (i) taking advantage of the nanostructure to provide anti-reflection (AR) properties to the PV cells, and (ii) use the p-AAO as a template to conform geometrically the PV cell components, and simultaneously to improve their ability to trap incoming photons. An early example of the production of an antireflecting nanostructure is the use of p-AAO with tapered holes [136] as template and replicate their structure with PMMA [77]. The same structures were further optimized by a parametric variation of the angle of the p-AAO tapered pores, with the conclusion that the AR structures with a gradually changing slope exhibited the lowest reflectance [137]. A more elaborated evolution of this idea, with a view to industrial up-scaling is found in [138], in which p-AAO nanostructures with tapered and cylindrical pores are fabricated with spacing of 100 and 200 nm and depth of 180–500 nm, and then replicated into polymer films by a roll-to-roll process. By means of angle-dependent optical transmission measurements, they concluded that the structures provide up to a 2 % increase in transmission at normal incidence and up to a 5 % increase at 70° incidence, with respect to an equivalent flat surface. In this work, it is also remarkable that no differences in transmittance increase between tapered and cylindrical nanopores can be registered. Besides using p-AAO as a template to shape other materials such as polymers, a graded p-AAO layer can be used as an AR coating itself, as demonstrated by Chen et al. [139]. In this work, an AR coating based on graded p-AAO is obtained on a glass substrate with large area (4 cm × 4 cm). By electron-beam evaporation of the aluminum and sequential electrochemical etching to grow pores and chemical etching to modulate their diameter, five-layered graded-index broadband p-AAO coatings following a Gaussian profile of refractive index in depth are obtained. With this, reflectance of glass can be reduced to values as low as 0.64 % at a wavelength of 534 nm.

As stated above, a second path for application to p-AAO to PV is to conform the materials composing the cells with the additional consequence that their light trapping properties are increased. This approach can be found in any of the different kinds of PV cells that exist. For instance, Kato et al. use p-AAO as scaffolds for quasi-solid dye sensitized solar cells [140]. The straight channels of p-AAO offer straight paths for the $I_2(I_3^-)$ ionic species that mediate the PV conversion what results in a better diffusion and an increase in efficiency. Furthermore, the nanostructure permits a deeper penetration of light into the material with a consequent further increase in quantum efficiency around the wavelength of 400 nm. In the work by Sheng et al. [97], a p-AAO grown on the backside of a silicon-based PV device combined with a DBR structure composed of alternating layers of deposited Si and SiO_2, leads to an increase in absorption at the near-IR. The p-AAO acts as a subwavelength grating that scatters the light back-reflected by the Si–SiO_2 DBR generating the increase in absorption. Finally, an example of templating polymeric solar cells with p-AAO can be found in [78] where p-AAO is used as a template to conform Poly(3-hexylthiophene) (P3HT) nanopillars to be used as donor material in organic solar cells. By means of numerical simulation [141, 142] it has been shown that such nanopillars are a basic building block to obtain an ordered nanostructured interface with additional properties of light trapping.

6.3.6 Alternative Applications

The interesting optical properties of p-AAO structures owing to the dimensions and geometry of its self-ordered pores, has generated an increasing interest on replicating p-AAO morphologies with other materials. Recently, p-AAO has been successfully used as a template to endow complex structures to materials with limitations in the assembly of pores or the formation of high surface area layers and multilayers [77, 143]. In contrast to other methods used for the formation of two-dimensional photonic crystals (2D-PC) like electron-beam lithography (EBL) [144], holographic lithography (HL) [145] or focused ion beam milling (FIBM) [146], which require expensive equipment and have low yield for mass production, the easily controllable structure of p-AAO has made of this material a cost-effective tool for the formation of large-area wavelength-scale PC structures.

The use of p-AAO as a template is studied by Fu el al. for fabricating an improved light transmission surface [79]. In this work, a complex geometry of the p-AAO pores where the pore diameter decreases with depth is fabricated. The p-AAO structure was replicated onto the desired material (epoxy resin in this case) by hot embossing technique. Thus, highly ordered epoxy conical pillars with a diameter continually reduced from the bottom to the surface where obtained. The optical properties of the resulting complex epoxy layer are especially interesting, considering that the refractive index of the conical pillar arrays varies gradually from the refractive index of the epoxy substrate to the refractive index of air. The patterning of the surface with these two dimensional pillar arrays has demonstrated to be an interesting alternative to conventional thin film stack coating as the integrated transmission of conical pillar arrays presented in this work is 22.3 and 11.3 % higher than that of planar epoxy and the cylindrical pillar arrays, respectively.

Furthermore, this study is completed by the use of the conical pillar arrays of epoxy as encapsulant of a GaN based flip-chip LED. Compared to LED encapsulated by planar epoxy, the enhancement of light extraction from LED covered with conical and cylindrical pillar arrays is 46.8 and 34.9 %, respectively.

The use of p-AAO as a template for replicating complex structures is not the only way to transfer its 2D pore geometry to other materials. p-AAO used as a selective dry etching mask has been reported in [79], where the p-AAO structure is transferred onto the surfaces of a GaN-based LED chip to improve its light output power. For this purpose, p-AAO was adopted as a pattern transfer and selective etching mask of reactive ion etching (RIE). Thereafter the p-AAO structures with different pore depths were obtained at the entire surfaces of GaN-based LED. The pattern of the LED surfaces takes advantage of the p-AAO self-organizing structure and cost-effective formation process, in contrast to other expensive and technological complex methods like EBL, FIBM, laser holographic lithography and nanoimprint lithography.

The aim of this patterning is to increase the light extraction efficiency of conventional planar LED, limited by total internal reflection (TIR) due to the large index contrast between the GaN film and air. The patterning or roughening of the entire surfaces of LED is proposed to suppress the limitations of TIR by coupling guided modes trapped inside GaN layers [147, 148]. The light output power improvement of this PC-based GaN LED was up to 94 % compared to that of conventional GaN-based LED. This improvement is attributed to the diffraction and scattering effects of guided modes by the PC transferred structures on the whole emitting surface of the device.

Previous studies have demonstrated the 2D photonic bandgap of p-AAO [149, 150]. This interesting property of p-AAO was used by Masuda et al. to observe the laser emission of an optically pumped thin layer of dendrimer-encapsulated fluorescent dye (pyrromethene 597) covering the surface of p-AAO [76]. In this work, a sharp intense emission peak is observed at 547.26 nm, originated from the laser action that relies on the low group velocity at the photonic band edge in the p-AAO. The wavelength position of this emission peak was obtained for an interpore distance of 200 nm and a filling factor of 0.3, but different filling factors resulted in an emission peak shift. Therefore, the position of the emission peak is dependent on the filling factor, being its wavelength position tunable by changing the geometrical structure of the p-AAO. The sharp emission peak could only be observed for transverse electric (TE)-polarized light but not for transverse magnetic (TM)-polarized light.

Recently, an optical technique has been developed to investigate the dynamics of capillary-driven fluid imbibition in nanoporous substrates [151, 152]. This method is used by Urteaga et al. to characterize the internal structure of p-AAO in a nondestructive manner [153]. The nanometric dimension of p-AAO pores hinders the study and elucidation of their geometry with classical techniques such as electron microscopy. However, the technique proposed in this work overcomes these dimension limitations and analyze the flow of liquids through the nanochannel arrays which behave as effective media when interacting with visible electromagnetic waves. Laser interferometry with submillisecond resolution is used to follow the dynamics of capillary driven fluid imbibition into the substrates. The method deals with reflectance interference and therefore can be considered a variation of the well-known Fabry–Pérot interferometry, but at a single wavelength.

This work concludes that the p-AAO pores are axisymmetric conical channels with a smallest radius (R_S) at one end, and a largest radius (R_L) at the other end. Both radii are precisely determined by measuring the filling times on each side of the membrane. It is worthy to mention that the difference detected between R_L and R_S is as small as 5 nm in pores of 75 μm of length, what denotes the high accuracy and reduced error of this method compared to classical microscopy. The optical thickness variation is also determined with the total number of oscillations and therefore the average refractive index of the membrane for a given physical thickness. Moreover, this method determines that p-AAO conical nanochannels present a clear anisotropy of capillary filling times, even if they are extremely slim. This innovative method for the identification of the internal geometry of p-AAO nanochannels has interesting potential applications in the fields of nanofiltration, flow rectification and biosensing.

6.4 Concluding Remarks

In this chapter we have given an overview of the optical properties of porous anodic alumina and of different examples of application of such optical properties. The most interesting intrinsic optical property is the photoluminescence of the material, caused by the electrochemical production process. This property can be modulated in several ways and applied in several ways such as in sensing or labeling. However, it is the porosity of the material and the possibility of modulating it both in the whole volume of the material or in its depth what confers a great versatility in the optical and photonic properties that can be obtained. These structure-based photonic properties lead to structures such as Fabry–Pérot cavities, distributed-Bragg reflectors, photonic pseudo-random crystals or rugate filters. Most of the applications of the optical and photonic properties are found in the biosensing field, as light is an excellent transduction mechanism to obtain information about the changes that happen in the pores and the pores inner surface when the different sensing reactions take place. However, other applications are also reported in literature, specifically in light generation and in photon-energy conversion. In conclusion, porous anodic alumina has demonstrated to date a great applicability as an optical nanomaterial, which possesses still a great potential for enlarging the list of applications.

References

1. E.D. Palik, *Handbook of Optical Constants of Solids* (Academic Press, Orlando, 1985)
2. G.E. Thompson, G.C. Wood, in *Treatise on Materials Science and Technology*, ed. by J.C. Scully (Academic Press, New York, 1983), pp. 205–329
3. G.D. Sulka, in *Nanostructured Materials in Electrochemistry*, ed. by A. Eftekhari (Wiley-VCH Verlag GmbH & Co. KGaA, Weinheim, 2008), pp. 1–116
4. Y. Li, G.W. Meng, L.D. Zhang, F. Phillipp, Ordered semiconductor ZnO nanowire arrays and their photoluminescence properties. Appl. Phys. Lett. **76**, 2011 (2000)
5. Y. Du, W.L. Cai, C.M. Mo, J. Chen, L.D. Zhang, X.G. Zhu, Preparation and photoluminescence of alumina membranes with ordered pore arrays. Appl. Phys. Lett. **74**, 2951 (1999)
6. X. Sun, F. Xu, Z. Li, W. Zhang, Photoluminescence properties of anodic alumina membranes with ordered nanopore arrays. J. Lumin. **121**, 588–594 (2006)
7. S. Chan, Y. Li, L.J. Rothberg, B.L. Miller, P.M. Fauchet, Nanoscale silicon microcavities for biosensing. Mater. Sci. Eng., C **15**, 277–282 (2001)
8. L.F. Marsal, L. Vojkuvka, P. Formentin, J. Pallarés, J. Ferré-Borrull, Fabrication and optical characterization of nanoporous alumina films annealed at different temperatures. Opt. Mater. (Amst) **31**, 860–864 (2009)
9. S. Nakamura, M. Saito, L.-F. Huang, M. Miyagi, K. Wada, Infrared optical constants of anodic alumina films with micropore arrays. Jpn. J. Appl. Phys. **31**, 3589–3593 (1992)
10. V.A. Yakovlev, E.A. Vinogradov, N.N. Novikova, G. Mattei, M.-P. Delplancke-Ogletree, Infrared reflectivity spectra of thin porous aluminum oxide films. Phys. Status Solidi **6**, 1697–1699 (2009)
11. Y. Li, G.H. Li, G.W. Meng, L.D. Zhang, F. Phillipp, Photoluminescence and optical absorption caused by the F^+ centres in anodic alumina membranes. J. Phys.: Condens. Matter **13**, 2691–2699 (2001)
12. J.H. Wu, X.L. Wu, N. Tang, Y.F. Mei, X.M. Bao, Strong ultraviolet and violet photoluminescence from Si-based anodic porous alumina films. Appl. Phys. A Mater. Sci. Process. **72**, 735–737 (2001)
13. W.L. Xu, M.J. Zheng, S. Wu, W.Z. Shen, Effects of high-temperature annealing on structural and optical properties of highly ordered porous alumina membranes. Appl. Phys. Lett. **85**, 4364 (2004)
14. A. Rauf, M. Mehmood, M. Ahmed, M. ul Hasan, M. Aslam, Effects of ordering quality of the pores on the photoluminescence of porous anodic alumina prepared in oxalic acid. J. Lumin. **130**, 792–800 (2010)
15. G. Shi, C.M. Mo, W.L. Cai, L.D. Zhang, Photoluminescence of ZnO nanoparticles in alumina membrane with ordered pore arrays. Solid State Commun. **115**, 253–256 (2000)
16. N.I. Mukhurov, S.P. Zhvavyi, S.N. Terekhov, A.Y. Panarin, I.F. Kotova, P.P. Pershukevich, I.A. Khodasevich, I.V. Gasenkova, V.A. Orlovich, Influence of electrolyte composition on photoluminescent properties of anodic aluminum oxide. J. Appl. Spectrosc. **75**, 214–218 (2008)
17. A. Nourmohammadi, S.J. Asadabadi, M.H. Yousefi, M. Ghasemzadeh, Photoluminescence emission of nanoporous anodic aluminum oxide films prepared in phosphoric acid. Nanoscale Res. Lett. **7**, 689 (2012)
18. T. Gao, G. Meng, L. Zhang, Blue luminescence in porous anodic alumina films: the role of the oxalic impurities. J. Phys.: Condens. Matter **15**, 2071–2079 (2003)
19. J.H. Chen, C.P. Huang, C.G. Chao, T.M. Chen, The investigation of photoluminescence centers in porous alumina membranes. Appl. Phys. A **84**, 297–300 (2006)
20. Y. Yamamoto, N. Baba, S. Tajima, Coloured materials and photoluminescence centres in anodic film on aluminium. Nature **289**, 572–574 (1981)
21. G.S. Huang, X.L. Wu, Y.F. Mei, X.F. Shao, G.G. Siu, Strong blue emission from anodic alumina membranes with ordered nanopore array. J. Appl. Phys. **93**, 582 (2003)

22. W.J. Stępniowski, M. Norek, M. Michalska-Domańska, A. Bombalska, A. Nowak-Stępniowska, M. Kwaśny, Z. Bojar, Fabrication of anodic aluminum oxide with incorporated chromate ions. Appl. Surf. Sci. **259**, 324–330 (2012)
23. K. Nielsch, J. Choi, K. Schwirn, R.B. Wehrspohn, U. Gösele, Self-ordering regimes of porous alumina: the 10 % porosity rule. Nano Lett. **2**, 677–680 (2002)
24. A. Santos, M. Alba, M.M. Rahman, P. Formentín, J. Ferré-Borrull, J. Pallarès, L.F. Marsal, Structural tuning of photoluminescence in nanoporous anodic alumina by hard anodization in oxalic and malonic acids. Nanoscale Res. Lett. **7**, 228 (2012)
25. J. Choi, Y. Luo, R.B. Wehrspohn, R. Hillebrand, J. Schilling, U. Gösele, Perfect two-dimensional porous alumina photonic crystals with duplex oxide layers. J. Appl. Phys. **94**, 4757 (2003)
26. T. Iijima, S. Kato, R. Ikeda, S. Ohki, G. Kido, M. Tansho, T. Shimizu, Structure of duplex oxide layer in porous alumina studied by 27Al MAS and MQMAS NMR. Chem. Lett. **34**, 1286–1287 (2005)
27. R. Dronov, A. Jane, J.G. Shapter, A. Hodges, N.H. Voelcker, Nanoporous alumina-based interferometric transducers ennobled. Nanoscale **3**, 3109–3114 (2011)
28. A. Santos, T. Kumeria, D. Losic, Nanoporous anodic aluminum oxide for chemical sensing and biosensors. TrAC Trends Anal. Chem. **44**, 25–38 (2013)
29. S.D. Alvarez, C.-P. Li, C.E. Chiang, I.K. Schuller, M.J. Sailor, A label-free porous alumina interferometric immunosensor. ACS Nano **3**, 3301–3307 (2009)
30. G. Macias, L.P. Hernández-Eguía, J. Ferré-Borrull, J. Pallares, L.F. Marsal, Gold-coated ordered nanoporous anodic alumina bilayers for future label-free interferometric biosensors. ACS Appl. Mater. Interfaces **5**, 8093–8098 (2013)
31. K.H.A. Lau, L.-S. Tan, K. Tamada, M.S. Sander, W. Knoll, Highly sensitive detection of processes occurring inside nanoporous anodic alumina templates: a waveguide optical study. J. Phys. Chem. B **108**, 10812–10818 (2004)
32. A. Santos, T. Kumeria, D. Losic, Optically optimized photoluminescent and interferometric biosensors based on nanoporous anodic alumina: a comparison. Anal. Chem. **85**, 7904–7911 (2013)
33. S. Bosch, J. Ferré-Borrull, J. Sancho-Parramon, A general-purpose software for optical characterization of thin films: specific features for microelectronic applications. Solid State Electron. **45**, 703–709 (2001)
34. S. Bosch, J. Ferré-Borrull, N. Leinfellner, A. Canillas, Effective dielectric function of mixtures of three or more materials: a numerical procedure for computations. Surf. Sci. **453**, 9–17 (2000)
35. D.A.G. Bruggeman, Berechnung verschiedener physikalischer Konstanten von heterogenen Substanzen. I. Dielektrizitätskonstanten und Leitfähigkeiten der Mischkörper aus isotropen Substanzen. Ann. Phys. **416**, 636–664 (1935)
36. D. Aspnes, J. Theeten, F. Hottier, Investigation of effective-medium models of microscopic surface roughness by spectroscopic ellipsometry. Phys. Rev. B **20**, 3292–3302 (1979)
37. S. Berthier, *Optique des milieux composites* (Polytechnica, Paris, 1993)
38. R. Swanepoel, Determination of the thickness and optical constants of amorphous silicon. J. Phys. E. **16**, 1214–1222 (1983)
39. A.C. Gâlcă, E.S. Kooij, H. Wormeester, C. Salm, V. Leca, J.H. Rector, B. Poelsema, Structural and optical characterization of porous anodic aluminum oxide. J. Appl. Phys. **94**, 4296 (2003)
40. P.H. Berning, Theory and calculations of optical thin films, in *Physics of Thin Films*, ed. by G. Hass (Academic Press, New York, 1963), pp. 69–121
41. H.A. Macleod, *Thin-film optical filters* (CRC Press Taylor, Boca Raton, 2010)
42. Q. Xu, Y. Yang, J. Gu, Z. Li, H. Sun, Influence of Al substrate on the optical properties of porous anodic alumina films. Mater. Lett. **74**, 137–139 (2012)
43. Q. Xu, H.-Y. Sun, Y.-H. Yang, L.-H. Liu, Z.-Y. Li, Optical properties and color generation mechanism of porous anodic alumina films. Appl. Surf. Sci. **258**, 1826–1830 (2011)

44. S. Garabagiu, G. Mihailescu, Thinning anodic aluminum oxide films and investigating their optical properties. Mater. Lett. **65**, 1648–1650 (2011)
45. K. Huang, L. Pu, Y. Shi, P. Han, R. Zhang, Y.D. Zheng, Photoluminescence oscillations in porous alumina films. Appl. Phys. Lett. **89**, 201118 (2006)
46. L.D. Zeković, V.V. Urošević, B.R. Jovanić, Determination of the refractive index of porous anodic oxide films on aluminium by a photoluminescence method. Thin Solid Films **139**, 109–113 (1986)
47. S. Gardelis, A.G. Nassiopoulou, V. Gianneta, M. Theodoropoulou, Photoluminescence-induced oscillations in porous anodic aluminum oxide films grown on Si: Effect of the interface and porosity. J. Appl. Phys. **107**, 113104 (2010)
48. C. Yang, W. Shen, Y. Zhang, Z. Ye, X. Zhang, K. Li, X. Fang, X. Liu, Color-tuning method by filling porous alumina membrane using atomic layer deposition based on metal-dielectric-metal structure. Appl. Opt. **53**, A142–A147 (2014)
49. J. Wang, Y. Li, D.-S. Wang, C.-W. Wang, The optical responses of one-dimensional photonic crystals based on the transparent Ag-anodic aluminum oxide composites with super low-refractive index. Thin Solid Films **520**, 6970–6974 (2012)
50. M.M. Rahman, E. Garcia-Caurel, A. Santos, L.F. Marsal, J. Pallarès, J. Ferré-Borrull, Effect of the anodization voltage on the pore-widening rate of nanoporous anodic alumina. Nanoscale Res. Lett. **7**, 474 (2012)
51. R.M.A. Azzam, N.M. Bashara, *Ellipsometry and polarized light* (North-Holland, Amsterdam, 1987)
52. A. De Martino, Y.K. Kim, E. Garcia-Caurel, B. Laude, B. Drévillon, Optimized Mueller polarimeter with liquid crystals. Opt. Lett. **28**, 616–618 (2003)
53. E. Garcia-Caurel, A. De Martino, B. Drévillon, Spectroscopic Mueller polarimeter based on liquid crystal devices. Thin Solid Films **455–456**, 120–123 (2004)
54. E. Garcia-Caurel, J. Nguyen, L. Schwartz, B. Drévillon, Application of FTIR ellipsometry to detect and classify microorganisms. Thin Solid Films **455–456**, 722–725 (2004)
55. E. Garcia-Caurel, B. Drévillon, A. De Martino, L. Schwartz, Application of Fourier transform infrared ellipsometry to assess the concentration of biological molecules. Appl. Opt. **41**, 7339–7345 (2002)
56. N. Leinfellner, J. Ferré-Borrull, S. Bosch, A software for optical characterization of thin films for microelectronic applications. Microelectron. Reliab. **40**, 873–875 (2000)
57. J. Sancho-Parramon, J. Ferré-Borrull, S. Bosch, M.C. Ferrara, Use of information on the manufacture of samples for the optical characterization of multilayers through a global optimization. Appl. Opt. **42**, 1325–1329 (2003)
58. W. Lee, R. Ji, U. Gösele, K. Nielsch, Fast fabrication of long-range ordered porous alumina membranes by hard anodization. Nat. Mater. **5**, 741–747 (2006)
59. W. Lee, K. Schwirn, M. Steinhart, E. Pippel, R. Scholz, U. Gösele, Structural engineering of nanoporous anodic aluminium oxide by pulse anodization of aluminium. Nat. Nanotechnol. **3**, 234–239 (2008)
60. D. Losic, M. Lillo, Porous alumina with shaped pore geometries and complex pore architectures fabricated by cyclic anodization. Small **5**, 1392–1397 (2009)
61. B. Wang, G.T. Fei, M. Wang, M.G. Kong, L.De Zhang, Preparation of photonic crystals made of air pores in anodic alumina. Nanotechnology **18**, 365601 (2007)
62. M.M. Rahman, L.F. Marsal, J. Pallarès, J. Ferré-Borrull, Tuning the photonic stop bands of nanoporous anodic alumina-based distributed bragg reflectors by pore widening. ACS Appl. Mater. Interfaces **5**, 13375–13381 (2013)
63. G. Macias, J. Ferré-Borrull, J. Pallarès, L.F. Marsal, 1-D nanoporous anodic alumina rugate filters by means of small current variations for real-time sensing applications. Nanoscale Res. Lett. **9**, 315 (2014)
64. H. Masuda, K. Fukuda, Ordered metal nanohole arrays made by a two-step replication of honeycomb structures of anodic alumina. Science **268**(80), 1466–1468 (1995)
65. K. Sakoda, *Optical Properties of Photonic Crystals*, 2nd edn. (Springer, Berlin, 2005)

66. J.D. Joannopoulos, R.D. Meade, J.N. Winn, *Photonic crystals: molding the flow of light* (Princenton University Press, Princenton, 1995)
67. H. Masuda, M. Satoh, Fabrication of gold nanodot array using anodic porous alumina as an evaporation mask. Jpn. J. Appl. Phys. **35**, L126–L129 (1996)
68. A.P. Li, F. Müller, A. Birner, K. Nielsch, U. Gösele, Hexagonal pore arrays with a 50–420 nm interpore distance formed by self-organization in anodic alumina. J. Appl. Phys. **84**, 6023 (1998)
69. H. Masuda, H. Asoh, M. Watanabe, K. Nishio, M. Nakao, T. Tamamura, Square and triangular nanohole array architectures in anodic alumina. Adv. Mater. **13**, 189–192 (2001)
70. H. Masuda, H. Yamada, M. Satoh, H. Asoh, M. Nakao, T. Tamamura, Highly ordered nanochannel-array architecture in anodic alumina. Appl. Phys. Lett. **71**, 2770 (1997)
71. X. Wang, S. Xu, M. Cong, H. Li, Y. Gu, W. Xu, Hierarchical structural nanopore arrays fabricated by pre-patterning aluminum using nanosphere lithography. Small **8**, 972–976 (2012)
72. I. Mikulskas, S. Juodkazis, R. Tomasiūnas, J.G. Dumas, Aluminum oxide photonic crystals grown by a new hybrid method. Adv. Mater. **13**, 1574 (2001)
73. H. Masuda, M. Yotsuya, M. Asano, K. Nishio, M. Nakao, A. Yokoo, T. Tamamura, Self-repair of ordered pattern of nanometer dimensions based on self-compensation properties of anodic porous alumina. Appl. Phys. Lett. **78**, 826 (2001)
74. J.M. Montero Moreno, M. Waleczek, S. Martens, R. Zierold, D. Görlitz, V.V. Martínez, V. M. Prida, K. Nielsch, Constrained order in nanoporous alumina with high aspect ratio: smart combination of interference lithography and hard anodization. Adv. Funct. Mater. **24**, 1857–1863 (2014)
75. A. Rodríguez, A. Arriola, T. Tavera, N. Pérez, S.M. Olaizola, Enhanced depth control of ultrafast laser micromachining of microchannels in soda-lime glass. Microelectron. Eng. **98**, 672–675 (2012)
76. H. Masuda, M. Yamada, F. Matsumoto, S. Yokoyama, S. Mashiko, M. Nakao, K. Nishio, Lasing from two-dimensional photonic crystals using anodic porous alumina. Adv. Mater. **18**, 213–216 (2006)
77. T. Yanagishita, K. Yasui, T. Kondo, Y. Kawamoto, K. Nishio, H. Masuda, Antireflection polymer surface using anodic porous alumina molds with tapered holes. Chem. Lett. **36**, 530–531 (2007)
78. A. Santos, P. Formentín, J. Pallarés, J. Ferré-Borrull, L.F. Marsal, Fabrication and characterization of high-density arrays of P3HT nanopillars on ITO/glass substrates. Sol. Energy Mater. Sol. Cells **94**, 1247–1253 (2010)
79. X. Fu, B. Zhang, X. Kang, J. Deng, C. Xiong, T. Dai, X. Jiang, T. Yu, Z. Chen, G.Y. Zhang, GaN-based light-emitting diodes with photonic crystals structures fabricated by porous anodic alumina template. Opt. Express **19**(Suppl 5), A1104–A1108 (2011)
80. Y.-W. Cheng, K.-M. Pan, C.-Y. Wang, H.-H. Chen, M.-Y. Ke, C.-P. Chen, M.-Y. Hsieh, H.-M. Wu, L.-H. Peng, J. Huang, Enhanced light collection of GaN light emitting devices by redirecting the lateral emission using nanorod reflectors. Nanotechnology **20**, 035202 (2009)
81. T. Dai, B. Zhang, X.N. Kang, K. Bao, W.Z. Zhao, D.S. Xu, G.Y. Zhang, Z.Z. Gan, Light extraction improvement from GaN-based light-emitting diodes with nano-patterned surface using anodic aluminum oxide template. IEEE Photonics Technol. Lett. **20**, 1974–1976 (2008)
82. O. Takayama, M. Cada, Two-dimensional metallo-dielectric photonic crystals embedded in anodic porous alumina for optical wavelengths. Appl. Phys. Lett. **85**, 1311 (2004)
83. Z. Král, L. Vojkůvka, E. Garcia-Caurel, J. Ferré-Borrull, L.F. Marsal, J. Pallarès, Calculation of angular-dependent reflectance and polarimetry spectra of nanoporous anodic alumina-based photonic crystal slabs. Photonics Nanostruct. Fundam. Appl. **7**, 12–18 (2009)
84. O. Jessensky, F. Müller, U. Gösele, Self-organized formation of hexagonal pore arrays in anodic alumina. Appl. Phys. Lett. **72**, 1173 (1998)

85. I. Maksymov, J. Ferré-Borrull, J. Pallarès, L.F. Marsal, Photonic stop bands in quasi-random nanoporous anodic alumina structures. Photonics Nanostruct. Fundam. Appl. **10**, 459–462 (2012)
86. M.M. Rahman, J. Ferré-Borrull, J. Pallarès, L.F. Marsal, Photonic stop bands of two-dimensional quasi-random structures based on macroporous silicon. Phys. Status Solidi **8**, 1066–1070 (2011)
87. J. Ferré-Borrull, J. Pallarès, G. Macías, L. Marsal, Nanostructural engineering of nanoporous anodic alumina for biosensing applications. Materials (Basel). **7**, 5225–5253 (2014)
88. M. Noormohammadi, M. Moradi, Structural engineering of nanoporous alumina by direct cooling the barrier layer during the aluminum hard anodization. Mater. Chem. Phys. **135**, 1089–1095 (2012)
89. W.J. Zheng, G.T. Fei, B. Wang, Z. Jin, L.De Zhang, Distributed Bragg reflector made of anodic alumina membrane. Mater. Lett. **63**, 706–708 (2009)
90. C. Papadopoulos, A. Rakitin, J. Li, A. Vedeneev, J. Xu, Electronic transport in Y-junction carbon nanotubes. Phys. Rev. Lett. **85**, 3476–3479 (2000)
91. W. Zheng, G. Fei, B. Wang, L. De Zhang, Modulation of transmission spectra of anodized alumina membrane distributed bragg reflector by controlling anodization temperature. Nanoscale Res. Lett. **4**, 665–667 (2009)
92. Y. Su, G.T. Fei, Y. Zhang, H. Li, P. Yan, G.L. Shang, L. De Zhang, Anodic alumina photonic crystal heterostructures. J. Opt. Soc. Am. B **28**, 2931 (2011)
93. G.D. Sulka, K. Hnida, Distributed Bragg reflector based on porous anodic alumina fabricated by pulse anodization. Nanotechnology **23**, 075303 (2012)
94. G.L. Shang, G.T. Fei, S.H. Xu, P. Yan, L. De Zhang, Preparation of the very uniform pore diameter of anodic alumina oxidation by voltage compensation mode. Mater. Lett. **110**, 156–159 (2013)
95. G.L. Shang, G.T. Fei, Y. Zhang, P. Yan, S.H. Xu, H.M. Ouyang, L.De Zhang, Fano resonance in anodic aluminum oxide based photonic crystals. Sci. Rep. **4**, 3601 (2014)
96. X.-X. Fu, X.-N. Kang, B. Zhang, C. Xiong, X.-Z. Jiang, D.-S. Xu, W.-M. Du, G.-Y. Zhang, Light transmission from the large-area highly ordered epoxy conical pillar arrays and application to GaN-based light emitting diodes. J. Mater. Chem. **21**, 9576 (2011)
97. X. Sheng, J. Liu, N. Coronel, A.M. Agarwal, J. Michel, L.C. Kimerling, Integration of self-assembled porous alumina and distributed bragg reflector for light trapping in Si photovoltaic devices. IEEE Photonics Technol. Lett. **22**, 1394–1396 (2010)
98. A. Santos, V.S. Balderrama, M. Alba, P. Formentín, J. Ferré-Borrull, J. Pallarès, L.F. Marsal, Nanoporous anodic alumina barcodes: toward smart optical biosensors. Adv. Mater. **24**, 1050–1054 (2012)
99. M. Saito, M. Shibasaki, S. Nakamura, M. Miyagi, Optical waveguides fabricated in anodic alumina films. Opt. Lett. **19**, 710 (1994)
100. A. Yamaguchi, K. Hotta, N. Teramae, Optical waveguide sensor based on a porous anodic alumina/aluminum multilayer film. Anal. Chem. **81**, 105–111 (2009)
101. F. Trivinho-Strixino, H.A. Guerreiro, C.S. Gomes, E.C. Pereira, F.E.G. Guimarães, Active waveguide effects from porous anodic alumina: An optical sensor proposition. Appl. Phys. Lett. **97**, 011902 (2010)
102. A. Gitsas, B. Yameen, T.D. Lazzara, M. Steinhart, H. Duran, W. Knoll, Polycyanurate nanorod arrays for optical-waveguide-based biosensing. Nano Lett. **10**, 2173–2177 (2010)
103. K. Hotta, A. Yamaguchi, N. Teramae, Deposition of polyelectrolyte multilayer film on a nanoporous alumina membrane for stable label-free optical biosensing. J. Phys. Chem. C **116**, 23533–23539 (2012)
104. T.D. Lazzara, I. Mey, C. Steinem, A. Janshoff, Benefits and limitations of porous substrates as biosensors for protein adsorption. Anal. Chem. **83**, 5624–5630 (2011)
105. C. Fu, Y. Gu, Z. Wu, Y. Wang, S. Xu, W. Xu, Surface-enhanced Raman scattering (SERS) biosensing based on nanoporous dielectric waveguide resonance. Sensors Actuators B Chem. **201**, 173–176 (2014)

106. Y. Du, L. Shi, T. He, X. Sun, Y. Mo, SERS enhancement dependence on the diameter and aspect ratio of silver-nanowire array fabricated by anodic aluminium oxide template. Appl. Surf. Sci. **255**, 1901–1905 (2008)
107. M. Shaban, A.G.A. Hady, M. Serry, A new sensor for heavy metals detection in aqueous media. IEEE Sens. J. **14**, 436–441 (2014)
108. D.-K. Kim, K. Kerman, H.M. Hiep, M. Saito, S. Yamamura, Y. Takamura, Y.-S. Kwon, E. Tamiya, Label-free optical detection of aptamer-protein interactions using gold-capped oxide nanostructures. Anal. Biochem. **379**, 1–7 (2008)
109. D.-K. Kim, K. Kerman, S. Yamamura, Y.-S. Kwon, Y. Takamura, E. Tamiya, Label-free optical detection of protein antibody-antigen interaction on Au capped porous anodic alumina layer chip. Jpn. J. Appl. Phys. **47**, 1351–1354 (2008)
110. S.-H. Yeom, O.-G. Kim, B.-H. Kang, K.-J. Kim, H. Yuan, D.-H. Kwon, H.-R. Kim, S.-W. Kang, Highly sensitive nano-porous lattice biosensor based on localized surface plasmon resonance and interference. Opt. Express **19**, 22882–22891 (2011)
111. E.J. Anglin, M.P. Schwartz, V.P. Ng, L.A. Perelman, M.J. Sailor, Engineering the chemistry and nanostructure of porous silicon Fabry-Pérot films for loading and release of a steroid. Langmuir **20**, 11264–11269 (2004)
112. K.-P.S. Dancil, D.P. Greiner, M.J. Sailor, A porous silicon optical biosensor: detection of reversible binding of IgG to a protein a-modified surface. J. Am. Chem. Soc. **121**, 7925–7930 (1999)
113. M.M. Orosco, C. Pacholski, M.J. Sailor, Real-time monitoring of enzyme activity in a mesoporous silicon double layer. Nat. Nanotechnol. **4**, 255–258 (2009)
114. E.C. Wu, J.S. Andrew, L. Cheng, W.R. Freeman, L. Pearson, M.J. Sailor, Real-time monitoring of sustained drug release using the optical properties of porous silicon photonic crystal particles. Biomaterials **32**, 1957–1966 (2011)
115. T. Kumeria, M.M. Rahman, A. Santos, J. Ferré-Borrull, L.F. Marsal, D. Losic, Structural and optical nanoengineering of nanoporous anodic alumina rugate filters for real-time and label-free biosensing applications. Anal. Chem. **86**, 1837–1844 (2014)
116. S. Pan, L.J. Rothberg, Interferometric sensing of biomolecular binding using nanoporous aluminum oxide templates. Nano Lett. **3**, 811–814 (2003)
117. J.-C. Lee, J.Y. An, B.-W. Kim, Application of anodized aluminium oxide as a biochip substrate for a Fabry–Perot interferometer. J. Chem. Technol. Biotechnol. **82**, 1045–1052 (2007)
118. T. Kumeria, D. Losic, Reflective interferometric gas sensing using nanoporous anodic aluminium oxide (AAO). Phys. Status Solidi Rapid Res. Lett **5**, 406–408 (2011)
119. T. Kumeria, L. Parkinson, D. Losic, A nanoporous interferometric micro-sensor for biomedical detection of volatile sulphur compounds. Nanoscale Res. Lett. **6**, 634 (2011)
120. T. Kumeria, A. Santos, D. Losic, Ultrasensitive nanoporous interferometric sensor for label-free detection of gold(III) ions. ACS Appl. Mater. Interfaces **5**, 11783–11790 (2013)
121. T. Kumeria, M.D. Kurkuri, K.R. Diener, L. Parkinson, D. Losic, Label-free reflectometric interference microchip biosensor based on nanoporous alumina for detection of circulating tumour cells. Biosens. Bioelectron. **35**, 167–173 (2012)
122. Y. He, X. Li, L. Que, A transparent nanostructured optical biosensor. J. Biomed. Nanotechnol. **10**, 767–774 (2014)
123. J. Álvarez, L. Sola, M. Cretich, M.J. Swann, K.B. Gylfason, T. Volden, M. Chiari, D. Hill, Real time optical immunosensing with flow-through porous alumina membranes. Sens. Actuators B Chem. **202**, 834–839 (2014)
124. A. Santos, J. Pallarès, P. Formentín, M. Alba, J. Ferré-Borrull, L.F. Marsal, V.S. Balderrama, Tunable Fabry-Pérot interferometer based on nanoporous anodic alumina for optical biosensing purposes. Nanoscale Res. Lett. **7**, 370 (2012)
125. T. Kumeria, D. Losic, Controlling interferometric properties of nanoporous anodic aluminium oxide. Nanoscale Res. Lett. **7**, 88 (2012)
126. D.-L. Guo, L.-X. Fan, F.-H. Wang, S.-Y. Huang, X.-W. Zou, Porous anodic aluminum oxide bragg stacks as chemical sensors. J. Phys. Chem. C **112**, 17952–17956 (2008)

127. P. Yan, G.T. Fei, G.L. Shang, B. Wu, L. De Zhang, Fabrication of one-dimensional alumina photonic crystals with a narrow band gap and their application to high-sensitivity sensors. J. Mater. Chem. C **1**, 1659 (2013)
128. G.L. Shang, G.T. Fei, Y. Zhang, P. Yan, S.H. Xu, L.De Zhang, Preparation of narrow photonic bandgaps located in the near infrared region and their applications in ethanol gas sensing. J. Mater. Chem. C **1**, 5285 (2013)
129. T. Kumeria, A. Santos, M.M. Rahman, J. Ferré-Borrull, L.F. Marsal, D. Losic, Advanced structural engineering of nanoporous photonic structures: tailoring nanopore architecture to enhance sensing properties. ACS Photonics **1**, 1298–1306 (2014)
130. T. Kumeria, A. Santos, D. Losic, Nanoporous anodic alumina platforms: engineered surface chemistry and structure for optical sensing applications. Sensors (Basel). **14**, 11878–11918 (2014)
131. A. Santos, T. Kumeria, D. Losic, Nanoporous anodic alumina: a versatile platform for optical biosensors. Materials (Basel). **7**, 4297–4320 (2014)
132. A. Santos, P. Formentín, J. Pallarès, J. Ferré-Borrull, L.F. Marsal, Structural engineering of nanoporous anodic alumina funnels with high aspect ratio. J. Electroanal. Chem. **655**, 73–78 (2011)
133. J. Pepper, R. Noring, M. Klempner, B. Cunningham, A. Petrovich, R. Bousquet, C. Clapp, J. Brady, B. Hugh, Detection of proteins and intact microorganisms using microfabricated flexural plate silicon resonator arrays. Sens. Actuators B Chem. **96**, 565–575 (2003)
134. A. Santos, G. MacÍas, J. Ferré-Borrull, J. Pallarès, L.F. Marsal, Photoluminescent enzymatic sensor based on nanoporous anodic alumina. ACS Appl. Mater. Interfaces **4**, 3584–3588 (2012)
135. X. Li, Y. He, L. Que, Fluorescence detection and imaging of biomolecules using the micropatterned nanostructured aluminum oxide. Langmuir **29**, 2439–2445 (2013)
136. L.F. Huang, M. Saito, M. Miyagi, K. Wada, Graded index profile of anodic alumina films that is induced by conical pores. Appl. Opt. **32**, 2039–2044 (1993)
137. T. Yanagishita, T. Kondo, K. Nishio, H. Masuda, Optimization of antireflection structures of polymer based on nanoimprinting using anodic porous alumina. J. Vac. Sci. Technol. B Microelectron. Nanom. Struct **26**, 1856–1859 (2008)
138. G. Hubbard, M.E. Nasir, P. Shields, C.R. Bowen, A. Satka, K.P. Parsons, N.H. Holmes, D. W.E. Allsopp, Angle dependent optical properties of polymer films with a biomimetic anti-reflecting surface replicated from cylindrical and tapered nanoporous alumina. Nanotechnology **23**, 155302 (2012)
139. J. Chen, B. Wang, Y. Yang, Y. Shi, G. Xu, P. Cui, Porous anodic alumina with low refractive index for broadband graded-index antireflection coatings. Appl. Opt. **51**, 6839–6843 (2012)
140. T. Kato, S. Hayase, Quasi-solid dye sensitized solar cell with straight ion paths. J. Electrochem. Soc. **154**, B117 (2007)
141. P. Granero, V.S. Balderrama, J. Ferré-Borrull, J. Pallarès, L.F. Marsal, Two-dimensional finite-element modeling of periodical interdigitated full organic solar cells. J. Appl. Phys. **113**, 043107 (2013)
142. P. Granero, V.S. Balderrama, J. Ferré-Borrull, J. Pallarès, L.F. Marsal, Light absorption modeling of ordered bulk heterojunction organic solar cells. Curr. Appl. Phys. **13**, 1801–1807 (2013)
143. K. Choi, S.H. Park, Y.M. Song, Y.T. Lee, C.K. Hwangbo, H. Yang, H.S. Lee, Nano-tailoring the surface structure for the monolithic high-performance antireflection polymer film. Adv. Mater. **22**, 3713–3718 (2010)
144. C.-F. Lai, C.-H. Chao, H.-C. Kuo, H.-H. Yen, C.-E. Lee, W.-Y. Yeh, Directional light extraction enhancement from GaN-based film-transferred photonic crystal light-emitting diodes. Appl. Phys. Lett. **94**, 123106 (2009)
145. D.-H. Kim, C.-O. Cho, Y.-G. Roh, H. Jeon, Y.S. Park, J. Cho, J.S. Im, C. Sone, Y. Park, W. J. Choi, Q.-H. Park, Enhanced light extraction from GaN-based light-emitting diodes with holographically generated two-dimensional photonic crystal patterns. Appl. Phys. Lett. **87**, 203508 (2005)

146. D. Tao, Z. Bei, Z. Zhen-Sheng, L. Dan, W. Xiao, B. Kui, K. Xiang-Ning, X. Jun, Y. Da-Peng, Z. Xing, Surface light extraction mapping from two-dimensional array of 12-fold photonic quasicrystal on current injected GaN-based LEDs. Chinese Phys. Lett. **24**, 979–982 (2007)
147. J.J. Wierer, A. David, M.M. Megens, III-nitride photonic-crystal light-emitting diodes with high extraction efficiency. Nat. Photonics **3**, 163–169 (2009)
148. C. Xiong, B. Zhang, X. Kang, T. Dai, G. Zhang, Diffracted transmission effects of GaN and polymer two-dimensional square-lattice photonic crystals. Opt. Express **17**, 23684–23689 (2009)
149. H. Masuda, M. Ohya, H. Asoh, M. Nakao, M. Nohtomi, T. Tamamura, Photonic crystal using anodic porous alumina. Jap. J. Appl. Phys., Part 2 Lett. **38**, (1999)
150. H. Masuda, M. Ohya, H. Asoh, K. Nishio, Photonic band gap in naturally occurring ordered anodic porous alumina. Jap. J. Appl. Phys., Part 2 Lett. **40**, (2001)
151. L.N. Acquaroli, R. Urteaga, C.L.A. Berli, R.R. Koropecki, Capillary filling in nanostructured porous silicon. Langmuir **27**, 2067–2072 (2011)
152. E. Elizalde, R. Urteaga, R.R. Koropecki, C.L.A. Berli, Inverse problem of capillary filling. Phys. Rev. Lett. **112**, 134502 (2014)
153. R. Urteaga, L.N. Acquaroli, R.R. Koropecki, A. Santos, M. Alba, J. Pallarès, L.F. Marsal, C. L.A. Berli, Optofluidic characterization of nanoporous membranes. Langmuir **29**, 2784–2789 (2013)

Chapter 7
Nanoporous Anodic Alumina for Optical Biosensing

Abel Santos and Tushar Kumeria

Abstract As a result of its physical and chemical properties, nanoporous anodic alumina (NAA) has been envisaged as a promising nanomaterials for developing optical biosensing system with outstanding capabilities and performances. Recent studies have demonstrated that the combination of NAA with different optical techniques can yield a variety of innovative optical sensing systems for a broad range of analytes and applications. In this context, this chapter aims to report on the recent advances and developments of NAA-based optical biosensing systems. The different optical detection techniques, principles and concepts of NAA-based optical biosensing systems will be described in detail. Our objective is to provide a simple but detailed overview about the different NAA-based optical biosensing systems, with especial emphasis on the performance and capabilities of these devices. Finally, we will provide a future perspective on challenges and developments for this promising research field.

7.1 Introduction

Current technological advances have made it possible to implement sensing devices into technological gadgets present in our ordinary lives such as smart phones or computers [1–5]. This has generated new opportunities towards the development of easy-to-use sensing systems for a broad range of applications, ranging from detection of levels of glucose in blood to water quality analysis [1–4]. Regardless of these outstanding technological advances, extensive fundamental research is still needed in order to make this technology reliable, feasible, cost-competitive and

A. Santos (✉) · T. Kumeria
School of Chemical Engineering, The University of Adelaide,
North Engineering Building, Adelaide 5005, Australia
e-mail: abel.santos@adelaide.edu.au

D. Losic and A. Santos (eds.), *Nanoporous Alumina*,
Springer Series in Materials Science 219, DOI 10.1007/978-3-319-20334-8_7

efficient. Optical sensors are composed of three basic parts: namely; (i) the light source which generates optical waves, (ii) the sensing platforms where the interaction light-matter occurs and (iii) the detector where the shift in the optical signal is measured [6]. These three components will establish the sensing characteristics and performance of the optical sensor and thus are critical parameters to consider for developing optical sensing system with optimised capabilities. In particular, the sensing platform is a key element in any optical sensor given that it will establish the interaction light-matter and thus how different analytes are detected and quantified. Note that the interaction between light and matter produces changes in the characteristic optical signal of the sensing platform. Shifts in the characteristic optical signal are measured by the detector and a quantitative (concentration) and/or qualitative (spectral signature) relationship between analyte molecules and optical signal shifts can be established.

Different materials can be used to produce optical sensing platforms. These can be optically active or passive materials with capability to guide, reflect, absorb, transmit, emit, confine and/or enhance incident optical waves emitted by the source. In the last decade, the use of nanomaterials as optical sensing platforms has made it possible to develop optical sensing systems with outstanding capabilities and performances in terms of sensitivity, selectivity, versatility and applicability. Some examples of optical techniques that have been successfully combined with nanomaterials are surface-enhanced Raman spectroscopy (SERS), surface plasmon resonance spectroscopy (SPR), localised surface plasmon resonance spectroscopy (LSPR), reflectometric interference spectroscopy (RIFS) and photoluminescence (PL). Among the different materials used to develop optical sensing platforms, nanoporous anodic alumina (NAA) has been recently envisaged as a promising nanomaterial due to its unique set of physical and chemical properties [7, 8]. NAA is produced by the electrochemical anodisation of aluminium and its nanopores can be used as nanocontainers to accommodate, immobilise and detect a broad range of analytes in a selective manner (Fig. 7.1). The geometry of nanopores in NAA can be engineered through different electrochemical approaches, making it possible to design and engineer the effective medium of this nanoporous material in order to produce a variety of optical structures such as distributed Bragg refractors, waveguides, microcavities, omnidirectional mirrors, rugate filters and so on [7, 8]. Additionally, surface chemistry in NAA can be tuned according to specific sensing purposes by well-established chemistry protocols [9, 10].

This chapter is aimed at describing the most relevant optical biosensing systems based on NAA combined with different optical techniques. First, we will introduce different electrochemical approaches used to produce photonic structures based on NAA. After this, we will provide detailed descriptions of NAA-based optical biosensing systems, explaining their sensing principles, performances and practical applications. Finally, we will conclude this chapter with a prospective outlook on this field and its future challenges.

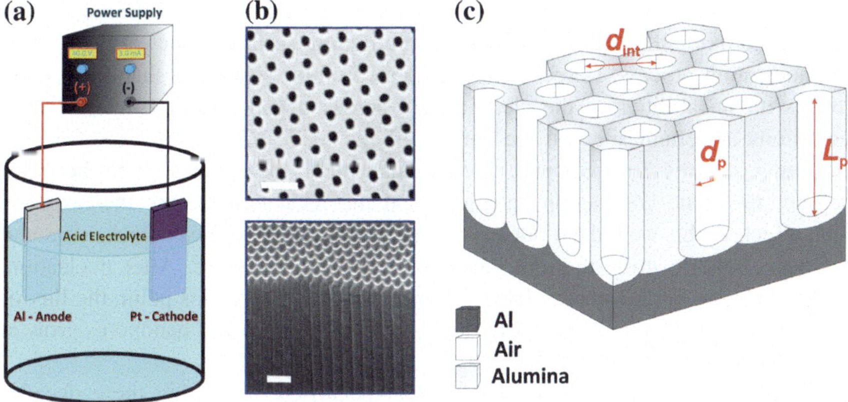

Fig. 7.1 Fabrication and geometric features of nanoporous anodic alumina. **a** Schematic illustration describing a basic electrochemical anodisation cell used to produce NAA. **b** Top and cross-section scanning electron microscopy (SEM) images of NAA (scale bars = 400 and 250 nm, respectively). **c** Scheme describing the most representative geometric features of NAA (i.e. d_p pore diameter, L_p pore length and d_{int} interpore distance)

7.2 Structural Engineering of Nanoporous Anodic Alumina

Electrochemical anodisation of high purity aluminium foils under specific conditions yields nanoporous anodic alumina structures. This nanofabrication technique, which was originally used in industrial processes such as metal surface finishing, machinery and automobile engineering, corrosion protection and so forth, was adopted by nanotechnology in the 1950s after electronic microscopes revealed the nanoporous nature of the aluminium oxide layers produced by anodisation of aluminium [11]. In particular, Keller characterised the structure of NAA for the first time by electron microscopy. This study revealed that NAA is composed of hexagonally arranged arrays of nanopores, in which the distance between adjacent nanopores (i.e. interpore distance) is directly proportional to the anodisation voltage [12]. This work was followed by a flood of studies aimed to characterise the physical and chemical properties of NAA structures produced under different conditions. Furthermore, pioneering theoretical models explaining the formation mechanism of NAA were proposed by Diggle in 1968 [13]. In the next years, Thompson and Wood made good use of new characterisation techniques such as microtome sectioning and transmission electron microscopy in order to understand the different phenomena playing a role in the formation mechanism of NAA [14, 15]. These studies established the anion and water incorporation mechanisms in the structure of NAA, which take place in the course of the anodisation process. After this, numerous theoretical models about the pore nucleation and growth mechanism in NAA were postulated [16–24]. However, the actual mechanism of NAA growth has yet to be completely clarified. In general, it is assumed that pore nucleation starts in the oxide thin film formed on the aluminium surface at the early stages of the anodisation process. Defects and pits located on the surface of that oxide layer along with instabilities in the electric field across it generate electric field concentrations at certain sites on the surface of the oxide film. These sites become nucleating centres for nanopore formation, where the electric field becomes focused, the ionic conduction enhanced and the local temperature increased by Joule effect. In this way, the oxide layer is preferentially dissolved at these sites and nanopores start to grow through the surface of the oxide film. After nucleation, nanopores grow until a steady state of growth is achieved. At this point, the flux of ionic species across the oxide barrier layer is in equilibrium and nanopores grow at constant rate (Fig. 7.2).

The flood of studies and intensive research carried out during these decades originated the seminal works of Masuda and co-workers, which are considered as the most important milestones in the use of NAA in nanotechnology [25–28]. Masuda and co-workers introduced the two-step anodisation process and reported on the self-organisation conditions in the most commonly used acid electrolytes

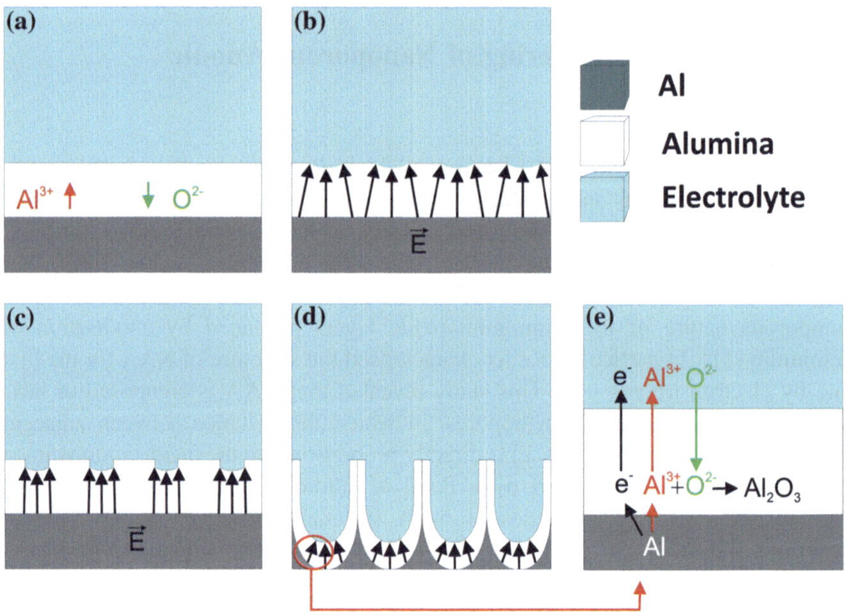

Fig. 7.2 Schematic cross-section diagram describing the first stages of growth of NAA. **a** Formation of a thin, compact layer of Al_2O_3. **b** Instabilities in the electric field across the oxide film dissolve partially the oxide at certain sites (i.e. nucleating centres). **c** Pore formation at the nucleating centres on the aluminium oxide surface. **d** Steady growth of pores (i.e. competition between formation and dissolution of aluminium oxide). **e** Magnified view of the *red circle* in (**d**) showing the transport of the main ionic species through the oxide barrier layer at the pore *bottom tips*

(i.e. sulphuric (H_2SO_4), oxalic ($H_2C_2O_4$) and phosphoric acids (H_3PO_4)). In this electrochemical approach, after the first anodisation step, the resulting NAA layer is chemically dissolved. The result is a pattern of hexagonally arranged hemispherical nanoconcavities on the surface of the aluminium substrate. These nanoconcavities enable the self-organised growth and propagation from top to bottom of cylindrical nanopores in the course of the second anodisation step, which is conducted under the same conditions as the first anodisation step. This simple but cost-effective and elegant anodisation approach makes it possible to produce highly organised NAA structures without the use of lithographic methods or expensive laboratory facilities. Self-organised NAA can be described as a nanoporous matrix of aluminium oxide (Al_2O_3—alumina) that features close-packed arrays of hexagonal cells containing a cylindrical nanopore at the centre. Nanopores grow perpendicularly to the underlying aluminium substrate. NAA is typically produced in electrolytes composed of aqueous solutions of acids, in which the anode (i.e. aluminium foil) and cathode (e.g. platinum) are immersed. The anodisation process starts when certain anodisation voltage or current is applied between anode and cathode. Then, nanopores nucleate and start to grow perpendicularly to the aluminium substrate. The growth of nanopores in steady state is a result of a competing electrochemical process between oxidation (i.e. formation of oxide) and dissolution (i.e. oxide dissolution) through the oxide barrier layer at the nanopore bottom tips. In this process, aluminium oxide grows at the interface aluminium-alumina as a result of the countermigration of ionic species through the oxide barrier layer. At the same time, alumina is dissolved at the interface alumina-electrolyte. Electrochemical anodisation of aluminium can be basically described by the following reduction-oxidation equations:

(i) Formation of alumina—aluminium-alumina interface (anode)

$$2Al(s) + 3H_2O(l) \rightarrow Al_2O_3(s) + 6H^+(aq) + 6e^- \tag{7.1}$$

(ii) Dissolution of alumina—alumina-electrolyte interface (anode)

$$Al_2O_3(s) + 6H^+(aq) \rightarrow 2Al^{3+}(aq) + 3H_2O(l) \tag{7.2}$$

(iii) Diffusion of aluminium cations—within oxide barrier layer (anode)

$$2Al(s) \rightarrow 2Al^{3+}(aq) + 6e^- \tag{7.3}$$

(iv) Hydrogen evolution—electrolyte-cathode interface (cathode)

$$6H^+(aq) + 6e^- \rightarrow 3H_2(g) \tag{7.4}$$

Note that, apart from the above electrochemical reaction, side reactions such as oxygen evolution at the anode take place in the course of the anodisation process. Therefore, the experimental current efficiency is always lower than 100 % during the electrochemical anodisation of aluminium. Aluminium can be anodised to produce NAA under two different regimes, so-called mild anodisation (MA) and hard anodisation (HA) [11, 25–28]. MA is conducted at moderate voltages and temperatures, while HA is carried out at high voltages and low temperatures. It is worth stressing that MA and HA conditions are dependent on the acid electrolyte. Another characteristic feature of these anodisation regimes is the growth rate, which is constant and slow under MA conditions (i.e. 3–8 μm h^{-1}) and exponentially decreasing and extremely fast at HA regime (i.e. 50–70 μm h^{-1}). As far as the geometric features of the resulting NAA structures, these can be defined by the pore diameter (d_p), pore length (L_p) and interpore distance (d_{int}). These geometric features can be precisely engineered by the anodisation conditions within the range of 10–400 nm for d_p, from a few nanometres to hundreds of micrometres for L_p and 50–600 nm for d_{int} [29, 30]. Table 7.1 compiles the geometric features of NAA structures produced under the most commonly used MA and HA anodisation conditions.

In terms of chemical composition, the structure of NAA presents an onion-like chemical distribution with an outer layer composed of aluminium oxide contaminated with impurities from the electrolyte and an inner layer basically composed of pure alumina [28] (Fig. 7.3). However, some studies have revealed that the actual chemical structure of NAA is divided into more than two chemical layers [31, 32]. In a study reported by Yamamoto et al., the chemical structure of NAA was analysed by studying its photoluminescence spectrum [31]. The obtained results revealed that the chemical structure of NAA is divided into three layers, with decreasing content of impurities from the central pore to the outer pore walls. Recently, Santos et al. analysed the chemical composition of NAA by studying its chemical dissolution in real-time by reflectometric interference spectroscopy [32]. These results revealed that the actual chemical structure of NAA is composed of up to four stratified chemical layers with increasing concentration of impurities from the inner side of the pores (i.e. layer close to the cell walls) to the outer side of the pores (i.e. layer close to the pore walls).

Note that the optical properties of NAA are directly related to its chemical composition and nanoporous structure. The structure of nanopores in NAA can be engineered by different electrochemical approaches in order to produce a variety of optical nanostructures with precisely designed optical properties. In this way, the

Table 7.1 Summary of the most widespread fabrication conditions used to produce NAA by mild (MA) and hard anodisation (HA) of high purity aluminium substrates along with the geometric features of the resulting nanostructures

Acid electrolyte	Anodisation regime	V (V)	T (°C)	d_p (nm)	d_{int} (nm)	Growth rate (μm h^{-1})	References
$H_2C_2O_4$ 0.3 M	HA	140	0–1	50	280	50	[11]
$H_2C_2O_4$ 0.3 M	MA	40	5–8	30	100	3.5	[25]
H_2SO_4 0.3 M	MA	25	5–8	25	63	7.5	[26]
H_3PO_4 0.1 M	MA	195	0–1	160	500	2	[27]
H_2SO_4 0.3 M	HA	40	0–1	30	78	85	[51]

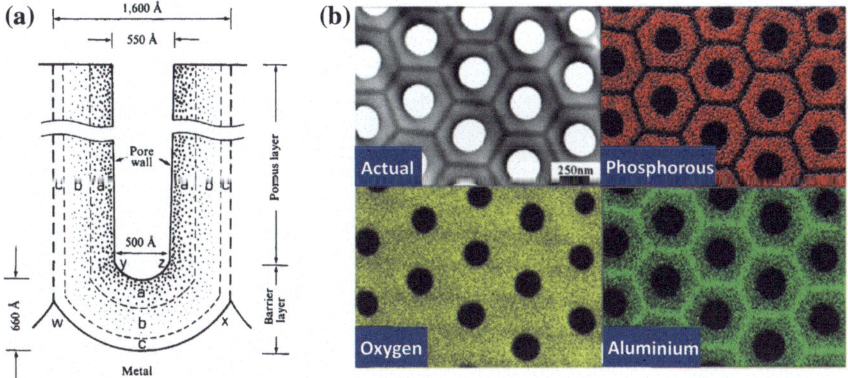

Fig. 7.3 Chemical structure of NAA. **a** Cross-section diagram showing the distribution of chemical impurities in the structure of NAA (adapted from [31]). **b** X-ray maps of elements distributed in the structure of NAA (adapted from [30])

Table 7.2 Compilation of the most representative pore geometries in NAA produced by different electrochemical approaches

Type of NAA	Electrochemical approach	References
Funnel-like	Combination of multiple anodisation and pore widening steps	[33–38]
Multilayered	Stepwise or pulse anodisation	[39–43]
Serrated	Anodisation at low voltage and high temperature	[44, 45]
Hierarchical	Asymmetric two-step anodisation	[46, 47]
Three-dimensional	Modulated anodisation profile and final wet chemical etching	[48, 49]
Modulated	Modulated anodisation profile	[50–52]

interaction between light and matter can be engineered in order to guide, confine, transmit, emit or reflect light in a selective manner for different purposes and applications. So far, optical nanostructures based on NAA have demonstrated a promising potential to be used as sensing platforms for developing optical biosensing systems. In that respect, it is worthwhile noting the versatility of NAA in terms of pore architecture, which can present nanopores with funnel-like [33–38], multilayered [39–43], serrated [44, 45], hierarchical [46, 47], three-dimensional [48, 49] and modulated [50–52]. Table 7.2 compiles some of the most common examples of NAA architectures produced by different electrochemical approaches.

Among these, the generation of pore diameter modulations in the course of the anodisation process has been intensively researched in the last years. Different electrochemical approaches have made it possible to modulate the pore diameter of NAA in depth and produce a variety of pore shapes with precisely controlled geometric features. Figure 7.4 summarises the most representative strategies used to modulate the pore diameter of NAA. Pore diameter modulations in NAA were

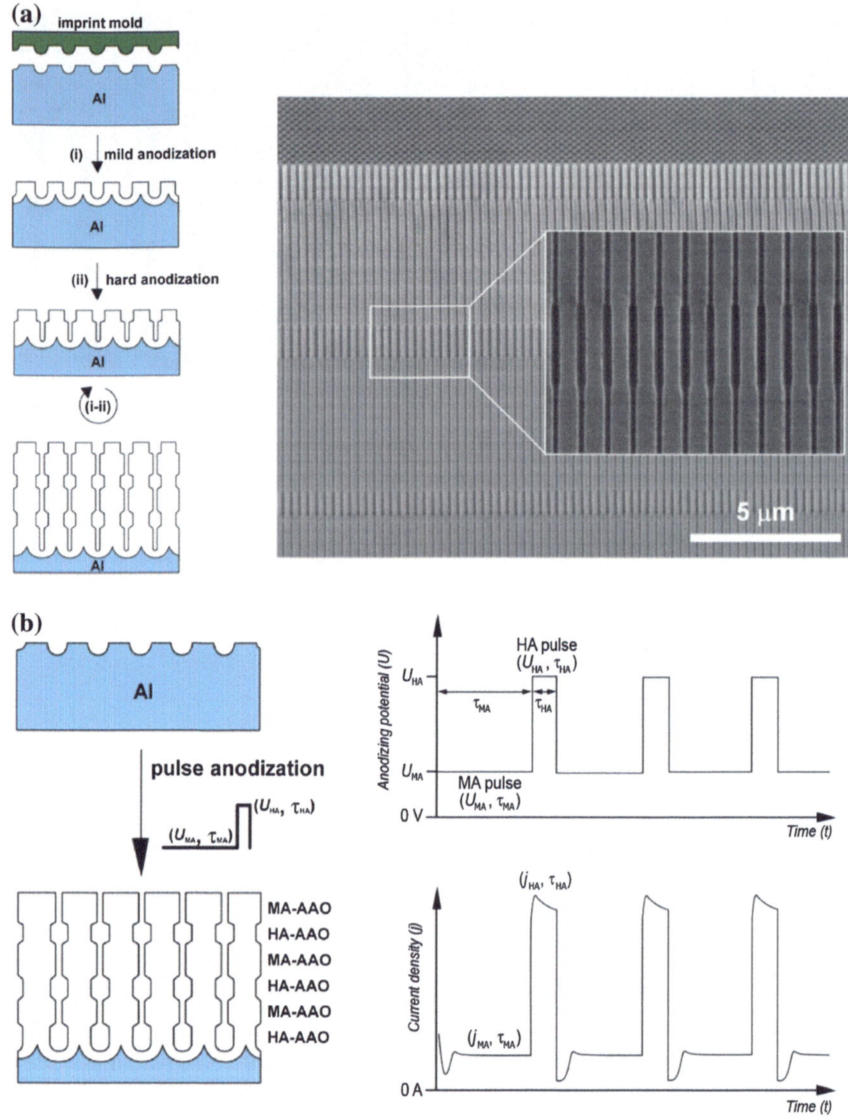

Fig. 7.4 Generation of pore diameter modulations in NAA. **a** Discontinuous nanopore modulation (adapted from [30]). **b** Continuous nanopore diameter by pulse anodisation (adapted from [30])

pioneered by Lee et al. by switching the anodisation conditions between MA to HA regimes [11]. To this end, the surface of electropolished aluminium substrates was patterned by a master stamp with hexagonally arranged arrays of nanopyramids prior to anodisation. The period of these nanopyramids (i.e. distance between the centre of adjacent nanopyramids) was 275 nm and the anodisation conditions of the first step were set to match that interpore distance under MA regime (i.e. H_3PO_4

0.4 M at 110 V and 10 °C for 15 min). As a result of that patterning stage nanopores grew in an organised arrangement from top to bottom in a single anodisation step. Then, the anodisation conditions were modified from MA to HA using an aqueous solution of oxalic acid 0.015 M at 137 V and 0.5 °C for 2 min. By repeating this process in a sequential cyclic fashion, they were able to modulate the diameter of nanopores in NAA from top to bottom. In this approach Lee et al. made good use of the different porosity levels between MA (10 %) and HA (3 %) conditions in oxalic acid as well as the fact that the interpore distance was the same under these conditions, preventing nanopores from branching after the anodisation conditions were changed. Furthermore, the length of each segment (i.e. MA or HA segments) was precisely controlled by the anodisation time, enabling the generation of NAA structures with exquisitely engineered geometric features from top to bottom. Although this work was the starting point of a flood of studies, this electrochemical approach has some inherent limitations given that a pre-patterning stage must be used and the acid electrolyte has to be changed after every anodisation step. In that respect, pulse anodisation, which is an electrochemical approaches based on voltage or current density pulses, made it possible to overcome these limitations and thus modulate the pore diameter in NAA in a simply manner [39, 50–52]. Among the different properties of NAA featuring pore diameter modulations, its optical properties are particularly interesting as they can be used to develop a variety of optical nanostructures for sensing applications.

Multilayered NAA is another type of structures with interesting optical properties, which can be used to develop sensing systems. These optical nanostructures are generated by changing the level of porosity between stacks of layers of NAA by switching the anodisation profile in the course of the anodisation process [39–43]. Note that different anodisation profiles such as saw-like, stepwise, sinusoidal or pseudosinusoidal can be used to design the effective medium of NAA by structural engineering. These electrochemical approaches have been used to produce NAA-based optical platforms such as microcavities, rugate filters, distributed Bragg reflectors, waveguides and omnidirectional mirrors. Lee et al. used a pulse anodisation approach under potentiostatic conditions in order to create multilayers of NAA in sulphuric acid electrolyte [39]. In this study, periodic stepwise voltage pulses between MA and HA regimes were used to design the nanoporous architecture of NAA along the pore length by introducing pore diameter modulations (Fig. 7.5). Making a good use of this approach, Sulka et al. demonstrated a NAA-based distributed Bragg reflector (DBR) structure [40]. In this approach, NAA was first produced by pulse anodisation in order to engineer its effective refractive index in depth. Then, the surface area of NAA was selectively infiltrated with polymer by using transmission electron microscopy grids as masks. The resulting NAA-based DBR mirrors showed enhanced reflection of light at two different ranges of wavelength. Zheng et al. produced NAA-based DBR structures by a pseudosinusoidal anodisation profile under voltage control [41]. This study demonstrated that the characteristic transmission peak of these DBR structures can be tuned within the visible spectrum by changing the temperature of the electrolyte. Furthermore, Santos et al. analysed the pore rearrangement phenomenon during the

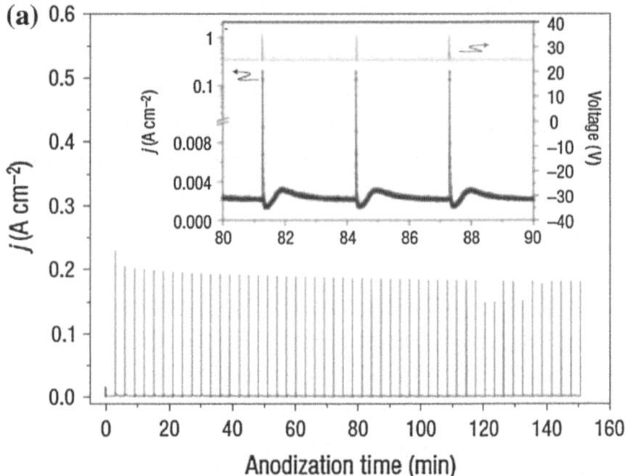

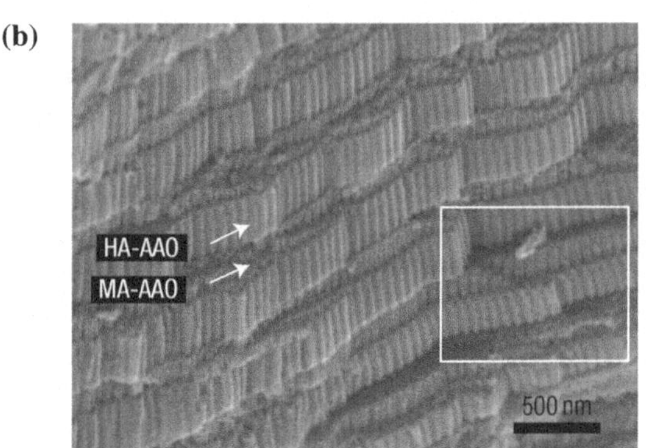

Fig. 7.5 Generation of multilayered nanostructures based on NAA by pulse anodisation. **a** Anodisation profile (adapted from [39]). **b** Cross-section SEM image of the resulting multilayered NAA structure (adapted from [39])

MA to HA regime transition studying the formation of bilayered NAA structures [42]. In this study, different parameters such as the anodisation voltage ramp and the HA voltage were systematically analysed in order to obtain a better understanding of the arrangement of nanopores when the anodisation regime is changed. These nanostructures can be used to develop bilayered NAA optical structures such as complex Fabry–Pérot interferometers. In a more recent work, Rahman et al. fabricated NAA-based DBR structures by a cyclic anodisation approach under voltage control [53]. In this study, a direct relationship between the photonic stop-band of the resulting NAA structures and the pore diameter was established, demonstrating that the photonic stop-band of these DBR structures can be engineered by modifying the pore diameter of NAA.

Funnel-like NAA is another type of structure with interesting optical properties, which can be used to develop optical biosensing systems [33–38]. Typically, this type of nanostructure is fabricated by combining sequential anodisation and pore widening steps in a discontinuous fashion (Fig. 7.6). The resulting NAA structures are composed of a stack of layers of NAA featuring decreasing pore diameter from top to bottom, which can range from a bilayered to a multilayered structure, depending on the number of anodisation and pore widening steps. As far as the

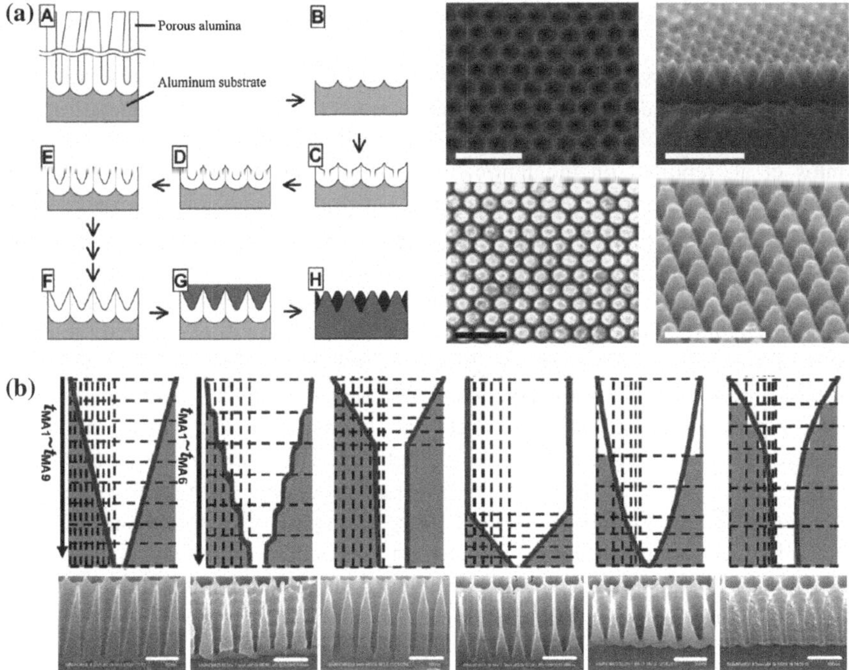

Fig. 7.6 Generation of funnel-like nanostructures based on NAA sequential combination of multiple anodisation and pore widening steps. **a** Fabrication process for low aspect ratio *funnel-like* NAA and SEM images of templates and nickel replicas (adapted from [35]). **b** Anodisation profiles and cross-section SEM image of *funnel-like* NAA structures featuring different geometries and shapes (adapted from [38])

fabrication process is concerned, this starts with an anodisation step, which is followed by a pore widening step. This process is repeated in a sequential fashion in order to engineer the nanoporous structure of NAA in depth. Note that funnel-like NAA structures can be used as templates to produce other nanostructures based on different materials such as polymers or metals by replica synthesis. The resulting nanostructures have interesting optical properties for different purposes such as antireflection coatings or optical biosensors. In that regard, He et al. replicated the inner structure of funnel-like NAA in order to produce shape-coded silica nanotubes, which were used as optical biosensing platforms for multiplexed immunoassays [33]. Another study by Nagaura et al. reported on the fabrication of funnel-like NAA featuring low aspect ratio (i.e. multi-short segments of NAA featuring smoothly decreased pore diameter from top to bottom) [34]. This work was extended by the same authors, who used these templates in order to create nickel-based replicas [35]. The resulting metallic films featured nanocones distributed across the surface according to the pattern established by the funnel-like NAA template. In another study, Yanagishita et al. made good use of funnel-like NAA structures in order to developed antireflection coatings based on polymers

[36]. These films were created by photoimprinting approach using funnel-like NAA templates featuring different pore geometry. After fabrication, the resulting polymer-based films were optically evaluated by studying their transmittance spectra. The obtained results revealed that smoother changes in the inner structure of the funnel-like NAA structure provide lower reflectance. Another study by Santos et al. reported on the fabrication of high aspect ratio funnel-like NAA structures (i.e. multi-long segments of NAA featuring smoothly decreased pore diameter from top to bottom) [37]. In this work, the length of each segment in the funnel-like structure was controlled by the total charge (i.e. total current charge passed through the system during each anodisation step), enabling a precise control over the length of each segment. In a more recent study, Li et al. tailored the inner shape of funnel-like NAA structures by combining anodisation and pore widening steps of different lengths [38]. They were able to generate whorl-embedded cones, funnels, pencils, linear cones, trumpets and parabolas in the inner nanoporous structure of NAA by this electrochemical approach. Traditionally, funnel-like NAA structures are produced featuring cylindrical nanopores with decreasing pore diameter from top to bottom. Nevertheless, Santos et al. reported on a unique electrochemical approach aimed to generate inverted funnel-like structures [32]. This type of NAA structure was produced by a sequential combination of anodisation and annealing steps with a final pore widening step. The resulting funnel-like structures feature increasing pore diameter from top to bottom. Note that, taking advantage of the optical properties of NAA, the formation process of inverted funnel-like structures was monitored in real-time by reflectometric interference spectroscopy, making it possible to precisely engineer their structure.

7.3 Optical Biosensors Based on NAA Structures

As mentioned above, nanostructures based on NAA have a unique set of optical properties that can be used to develop optical biosensing systems. As a result of its outstanding electronic and optical properties, porous silicon has been intensively researched in the last decades to produce a variety of optical sensing platforms. However, regardless of these properties, porous silicon has some inherent limitations such as disorganised nanoporous structure, fragility and poor chemical stability under harsh conditions. In that respect, NAA offers multiple advantages such as chemical stability under acidic or basic conditions, mechanical robustness, well-defined nanoporous structure and functional surface chemistry [54–56]. These assets make NAA a promising candidate for developing optical sensing systems and an alternative nanomaterial to traditional porous silicon sensing platforms. Recently, an extensive research work has been devoted to developing NAA-based optical sensing systems. Here, we will report on the most outstanding advances in the design, development and applicability of these optical sensing systems. Table 7.3 summarises the most representative examples of these optical systems and their sensing capabilities.

Table 7.3 Summary of optical biosensors based on NAA platforms

Sensing system	Analyte	Detection limit	References
SERS-NAA	p-Aminothiophenol	0.5 M	[58]
	4-Mercaptopyridine	$1 \cdot 10^{-6}$ M	[59]
	3-Mercaptobenzoic	3 mM	[61]
	Benzenethiol	500 ppb	[62]
	N-Methyl-4-nitroaniline	3 ppb	[62]
SPR-NAA	Avidin	10 μg mL^{-1}	[69]
	Anti-5-fluorouracil	100 mg mL^{-1}	[69]
	Melittin	100 ng mL^{-1}	[69]
	$Ru[BPhen_3]^{2+}$	2 μM	[64]
	$Fe[Phen_3]^{2+}$	1 μM	[64]
	BSA	60 nM	[65]
	Invertase	10 nM	[35]
RIfS-NAA	H_2S	0.5 v%	[83]
	DNA	2 nmol cm^{-2}	[81]
	Circulating tumour cells	1000 cells mL^{-1}	[82]
	Immunoglobulin	0.1 mg mL^{-1}	[79]
	Glucose	100 mM	[85]
	Cysteine	5 mM	[85]
	Gold(III)	0.1 μM	[86]
	BSA	1 mg mL^{-1}	[87]
	Glucose	0.01 M	[88]
	BSA	15 μM	[82]
	Human IgG	600 nM	[84]
PL-NAA	Morin	$5 \cdot 10^{-6}$ M	[97]
	Trypsin	40 μg mL^{-1}	[97, 100]
		0.1 mg mL^{-1}	
	DNA	100 mM	[98]
	Oxazine 170	$6.5 \cdot 10^{-3}$ M	[99]
	Glucose	0.1 M	[99]
	Glucose	10 mM	[85]
	Cysteine	5 mM	[85]

Detection technique, analyte and detection limit

7.3.1 Surface-Enhanced Raman Scattering (SERS)

Surface-enhanced Raman scattering spectroscopy (SERS) is an ultra-sensitive optical technique based on the enhancement of Raman scattering that takes place when biomolecules are absorbed on the surface of metal films featuring nano-structured roughness. This optical phenomenon has been associated with the excitation of localised surface plasmons on the surface of metallic nanostructures, which are known as "hot spots" or "hot junctions". However, a complete explanation of the actual mechanism involved in SERS is still to come. In the last decade, many studies have made good use of SERS as a powerful analytical technique in order to detect a broad range of analytes with implications in many research fields. Raman signals from biomolecules adsorbed onto the surface of SERS substrates can be enhanced around six orders of magnitude as compared to other analytical techniques, making it possible to achieve limits of detection as low as single molecules. Furthermore, SERS is not only a quantitative technique but also qualitative as biomolecules present characteristic SERS signals, which can be used as fingerprints for identification (i.e. spectroscopy signature). Among the different nanomaterials used to develop SERS substrates, NAA has showed a very promising potential as a result of its well-organised nanometric structure, which can be precisely replicated to produce metallic substrates with nanometric features, and its scalable and cost-efficient fabrication process.

NAA-based SERS substrates can be fabricated by depositing metals such as gold or silver on the top or bottom surface of NAA templates by thermal evaporation or sputtering. Nonetheless, other approaches have been used to produce NAA-based SERS substrates. Some examples of these are metallic membranes, decoration of nanopores with nanoparticles and growth of nanowires (Fig. 7.7). As mentioned before, the geometric features of NAA structures can be precisely engineered by the anodisation conditions, which along with the metal deposition conditions enable the fabrication of SERS substrates with optimised optical signals for specific sensing

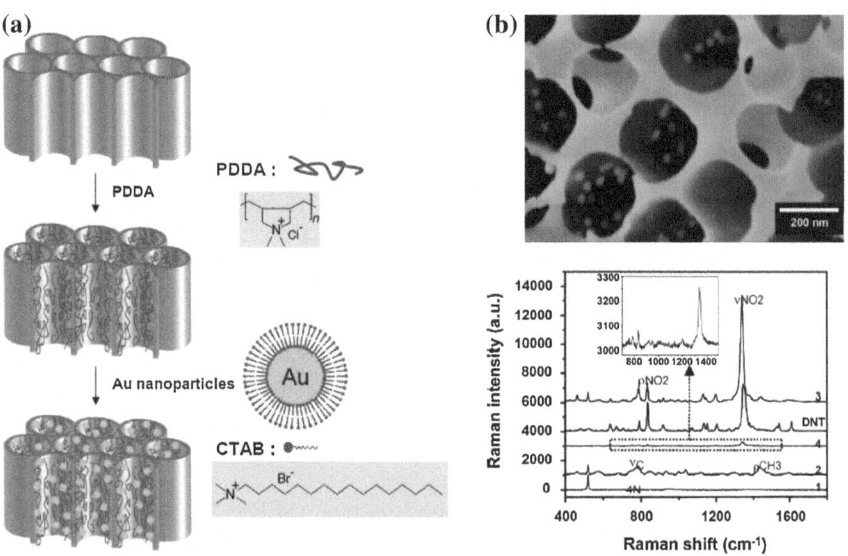

Fig. 7.7 NAA-based SERS systems. **a** Fabrication process for a NAA-based SERS substrate based on gold nanoparticles (adapted from [57]). **b** SEM image of the resulting SERS system and its sensing performance (adapted from [57])

purposes. For example, the inner surface of nanopores in NAA substrates was decorated with gold nanoparticles by Ko et al. [57]. The resulting NAA-based SERS substrates were used to detect trace amounts of an aromatic compound of TNT-based plastic explosive (i.e. 2,4-dinitrotoluene), which cannot be detected by conventional Raman spectroscopy. A similar approach was used by Lu et al., who decorated the inner surface of nanopores in NAA substrates with silver nanoparticles [58]. In this approach, silver nanoparticles were grown by incubation in an electrolyte composed of $Ag(NH_3)^{2+}$, which was subsequently reduced to silver nanoparticles.

Then, SERS signals were optimised as a function of the size and density of silver nanoparticles when detecting p-aminothiophenol. Note that a point-by-point SERS mapping of analyte molecules immobilised onto these NAA-based SERS substrates was used to characterise the homogeneity of adsorbed molecules across these substrates. In another approach, Ji et al. developed NAA-based SERS substrates by electrodepositing silver nanoparticles on the inner surface of NAA substrates [59]. The homogeneity and stability of the resulting SERS substrates were optimised by studying the effect of different electrodeposition times. The sensing performance of these NAA-based SERS substrates was evaluated by detecting 4-mercaptopyridine molecules. Using a different approach, Lee et al. developed NAA-based SERS substrates by electrodepositing silver nanowires inside NAA templates [60]. The sensing performance of the resulting optical sensing platforms was optimised by analysing the effect of the distance between adjacent nanowires within the alumina matrix when detecting 4-aminobenzenethiol. Their results revealed that the intensity of SERS signals can be enhanced up to 200-fold by reducing the intergap distance from 35 to 10 nm. Valleman et al. made good use of an electroless deposition approach inside NAA membranes [61]. The resulting gold-NAA composite membranes were used as SERS substrates to characterise the formation of monolayers of 3-mercaptobenzoic acid. It was reported that the resulting gold structure featured a slight curvature along the nanopores, which increased the SERS effect from the middle to the bottom of the nanopores. Recently, the inner surface of NAA substrates was decorated with silver nanocubes by Kodiyath et al., who used a layer-by-layer deposition approach based on polyelectrolytes [62]. The sensing performance of these NAA-based SERS substrates was evaluated by detecting trace levels of organic vapours of benzenethiol and N-methyl-4-nitroaniline. The obtained results demonstrated that these nanostructures can provide efficient, reliable and versatile sensing capabilities and that the limit of detention and saturation level are directly related to the number of polyelectrolyte bilayers deposited onto the inner surface of NAA nanopores. Furthermore, it was reported a direct and strong relationship between the density and distribution of silver nanocubes and the sensing performance. Finally, this study compared the sensing performance of silver nanocubes and nanoparticles, demonstrating that the former metallic nanostructures can provide better sensing capabilities and performances for the above-mentioned vapour analytes (Fig. 7.8).

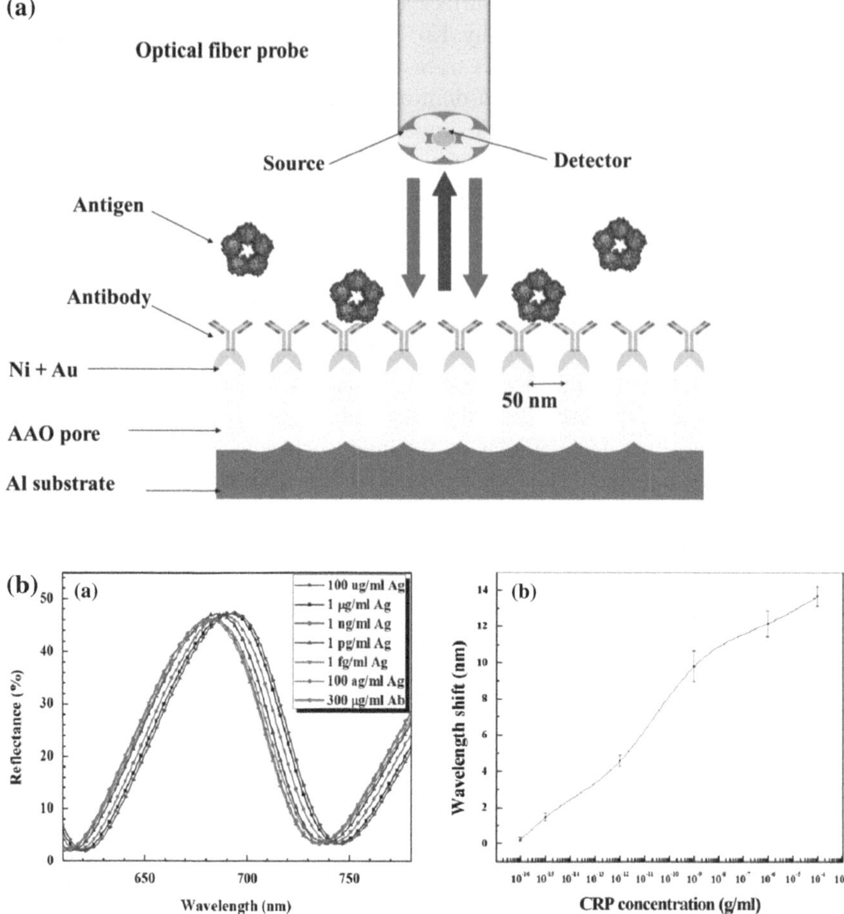

Fig. 7.8 NAA-based LSPR systems. **a** Structure of a LSPR system based on NAA coated with *gold* and chemically functionalised with antibodies to target C-reactive protein antigens (adapted from [70]). **b** Sensing performance of that system when detecting C-reactive protein (adapted from [70])

7.3.2 *Surface Plasmon Resonance (SPR)*

Similarly to SERS, surface plasmon resonance (SPR) is based on the generation of surface plasmons from an evanescent electromagnetic wave. This phenomenon occurs when a light beam is focused on the surface of a prism coated with a thin metallic film. This optical technique is typically implemented in a Kretschmann configuration. NAA structures can be used to develop unique SPR sensing systems. The most widespread method to produce NAA-based SPR systems is to grow a layer of NAA on the surface of a thin aluminium film deposited on the surface of the prism [63]. In this system, the plasmonic properties depend on the effective medium of the adjacent nanoporous matrix within distances of several hundreds of nanometres. For this reason, any change in the effective medium inside the nanopores of the NAA film is translated into a shift in the SPR signal, making it possible to quantify and detect trace levels of analyte molecules. Furthermore, NAA-based SPR systems can be used to monitor in real time and in situ biological events taking place inside the nanopores such as adsorption or desorption of molecules, association or dissociation of specific interactions between ligands or digestion of proteins. It is worthwhile mentioning that NAA-based SPR systems can be designed to provide improved sensing performances by engineering the nanoporous structure of the NAA film. In this way, light can be guided and confined in a selective manner, minimising losses associated with scattering effects [64]. Lau et al. and Koutsioubas et al. developed NAA-based SPR systems capable of monitoring the adsorption and desorption of bovine serum albumin (BSA) molecules inside the nanopores of NAA films in real-time and establishing the effect of pH on these processes [65, 66]. In another study, Dhathathreyan used a NAA-based SPR system in order to determine the kinetics of digestion of sucrose by enzyme invertase [67]. First, enzyme molecules were immobilised onto the inner surface of NAA nanopores and the digestion of sucrose to produce glucose and fructose was characterised in real-time as a function of pH by SPR. The obtained results demonstrated that this digestion process presents a biphasic kinetic for both the absorption of enzyme molecules and the digestion of sucrose. Lau et al. reported on the use of a NAA-based SPR system to characterise the grafting of poly (g-benzyl-L-glutamate) inside the nanopores, taking as a reference the surface of planar silicon dioxide substrates [68]. The obtained results confirmed that the conformation of the polymeric nanostructure inside the nanopores of NAA was a result of polymer chains confinement. Hotta et al. demonstrated that the sensing performance of NAA-based SPR systems can be significantly improved by engineering the structure of the NAA film [64]. To this end, they optimised the sensing performance of this system as a function of the pore geometry when detecting the adsorption of BSA molecules. They reported an enhanced red shift of more than 300 nm as compared to other NAA-based SPR system without optimised nanoporous structure.

It is worth stressing that SPR systems based on NAA structures are not only limited to a Kretschmann configuration. Localised surface plasmon resonance (LSPR) systems based on NAA platforms have been intensively researched as well. These systems are composed of highly ordered arrays of metallic nanoparticles such as gold or silver, which are deposited on the top or bottom surfaces of NAA substrates. The resulting metallic nanostructures feature organised arrangement at nanometric scale and can provide unique sensing performances for a broad range of analytes and applications. Hiep et al. developed a NAA-based LSPR system by depositing gold nanoparticles on the top surface of NAA substrates [69]. The sensing capabilities of that system were evaluated by detecting specific interactions between biotin/avidin and 5-fluorouracil/anti-5-fluorouracil. The obtained results demonstrated that this system provides a significant enhancement in terms of detection as a result of combined SPR and interference effects. Making good use of this, Yeom et al. fabricated an immunosensor to detect C-reactive protein [70]. In this system, gold nanoparticles were first deposited on the surface of NAA substrates and C-reactive protein antibodies immobilised onto theses LSPR platforms. Subsequently, C-reactive protein antigen molecules were selectively detected, achieving a low limit of detection as low as 1 fg mL^{-1}. Kim et al. developed a similar LSPR system, which was used to detect picomolar concentrations of untagged oligonucleotides when they hybridise with synthetic and PCR-amplified DNA molecules [71].

7.3.3 Reflectometric Interference Spectroscopy (RIfS)

Another highly sensitive and intensively researched optical technique is reflectometric interference spectroscopy (RIfS), which relies on the interaction between a white light beam and a thin film [72]. Gauglitz pioneered the use of RIfS for biosensing applications by combining RIfS with sensing platforms composed of polymer films based on polyethylene glycols and dextrans [73]. Typically, thin films of these polymers are applied onto different substrates such as silicon dioxide or glass and, subsequently, recognition elements for biomolecules such as antibodies, aptamers, esterone and phospholipid membranes are immobilized onto them. The resulting molecular interaction causes a quantifiable change in the effective optical thickness of the film, which results in a modulation of the interference spectrum. Therefore, to monitor this change over time makes it possible to characterise the binding behaviour of target molecules. Another important contribution to the development of biosensing systems based on RIfS was made by Sailor, who combined porous silicon platforms with RIfS in order to develop a variety of highly sophisticated sensing systems for a broad range of applications [74, 75]. Porous silicon-based RIfS systems have emerged as an outstanding alternative to traditional polymer thin films [76, 77]. Recently, some studies have reported on the development of NAA-based RIfS systems, which have showed a very promising potential. These systems could become an complementary alternative to porous silicon systems, providing some superior advantages in terms of chemical and mechanical stability, controllable pore geometry, scalable and cost-competitive fabrication process [7, 8] (Fig. 7.9). The RIfS spectrum of NAA structures presents a well-resolved interference pattern produced by the Fabry–Pérot effect, which is defined by $2n_{eff}L_p = m\lambda$, where n_{eff} is the effective refractive index of NAA, L_p is the physical thickness of the NAA film, and m is the order of the RIfS fringe, the maximum of which is located at the wavelength λ [78]. Note that RIfS is a label-free technique and thus makes it possible to follow biological events in real-time and in situ without the use of fluorescence or radioactive labels. NAA-based RIfS systems are able to perform ultra-sensitive detection of analytes such as organic gases and biomolecules in a quantitative and qualitative manner [79, 80]. Pan et al. developed a NAA-based RIfS system for label-free detection of

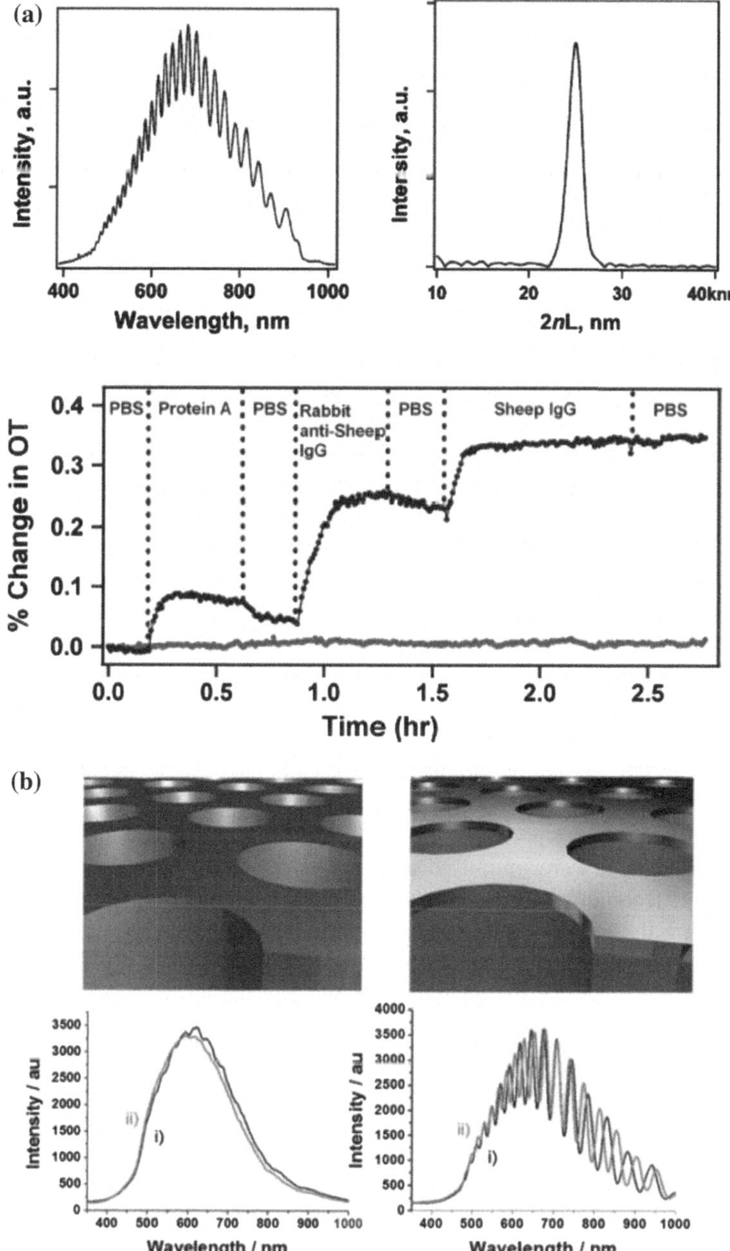

Fig. 7.9 NAA-based RIfS systems. **a** Sensing principle of a NAA-RIfS system and real-time monitoring of antibody-antigen interactions (adapted from [79]). **b** Enhancement of RIfS signals in NAA by coating its *top surface* with thin films of metal (adapted from [84])

complementary DNA molecules [81]. In another interesting study, Alvarez et al. fabricated a label-free immunosensor based on the combination of RIfS with NAA platforms [79]. This system was capable of selectively monitor binding interactions between antibodies and antigens, confirming that these produce a significative change in the effective optical thickness when the interaction is specific. Kumeria et al. made good use of this characteristic in order to develop a microchip biosensor based on RIfS and NAA platforms to detect and quantify circulating tumour cells (CTCs) [82]. NAA substrates were first coated with a thin layer of gold, which was subsequently functionalised with biotinylated anti-EpCAM antibodies aimed at capturing and detecting CTCs in a single step. Furthermore, these authors used a similar system to detect volatile sulphur and hydrogen sulphide compounds associated with malodour [83]. First, NAA platforms were coated with thin films of metals and monolayers of functional molecules were immobilised onto them in order to endow these sensors with chemical selectivity towards target gas molecules. Another NAA-based RIfS system was developed by Dronov et al., who fabricated an interferometric transducer by coating NAA platforms with platinum [84]. This study demonstrated that thin metal films on the top surface of NAA platforms enhance the interferometric pattern in RIfS, increasing the signal-to-noise ratio by LSPR effect. Furthermore, this study compared the sensing performance of these platforms with traditional porous silicon. The obtained results revealed that NAA is more sensitive than porous silicon, providing more stable and reliable optical signals. Note that the sensing performance of the proposed system was evaluated by detecting two types of immunoglobulin antibodies, anti-BSA and anti-human IgG. In another study, Santos et al. compared the sensing performance of NAA-based RIfS and photoluminescence sensors based on the same nanoporous sensing platform (vide infra) [85]. These sensors were assessed when detecting different analytes under non-specific and specific adsorption conditions. To this end, optical signals were optimised as a function of the pore geometry in both spectroscopic techniques. Then, the most optimal sensing platform was established by comparing the sensing performance when detecting D-glucose and L-cysteine under non-specific and specific adsorption conditions, respectively. These results showed that the nature of the analyte molecule and the adsorption conditions are key parameters in the resulting sensing performance of these optical systems. In addition, these results revealed that NAA platforms combined with photoluminescence present a better sensing performance than that of NAA-based RIfS platforms, with a better linearity, higher sensitivity and lower limit of detection. Recently, Kumeria et al. developed a NAA-based RIfS system capable of performing ultra-sensitive detection of gold ions in aqueous solutions [86]. First, NAA substrates were chemically functionalised with 3-mercaptopropyl-tirethoxysilane (MPTES) in order to endow these sensing platforms with chemical selectivity towards gold ions. The sensing principle in this system relies on the effective optical thickness change that takes place when gold ions are immobilised by thiol groups present of the inner surface of NAA nanopores. The sensing performance of this system was assessed by establishing the linear range, saturation concentration, sensitivity and low limit of detection. Furthermore, these authors demonstrated the applicability of this system for real-life scenarios through a series of experiments using different media such as buffer solution and tap water. In another study, Macías et al. made good use of a bilayered funnel-like NAA structure to develop a NAA-based RIfS system [87]. The RIfS spectrum of these sensing platforms presented a complex series of Fabry–Pérot interference fringes, which were studied when detecting BSA molecules. NAA-based rugate filters were developed by Kumeria et al. in order to demonstrate the direct relationship between sensing performance and the nanoporous structure of NAA [88]. This system, the sensing principle of which was based on shifts in the characteristic peak position of the rugate filter structure, was assessed when detecting different levels of glucose in a non-specific manner. Small changes in the effective medium of the nanoporous material produced shifts in the characteristic peak position, making it possible to discern the most sensitive NAA structure as a function of the nanoporous structure. Finally, the obtained results were verified with the Looyenga–Landau–Lifshitz model. An extension of this work was reported by the same authors, who demonstrated these NAA-based DBR structures are capable of performing detection of low concentrations of mercury ions in tap and river water [89].

In addition to the above mentioned sensing capabilities, NAA-based RIfS systems can be used to monitor the release of molecules from these nanoporous substrates. Making good use of this property, Kumeria et al. were able to study the release of indomethacin, an anti-inflammatory drug, from NAA substrates under dynamic flow conditions through changes in the effective optical thickness of the nanoporous platform [90]. This study demonstrated that this process is controlled by the diffusion of drug molecules from the nanopores to the eluting medium and that this is directly related with the flow rate. These results revealed that the faster the flow rate the faster the release of drug molecules from the nanoporous substrate. In contrast to traditional static conditions, this system provides more reliable information that can help to engineer drug-releasing implants with improved capabilities for in vivo applications.

7.3.4 Photoluminescence Spectroscopy (PLS)

Nanoporous anodic alumina is a well-known photoluminescent material, the properties of which have been intensively studied during the last decades. However, the actual origin of its photoluminescent properties is still a matter of debate [91–93]. It is generally accepted that PL in NAA relies on two types of photoluminescent centres. F-centres are associated with the total amount of impurities incorporated into the chemical structure of NAA from the acid electrolyte in the course of the anodisation process [94]. F^{+}-centres are related to ionised oxygen vacancies present in the amorphous structure of NAA [92, 93]. Therefore, the PL properties of NAA are highly dependent on the fabrication conditions such as the acid electrolyte, the anodisation voltage/current density, the anodisation regime and thermal treatments [91, 95, 96]. As far as the use of these PL properties for sensing

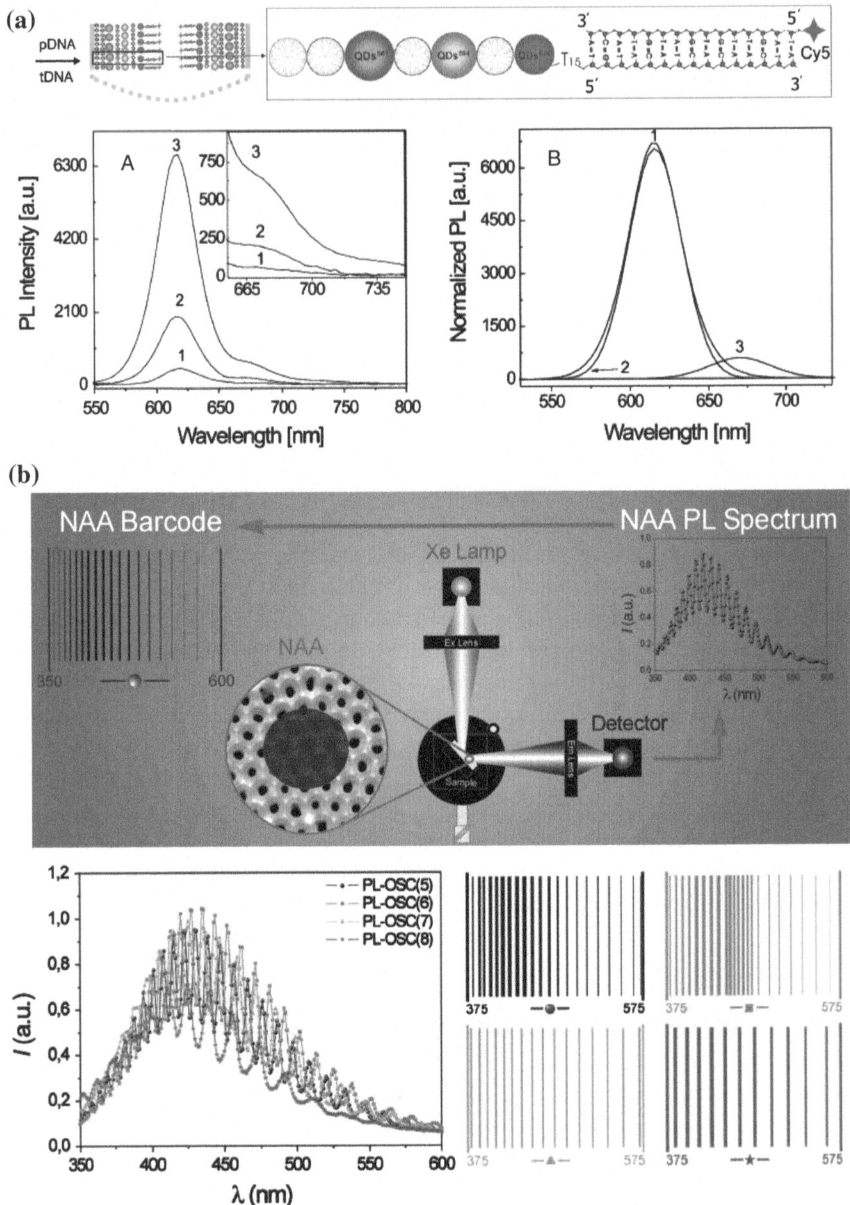

Fig. 7.10 NAA-based PLS systems. **a** PL-NAA biosensing platform developed to monitor DNA hybridisation through changes in the PL spectrum of NAA (adapted from [98]). **b** Optical barcode system based on PL-NAA biosensing platforms (adapted from [99])

purposes, pioneering studies used NAA substrates to accommodate, immobilise and detect analyte molecules through changes in the PL spectrum of NAA platforms. Ji et al. detected non-fluorescent morin and morin-trypsin infiltrated in the nanoporous network of NAA substrates using as a sensing principle shifts in the PL spectrum of these platforms [97]. The obtained results revealed chemical interactions between morin molecules and alumina. Feng et al. developed a more sophisticated system capable of monitoring the hybridisation of DNA molecules immobilised onto the nanoporous structure of NAA [98]. In this system, the inner surface of NAA nanopores was chemically functionalised with mercaptoundecanoic acid. This was followed by layer-by-layer deposition of positively and negatively charged dendrimers, intercalating layers of ZnCdSe quantum dots.

This NAA-based PL system demonstrated enhanced sensitivity afforded by the use of photoluminescent quantum dots, which enable the detection of the hybridisation of DNA molecules (Fig. 7.10a). Furthermore, a direct relationship between the assemblies of quantum dots and the sensitivity of the system was established. Santos et al. pioneered a barcode system for sensing applications based on the PL spectrum of NAA substrates in the UV-visible region [99]. Likewise in RIfS, the PL spectrum of NAA presents well-resolved and narrow oscillations as a result of interferometric effect, which amplifies and enhances the emission of light from the NAA structure (Fig. 7.10b). These oscillations correspond to the optical modes of the optical cavity formed by the NAA film and can be used to develop optical sensors as any change in the effective medium is translated into shifts in these characteristic oscillations. This principle was used to develop a barcode system for sensing applications, which was demonstrated by detecting organic molecules such as organic dyes, enzymes and glucose [99, 100].

7.4 Conclusions

As this chapter has shown, NAA has become the nanomaterial of choice to develop a variety of nanostructures with outstanding physical and chemical properties. Particularly interesting are the optical properties of these nanostructures as nanopores in NAA can be engineered through different electrochemical approaches in order to reflect, guide, emit, transmit and confine light in a selective manner and create a variety of optical sensing platforms with unique capabilities and performances. Here, we have summarised the most representative fundamental aspects related with the structural engineering of NAA aimed at producing optical nanostructures. Furthermore, we have presented in detail some outstanding combinations of optical sensing techniques NAA photonic structures such as SERS, SPR, LSPR, RIfS and PLS. These studies have demonstrated that the key features of NAA such as scalable and cost-efficient fabrication process, versatile nanopore geometry, chemical and mechanical stability and optical properties make this nanomaterial a promising alternative to conventional sensing platforms.

There are still excellent opportunities for further developments of optical biosensing systems based on NAA structures. Particularly interesting are the implementation of these systems into lab-on-a-chip devices with separation, detection and quantification capabilities for multiple analytes. Furthermore, the development of implantable biosensors for in vivo applications and the development of point-of-care systems for medical and environmental diagnosis are thought to be two areas of focus for future applications of NAA-based optical sensors.

Acknowledgments Financial support from Australian Research Council (DE14010054) and the School of Chemical Engineering of The University of Adelaide are greatly acknowledged.

References

1. Q. Wei, R. Nagi, K. Sadeghi, S. Feng, E. Yan, S.J. Ki, R. Caire, D. Tseng, A. Ozcan, Detection and spatial mapping of mercury contamination in water samples using a smart-phone. ACS Nano **2**, 1121–1129 (2014)
2. Q. Wei, H. Qi, W. Luo, D. Tseng, S.J. Ki, Z. Wan, Z. Göröcs, L.A. Bentolila, T.T. Wu, R. Sun, A. Ozcan, Fluorescent imaging of single nanoparticles and viruses on a smart phone. ACS Nano **10**, 9147–9155 (2013)
3. S. Khatua, M. Orrit, Toward single-molecule microscopy on a smart phone. ACS Nano **10**, 8340–8343 (2013)
4. S. Ayas, A. Cupallari, O.O. Ekiz, Y. Kaya, A. Dana, Counting molecules with a mobile phone camera using plasmonic enhancement. ACS Photonics **1**, 17–26 (2014)
5. Y.D. Ivanov, T.O. Pleshakova, N.V. Krohin, A.L. Kaysheva, S.A. Usanov, A.I. Archakov, Registration of the protein with compact disk. Biosens. Bioelectron. **43**, 384–390 (2013)
6. J. Homola, S.S. Yee, G. Gauglitz, Surface plasmon resonance sensors: review. Sens. Actuators, B **54**, 3–15 (1999)
7. A. Santos, T. Kumeria, D. Losic, Nanoporous anodic alumina: a versatile platform for optical biosensors. Materials **7**, 4297–4320 (2014)
8. T. Kumeria, A. Santos, D. Losic, Nanoporous anodic alumina platforms: engineered surface chemistry and structure for optical sensing applications. Sensors **14**, 11878–11918 (2014)
9. F. Rusmini, Z. Zhong, J. Feijen, Protein immobilization strategies for protein biochips. Biomacromolecules **8**, 1775–1789 (2007)
10. C.J. Ingham, J. ter Maat, W.M. de Vos, Where bio meets nano: the many used of nanoporous aluminium oxide in biotechnology. Biotechnol. Adv. **30**, 1089–1099 (2012)
11. W. Lee, R. Ji, U. Gösele, K. Nielsch, Fast fabrication of long-range ordered porous alumina membranes by hard anodization. Nat. Mater. **5**, 741–747 (2006)
12. F. Keller, M.S. Hunter, D.L. Robinson, Structural features of oxide coatings on aluminum. J. Electrochem. Soc. **100**, 411–419 (1953)
13. J.W. Diggle, T.C. Downie, C.W. Coulding, Anodic oxide films on aluminum. Chem. Rev. **69**, 365–405 (1969)
14. J.P. O'Sullivan, G.C. Wood, The morphology and mechanism of formation of porous anodic films on aluminium. Proc. R. Soc. London, Ser. A **317**, 511–543 (1970)
15. G.E. Thompson, G.C. Wood, Anodic films on aluminum. *Treatise on materials science and technology*, vol. 23 (Academic Press, New York, 1983)
16. T.P. Hoar, N.F. Mott, A mechanism for the formation of porous anodic oxide films on aluminium. J. Phys. Chem. Solids **9**, 97–99 (1959)
17. V.P. Parkhutik, V.I. Shershulsky, Theoretical modelling of porous oxide growth on aluminum. J. Phys. D Appl. Phys. **25**, 1258–1263 (1992)

18. G. Patermarakis, P. Lenas, C.H. Karavassilis, G. Papayiannis, Kinetics of growth of porous anodic Al_2O_3 films on Al metal. Electrochim. Acta **36**, 709–725 (1991)
19. G. Patermarakis, N. Papandreadis, Study on the kinetics of growth of porous anodic Al_2O_3 films on Al metal. Electrochim. Acta **38**, 2351–2361 (1993)
20. G. Patermarakis, D. Tzouvelekis, Development of a strict kinetic model for the growth of porous anodic Al_2O_3 films on aluminium. Electrochim. Acta **39**, 2419–2429 (1994)
21. G. Patermarakis, H.S. Karayannis, The mechanism of growth of porous anodic Al_2O_3 films on aluminium at high film thicknesses. Electrochim. Acta **40**, 2647–2656 (1995)
22. G. Patermarakis, Transformation of the overall strict kinetic model governing the growth of porous anodic Al_2O_3 films on aluminium to a form applicable to the non-stirred bath film growth. Electrochim. Acta **41**, 2601–2611 (1996)
23. J. Randon, P.P. Mardilovich, A.N. Govyadinov, R. Paterson, Computer-simulation of inorganic membrane morphology Part 3 anodic alumina films and membranes. J. Colloid Interface Sci. **169**, 335–341 (1995)
24. S.K. Thamida, H.C. Chang, Nanoscale pore formation dynamics during aluminum anodization. Chaos **12**, 240–251 (2002)
25. H. Masuda, K. Fukuda, Ordered metal nanohole arrays made by a two-step replication of honeycomb structures of anodic alumina. Science **268**, 1466–1468 (1995)
26. H. Masuda, F. Hasegwa, Self-ordering of cell arrangement of anodic porous alumina formed in sulfuric acid solution. J. Electrochem. Soc. **144**, L127–L130 (1997)
27. H. Masuda, K. Yada, A. Osaka, Self-ordering of cell configuration of anodic porous alumina with large-size pores in phosphoric acid solution. Jpn. J. Appl. Phys. **37**, L1340–L1342 (1998)
28. K. Nielsch, J. Choi, K. Schwirn, R.B. Wehrspohn, U. Gösele, Self-ordering regimes of porous alumina: the 10 % porosity rule. Nano Lett. **2**, 677–680 (2002)
29. G.D. Sulka, Highly ordered anodic porous alumina formation by self-organized anodizing, in *Nanostructured Materials in Electrochemistry*, vol. 1, ed. by A. Eftekahari (Wiley-VCH Verlag GmbH and Co. KGaA, Weinhem, Germany, 2008), pp. 1–116
30. W. Lee, S.J. Park, Porous anodic aluminium oxide: anodization and template synthesis of functional nanostructures. Chem. Rev. **114**, 7487–7556 (2014)
31. Y. Yamamoto, N. Baba, S. Tajima, Coloured materials and photoluminescence centres in anodic film on aluminium. Nature **289**, 572–574 (1981)
32. A. Santos, T. Kumeria, Y. Wang, D. Losic, *In situ* monitored engineering of inverted nanoporous anodic alumina funnels: on the precise generation of 3D optical nanostructures. Nanoscale **6**, 9991–9999 (2014)
33. B. He, S.J. Son, S.B. Lee, Suspension array with shape-coded silica nanotubes for multiplexed immuno assays. Anal. Chem. **79**, 5257–5263 (2007)
34. T. Nagaura, F. Takeuchi, S. Inoue, Fabrication and structural control of anodic alumina films with inverted cone porous structure using multi-step anodizing. Electrochim. Acta **53**, 2109–2114 (2008)
35. T. Nagaura, F. Takeuchi, Y. Yamauchi, K. Wada, S. Inoue, Fabrication of ordered Ni nanocones using porous anodic alumina template. Electrochem. Commun. **10**, 681–685 (2008)
36. T. Yanagishita, T. Kondo, K. Nishio, H. Masuda, Optimization of antireflection structures of polymer based on nanoimprinting using anodic porous alumina. J. Vac. Sci. Technol., B **26**, 1856–1859 (2008)
37. A. Santos, P. Formentín, J. Pallarès, J. Ferré-Borrull, L.F. Marsal, Structural engineering of nanoporous anodic alumina funnels with high aspect ratio. J. Electroanal. Chem. **655**, 73–78 (2011)
38. J. Li, C. Li, C. Chen, Q. Hao, Z. Wang, J. Zhu, X. Gao, Facile method for modulating the profiles and periods of self-ordered three-dimensional alumina taper-nanopores. ACS Appl. Mater. Interfaces **4**, 5678–5683 (2012)

39. W. Lee, K. Schwirn, M. Steinhart, E. Pippel, R. Scholz, U. Gösele, Structural engineering of nanoporous anodic aluminium oxide by pulse anodization of aluminium. Nat. Nanotechnol. **3**, 234–239 (2008)
40. G.D. Sulka, K. Hnida, Distributed Bragg reflector base on porous anodic alumina fabricated by pulse anodization. Nanotechnology **23**, 075303 (2012)
41. W.J. Zheng, G.T. Fei, B. Wang, L.D. Zhang, Modulation of transmission spectra of anodized alumina membrane distributed Bragg reflector by controlling anodization temperature. Nanoscale Res. Lett. **4**, 665–667 (2009)
42. A. Santos, J.M. Montero-Moreno, J. Bachmann, K. Nielsch, P. Formentín, J. Ferré-Borrull, J. Pallarès, L.F. Marsal, Understanding pore rearrangement during mild to hard transition in bilayered porous anodic alumina membranes. ACS Appl. Mater. Interfaces **3**, 1925–1932 (2011)
43. M.M. Rahman, L.F. Marsal, J. Pallarès, J. Ferré-Borrull, Tuning the photonic stop bands of nanoporous anodic alumina-based distributed Bragg reflectors by pore widening. ACS Appl. Mater. Interfaces **5**, 13375–13381 (2013)
44. D. Li, L. Zhao, C. Jiang, J.G. Lu, Formation of anodic aluminum oxide with serrated nanochannels. Nano Lett. **10**, 2766–2771 (2010)
45. X. Zhu, L. Liu, Y. Song, H. Jia, H. Yu, X. Xiao, X. Yang, Oxygen bubble mould effect: serrated nanopore formation and porous alumina growth. Monatshefte Chem. **139**, 999–1003 (2008)
46. J.M. Montero-Moreno, M. Sarret, C. Müller, Some considerations on the influence of voltage in potentiostatic two-step anodizing of AA1050. J. Electrochem. Soc. **154**, C169–C174 (2007)
47. A. Santos, J. Ferré-Borrull, J. Pallarès, L.F. Marsal, Hierarchical nanoporous anodic alumina templates by asymmetric two-step anodization. Phys. Status Solidi A **208**, 668–674 (2011)
48. D. Losic, D. Losic Jr, Preparation of porous anodic alumina with periodically perforated pores. Langmuir **25**, 5426–5431 (2009)
49. A. Santos, L. Vojkuvka, M. Alba, V.S. Balderrama, J. Ferré-Borrull, J. Pallarès, L.F. Marsal, Understanding and morphology control of pore modulations in nanoporous anodic alumina by discontinuous anodization. Phys. Status Solidi A **209**, 2045–2048 (2012)
50. K. Pitzschel, J.M. Montero-Moreno, J. Escrig, O. Albrecht, K. Nielsch, J. Bachmann, Controlled introduction of diameter modulations in arrayed magnetic iron oxide nanotubes. ACS Nano **3**, 3463–3468 (2009)
51. K. Schwirn, W. Lee, R. Hillebrand, M. Steinhart, K. Nielsch, U. Gösele, Self-ordered anodic aluminum oxide formed by H_2SO_4 hard anodization. ACS Nano **2**, 302–310 (2008)
52. D. Losic, M. Lillo, D. Losic Jr, Porous alumina with shaped pore geometries and complex pore architectures fabricated by cyclic anodization. Small **5**, 1392–1397 (2009)
53. M.M. Rahman, E. García-Caurel, A. Santos, L.F. Marsal, J. Pallarès, J. Ferré-Borrull, Effect of the anodization voltage on the pore-widening rate of nanoporous anodic alumina. Nanoscale Res. Lett. **7**, 474 (2012)
54. A. Santos, T. Kumeria, D. Losic, Nanoporous anodic aluminum oxide for chemical sensing and biosensors. Trends Anal. Chem. **44**, 25–38 (2013)
55. M. Stephan, I. Mey, C. Steinem, A. Janshoff, Combining reflectometry and fluorescence microscopy: an assay for the investigation of leakage processes across lipid membranes. Anal. Chem. **86**, 1366–1371 (2014)
56. H. Neubacher, I. Mey, C. Carnarius, T.D. Lazzara, C. Steinem, Permeabilization assay for antimicrobial peptides based on pore-spanning lipid membranes on nanoporous alumina. Langmuir **30**, 4767–4774 (2014)
57. H. Ko, V.V. Tsukruk, Nanoparticle-decorated nanocanals for surface-enhanced Raman scattering. Small **4**, 1980–1984 (2008)
58. Z. Lu, W. Ruan, J. Yang, W. Xu, C. Zhao, B. Zhao, Deposition of Ag nanoparticle on porous anodic alumina for surface enhanced Raman scattering substrate. J. Raman Spectrosc. **40**, 112–116 (2009)

59. N. Ji, W. Ruan, C. Wang, Z. Lu, B. Zhao, Fabrication of silver decorated anodic aluminum oxide substrate and its optical properties on surface-enhanced Raman scattering and thin film interference. Langmuir **25**, 11869–11873 (2009)
60. S.J. Lee, Z. Guan, H. Xu, M. Moskovits, Surface-enhanced Raman spectroscopy and nanogeometry: the plasmonic origin of SERS. J. Phys. Chem. C **111**, 17985–17988 (2007)
61. L. Velleman, J.L. Bruneel, F. Guillaume, D. Losic, J.G. Shapter, Raman spectroscopy probing of self-assembled monolayers inside the pores of gold nanotube membranes. Phys. Chem. Chem. Phys. **13**, 19587–19593 (2011)
62. R. Kodiyath, S.T. Malak, Z.A. Combs, T. Koenig, M.A. Mahmoud, M.A. El-Sayed, V.V. Tsukruk, Assemblies of silver nanocubes for highly sensitive SERS chemical vapor detection. J. Mater. Chem. A **1**, 2777–2788 (2013)
63. R.J. Green, R.A. Frazier, K.M. Shakesheff, M.C. Davies, C.J. Roberts, S.J.B. Tendler, Surface plasmon resonance analysis of dynamic biological interactions with biomaterials. Biomaterials **21**, 1823–1835 (2000)
64. K. Hotta, A. Yamaguchi, N. Teramae, Nanoporous waveguide sensor with optimized nanoarchitectures for highly sensitive label-free biosensing. ACS Nano **6**, 1541–1547 (2012)
65. K.H.A. Lau, L.S. Tan, K. Tamada, M.S. Sander, W. Knoll, Highly sensitive detection of processes occurring inside nanoporous anodic alumina templates: a waveguide optical study. J. Phys. Chem. B **108**, 10812–10818 (2004)
66. A.G. Koutsioubas, N. Spiliopoulos, D. Anastassopoulos, A.A. Vradis, G.D. Priftis, Nanoporous alumina enhanced surface plasmon resonance sensors. J. Appl. Phys. **103**, 094521 (2008)
67. A. Dhathathreyan, Real-time monitoring of invertase activity immobilized in nanoporous aluminium oxide. J. Phys. Chem. B **115**, 6678–6682 (2011)
68. K.H.A. Lau, H. Duran, W. Knoll, *In situ* characterization of N-carboxy anhydride polymerization in nanoporous anodic alumina. J. Phys. Chem. B **113**, 3179–3189 (2009)
69. H.M. Hiep, H. Yoshikawa, E. Tamiya, Interference localized surface plasmon resonance nanosensor tailored for the detection of specific biomolecular interactions. Anal. Chem. **82**, 1221–1227 (2010)
70. S.H. Yeom, O.G. Kim, B.H. Kang, K.J. Kim, H. Yuan, D.H. Kwon, H.R. Kim, S.W. Kang, Highly sensitive nano-porous lattice biosensor based on localized surface plasmon resonance and interference. Opt. Express **19**, 22882–22890 (2011)
71. D.K. Kim, K. Kerman, M. Saito, R.R. Sathuluri, T. Endo, S. Yamamura, Y.S. Kwon, E. Tamiya, Label-free DNA biosensor based on localized surface plasmon resonance coupled with interferometry. Anal. Chem. **79**, 1855–1864 (2007)
72. G. Gauglitz, J. Ingenhoff, Design of new integrated optical substrates for immuno-analytical applications. Fresenius' J. Anal. Chem. **349**, 355–359 (1994)
73. G. Gauglitz, A. Brecht, G. Kraus, W. Nahm, Chemical and biochemical sensors based on interferometry at thin (multi-) layers. Sensor. Actuat. B **11**, 21–27 (1993)
74. C. Pacholski, M. Sartor, M.J. Sailor, F. Cunin, G.M. Miskelly, Biosensing using porous silicon double-layer interferometers: reflective interferometric Fourier transform spectroscopy. J. Am. Chem. Soc. **127**, 11636–11645 (2005)
75. C. Pacholski, C. Yu, G.M. Miskelly, D. Godin, M.J. Sailor, Reflective interferometric Fourier transform spectroscopy: a self-compensating label-free immunosensor using double-layers of porous SiO_2. J. Am. Chem. Soc. **128**, 4250–4252 (2006)
76. F. Cunin, T.A. Schmedake, J.R. Link, Y.Y. Li, J. Koh, S.N. Bhatia, M.J. Sailor, Biomolecular screening with encoded porous silicon photonic crystals. Nat. Mater. **1**, 39–41 (2002)
77. A. Jane, R. Dronov, A. Hodges, N.H. Voelcker, Porous silicon biosensors on the advance. Trends Biotechnol. **27**, 230–239 (2009)
78. T. Kumeria, D. Losic, Controlling interferometric properties of nanoporous anodic aluminium oxide. Nanoscale Res. Lett. **7**, 88 (2012)
79. S.D. Alvarez, C.P. Li, C.E. Chiang, I.K. Schuller, M.J. Sailor, A label-free porous alumina interferometric immunosensor. ACS Nano **3**, 3301–3307 (2009)

80. F. Casanova, C.E. Chiang, C.P. Li, I.V. Roshchin, A.M. Ruminski, M.J. Sailor, Gas adsorption and capillary condensation in nanoporous alumina films. Nanotechnology **19**, 315709 (2008)
81. S. Pan, L.J. Rothberg, Interferometric sensing of biomolecular binding using nanoporous aluminum oxide templates. Nano Lett. **3**, 811–814 (2003)
82. T. Kumeria, M.D. Kurkuri, K.R. Diener, L. Parkinson, D. Losic, Label-free reflectometric interference microchip biosensor based on nanoporous alumina for detection of circulating tumour cells. Biosens. Bioelectron. **35**, 167–173 (2012)
83. T. Kumeria, D. Losic, Reflective interferometric gas sensing using nanoporous anodic aluminium oxide (AAO). Phys. Status Solidi RRL **5**, 10–11 (2011)
84. R. Dronov, A. Jane, J.G. Shapter, A. Hodges, N.H. Voelcker, Nanoporous alumina-based interferometric transducers ennobled. Nanoscale **3**, 3109–3114 (2011)
85. A. Santos, T. Kumeria, D. Losic, Optically optimized photoluminescent and interferometric biosensors base on nanoporous anodic alumina: a comparison. Anal. Chem. **85**, 7904–7911 (2013)
86. T. Kumeria, A. Santos, D. Losic, Ultrasensitive nanoporous interferometric sensor for label-free detection of gold(III) ions. ACS Appl. Mater. Interfaces **5**, 11783–11790 (2013)
87. G. Macias, L.P. Hernández-Eguía, J. Ferré-Borrull, J. Pallares, L.F. Marsal, Gold-coated ordered nanoporous anodic alumina bilayers for future label-free interferometric biosensors. ACS Appl. Mater. Interfaces **5**, 8093–8098 (2013)
88. T. Kumeria, M.M. Rahman, A. Santos, J. Ferré-Borrull, L.F. Marsal, D. Losic, Structural and optical nanoengineering of nanoporous anodic alumina rugate filters for real-time and label-free biosensing applications. Anal. Chem. **86**, 1837–1844 (2014)
89. T. Kumeria, M.M. Rahman, A. Santos, J. Ferré-Borrull, L.F. Marsal, D. Losic, Nanoporous anodic alumina rugate filters for sensing of ionic mercury: toward environmental point-of-analysis systems. ACS Appl. Mater. Interfaces **6**, 12971–12978 (2014)
90. T. Kumeria, K. Gulati, A. Santos, D. Losic, Real-time and *in situ* drug release monitoring from nanoporous implants under dynamic flow conditions by reflectometric interference spectroscopy. ACS Appl. Mater. Interfaces **5**, 5436–5442 (2013)
91. A. Santos, M. Alba, M.M. Rahman, P. Formentín, J. Ferré-Borrull, J. Pallarès, L.F. Marsal, Structural tuning of photoluminescence in nanoporous anodic alumina by hard anodization in oxalic and malonic acids. Nanoscale Res. Lett. **7**, 228 (2012)
92. Y.B. Li, M.J. Zheng, L. Ma, High-speed growth and photoluminescence of porous anodic alumina with controllable interpore distances over a large range. Appl. Phys. Lett. **91**, 073109 (2007)
93. Y. Du, W.L. Cai, C.M. Mo, J. Chen, L.D. Zhang, X.G. Zhu, Preparation and photoluminescence of alumina membranes with ordered pore arrays. Appl. Phys. Lett. **74**, 2951–2953 (1999)
94. G.S. Huang, X.L. Wu, Y.F. Mei, X.F. Shao, G.C. Siu, Strong blue emission from anodic alumina membranes with ordered nanopore array. J. Appl. Phys. **93**, 582–585 (2003)
95. N.I. Mukhurov, S.P. Zhvavyi, S.N. Terekhov, A.Y. Panarin, I.F. Kotova, P.P. Pershukevich, I.A. Khodasevich, I.V. Gasenkova, V.A. Orlovich, Influence of electrolyte composition on photoluminescent properties of anodic aluminum oxide. J. Appl. Spectrosc. **75**, 214–218 (2008)
96. Z. Li, K. Huang, Optical properties of alumina membranes prepared by anodic oxidation process. J. Lumin. **127**, 435–440 (2007)
97. R.P. Jia, Y. Shen, H.Q. Luo, X.G. Chen, Z.D. Hu, D.S. Xue, Enhanced photoluminescence properties of morin and trypsin absorbed on porous alumina films with ordered pore array. Solid State Commun. **130**, 367–372 (2004)
98. C.L. Feng, X. Zhong, M. Steinhart, A.M. Caminade, J.P. Majoral, W. Knoll, Graded-bandgap quantum-dot-modified nanotubes: a sensitive biosensor for enhanced detection of DNA hybridization. Adv. Mater. **19**, 1933–1936 (2007)

99. A. Santos, V.S. Balderrama, M. Alba, P. Formentín, J. Ferré-Borrull, J. Pallarès, L.F. Marsal, Nanoporous anodic alumina barcodes: toward smart optical biosensors. Adv. Mater. **24**, 1050–1054 (2012)
100. A. Santos, G. Macías, J. Ferré-Borrull, J. Pallarès, L.F. Marsal, Photoluminescent enzymatic sensor based on nanoporous anodic alumina. ACS Appl. Mater. Interfaces **4**, 3584–3588 (2012)

Chapter 8
Nanoporous Anodic Alumina for Optofluidic Applications

Raúl Urteaga and Claudio L.A. Berli

Abstract The dynamics of liquid imbibition in nanoporous anodic alumina (NAA) has been recently studied by using the integration of optics and microfluidics. Here the latest results on the subject are revised, and some perspectives for using NAA in optofluidics applications are considered, notably the characterization of the membranes itself.

8.1 Introduction

The term 'optofluidics' is relatively new, it has around 10 years old, and comes to define the field of research that couples optics and fluidics at the microscale to advance diverse areas of science and engineering [1, 2]. Actually, the integration of photonics and microfluidics has rapidly evolved and applications are continuously growing, from chemical and biological sensing [3, 4] to the development of new healthcare and energy systems [5, 6]. One of the capabilities of optofluidics is the possibility to modify, and sense, the optical properties of micro/nanofluidic devices by changing the amount of fluid contained within, such as the fluidic tuning of optics made in liquid-infiltrated photonic crystals [7, 8].

In this framework, here we discuss the fundamentals and the practical implementation of an optofluidic technique to examine the imbibition dynamics of fluids in nanoporous anodic alumina (NAA), an information that is directly related to the inner morphology of the membrane. The method takes advantage of the fact that the nanopores array behaves as an effective media in response to visible electromagnetic waves, since nanopore radii are much smaller than the light wavelength. Thus laser

R. Urteaga
IFIS-Litoral (UNL-CONICET) Güemes 3450, 3000 Santa Fe, Argentina
e-mail: urteagar@santafe-conicet.gov.ar

C.L.A. Berli (✉)
INTEC, UNL-CONICET, Güemes 3450, 3000 Santa Fe, Argentina
e-mail: cberli@santafe-conicet.gov.ar

D. Losic and A. Santos (eds.), *Nanoporous Alumina*,
Springer Series in Materials Science 219, DOI 10.1007/978-3-319-20334-8_8

interferometry with sub-millisecond resolution can be used to follow optical property variations associated to capillary imbibition into NAA membranes [9, 10]. The measurement system is a variation of the well-known Fabry-Pérot interferometry, which has been also used to study nanoscale fluid dynamics [11].

The method has promising applications in lab-on-a-chip and microfluidic technology, for instance, it can be used as nanofluidic sensor to determine fluid properties of a small sample of liquid. In addition, knowing the imbibition kinematics of a given fluid allows one to elucidate the nanopores morphology in a non-destructive manner, which is of great interest for the characterization and specific design of NAA matrices. It is worth noting here that determining the detailed internal structure of NAA by using conventional techniques like microscopy imaging demands an irreversible alteration of the sample. More precisely, the optofluidic measurement together with the solution of mathematical model [12] enables the determination of the pore radius profile in samples whose structures involve regular arrays of nanopores with both very large aspect ratios and narrow pore-size distributions.

It is relevant to observe that scientific reports on liquid flow through NAA driven by capillary forces are barely found in the literature. Related studies refer to liquid spreading onto NAA surfaces and imbibition of sessile drops into the substrate [13, 14], where experiments consisted in capturing the shape evolution of drops with a video camera and posterior analysis of droplet volume conservation. The wetting behavior was also investigated by measuring the macroscopic contact angle in both open and close-pore structures [15, 16]; these works reveal that the wettability behavior of NAA can be controlled from the pore size and depth, which has much interest in industry applications. In addition, experiments of equilibrium wetting and capillary condensation were carried out with NAA [17], which can be sensitive to pore structure and morphology as well [18, 19]. It is worth noting here that the fluid imbibition we are dealing with involves liquid flowing throughout the membrane, in contrast to the experiments mentioned above.

Concerning the optical properties of NAA, Chap. 7 provides a brief review and detailed descriptions. Therefore, the present Chapter is organized as follows: firstly, the fundamentals and the formulation of fluid dynamic and optic models are described, then the setup of the optofluidic experiment is presented, the experimental results are discussed, and finally the main conclusions are outlined.

8.2 Theory

8.2.1 Fluid Dynamic Model

8.2.1.1 Capillary-Driven Fluid Imbibition

In this section we derive a mathematical model that depicts the spontaneous imbibition of liquids in NAA driven by capillary forces. The dynamics of capillary filling is described in the theoretical context of continuum fluid mechanics, which is known to provide a suitable description of the phenomenon in milli- and micro-channels [20]. In fact, although the continuum hypothesis is expected to breakdown when fluids are confined to spaces containing relatively few fluid molecules, several recent studies prove the validity of the approach in tubes with cross-sections smaller than 10 nm [11, 21–23].

Given the well ordered array of pores in NAA (Fig. 8.1a), the membrane can be considered as an assembly of straight and rigid nanochannels, not interconnected, aligned in the flow direction, and all of them with the same length and pore radius profile. Under these conditions, the velocity at which the liquid invades the membrane is that of the capillary flow in each nanopore, hence the fluid dynamics of a single capillary represents the imbibition of the whole membrane. This ideal model is a reasonable approximation that greatly simplifies calculations, but it might be exercised with caution in cases where a certain distribution of pore radii could be present.

Further, nanopores are assumed to be axisymmetric, with circular cross-section, where the radius $r(x)$ varies smoothly along the axial direction x, as shown in Fig. 8.1b. Incompressible Newtonian fluids of viscosity μ, density ρ, and surface tension σ are considered. Temperature is assumed to be uniform in the entire flow domain. Given the strong confinement, liquid flow through the membrane takes place in the regime of very low Reynolds numbers, meaning that inertial effects are irrelevant in front of viscous dissipation [24]. Under these conditions, the relation

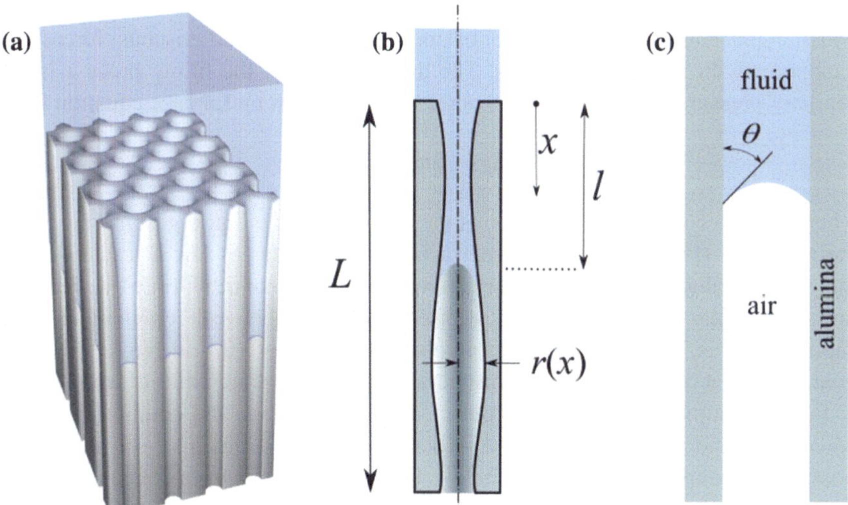

Fig. 8.1 Schematic representation of **a** the regular array of nanopores considered to model the fluid dynamics in NAA, **b** a single nanopore with the imbibing fluid and the coordinate systems used in calculations, and **c** details of the meniscus formed by the fluid in contact with alumina and air. The drawing is deliberately out of scale; note that actual aspect ratio of nanopores is $L/r \sim 10^3$

between the pressure gradient and the hydrodynamic resistance of the penetrating fluid is given by Poiseuille equation,

$$-\frac{dp}{dx} = \frac{8\mu Q}{\pi r(x)^4} \tag{8.1}$$

where Q is the volumetric flow rate. Here the hydrodynamic resistance of the displaced air is disregarded, considering that relatively low viscosity coefficient of air. Equation (8.1) is valid for fully developed flows, which is largely satisfied in NAA membranes because of the large aspect ratio of nanopores ($L/r \sim 10^3$). This fact also ensures that the assumption of one-dimensional flow is accurate.

It is also implicit in (8.1) that the liquid satisfies the no-slip condition at the solid-liquid interface [24], a classic assumption that lately appears to be violated in microfluidic experiments [25, 26], notably in hydrophobic nanotubes [27], but also in hydrophilic nanochannels [28]. In particular, the effect has been recently reported in NAA [29]. Despite these facts, here we use (8.1) in its present form, taking into consideration that the possible wall slip is a linear effect that modify the effective fluid viscosity only (the functionality of the meniscus velocity with time would not be altered).

Considering that Q is instantly uniform due to mass conservation, the integration of (8.1) from $x = 0$ to the meniscus position $x = l$ yields

$$\frac{8\mu Q}{\pi}\int_0^l \frac{dx}{r(x)^4} = \Delta p \tag{8.2}$$

which represents the steady state balance of capillary and viscous forces for nanopores with nonuniform radius. The driving force of the filling process is the capillary pressure generated in each pore, which is given by Laplace equation [20],

$$\Delta p = 2\sigma\cos\theta/r \tag{8.3}$$

where θ is the contact angle that forms the meniscus at the liquid-gas interface with the pore wall (Fig. 8.1c). As the interfacial tension scales as $1/r$, it reaches extremely large values in nano-scale curvatures, hence the contribution of hydrostatic pressure results negligible. More precisely, the ratio between gravitational and capillary forces, the so-called Bond number, is $\rho gLr/\sigma \sim 10^{-9}$ in NAA. This also guarantees that the meniscus is hemispherical, with a constant radius of curvature. It is further assumed here that the contact angle θ preserves the equilibrium value when the liquid flows. For the fluid velocities of our experiments ($\sim$1 mm/s), the possible error in the factor $\cos\theta$ is below 1 %, as calculated from equations for the dynamic contact angle reported elsewhere [30]. Concerning the pressure, it is worth to remark that pore ends are open to the atmosphere; otherwise the compression of entrapped air would cause an additional resistance (backpressure) that needs to be included in the model [9].

Eliminating Δp from (8.2) and (8.3), and introducing $Q = v(l)\pi r(l)^2$, where $v(l) = dl/dt$ is the average fluid velocity at the meniscus position, leads to the following expression for the meniscus velocity [10],

$$\frac{dl}{dt} = \alpha \left[4r(l)^3 \int_0^l \frac{dx}{r(x)^4} \right]^{-1} \tag{8.4}$$

where $\alpha = \sigma\cos\theta/\mu$ is a constant coefficient that characterizes fluid properties. This equation governs the kinematics of capillary flow in tubes of non-uniform cross section, and has been used to predict the instantaneous meniscus position $l(t)$, for a given function $r(x)$ specified beforehand [31–34]. That procedure is designated direct calculation, and it is described next. In contrast, the inverse calculation consists in determining a completely unknown function $r(x)$ from the curve of experimental data $l(t)$. This possibility has been just reported in the literature [12], and it is discussed subsequently.

To end this section it is relevant to note that (8.4) actually describes a quasi-steady state: it is assumed that, at any given time, the velocity profile instantaneously relaxes towards Poiseuille flow. This approximation is accurate if the characteristic time $\tau = \rho r^2/\mu$ is sufficiently short [24]. For ordinary fluids in NAA, $\tau \sim 10^{-10}$ s, which is much smaller than the time required to fill the membrane ($\sim 10^{-3}$ s).

8.2.1.2 Direct Calculation

In the trivial case of cylindrical pores of uniform radius R and length L, the integration of (8.4) yields

$$\bar{l}^2 = t/t_{\mathrm{F}} \tag{8.5}$$

which is the classical Lucas-Washburn equation [35, 36] for the dimensionless distance $\bar{l} = l/L$ traveled by the fluid in the axial direction of the tube, as a function of time t. In (8.5), $t_{\mathrm{F}} = 2L^2/R\alpha = 2L^2\mu/R\sigma\cos\theta$ is the characteristic filling time.

The aim here is to predict the imbibition of nanopores with a monotonic variation of $r(x)$. The simplest case corresponds to conical channels with the smallest radius R_{S} at one end, the largest radius R_{L} at the other and opening angle $\beta \approx (R_{\mathrm{L}}-R_{\mathrm{S}})/L$, which is around 10^{-4} in NAA membranes. Therefore, if the fluid enters through R_{S} and flows towards R_{L}, the pore radius opens downstream as $r(x) = R_{\mathrm{S}} + \beta x$. Solving (8.4) with this function yields

$$\frac{1}{6}(H-1)^2\bar{l}^4 + \frac{2}{3}(H-1)\bar{l}^3 + \bar{l}^2 = t/t_{\mathrm{F,S}} \tag{8.6}$$

where $H = R_L/R_S$, and $t_{F,S} = 2L^2/\alpha R_S$, which represents the filling time of a cylindrical channel of uniform radius R_S. The time to fill the conical pore is

$$t_O = t_{F,S}(H^2/6 + H/3 + 1/2) \tag{8.7}$$

In contrast, if the fluid enters through R_L and flows towards R_S, the pore radius shrinks in the direction of the flow as $r(x) = R_L - \beta x$, and (8.4) yields

$$\frac{1}{6}(1/H - 1)^2\bar{l}^4 + \frac{2}{3}(1/H - 1)\bar{l}^3 + \bar{l}^2 = t/t_{F,L} \tag{8.8}$$

where $t_{F,L} = 2L^2/\alpha R_L$ represents the filling time of a cylindrical channel with uniform radius R_L. The time to fill this conical pore is

$$t_C = t_{F,L}(H^{-2}/6 + H^{-1}/3 + 1/2) \tag{8.9}$$

Equations (8.6) and (8.8) indicate that the kinematics of fluid imbibition in conical pores no longer satisfies the scaling of cylindrical pores predicted by (8.5), except at very short times. When the meniscus advance into the conical geometry, the fact that viscous force varies as $r(x)^{-2}$ while capillary force varies as $r(x)^{-1}$ yields the polynomial relationships. Consistently, Lucas-Washburn law is recovered when $H = 1$. These results are illustrated in Fig. 8.2, where a cylindrical pore of radius R_0 is taken as reference.

It is seen in Fig. 8.2 that pores of uniform cross section have filling times inversely proportional to the tube radius. In conical pores however, as the radius increases in the flow direction from R_0 to $2R_0$, the filling time becomes larger than that corresponding to a cylinder with radius $R < R_0$. In other words, if one regards the conical pore as an equivalent cylinder, the equivalent cylinder radius is not between R_0 and $2R_0$, but it is smaller than R_0. And the opposite happens in converging pores. This rather counterintuitive behavior can be understood as follows. In a cylindrical capillary, when the fluid advances along the channel, the capillary pressure keeps constant, whereas the hydrodynamic resistance increases with the wetted length. In a conical capillary that opens in the flow direction, the capillary pressure progressively decreases with $r(x)$. At the same time, the hydrodynamic resistance increases with x, but at a rate that is slower than that in a cylinder with uniform radius. In a conical capillary that shrinks along the direction of the flow, the capillary pressure increases as $r(x)$ decreases. The hydrodynamic resistance also increases, but at a rate that is larger than that in a cylinder.

The overall consequence of predictions above is a strong asymmetry of the filling times, depending on the flow direction in the membrane. Actually, the asymmetric filling had been predicted by model simulation of capillary imbibition of porous substrates consisting of two layers with different permeabilities [37], thus filling times are shorter from wide to narrow pores, and vice versa. This behaviour has been also observed for NAA membranes [14], though associated to another cause: the presence of different pore structures on each side of the membrane. In the

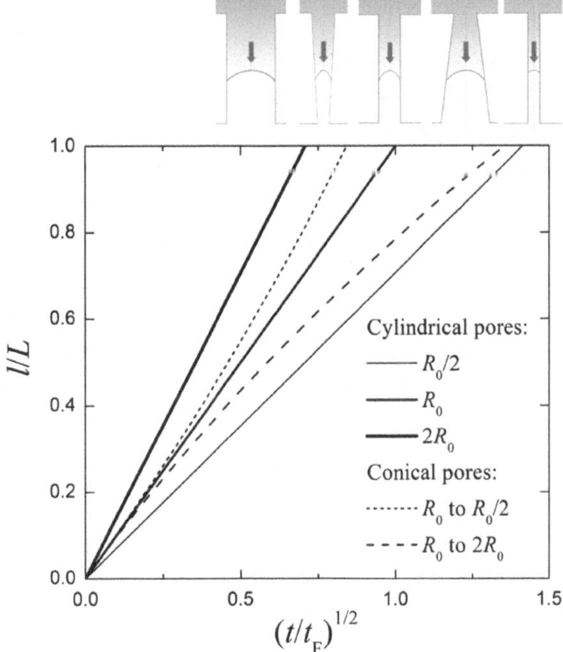

Fig. 8.2 Relative meniscus position as a function of the square root of the relative time for conical nanopores (8.6) and (8.8), in comparison to cylindrical pores (8.5). The characteristic filling time used is $t_F = 2L^2/R_0\alpha$. The drawings on the *top* are in the sequence of the filling times (increasing from *left* to *right*) of the corresponding curves

case considered here (conical nanopores), the relation between the time required to fill the membrane in each direction is defined by H only. In fact, taking the ratio of (8.7) and (8.9) leads to the following expression,

$$\frac{t_O}{t_C} = \frac{H^5 + 2H^4 + 3H^3}{3H^2 + 2H + 1} \approx H^{7/3} \tag{8.10}$$

which quantitatively predicts the asymmetry of filling times from different sides of the membrane. This prediction is illustrated in Fig. 8.3. The second equality of (8.10) is obtained from a Taylor series expansion at $H \approx 1$. It is however valid for a relatively wide range of H around 1. Furthermore, the prediction of (8.10) has been recently extended to non-Newtonian fluids in conical microchannels [38]. I was found that, for power law fluids of index m, the ratio of filling times obeys $t_O/t_C \approx H^{f(m)}$, where $f(m) = 7/3 + 1/m - 1$ represents reasonably well the curves of Fig. 8.3 with an error lower than 3 % for the range $1/2 \leq m \leq 3/2$. The rectification properties associated to the asymmetric filling open new potential applications in microfluidics and optofluidics. Later in Sect. 8.4.1 we employ these concepts to interpret experimental results with NAA membranes.

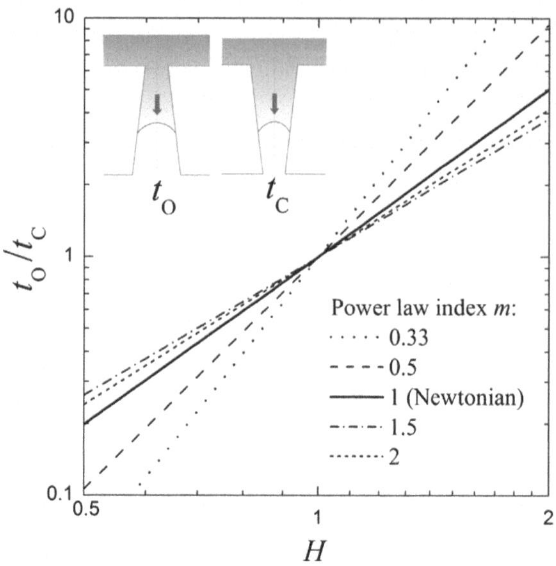

Fig. 8.3 Asymmetric capillary filling in conical nanopores, where $H = R_L/R_S$ is the ratio between pore ends radii. The full line is the prediction of (8.10). *Dotted* and *dashed lines* represents capillary filling of the same cones with non-Newtonian power law fluids for different indexes m, as reported in [38]. In the log-log representation, the respective slopes are $7/3 + 1/m - 1$

8.2.1.3 Inverse Calculation

It is relevant to note that solving the inverse calculations from integral equations is not trivial from the mathematical point of view. Actually, extracting $r(x)$ from (8.4) is an ill-posed problem, meaning that it has multiple solutions, and that choosing the best one presents several difficulties. In this section we outline a strategy recently reported for the purposes [12]. Firstly, (8.4) is converted into a differential equation by using the Leibnitz rule,

$$\frac{d}{dl}\left[r(l)u(l)^{1/3}\right] = -\frac{4u(l)^{4/3}}{3\alpha} \tag{8.11}$$

Subsequent integration leads to an explicit expression of the unknown function $r(l)$,

$$r(l) = \left[r(l_0)u(l_0)^{1/3} + \frac{4}{3\alpha}\int_l^{l_0} u(l)^{4/3}dl\right]u(l)^{-1/3} \tag{8.12}$$

where l_0 is an arbitrary limit of integration. The multiplicity of solutions of this equation can be readily seen: if one introduces $u(l)$ as the meniscus velocity corresponding to a given capillary, then (8.12) yields an infinite family of curves $r(x)$ parameterized by $r(l_0)$.

Concerning the input data, the meniscus position $l(t)$, hence $u(t)$, can be experimentally assessed by using the reflected light interference, as describe below in Sect. 3.1.2. Nevertheless, in the study of nanoporous matrices, the variable usually measured is the amount of imbibed fluid as a function of time, by means of gravimetric methods [39], neutrons absorption [40], or simply by geometric analysis of the drop volume [13, 14]. To take into account these possibilities, where the instantaneous position of the meniscus can be attained only indirectly, the model can be extended to the case where experimental data are proportional to the volume of imbibed fluid $v(l)$. Introducing $dv(x) = \pi r(x)^2 dx$ into (8.4) and reproducing the procedure to derive (8.12) yields

$$r(v) = \left[r(v_0)^5 Q(v_0)^5 + \frac{20}{\alpha \pi^2} \int_{v}^{v_0} Q(v)^6 dv \right]^{1/5} Q(v)^{-1} \tag{8.13}$$

where $Q = dv/dt$ is the volume flow rate [12]. It is worth noting that this change of variables does not avoid the ill-posed nature of the problem. The multiplicity of solutions of (8.13) is related to the arbitrary constant $r(v_0)$, which is the boundary condition required to confer uniqueness to the solution, analogously to $r(l_0)$ in (8.13).

Identifying the right $r(x)$ solution among the parameterized family of curves demands additional information from the problem, namely the pore radius at a given position, or the total pore volume. Alternatively, the method recently proposed [12] takes advantage of the possibility to measure capillary filling in both pore directions, i.e. from different sides of the membrane. The second measurement, when processed through (8.13), generates a new family of $r(x)$ curves, where one of them must reproduce the meniscus velocity of the first measurement. Thus the procedure allows one to identify the best function $r(x)$ representing the pore geometry. It is worth to add that, since this method generates two solutions, now the inverse problem is over determined, and cross-checking the two independent solutions serves as a consistency test for the model. The experimental proofs are described below in Sect. 8.4.2.

8.2.2 Optical Model

8.2.2.1 Effective Refractive Index. Linear Approximation

The alumina membranes can be regarded as a two dimensional periodic composite formed by a host matrix of bulk alumina with perfectly aligned cylindrical inclusions of air perpendicular to the substrate. In the long-wavelength limit, i.e. when the structure size is much smaller than the light wavelength used in experiments, an effective-medium theory can be used to describe the optical properties of this photonic crystal, considering that it as a homogeneous material with intermediate properties regarding its components [41]. In general, averaging the respective dielectric functions is non–trivial since it depends on the details of the topology, and only can be assessed analytically for some special cases [42].

Given the symmetry of the pore distribution in NAA, the resulting material is uniaxial, i.e. the effective dielectric constant of the membrane has two different values depending on whether the wave propagation is perpendicular or parallel to the pores [42]. In particular, it can be shown that the effective dielectric constant in the perpendicular direction can be obtained has the simple volumetric average of the dielectric constants of its components [42]. Several effective medium models have been developed to take into account the particular microstructure of the film. Among these models, Maxwell-Garnet and Bruggeman formulations [43] are considered here; (8.14) and (8.15), respectively. Both formulations can be used to calculate the effective dielectric constant (ε_{eff}) of ellipsoidal inclusions of dielectric constant ε_2 in a host matrix of dielectric constant ε_1 with a volume fraction P.

$$\frac{\varepsilon_{eff} - \varepsilon_1}{\varepsilon_{eff} + (\varepsilon_1 - \varepsilon_{eff})q} - \frac{\varepsilon_2 - \varepsilon_1}{\varepsilon_1 + (\varepsilon_2 - \varepsilon_1)q} P = 0 \tag{8.14}$$

$$\frac{\varepsilon_1 - \varepsilon_{eff}}{\varepsilon_{eff} + (\varepsilon_1 - \varepsilon_{eff})q}(1 - P) + \frac{\varepsilon_2 - \varepsilon_{eff}}{\varepsilon_{eff} + (\varepsilon_2 - \varepsilon_{eff})q} P = 0 \tag{8.15}$$

Here, the depolarization factor q can be calculated from the geometry parameters of the ellipsoids [44] having different values depending upon the relative orientation of the ellipsoids with the electric field. In particular, for spherical inclusions (isotropic) $q = 1/3$, for cylinders with propagation perpendicular to the axis $q = q_z = 0$, and for wave propagation parallel to the cylinders $q = q_{xy} = 1/2$ [44]. In fact, this depolarization factors have been used as fitting parameters to characterize the optical properties of NAA obtaining $q_{xy} \sim 1/2$ for the best fit [45, 46]. In general, the Maxwell-Garnet formulation gives better results for dilute fractions of inclusions, while Bruggeman's formula is more suitable for high porosities or interconnected pores morphology [44]. Given that absorption in alumina can be neglected in the visible and near infrared range [47], the effective refractive index of the film can be calculated simply as $n_{eff} = \varepsilon_{eff}^{1/2}$. The resulting refractive index calculated with both models for dry porous alumina and alumina completely filled with isopropyl alcohol are shown in Fig. 8.4. In this case, a depolarization factor of $q_{xy} = 1/2$ was used.

It can be observed in Fig. 8.4 that the effective refractive index of the porous film results in an almost linear relationship with the film porosity. Explicitly, the refractive index of the composite can be approximated by $n_d = n_{Al}(1 - P) + P$, for the dry alumina, where n_{Al} the refractive index of the bulk alumina, and $n_w = n_{Al}(1 - P) + n_{IP}P$ for the wet alumina, being n_{IP} the refractive index of isopropyl alcohol. The departure from linearity is more pronounced when the contrast between the individual refractive indexes is large, i.e. for dry alumina, however the difference is lower than 5 % in the worst condition (Bruggeman's model at $P \sim 0.6$).

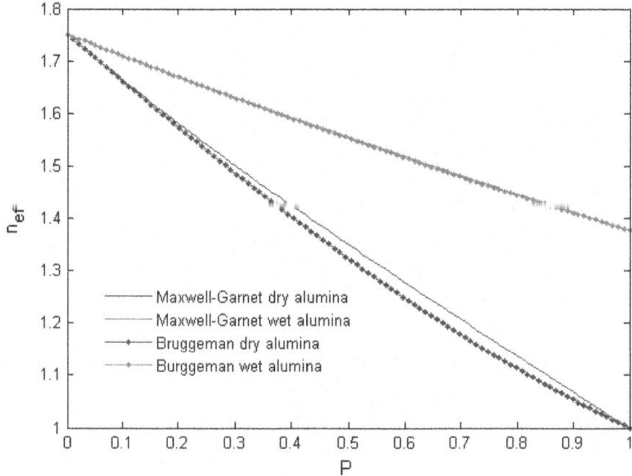

Fig. 8.4 Effective refractive index of dry (n_d) and fully wet alumina (n_w) with isopropyl alcohol, as a function of porosity, using different effective medium theories. Refractive indexes of pure substances were taken at $\lambda = 1000$ nm, and the depolarization factor is $q = 1/2$

8.2.2.2 Thin Film Interference

During liquid imbibition experiments on NAA films, the total reflected light intensity can be calculated using the simple theory of light interference in thin films [48]. The reflected intensity depends upon the contribution of reflection from three interfaces (Fig. 8.5).

The reflectance at each interface depends on the refractive index of the two involved media (n_i and n_{i+1}) and can be calculated as [48],

$$\mathcal{R} = \frac{(n_i - n_{i+1})^2}{(n_i + n_{i+1})^2} \tag{8.16}$$

In the particular case of the moving interface (air-liquid) inside the film, the surface is not perfectly defined due to small filling differences in individual channels. In these conditions the interface is blurred and can be intended as a smooth index profile that matches the refractive index of the two media at the interface. Even if a small region is smoothed at the interface, a big reduction in the reflectance is produced, in fact, a transition layer of only half wavelength thick can have a reflectance as low as 10^{-4} [48]. Therefore, the reflected intensity during the imbibition will be completely determined by the interference of the light reflected from the two fixed interfaces: dry alumina-air and wet alumina-liquid (Fig. 8.5). The interference will be constructive when the phase shift $\Delta\varphi$ is a multiple of 2π and destructive when $\Delta\varphi$ is an odd multiple of π. Taking into account the phase shift at the first interface, the total phase shift will be

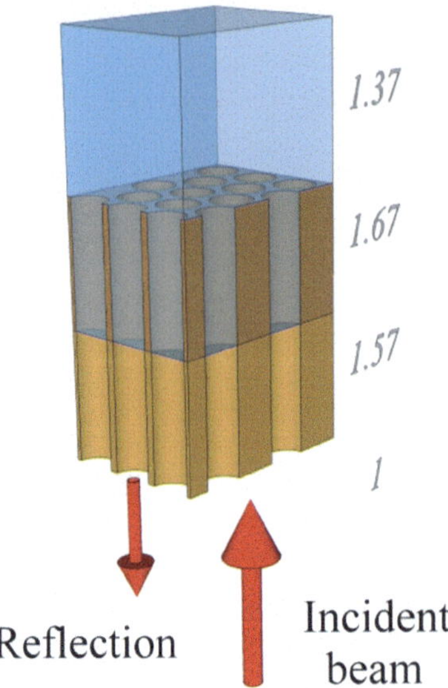

Fig. 8.5 Schematic representation of the dielectric layer sequence formed during fluid imbibitions in a NAA film. The effective refractive indexes indicated on the *right* (from *top* to *bottom*) correspond to the liquid, the NAA matrix plus the liquid, the NAA matrix plus the air, and simply air. Numerical values were calculated for isopropyl alcohol, NAA film with P = 0.2, and λ = 1000 nm

$$\Delta\varphi = \pi + 4\pi \frac{e_0}{\lambda} \tag{8.17}$$

where $e_0 = \int_0^L n_{eff}(x)dx$ is the optical path thickness of the membrane. As the liquid penetrates into the porous layer, $n_{eff}(x)$ increases, so does the optical path, and the reflectance reaches extremes every time e_0 is a multiple of $\lambda/4$. Therefore, the occurrence of maxima and minima in the reflected intensity is due to the variation of $n_{eff}(x)$ during liquid imbibitions, while the gap between the main reflecting interfaces is fixed.

Using the simple linear relationship for the effective refractive index we can write $n_{eff} = \sum_i n_i f_i$, where n_i and f_i are, respectively, the refractive index and volume fraction of each component: i = 1, alumina; i = 2, liquid; i = 3, air. Therefore the optical path can be calculated as

$$e_0 = \int_0^L n_1[1 - P(x)]dx + \int_0^l n_2 P(x)dx + \int_l^L n_3 P(x)dx \tag{8.18}$$

where $P(x) = \pi r^2(x)\delta$ is the local membrane porosity, δ being the pores surface density. Differentiating (8.18) with respect to l yields

$$\frac{de_0}{dl} = (n_2 - n_3)\pi r^2(l)\delta \tag{8.19}$$

Then integrating (8.19) with the particular function $r(x)$ yields an explicit relationship between the optical thickness and the relative position of the liquid front. In the case of conical nanochannels, when the fluid enters through the narrowest end and flows toward the widest end, $r(x) = R_S + \beta x$, thus

$$e_0(\bar{l}) = (n_2 - n_3)\pi R_S^2(\bar{l})\delta L\left[\bar{l} + (H-1)\bar{l}^2 + \frac{1}{3}(H-1)^2\bar{l}^3\right] \tag{8.20}$$

In addition, when the fluid enters through the widest end and flows toward the narrowest end, $R(x) = R_L - \beta x$, and the integration of (8.19) yields

$$e_0(\bar{l}) = (n_2 - n_3)\pi R_L^2(\bar{l})\delta L\left[\bar{l} + \left(H^{-1} - 1\right)\bar{l}^2 + \frac{1}{3}\left(H^{-1} - 1\right)^2\bar{l}^3\right] \tag{8.21}$$

Therefore, using (8.20) and (8.21) in (8.17) one obtains the respective expressions of $\Delta\varphi(\bar{l})$, i.e. one associates the extremes of the measured light reflectance to the instantaneous position of the moving liquid front, taking into account the pore radius variation across the membrane. Of course, a lineal relation for $\Delta\varphi(\bar{l})$ is obtained if $H = 1$.

8.3 Experiments

8.3.1 Alumina Preparation

NAA membranes were fabricated by the two-step anodization process [49]. Before anodization takes place, aluminum (Al) foils were electropolished under constant potential in a mixture of perchloric acid ($HClO_4$) and ethanol (EtOH). After this, the first electrochemical anodization was performed by applying direct voltage to the electropolished Al films in acidic solutions of sulfuric acid (H_2SO_4) and oxalic acid ($H_2C_2O_4$). Then, the aluminum oxide film with disordered pores was dissolved by wet chemical etching using a mixture of phosphoric acid (H_3PO_4) and chromic acid (H_2CrO_4). Then, the second anodization step was started under the same anodization conditions and carried out until the NAA membrane reached the desired thicknesses (about 75 μm, as inspected by optical microscopy). Thereafter, the remaining Al substrate was removed from the bottom side by wet chemical etching in a saturated solution of cupric chloride and hydrochloric acid ($CuCl_2$/HCl).

Subsequently, a pore opening process was performed by wet chemical etching under current control [50]. In order to modify the porosity of the NAA (i.e., the pore diameter), a pore widening process was performed by wet chemical etching in a phosphoric acid solution. Thereby, a hexagonally distributed pore array is obtained with pore diameter between 20 and 50 nm and center-to-center pore distance ranging from 50 to 100 nm, as estimated from SEM image analysis [10]. Furthermore, pore radii values differ from one side ($\sim$28 nm) to the other ($\sim$34 nm), which indicates that nanopores present a sort of conical shape.

8.3.2 Optofluidic Measurements

The experiment designed to follow capillary filling in nanoporous substrates was described in [9, 10], and it is similar to that reported in [11]. A liquid drop released from a needle impinges over a NAA film, and the liquid enter the pores by capillary action. Light reflectance as a function of time is measured from the rear, as shown in Fig. 8.6a. A laser beam (λ = 980 nm) is focused onto the membrane by employing optical fibers. The light reflected from the NAA film at low angle (5°) is measured by using a photodetector with linear response. Data is recorded at 48 kHz with a 14 bits computer controlled data acquisition card. The sample holder (Fig. 8.6a) has a hole that allows light to interact only with the porous membrane. This also ensures that pores are open to the atmosphere at both ends.

Figure 8.6b presents the normalized reflectance of a 75 μm thick NAA film when it is filled with isopropanol at room temperature from both sides. At the beginning of the experiment, when the liquid reaches the surface, the reflectance decreases

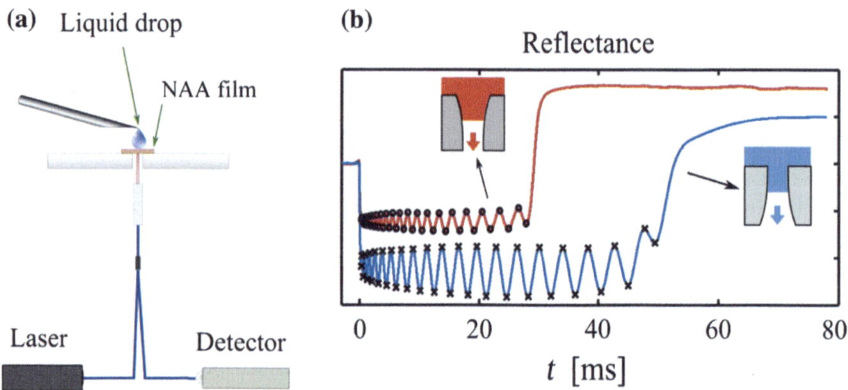

Fig. 8.6 **a** Diagram of the experimental setup. A liquid drop released from a needle reaches the NAA film, and the liquid penetrates into the porous membrane by capillary forces. A laser diode (λ = 980 nm) is focused on the porous membrane from the rear, and the time evolution of the laser reflectance is measured at 48 kHz. **b** Normalized reflectance measured from both sides of the same NAA film. The insets show schematically the flow direction of each measurement

steeply due to the sudden change in the refractive index contrast at the upper interface. Then small oscillations are observed as a result of the interference related to the change of $n_{eff}(\bar{l})$ with fluid imbibition. When pores are completely filled, the reflectance grows again due to the increases of the refractive index contrast at the lower interface. The second intensity step is less abrupt than the first one because not all the pores are filled at exactly the same time, considering that a narrow distribution of pore sizes is present. In what follows we will identify with the red color the filling process in the closing pore direction (red curve with dots, Fig. 8.6b) and with blue the filling in the opening pore direction (blue curve with crosses, Fig. 8.6b).

8.4 Results and Discussions

8.4.1 Direct Calculation: Prediction of Imbibition Kinematics

In this section we confront the experimental results against the straightforward prediction of fluid dynamics and optical models. For this purpose, we consider the imbibition dynamics determined from the measurements of reflectance evolution (Fig. 8.6b). Between two consecutive extremes of the reflectance oscillations, the optical thickness changes a quarter wave (see (8.17) above). If one assumes that nanopores are conical, then each extreme of the oscillations can be related to the instantaneous position of the liquid front by using (8.20) and (8.21).

Figure 8.7 shows the normalized squared position of the liquid front as a function of time (symbols), for imbibitions from both sides of the membrane [10]. The insets schematically show the flow direction for each measurement. Full lines in Fig. 8.7 are the theoretical predictions for each case, according to (8.6) and (8.8). Additionally, dashes lines are the expected dynamics of capillary filling of cylinders of radii R_S and R_L, according to (8.5).

Firstly, one may observe that the filling times corresponding to the imbibition from different sides of the membrane are quite distinctive, despite the relative slight variation in pore radius across the membrane: a few nanometers in 75 μm. Secondly, as discussed above in relation to Fig. 8.2, experiments show that the filling time for the opening cone t_O is larger than that of a cylindrical cone of radius R_S (Fig. 8.7), while the filling time for the closing cone t_C is shorter than that of a cylinder with radius R_L. A relevant corollary of this behavior is that the only way to fit the experimental curves to (8.5) is to accept that the equivalent cylindrical pore has a radius smaller than R_S for red data and larger than R_L for blue data. Finally, a reasonable agreement between experiments and the prediction of (8.6) and (8.8) is observed, which means that nanopores present a shape quite similar to cones. It is

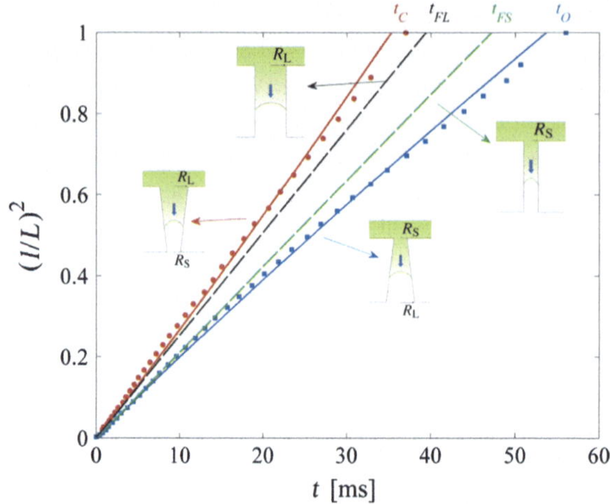

Fig. 8.7 Filling dynamics of isopropanol from both sides of the membrane (symbols) in the direction showed in the insets. Full lines are the theoretical predictions according to (8.6) and (8.8). Dashed lines are the expected filling dynamics for cylindrical channels with radius R_S and R_L, according to (8.5). Adapted from [10]

worth to remark that the optofluidic technique captures the phenomenon, even with very small cone angles. In this example the opening angle is around 0.007°, taking into consideration that the pore radius changes from ~28 to ~34 nm in 75 μm.

8.4.2 *Inverse Calculation: Determining Pore Morphology*

In this section we deal with the objective of determining the radius profile $r(x)$ of nanopores from experimental results of capillary imbibitions, by using the procedure described above in Sect. 8.2.1.3. Experimental results of reflectance variation due to fluid imbibitions are considered again (Fig. 8.6b). Now the extreme positions of reflectance oscillations are converted into the imbibed fluid volume fraction v/V by using the linear approximation of effective medium discussed in Sect. 8.2.2.1; results firstly reported in [12] are shown in Fig. 8.8. To determine the membrane pore radius profile from these data, the following procedure was implemented. First, v/V data were numerically differentiated by central differences to obtain $Q(v)/V$. Using these data as input in (8.13), the radius profile $r(v)$ was obtained for measurements in each direction (red and blue data). More precisely, taking the red $Q(v)$ data, (8.3) yields a family of possible solutions for $r(v)$. With these functions, the expected blue filling data is simulated with the direct calculation. The simulated curves are compared to the measured blue curve in Fig. 8.8, to identify the right $r(v)$ function by minimum least squares fit. Then the best $r(v)$ curve is converted into $r(x)$ data with the relation $dv(x) = \pi r(x)^2 dx$, which is plotted in Fig. 8.9. The whole

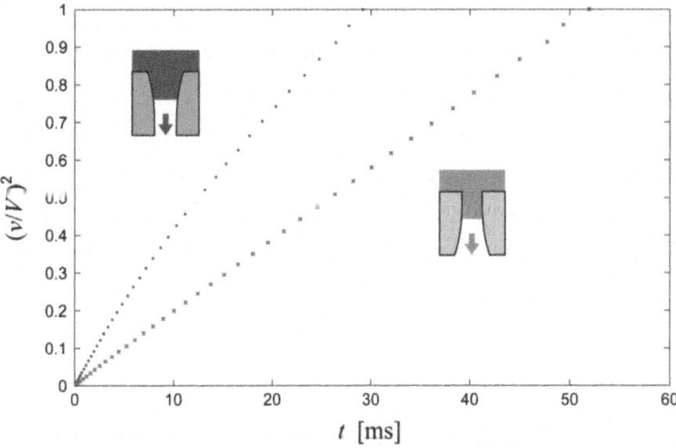

Fig. 8.8 Squared volume fraction of the imbibed fluid as a function of time, obtained from the extremes of the reflectance oscillations (Fig. 8.6b), measured from different sides of the membrane. Adapted from [12]

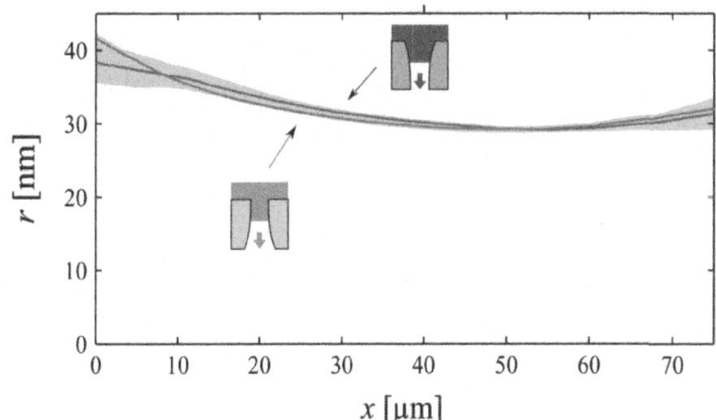

Fig. 8.9 Pore radius as a function of the axial distance. The *red* and *blue curves* are the predicted $r(x)$ functions. The *shaded area* represents the confidence bounds obtained from five different trials in each direction. Adapted from [12]

procedure is then repeated starting with the blue $Q(v)$ data, which yields the blue $r(x)$ curve in Fig. 8.9. All calculations were made by using $\alpha = 0.18$ m/s.

The solutions $r(x)$ thus obtained fall within the shaded area that represents the error in Fig. 8.9 (uncertainties of the experimental data from five measurements in each direction). This overlapping of two solutions shows the consistency of the method. Moreover, the radius values at pore ends are in agreement with those obtained from SEM in both membrane faces. These results confirm that the solution of the inverse problem proposed in [12] is suitable to determine the pore radius profile in NAA films. The method is also self-sufficient in the sense that no additional data is needed to identify the right solution.

Finally one may summarize the optofluidic characterization of NAA films: in the previous section, the so-called direct calculation was used to predict the kinematics $l(t)$ for a given $r(x)$ variation of pore radii. The conical geometry was prescribed for nanopores, and predictions agree rather well with experiments (Fig. 8.7). In this section, the inverse calculation is implemented to determine the precise $r(x)$ function from experimental data $l(t)$. It is now observed (Fig. 8.9) that the resulting $r(x)$ profile is not perfectly linear, as expected for conical pores. However the net change of pore radius is about 10 nm along 75 μm, with a very smooth variation, thus nanopores appear roughly similar to cones in practical terms. Moreover, this result indicates that the optofluidic technique is sensitive enough to capture subtle differences in nanopore morphology.

8.4.3 Further Applications: Pore Opening Control

As it was mentioned in Sect. 8.3.1, at the end of the anodization process, the bottom of nanopores is closed by a thin barrier oxide layer. The removal of this barrier is required for most applications of NAA, hence the pore opening (PO) is made by wet-chemical etching from the closed pores side of the membrane. Once the thin barrier oxide layer is dissolved, the etching process extends to the inner walls of the pores, with the consequent pore widening (PW). It is crucial to determine precisely the PO point in order to differentiate one process from the other. The usual technique for monitoring the PO follows the conductivity through the film [51]. The method has been recently impressved by measuring the current at only one side of the membrane [52]. In any case, the response involves a delay associated to ion diffusion in the liquid after PO. Alternatively, the optofluidic technique discussed above has been successfully employed to precisely determine the onset of PO and, simultaneously, to control the PO process in real time [53].

The procedure is briefly as follows. When a drop of the wet etching solution (phosphoric acid 5 % w/w) is released over the closed side of a membrane ($t = 0$ in Fig. 8.10), the reflectance immediately falls due to the reduction in the refractive

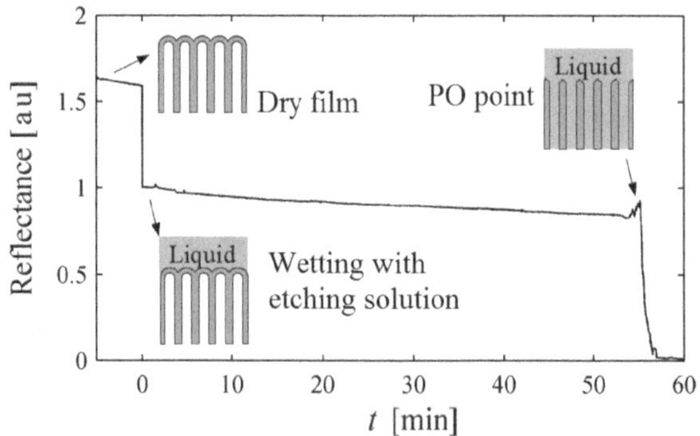

Fig. 8.10 Light reflectance measured during PO process, where the penetrating liquid is the wet-etching solution. The PO point is clearly determined in real time. Adapted from [53]

index contrast. At this step no oscillations are detected, indicating that the liquid does not penetrate the membrane. Then the etching proceeds, and at a given time (t = 55 min. in Fig. 8.10) a series of oscillations are observed, which indicates the penetration of the liquid, hence the PO point. Further, as the etching rate strongly depends on temperature, a small spot of increased temperature can be attained at a specific region of the membrane by tuning the laser power and focus, as well as by adding and adsorbing dye in solution, thus improving the control of the process.

8.5 Concluding Remarks

Optofluidics, the integration of optics and microfluidics, is a relatively new field of research with enormous potentials in both basic science and technological applications. In this context, NAA membranes are studied by measuring the dynamics of capillary-driven fluid imbibition by laser interferometry, and a theoretical model is derived to rationalize experimental results. On the one hand, the measurement system can be used as a nanofluidic sensor to determine fluid properties of a small sample of liquid, which has promising applications in lab-on-a-chip technology. On the other hand, the spontaneous filling of the membranes with a known fluid is used to elucidate the nanopores morphology, which is of interest for the characterization and specific design of NAA matrices.

More precisely, if the pore geometry is specified beforehand, the model predicts both the meniscus position and the variation of the effective refractive index to be obtained in an imbibition experiment with a given fluid. In addition, the inverse calculation with the model enables the identification of pore geometry from the experimental curves of reflectance data. Due to the mathematical complexity, this calculation is currently limited to well ordered arrays of straight and non-connected pores; nevertheless further efforts are being made to expand the implementation of the optofluidic technique to NAA matrices with arbitrary morphologies.

Concerning characterization, it is worth noting that the optofluidic technique presents some advantages in relation to microscopy. In fact, quantification of objects size from image analysis of SEM requires a crucial step that consists in defining a threshold, which is normally done by grey levels. The cut off value for pixel intensity may be rather arbitrary, thus pore radii are difficult to determine precisely. The relative errors take relevance when pores sizes are about few nanometers. In contrast, the optofluidic technique we are discussing here allows one to precisely determine both end radii and, notably, the inner profile across NAA membranes.

Another relevant aspect of the modeling is the theoretical prediction of anisotropic filling. The phenomenon is clearly seen in imbibition experiments, even when nanopores are extremely slim. The simplest case of conical pores was considered here, but any radius profile can be in principle included in calculations. These are a novel results in nanoscale fluid dynamics, which also open new potential applications for NAA arrays, for instance in the attractive fields of nanofiltration and flow rectification.

Acknowledgments The authors wish to acknowledge the financial support from the CONICET and the Universidad Nacional del Litoral, Argentina.

References

1. D. Psaltis, S.R. Quake, C. Yang, Nature **442**, 381 (2006)
2. Y. Fainman, L. Lee, D. Psaltis, C. Yang, *Optofluidics: Fundamentals, Devices, and Applications* (McGraw-Hill, New York, 2010)
3. X. Fan, I.M. White, Nat. Photonics **5**, 591 (2011)
4. L. Pang, H.M. Chen, L.M. Freeman, Y. Fainman, Lab Chip **12**, 3543 (2012)
5. D. Erickson, D. Sinton, D. Psaltis, Nat. Photonics **5**, 583 (2011)
6. Y.-F. Chen, L. Jiang, M. Mancuso, A. Jain, V. Oncescu, D. Erickson, Nanoscale **4**, 4839 (2012)
7. N.A. Mortensen, S. Xiao, J. Pedersen, Microfluid. Nanofluid. **4**, 117 (2008)
8. G. Barillaro, S. Merlo, S. Surdo, L.M. Strambini, F. Carpignano, Microfluid. Nanofluid. **12**, 545 (2012)
9. L.N. Acquaroli, R. Urteaga, C.L.A. Berli, R.R. Koropecki, Langmuir **27**, 2067 (2011)
10. R. Urteaga, L.N. Acquaroli, R.R. Koropecki, A. Santos, M. Alba, J. Pallares, L.F. Marsal, C.L. A. Berli, Langmuir **29**, 2784 (2013)
11. K.M. Van Delft, J.C.T. Eijkel, D. Mijatovic, T.S. Druzhinina, H. Rathgen, N.R. Tas, A. Van den Berg, F. Mugele, Nano Lett. **7**, 345 (2007)
12. E. Elizalde, R. Urteaga, R.R. Koropecki, C.L.A. Berli, Phys. Rev. Lett. **112**, 134502 (2014)
13. H. Haidara, B. Lebeau, C. Grzelakowski, L. Vonna, F. Biguenet, L. Vidal, Langmuir **24**, 4209 (2008)
14. C. Grzelakowski, D. Ben Jazia, B. Lebeau, L. Vonna, D. Dupuis, H. Haidara, Langmuir **25**, 5855 (2009)
15. C. Ran, G. Ding, W. Liu, Y. Deng, W. Hou, Langmuir **24**, 9952 (2008)
16. Z. Li, J. Wang, Y. Zhang, J. Wang, L. Jiang, Y. Song, Appl. Phys. Lett. **97**, 233107 (2010)
17. K.J. Alvine, O.G. Shpyrko, P.S. Pershan, K. Shin, T.P. Russell, Phys. Rev. Lett. **97**, 175503 (2006)
18. L. Bruschi, G. Mistura, L. Liu, W. Lee, U. Gösele, B. Coasne, Langmuir **26**, 11894 (2010)
19. F. Casanova, C.E. Chiang, A.M. Ruminski, M.J. Sailor, I.K. Schuller, Langmuir **28**, 6832 (2012)
20. P.G. de Gennes, F. Brochard-Wyart, D. Quére, *Capillarity and Wetting Phenomena* (Springer-Verlag, New York, 2004)
21. A. Han, G. Mondin, N.G. Hegelbach, N.F. de Rooij, U.J. Staufer, J. Colloid Interface Sci. **293**, 151 (2006)
22. D.I. Dimitrov, A. Milchev, K. Binder, Phys. Rev. Lett. **99**, 054501 (2007)
23. J. Haneveld, N.R. Tas, N. Brunets, H.V. Jansen, M.J. Elwenspoek, J. Appl. Phys. **104**, 014309 (2008)
24. P. Tabeling, *Introduction to Microfluidics* (Oxford University Press, New York, 2005)
25. S. Granick, Y. Zhu, H. Lee, Nat. Mater. **2**, 221 (2003)
26. E. Lauga, M.P. Brenner, H.A. Stone, in *Handbook of Experimental Fluid Dynamics* ed. by C. Tropea, A. Yarin, J. Foss (Springer, New-York, 2007), pp. 1219–1240
27. M. Majumder, N. Chopra, R. Andrews, B.J. Hinds, Nature **438**, 44 (2005)
28. K.P. Lee, H. Leese, D. Mattia, Nanoscale **4**, 2621 (2012)
29. C. Wu, H.S. Leese, D. Mattia, R.R. Dagastine, D.Y.C. Chan, R.F. Tabor, Langmuir **29**, 8969 (2013)
30. M.N. Popescu, J. Ralston, R. Sedev, Langmuir **24**, 12710 (2008)
31. D. Erickson, D. Li, C.B. Park, J. Colloid Interface Sci. **250**, 422 (2002)
32. T.L. Staples, D.G. Shaffer, Colloids Surf. A: Physicochem. Eng. Aspects **204**, 239 (2002)

33. M. Reyssat, L. Courbin, E. Reyssat, H.A. Stone, J. Fluid Mech. **615**, 335 (2008)
34. Liou W.W. Liou, Y. Peng, P.E. Parker, J. Colloid Interface Sci. **333**, 389 (2009)
35. R. Lucas, Kolloid Z. **23**, 15 (1918)
36. E.W. Washburn, Phys. Rev. **17**, 273 (1921)
37. N. Alleborn, H.J. Raszillier, J. Colloid Interface Sci. **280**, 449 (2004)
38. C.L.A. Berli, R. Urteaga, Microfluid Nanofluid, in press (2014). doi:10.1007/s10404-014-1388-9
39. S. Gruener, P. Huber, Phys. Rev. Lett. **103**, 174501 (2009)
40. S. Gruener, Z. Sadjadi, H.E. Hermes, A.V. Kityk, K. Knorr, S.U. Egelhaaf, H. Rieger, P. Huber, PNAS **109**, 10245 (2012)
41. L.D. Landau, E.M. Lifshitz, *Electrodynamics of Continuous Media*, vol. 8, 2nd edn. (Elsevier, Burlington, 1984)
42. A.A. Krokhin, P. Halevi, J. Arriaga, Phys. Rev. B. **65**, 115208 (2002)
43. D.E. Aspnes, Thin Solid Films **89**, 249 (1982)
44. O. Stenzel, *The Physics of Thin Film Optical Spectra*, vol. 44 (Springer Series in Surface Sciences, Springer, Heidelberg, 2005)
45. C. Galca, E.S. Kooij, H. Wormeester, C. Salm, V. Leca, J.H. Rector, B. Poelsema, J. Appl. Phys. **94**, 4296 (2003)
46. D.W. Thompson, P.G. Snyder, L. Castro, L. Yan, P. Kaipa, J.A. Woollam, J. Appl. Phys. **97**, 113511 (2005)
47. H. Zhuo, F. Peng, L. Lin, Y. Qu, F. Lai, Thin Solid Films **519**, 2308 (2011)
48. Z. Knittl, *Optics of thin films. An optical multilayer theory* (John Wiley & Sons, Czechoslovakia, 1976)
49. H. Masuda, K. Fukuda, Science **268**, 1466 (1995)
50. D. Losic, D. Losic Jr, Langmuir **25**, 5426 (2009)
51. M. Lillo, D. Losic, J. Membrane Sci. **327**, 11 (2009)
52. H. Han, S.J. Park, J.S. Jang, H. Ryu, K.J. Kim, S. Baik, W. Lee, ACS Applied Materials & Interfaces **5**, 3441 (2013)
53. E. Elizalde, F.A. Garcés, R. Urteaga, R.R. Koropecki, C.L.A. Berli *9th International Conference of the Porous Semiconductors—Science and Technology, PSST.* (Alicante-Benidorm, Spain, 9–14 March 2014)

Chapter 9
Protein and DNA Electrochemical Sensing Using Anodized Aluminum Oxide Nanochannel Arrays

Alfredo de la Escosura-Muñiz, Marisol Espinoza-Castañeda and Arben Merkoçi

Abstract This chapter shows the recent trends on the use of anodized aluminum oxide (AAO) nanochannel arrays for electrochemical sensing of proteins and DNA with a special focus on those based on voltammetric detections. Some general considerations on nanochannels and especially on AAO nanoporous membranes are given first, followed by the receptors (antibody, DNA) immobilization as well as the set-up configuration. The typical optimization procedures as well as the detection principles ranging from the use of ionic electroactive indicators and nanoparticles (used also as blockers) are discussed. Aspects related to the analytical performance of the developed devices while applied in diagnostics including cancer biomarker detection are also given. Finally an overview for future improvements and applications of this technology are included.

9.1 General Introduction: Stochastic Sensing Using Biological Single Nanochannels

The nowadays's nanochannels-based sensing systems [14] are inspired by the pioneer microparticle counter device patented by Wallace Coulter in 1953 [8, 9]. This simple microparticle counting system consists of a single microchannel that separates two electrolyte-filled chambers. Changes in the electrical conductance between both chambers are recorded as electric current or voltage pulse when a microparticle enters the nanochannel, allowing to determine its size, mobility, surface charge, and concentration. Devices for particle counting based on this concept are nowadays commercially available and commonly used in hospitals for

A. de la Escosura-Muñiz · M. Espinoza-Castañeda · A. Merkoçi (✉)
ICN2—Nanobioelectronics & Biosensors Group, Institut Catala de Nanociencia i Nanotecnologia, Campus UAB, 08193 Bellaterra (Barcelona), Spain
e-mail: arben.merkoci@icn.cat

A. Merkoçi
ICREA, Institucio Catalana de Recerca i Estudis Avançats, 08010 Barcelona, Spain

D. Losic and A. Santos (eds.), *Nanoporous Alumina*,
Springer Series in Materials Science 219, DOI 10.1007/978-3-319-20334-8_9

cells counting/determination as well as in the industry of painting and ceramics between other applications (Fig. 9.1).

This device evolved from micro- to nanochannels based ones, inspired by natural ion channels, able to detect molecules such as ssDNA and proteins [1, 4, 55]. The insertion of the α-hemolysin bacterial protein pore in artificial lipid bilayers was the pioneer approach to build these biomimetic nanochannels, opening the way of extensive sensing applications based on the so-called stochastic sensing [3] (Fig. 9.2). The selectivity of the system is given by specific receptors that are inserted inside the nanochannel by genetic engineering techniques, allowing to detect a variety of analytes such as DNA [41], proteins [43] and others [40].

It must be highlighted the potential ability of the α-hemolysin nanochannel for DNA sequencing [32]. DNA single strands can enter through the α-hemolysin channel by electrophoretic forces. Once inside the channel, each of the four bases (adenine, guanine, cytosine and thymine) interacts in a specific way with the walls of the channels which leads to a different stay-time and consequently giving specific pulses of current, constituting a "fingerprint" signal that can be related with the length and base composition of the nucleic acid [5, 25, 33, 54]. This ideal system could directly detect nucleotides, without reagents or labeling being also able to do sequencing of long strands of DNA. Some companies such as Oxford Nanopore

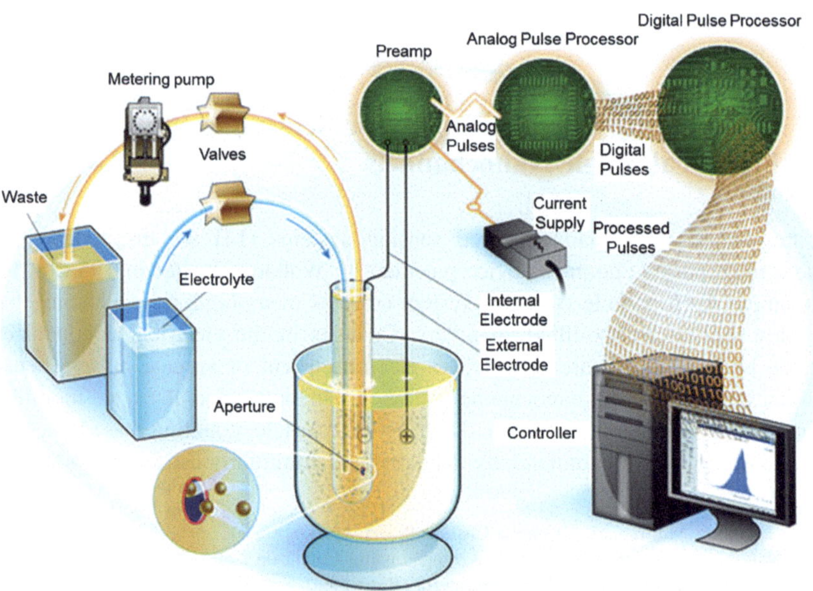

Fig. 9.1 Scheme of a commercial microparticle counting device based on the Coulter principle. An aperture (microchannel) separating two conductive chambers allows the flux of microparticles and its detection by measuring the changes in the electrical conductance between the chambers. Reproduced with permission of Beckman Coulter Inc.

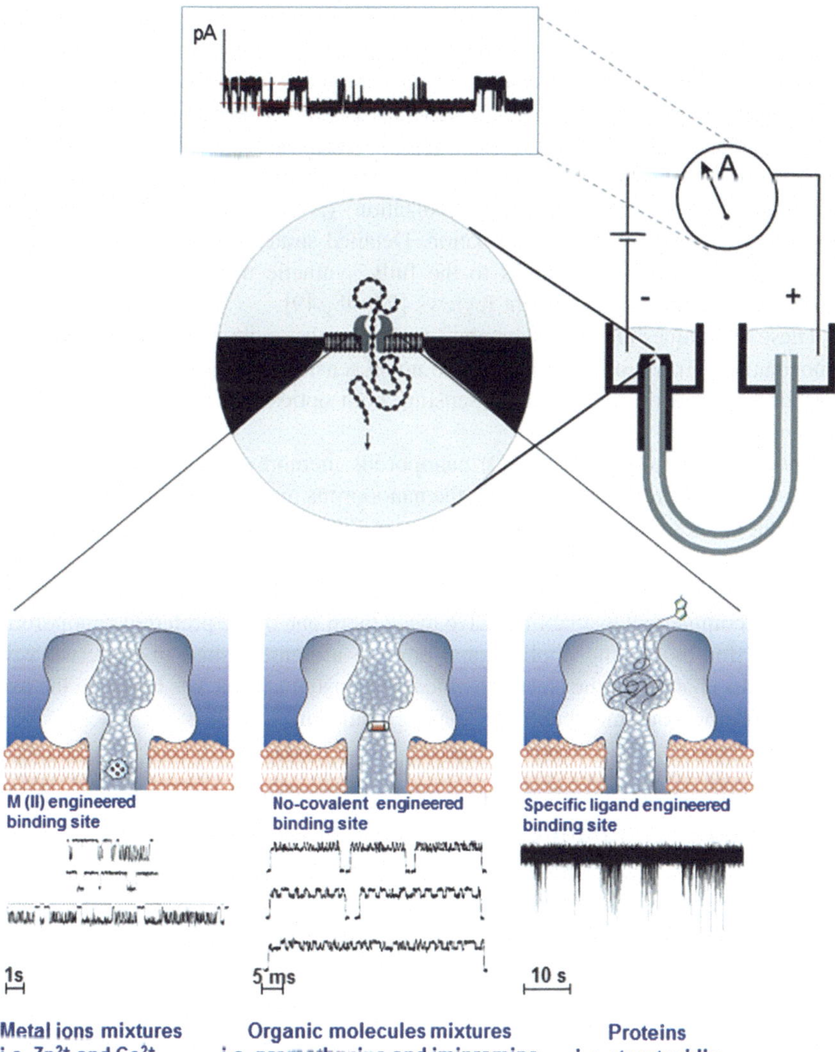

Fig. 9.2 Sensing using α-hemolysin single nanochannels: stochastic sensing. (*Up*) Scheme of a nanochannel based sensor for DNA evaluation based on the concept of the Coulter counter. Adapted from [55] with permission. (*Down*) Engineered protein nanochannels allow the detection of a variety of analytes by stochastic sensing after the insertion of specific receptors inside the nanochannels by genetic engineering. Adapted with permission from [3]. Copyright 2001 Nature Publishing Group

Technologies have recently proposed commercial approaches taking advantage of this sensing principle [7, 39].

However, the low stability, time consuming analysis and the need of genetic engineering techniques for the insertion of specific receptors have limited the implementation of these biological ion channels as robust sensing platforms. In this context, solid-state single nanochannels prepared by a variety of techniques have been proposed as alternative systems in the last years. The most successful strategies for the preparation of solid-state nanochannels for biosensing applications are: (i) track-etching on polymeric membranes; (ii) electron-beam lithography and ion beam sculpting: focused ion beam (FIB) and dual FIB/SEM systems on $SiN–SiN_2$ membranes; (iii) metallic substrate anodization; (iv) micromolding and (v) high ordered mesoporous thin films formation. Detailed strategies for the integration of nanochannels, from ion channels to the fully synthetic nanochannels have been extensively summarized in recent reviews [24, 48, 49].

These techniques allow the preparation of not only single nanochannels but also nanochannels arrays, opening the way to novel sensing approaches totally different from those based on the stochastic sensing, both optical [2, 35, 44, 46] and electrochemical [14].

Anodic aluminum oxide (AAO) nanoporous membranes prepared by metallic substrates anodization stand out from the nanoporous materials that can be prepared following the above mentioned methodologies. These membranes have a high pore density ($1 \times 10^9/cm^2$) and small pore diameters, which result in a substrate with high surface area that can be easily functionalized. These characteristics, together with their commercial availability, have made them one of the preferred nanoporous substrates for biosensing applications. For the aforementioned reasons, we will focus in this chapter on the electrochemical biosensing approaches based on the use of AAO nanoporous membranes as sensing platform.

9.2 Anodic Aluminum Oxide (AAO) Nanoporous Membranes Preparation and Functionalization

9.2.1 AAO Nanoporous Membranes Preparation

Anodic aluminum oxide (AAO) is chemically and thermally stable rigid and dense porous material. The main characteristics of AAO are their perfectly ordered and size- controlled nanopores [31]. The typical fabrication process consists of the anodization of aluminum of high purity in an acidic solution applying a constant high voltage [38, 45] (Fig. 9.3A). Under these conditions, aluminum oxide is generated, containing many surface irregularities (defects) where the electric field concentrates. This localized electric field enhances acidic dissolution of the oxide at the bottom of the pores while leaving the pore walls intact, resulting in the nanochannels formation. The barrier layer produced at the bottom of the pore channels is subsequently removed by immersion in dilute acid during which the pores are also enlarged. The anodization under optimized conditions gives rise to the production of porous alumina with a highly ordered cell configuration. The structure of AAO can be described as a close-packed hexagonal array of parallel cylindrical nanopore perpendicular to the surface on top of the underlying Al substrates (Fig. 9.3B) [47]. The rich content of hydroxyl groups on the alumina membrane surface allow them to be easily modified via modification with organic molecules with the desired functionality [34]. An exciting alternative of special interest also for later biosensing applications consists on the generation of AAO nanochannels directly on the surface of an electrotransducer surface as recently reported by Foong et al. [20] after depositing an aluminum layer by chemical vapor deposition (Fig. 9.3C).

9.2.2 AAO Nanoporous Membranes Functionalization

Binding ligands are commonly used for the modification of the inner nanochannel walls to provide recognition sites for analytes, playing a crucial role in achieving the desired sensing performance [10]. Wet chemical approaches for the nanochannel modification by using silanes or organic acids are commonly used. Layer-by-layer (LBL) assembly is also a very versatile method for incorporating functional groups

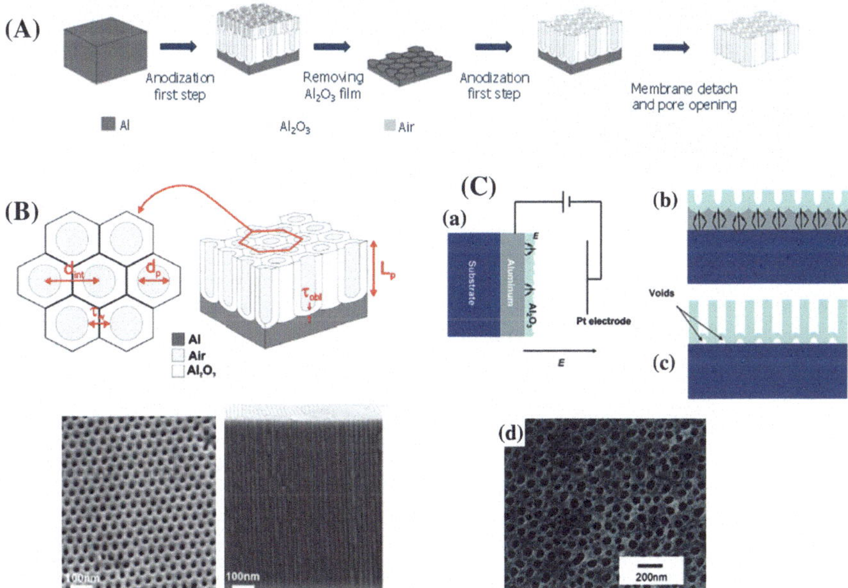

Fig. 9.3 Preparation of anodic aluminum oxide (AAO) nanoporous membranes by metallic substrates anodization. **A** Schematic of the main steps for AAO membranes preparation by the two-step aluminum anodization process under hard conditions. Adapted from [44] with permission. **B** AAO structure and SEM images showing the *top* and *cross-section* views of AAO structure. Adapted from [47]. **C** Schematic of AAO template formation on ITO glass. Shown are (*a*) schematic of the setup system and (*b*) process of acidic dissolution of the native aluminum oxide as well as (*c*) the formed layer after the anodization process and (*d*) SEM image of the obtained pores. Adapted from [20]. Copyright 2008 American Chemical Society

into nanochannels. This method avoids the need for complex chemical steps in the process [61].

The most common strategy for AAO nanoporous membranes functionalization is based on the formation of functional groups such as –COOH or $-NH_2$ inside the nanochannels able to react afterwards with antibodies or ssDNA through the formation of a peptide bond for later biosensing applications, as schematized in Fig. 9.4. The easy chemical route comprises the membranes silanization followed by reaction with glutaraldehyde. The efficient i.e. antibody immobilization inside nanochannels can be checked by using antibodies labelled with a fluorescent tag (i.e. FITC), observing in the confocal microscope the presence of fluorescence not only on the external surface of the membranes but also in the inner walls of the nanochannels (Fig. 9.4, upper left).

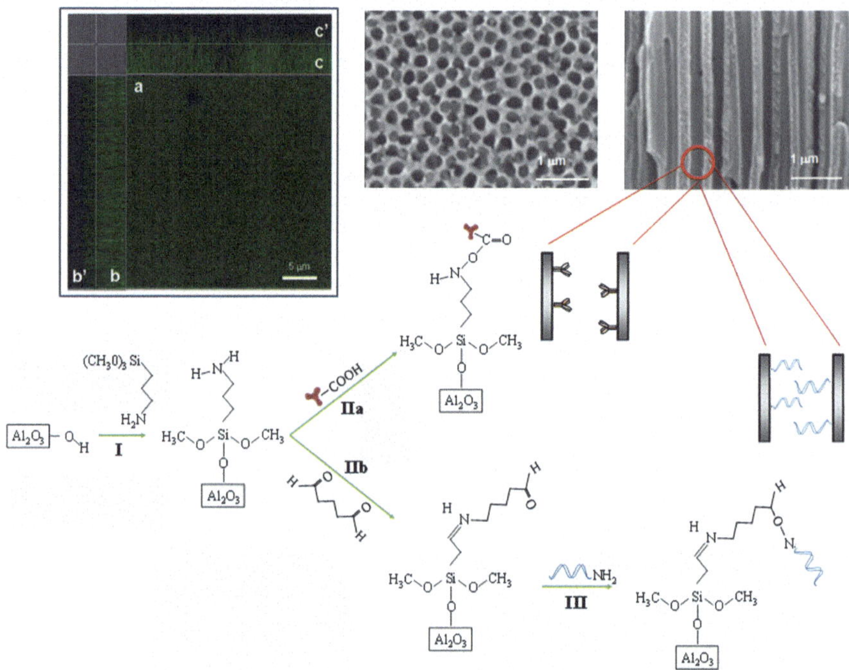

Fig. 9.4 Scheme of the biofunctionalization procedure for antibodies and ssDNA immobilization on AAO nanoporous membranes: generation of amino groups by silanization process (*I*) followed by covalent binding of antibodies using EDC/sulfo-NHS cross-linker (*IIa*). Carboxyl groups can be generated after the silanization by reaction with glutaraldehyde (*IIb*), followed by the immobilization of amino-modified ssDNA by the peptide bond (*III*). SEM images correspond to of a plan (*center*) and cross-sectional view (*right*) of a 200 nm pore AAO nanoporous membrane. Confocal image (*left*) corresponds to a 200 nm pore AAO membrane with antibody-FITC immobilized. Plan view (*a*) and planes until 5 μm in depth from the top surface (*b*, *c*) and from the *bottom* surface (b′, c′). Adapted from [11–13] with permission

9.3 Electrochemical Biosensing Systems Based on AAO Nanoporous Membranes

9.3.1 Voltammetric Sensing Systems

9.3.1.1 Detection Principle: Label-Free Protein and DNA Detection

The most recent and representative approaches of voltammetric biosensing using AAO nanoporous membranes are focused on their use as modifiers of conventional electrotransducer surfaces, measuring changes of the electrochemical response of an electroactive specie in solution due to the presence of the analyte inside the channels. The sensing principle for the detection of proteins and DNA is illustrated in Fig. 9.5. In the case of proteins, the formation of the immunocomplex inside the nanochannels produces a partial blockage in the diffusion of electroactive species through the nanoporous membranes toward the electrochemical transducer surface (working electrode) leading to a decrease of the electrochemical signal related to $[Fe(CN)_6]^{4-}$ oxidation to $[Fe(CN)_6]^{3-}$ [11]. The nanoporous cell set-up (see pictures of Fig. 9.5) is prepared by fixing the AAO membrane onto a screen-printed carbon electrode (SPCE) by physical attachment, placing the SPCE onto a methacrylate block and putting the membrane with the filtering side up over the three-electrode surface. Then, a second methacrylate block containing a hole of the same size as the working area of the SPCE is placed onto the AAO nanoporous membrane, with an insulating ring between them to avoid liquid leakage. Finally, the system is fixed with screws. In this way, a 300 μL electrolytic cell is defined, which is filled with a solution (100 μL) of 1 mM $K_3[Fe(CN)_6]$ in 0.1 M $NaNO_3$. Differential pulse voltammetry (DPV) oxidation peak is selected as analytical signal.

The blockage of the pores due to the formation of DNA hybridization complexes is also detected following the same principle [12]. Both systems are consistent with the relation between the pore size of the AAO membrane used (typically 200 nm) and the length of both antibody and antigen, that is, 14.5 nm × 8.5 nm × 4 nm in the case of human immunoglobulin G (HIgG) and also with the size of a i.e. 21-mer ssDNA (approximately diameter of 1.84 nm and length of 0.38 nm). Furthermore, AAO nanoporous membranes have the ability to act as filters of e.g. cells present in real samples (as illustrated in the figure), allowing to minimize matrix effects [13].

However, the poor sensitivity of both label-free bioassays (LOD: 100 μg/mL for HIgG protein and 5 μg/μL for a 21-mer ssDNA) does not fulfill the requirements for real diagnostic applications. To overcome this drawback two amplification strategies have been reported. The first one is based on the use of gold nanoparticle (AuNP) tags followed by silver enhancement so as to increase the blockage degree in the channels [13]. In the second strategy, the standard red-ox indicator ($[Fe(CN)_6]^{4-/3-}$) is changed by Prussian Blue nanoparticles (PBNPs) which diffusion inside the channels is hindered by lower quantities of analyte [19]. Both strategies and their application for biomarkers detection in real samples (whole human blood and cell cultures respectively) are described in the following sections.

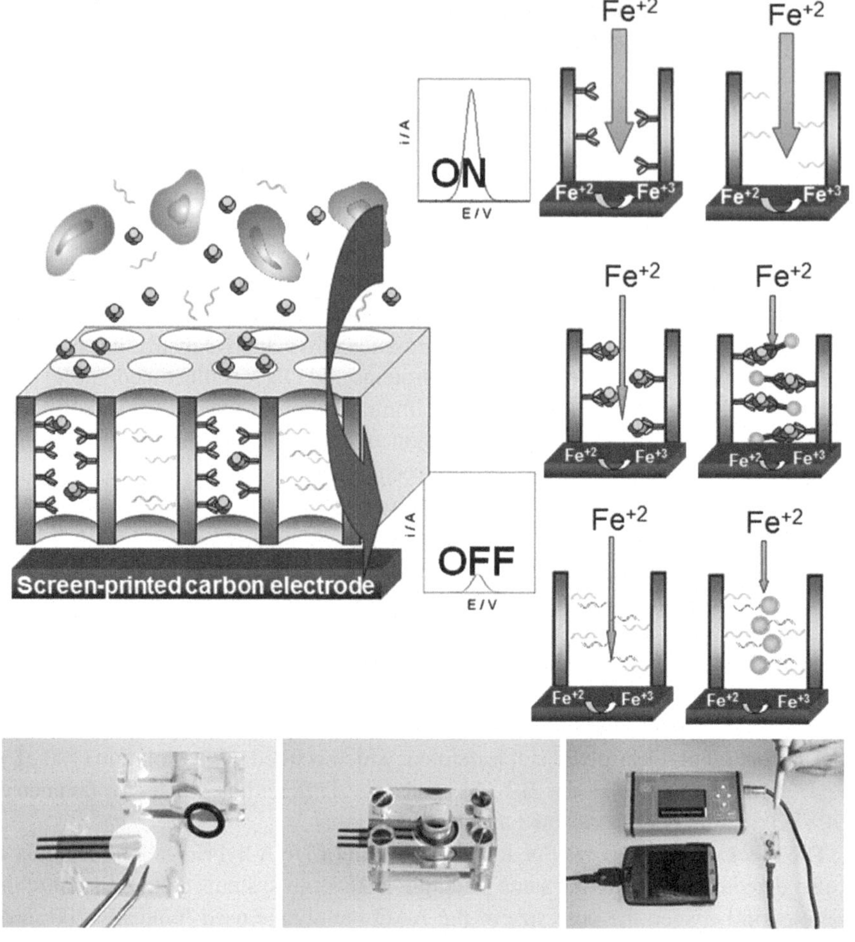

Fig. 9.5 Principle of electrochemical biosensing using AAO nanoporous membranes. *Left* Cells in the sample remain outside the pores while the proteins (or ssDNA) enter inside are recognized by specific antibodies (or complementary ssDNA). *Right* Sensing principle in the absence ("On" response) and presence ("Off" response) of the specific biomolecule in the sample. Gold nanoparticle tags increase the blockage in the channel, leading to an enhanced sensitivity. Pictures correspond to the electrochemical cell set-up. Adapted from [13] with permission

9.3.1.2 Sensitivity Enhancement Using AuNPs Tags and Silver Amplification: Application for Cancer Biomarkers Detection in Human Blood

The first amplification strategy relies in the use of AuNPs tags in sandwich assay approaches [13]. The presence of such AuNPs inside the channels increase in a high extent the blockage of the diffusion of the electroactive species achieving in this way lower detection limits (Fig. 9.5, right part). This system is AuNPs size-sensitive, being the LOD of proteins lower when larger is the size of AuNPs (20–80 nm range was assayed). Moreover, the catalytic activity of the AuNPs toward the silver deposition is approached for the selective formation of silver crystals around the AuNPs, increasing in this way the AuNP size and consequently the blockage of the channels and improving the LOD of proteins. The optimized assay using 80 nm-sized AuNPs and silver enhancement allows to obtain LODs of HIgG as low as 50 ng/mL.

The above detailed optimized system has been applied for the detection of protein biomarkers in whole human blood samples, without any sample preparation, taking advantage of the dual ability of the membranes to act not only as sensing platforms but also as filter of complex components such as red and blood cells. With this methodology, a new system for detection of CA15-3 cancer marker spiked in whole blood capable to detect 52 U/mL of CA15-3 with very low matrix effects was developed, as shown in Fig. 9.6c, d. In addition, the developed device presents the advantage of the quantitative analysis that can be performed with the low-cost electrochemical analyzers using a simple and low-cost detection technology. Electrochemical results are supported by SEM characterization of the nanoporous membranes. Figure 9.6a shows a SEM image corresponding to a top view of an AAO nanoporous membrane on which a drop of blood (50 μL) was deposited (and left to dry). It is observed that the constituents of the blood (probably red blood cells, white blood cells, platelets, and crystal salts, all on the micrometric scale) remain on the surface of the membrane which acts as a filter while free pores allow the penetration of the spiked cancer biomarker. When a membrane with immobilized anti-CA15-3 antibodies is left to react with a blood sample containing 240 U/mL of added CA15-3, upon following the optimized experimental procedure and using 80-nm AuNP tags and silver enhancement a cross-sectional SEM image as shown in Fig. 9.6b is obtained. It is observed that the presence of AuNPs inside the pores in the case of the specific reaction gives rise to the formation of white silver crystals with sizes around 150–200 nm, which contribute to the blockage in the pore.

The same sensing principle has been extended to the detection of proteins in human blood (thrombin as model one) using aptamers as bioreceptors [15] taking advantage of both aptamer-based recognition and the use of AAO nanoporous membrane platforms. The protocol involves sandwich format (aptamer/thrombin/antibody-AuNP) and silver amplification in the inner walls of the membranes. The effect of the electrostatic interactions between the aptamer/thrombin/antibody-AuNP complex and the redox indicator, in addition to the steric effects were also investigated here, concluding that the biosensing system is a result of the combination of both effects.

This system allows to detect thrombin spiked in whole blood at very low levels which are within the range of clinical interest for the diagnostic of coagulation abnormalities as well as pulmonary metastasis. In Fig. 9.7a, the filtering ability of the AAO membranes in the analysis of whole human blood is illustrated by the SEM image showing two red blood cells that remain out of the nanochannels. This image also shows that the solution containing the blood cells does not block the entire pores so thrombin biomarker can pass through the nanochannels to be in

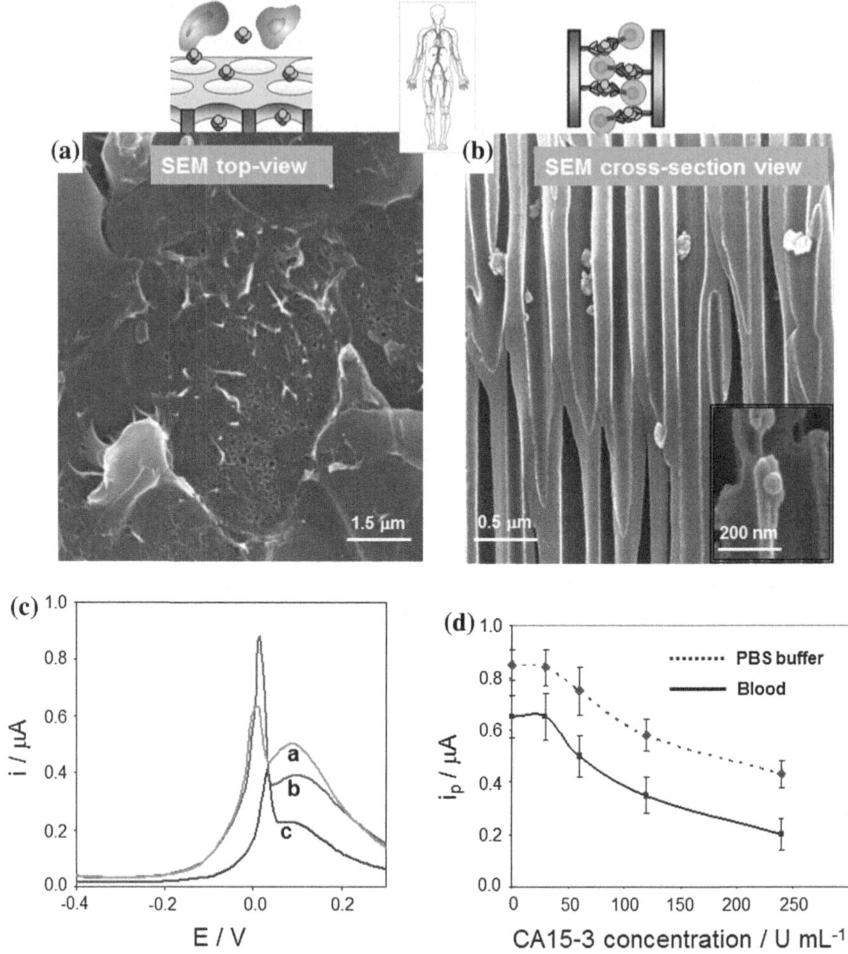

Fig. 9.6 Voltammetric detection of cancer biomarkers in blood using antibodies. **a** SEM *top view* of a 200-nm pore AAO membrane on which a drop of blood (50 µL) was deposited (and *left* to dry) on the filtering side. **b** SEM *cross-sectional view* of the membrane with immobilized anti-CA15-3 antibodies, *left* to react with a blood sample containing added CA15-3 (240 U/mL) and completing the sandwich using 80-nm AuNP tags and silver enhancement (silver crystals are observed). *Inset* Image of a single 80-nm AuNP attached to the nanochannel wall. **c** DPV results for blood samples containing different concentrations of added CA15-3: 60 (**a**), 120 (**b**), and 240 U/mL (**c**). **d** Comparison of the effect of the concentration of CA15-3 on the analytical signal obtained in PBS buffer (*dotted line*) and in the blood sample (*solid line*). Adapted from [13] with permission

contact with the working electrode surface. Figure 9.7b shows a cross-sectional SEM view of AAO membranes of 200 nm nanochannels size modified with the aptamer and left to react with a blood sample containing 100 ng/mL of spiked thrombin (inset cartoon is a schematic representation of the blockage occurring

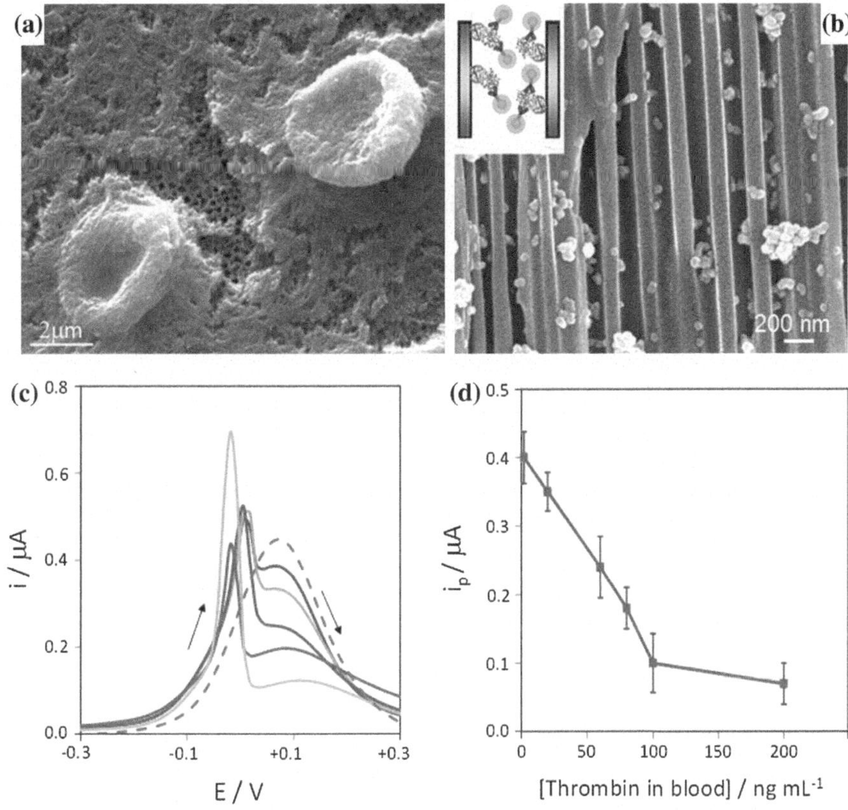

Fig. 9.7 Voltammetric detection of cancer biomarkers in blood using aptamers. **a** SEM top view of a 200-nm pore AAO membrane on which a drop of blood (30 μL) was deposited. Blood cells as well as the nanopores are observed. **b** SEM cross-sectional view of the membrane modified with the aptamer and left to react with a blood sample containing 100 ng/mL of spiked thrombin. The sandwich assay is then completed with anti-thrombin/AuNPs followed by silver enhancement (silver crystals are observed). *Inset* Cartoons correspond to scheme of the involved process. **c** DPV results for blood samples containing spiked thrombin at concentrations of (up to down): 0 (*dashed line*), 2, 20, 60, 80 and 100 ng/mL. **d** Effect of the concentration of thrombin spiked in blood on the analytical signal. Adapted from [15] with permission

inside the nanochannels after a sandwich assay and silver deposition onto AuNP tags). The sandwich assay is then completed with anti-thrombin/AuNPs followed by silver enhancement. The silver crystals observed in the SEM image are the main responsible of the blockage in the channel. The DPV signals and the quantitative analysis for different quantities of spiked thrombin are shown in Fig. 9.7c, d, respectively. This allows obtaining of a detection limit of 1.8 ng of thrombin per mL of human blood which is within the range of interest for the above mentioned clinical applications.

Finally, it is important to highlight that the use of AuNP tags have also been applied for improving the sensitivity of the DNA hybridization detection in AAO based sensing systems (Fig. 9.5, right part), reaching LODs of 42 ng/μL for a 21-mer ssDNA [12].

9.3.1.3 Sensitivity Enhancement Using PBNPs as Red-Ox Indicator

As stated before, the label-free system based on the monitoring of $[Fe(CN)_6]^{4-}$ indicator ions diffusion through AAO nanoporous membranes (attached onto a SPCE by DPV signal change after biomolecule recognition) lacks of sensitivity, making necessary sandwich based strategies that use AuNPs and silver enhancement. However, sandwich immunoassays have inherent drawbacks in terms of time of analysis and antibodies consumption besides the rather low efficiency while being applied for small proteins analysis.

One of the main possible reasons of the low sensitivity of the label-free assay probably relies on the small size of the $[Fe(CN)_6]^{4-}$ ions. For this reason, alternative red-ox indicators with sizes bigger than that of these ions would be potentially useful for the improvement of the label-free approach. Prussian blue nanoparticles (PBNPs) appear here as ideal candidates for this purpose. The Prussian blue (PB) is a mixed valence compound widely used as an electron transfer mediator for analytical applications [30, 37, 53].

Given the aforementioned context, a novel device that operates through PBNPs as red-ox indicator for sensitive label free immunodetection of a cancer biomarker is developed [19]. Stable and narrow-sized (around 4 nm) PBNPs, protected by polyvinylpyrrolidone (Fig. 9.8b), exhibit a well-defined and reproducible red-ox behavior and are successfully applied for the voltammetric evaluation of the AAO nanochannels (20 nm pore sized) blockage due to the immunocomplex formation. The bigger size of the PBNPs compared with ionic indicators such as the $[Fe(CN)_6]^{4-/3-}$ system leads to an increase in the steric effects hindering their diffusion toward the signaling electrode which in turn is transduced to an improvement of the detection limit from 200 μg/mL to 34 pg human IgG/mL (Fig. 9.8a).

The advantages of using this large sized nanochannel blocking indicator are related not only to a better sensitivity for the detection of a big protein (human IgG: 150 kDa; 6 nm approx.) but also to the detection of small proteins which would not produce any blockage for ionic red-ox indicators such as $[Fe(CN)_6]^{4-/3-}$. In this context, the developed label-free methodology appears as a very advantageous alternative for the detection of small proteins which are not easily detected using sandwich immunoassays. This is the case of PTHrP, a small protein of 173 aminoacids (MW: 18 kDa), the factor responsible for tumor-induced hypercalcemia [50]. Serum PTHrP determination is a useful indicator of not only hypercalcemia but also bone metastases and survival in lung carcinoma patients [22]. So, measurement of PTHrP is indeed of clinical relevance. The most challenging issue in the detection of this protein using a biosensing system is to find a pair of antibodies that recognize different aminoacids sequences and, consequently, are suitable to

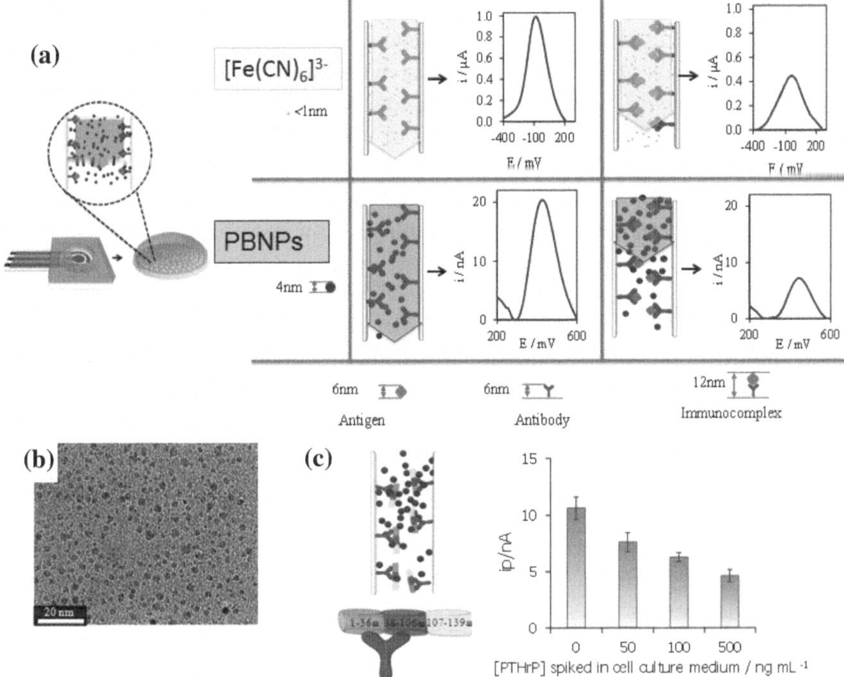

Fig. 9.8 Enhancement of the voltammetric sensing systems using PBNPs as red-ox indicators. **a** Schematic representation of the effect of the immunocomplex formation towards the voltammetric signal using $[Fe(CN)_6]^{3-}$ (*top*) and PBNPs (*bottom*) as red-ox indicator. DPV signals correspond to blank samples (*left*) and samples containing the antigen (HIgG) (*right*) at levels of 200 µg/mL and 100 ng/mL respectively. **b** Transmission electron micrograph (TEM) of a 20 µg/mL PBNPs suspension. **c** Schematic representation of the parathyroid hormone-related protein (PTHrP) capturing on the inner walls of the nanochannels and its blocking effect on the diffusion of the PBNPs to the electrode and effect of the concentration of PTHrP spiked in cell culture medium on the analytical signal. Adapted from [19] with permission

form a sandwich immunoassay. In the AAO-PBNPs based approach, a single antibody is used to capture the protein followed by its label-free detection avoiding the use of the sandwich format (Fig. 9.8c). In this way, PTHrP at levels of 50 ng/mL is detected in a real clinical scenario such as cell culture medium.

9.3.2 *Impedimetric, Capacitive, Conductometric and Resistive Sensing Systems*

In addition to the extensive voltammetric applications, impedance spectroscopy (EIS), capacitance, conductivity and resistivity measurements have also widely used in the last years in biosensing systems based on AAO nanoporous membranes, as recently reviewed by Santos and co-workers [47].

EIS takes advantage of the fact that the electrical behavior of AAO can be adjusted to an equivalent circuit, the components of which depend on the used model. In this way, AAO membranes can be used as impedimetric sensors by monitoring changes of conductance or impedance produced by bioreactions inside the nanochannels. The pioneer applications of impedance AAO biosensors focused on DNA hybridization detection [51, 56], study of protein and lipid membrane interactions [23] and detection of cancer cells and bacteria [52].

Recent approaches have also taken advantage of the use of AuNP tags and silver amplification for the improvement of the performance of these EIS-based sensing systems [60] reaching LODs of 50 pM for 25-mer ssDNA (Fig. 9.9).

Knowledge about electrochemical and electrical properties of AAO structures and the influence of channel dimensions on these properties is a crucial point for the development of such AAO based biosensing devices. Recent studies have explored the influence of AAO nanochannel dimensions (diameter and length) using non-faradic EIS [28]. The scheme of the proposed biosensing device and the experimental setup is shown in Fig. 9.10a–d. AAOs with channel diameters of 25–65 nm and thicknesses of 4–18 μm are first prepared and then a reduced working area of 20 × 20 μm^2 is defined by photolithography, followed by covalent immobilization of streptavidin in the inner walls of different sized nanochannels. Then a four-electrode cell is used to characterize the change in properties of the

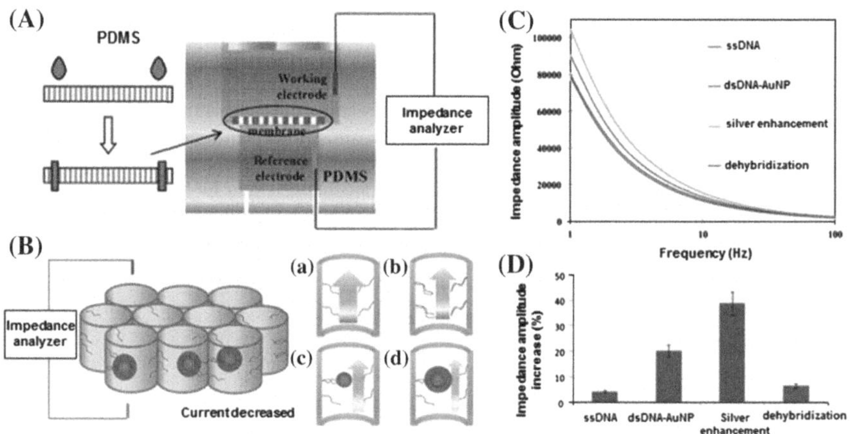

Fig. 9.9 Impedimetric detection of ssDNA: signal enhancement using AuNPs and silver amplification. **A** Schematic image of a PDMS chip integrated with a AAO membrane. **B** Sensing principle of AAO-based impendance sensor for DNA hybridization detection. (*a*) Relative large electrolyte current through AAO with immobilized ssDNA probes; (*b*) DNA hybridization causes partial blockage; (*c*) AuNP tags increase blocking degree; (*d*) silver enhancement on AuNPs matches the nanopore size better and further increases blocking degree. **C** Impedance spectra of dsDNA with AuNP tags, dsDNA with AuNP tags after silver enhancement, and dehybridization; **D** relative impedance amplitude change at 1 Hz relative to silane modified bare AAO membranes for dsDNA withAuNP tags, dsDNA with AuNP tags after silver enhancement, and dehybridization. Adapted from [60] with permission

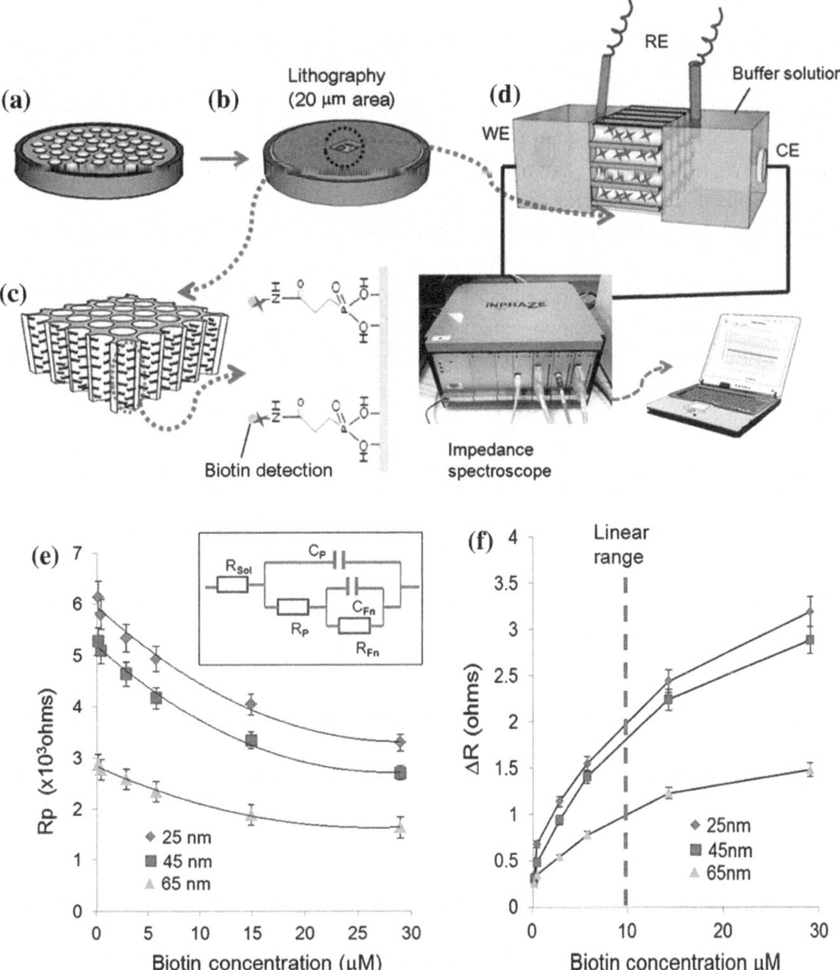

Fig. 9.10 Impedimetric detection of ssDNA: effect of nanochannel diameter and length. Schematic of experimental setup of the prepared AAO membrane before (**a**) and after (**b**) photolithography with a 20 × 20 μm^2 opening, after modification with streptavidin (**c**) and placed on a EIS cell for impedance characterization (**d**). **e** Relationship between the nanochannel diameter and biotin concentration and their influence on the nanochannel resistance Rp. The proposed equivalent circuit (*inset*) used for the simulation of experimental data is presented. **f** Calibration graph showing changes of nanochannel resistance ΔR with biotin concentration using AAO membranes with different nanochannel diameter. Adapted from [28] with permission

nanochannel array using EIS technique for the different sized nanochannels. The performance of this biosensing platform is evaluated for various concentrations of biotin (model analyte). The values of Rp show a decrease with the increasing concentration of analyte as consequence of the properties of binding biotin molecules inside the nanopores (Fig. 9.10e) adjusted to a linear relationship in the range from 0.2 to 10 mM (Fig. 9.10f). Moreover, it is found that lowering nanochannel diameters (<30 nm) provides better response and sensitivity. Regarding nanochannel length, it is observed that longer pores (>10 mm) are also not favourable for non-faradic EIS detection due to the higher resistance and long time for diffusion of analyte molecules inside nanopores. Hence a crucial point to achieve optimal performances of nanopore based electrochemical biosensing devices relies on the optimization of nanochannel dimensions.

Properties of AAO nanoporous membranes such as their stability at high temperatures, high surface area and affinity to adsorb gases and vapors make them very interesting platforms for capacitance-based gas-sensing devices. AAO-based relative humidity (RH) capacitance sensors have been reported in the last years [27]. Of special interest are the capacitance gas sensors able to detect important environmental contaminants such as polychlorinated biphenyls (PCBs) [26], nitrogen dioxide, ammonia and ethanol [29].

In addition to gas sensing, conductometric biosensors for the label-free detection of DNA molecules based on changes of ionic conductance of nanopores as result of DNA binding [27] and for the evaluation of enzymatic activities [62].

9.4 Conclusions and Prospects

The use of arrays of solid-state nanochannels has opened the way to novel and versatile electrochemical sensing systems, mainly based on the blockage of the channels by the affinity reaction (immunocomplex or DNA duplex formation). Anodized aluminum oxide (AAO) nanoporous membranes prepared by anodization of the aluminum metal substrate are the most used materials, due to their advantageous properties and mass production opportunities. These AAO nanoporous membranes have been shown to be excellent platforms for the analysis of real samples of, for example, human blood, due to their filtering properties, allowing minimization of matrix effects. These entire advantages make the AAO-based electrochemical biosensing technology a very promising research area that may bring significant scientific discoveries, with potential applications in fields such as diagnostics, safety-security, and other industries.

However, the practical application of the AAO nanoporous membranes based sensing systems are limited due to two main factors: (i) the membrane and the electrotransducer use to have to be assembled prior the detection (not integrated systems); and (ii) the large thickness of AAO membranes (in the micrometric scale) made necessary the use of amplification strategies for the enhancement of nanochannel blockage. For these reasons, nanochannels with nanosized thickness integrated on the electrotransducer surface are highly required so as to overcome such limitations. In this context, very interesting alternatives recently proposed for the preparation of nanochannels on solid substrates consist on the formation of highly ordered mesoporous thin film coatings, which has a great potential application in electrochemical analysis [58]. Strategies such as the use of self-assembled nanoparticle templates [42, 57], self-assembled surfactant templates [59, 63] and hard templating/nanocasting approaches [6] have been recently reported in order to get ordered nanochannel array films on electrodes. Of special relevance are the approaches based on the self-assembling of polystyrene nanospheres to build nanostructured surfaces on transparent conductive oxide (TCO) substrates [18, 21]. The recent report [16] on a novel, cheap, disposable and single-use assembled nanoparticles-based nanochannel platform onto a flexible substrate for label-free voltammetric immunosensing also is an interesting integration alternative. This sensing platform is formed by the dip-coating of a homogeneous and assembled monolayer of carboxylated polystyrene nanospheres (PS, 200 and 500 nm-sized) onto the working area of flexible screen-printed ITO/PET electrodes (Fig. 9.11a, b). The spaces between the self-assembled nanospheres generate well-ordered nanochannels, with inter-PS particles distances of around 65 and 24 nm respectively (Fig. 9.11c). The formed nanochannels are approached for the effectively immobilization of antibodies and subsequent protein detection based on the monitoring of $[Fe(CN)_6]^{4-}$ flow through diffusion and the decrease in the DPV signal upon immunocomplex formation (Fig. 9.11d). The obtained sensing system is

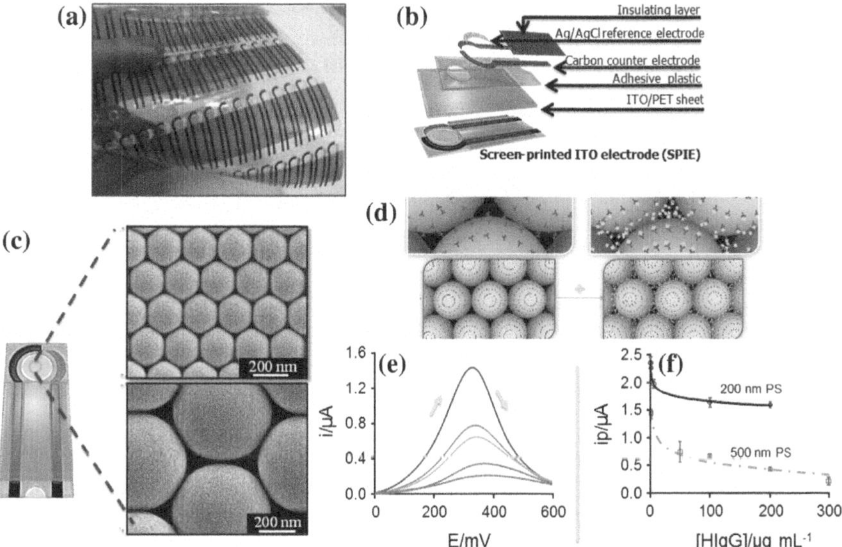

Fig. 9.11 Prospects: biosensing system based on monolayers of assembled nanoparticles. **a** Pictures of a sheet of screen-printed ITO electrodes (SPIEs). **b** Schematic representation of the different materials and layers which form the SPIE. **c** SEM images of a monolayer of 200 nm (*up*) and 500 nm (*down*) carboxylated polystyrene spheres (PS) on the working electrode of the SPIE. **d** Schematic representation (not in scale) of the sensing principle. **e** DPV obtained in 1 mM $K_3[Fe(CN)_6]$/0.1 M $NaNO_3$ for different concentrations of human IgG, from *top* to *bottom* 1, 50, 100, 200 and 300 µg/mL (SPIEs modified with 500 nm PS monolayer). **f** Relationship between the analytical signal and the concentration of HIgG for SPIEs modified with 200 nm PS monolayer (*solid line*) and 500 nm PS monolayer (*dashed line*). Adapted from [16] with permission

nanochannel-size dependent and allows to detect human IgG (chosen as model analyte) at levels of 580 ng/mL (Fig. 9.11e, f). The system also exhibits an excellent specificity against other proteins present in real sample and shows a good performance in a human urine sample.

In addition to the nanoparticle assembling, nanoimprinting technologies [17, 36] (i.e., roll-to-roll printing) that together with other fabrication techniques and emerging nanomaterials may bring new opportunities in terms of robustness, mass production, and better tailoring of nanochannel-based platforms toward various applications delivering this advantageous technology to more users interested in protein, DNA, and cell sensing.

Acknowledgments We acknowledge support of the Spanish MINECO under Project MAT2011–25870 and through the Severo Ochoa Centers of Excellence Program under Grant SEV-2013-0295.

References

1. T. Albrecht, J.B. Edel, M. Winterhalter, New developments in nanopore research-from fundamentals to applications. J. Phys. Condens. Matter. **22**, 450301 (1 pp) (2010)
2. J.O. Arroyo, J. Andrecka, K.M. Spillane et al., Label-free, all-optical detection, imaging, and tracking of a single protein. Nano Lett. **14**, 2065–2070 (2014)
3. H. Bayley, P.S. Cremer, Stochastic sensors inspired by biology. Nature **413**, 226–230 (2001)
4. S. Bezrukov, I. Vodyanoy, A. Parsegian, Counting polymers moving through a single ion channel. Nature **370**, 279–281 (1994)
5. D. Branton, D.W. Deamer, A. Marziali, H. Bayley, S.A. Benner, T. Butler, M. Di Ventra, S. Garaj, A. Hibbs, X. Huang et al., The potential and challenges of nanopore sequencing. Nat. Biotechnol. **26**, 1146–1153 (2008)
6. H. Chang, S.H. Joo, C. Pak, Synthesis and characterization of mesoporous carbon for fuel cell applications. J. Mater. Chem. **17**, 3078–3088 (2007)
7. J. Clarke, H.C. Wu, L. Jayasinghe, A. Patel, S. Reid, H. Bayley, Continuous base identification for single-molecule nanopore DNA sequencing. Nat. Nanotechnol. **4**, 265–270 (2009)
8. W.H. Coulter, Means for counting particles suspended in a fluid. U.S. Patent No. 2, 656:508 (1953)
9. W.H. Coulter, High speed automatic blood cell counter and cell size analyzer. Proc. Natl. Electronics Conf. **12**, 1034–1042 (1956)
10. A. Debrassi, A. Ribbera, Vos W.M. De et al., Stability of (bio)functionalized porous aluminum oxide. Langmuir **30**, 1311–1320 (2014)
11. A. de la Escosura-Muñiz, A. Merkoçi, Label-free voltammetric immunosensor using a nanoporous membrane based platform. Electrochem. Comm. **12**, 859–863 (2010)
12. A. de la Escosura-Muñiz, A. Merkoçi, Nanoparticle based enhancement of electrochemical DNA hybridization signal using nanoporous electrodes. Chem. Commun. **46**, 9007–9009 (2010)
13. A. de la Escosura-Muñiz, A. Merkoçi, A nanochannel/nanoparticle-based filtering and sensing platform for direct detection of a cancer biomarker in blood. Small **7**, 675–682 (2011)
14. A. de la Escosura-Muñiz, A. Merkoçi, Nanochannels preparation and application in biosensing. ACS Nano **6**(9), 7556–7583 (2012)

15. A. de la Escosura-Muñiz, W. Chunglok, W. Surareungchai, A. Merkoçi, Nanochannels for diagnostic of thrombin-related diseases in human blood. Biosens. Bioelectron. **40**, 24–31 (2013)
16. A. de la Escosura-Muñiz, M. Espinoza-Castañeda, M. Hasegawa, L. Philippe, A. Merkoçi, Nanoparticles-based nanochannels assembled on a plastic flexible substrate for label-free immunosensing. Nano Res. **8**(4), 1180–1184 (2015)
17. Y. Ding et al., Nanoimprint Lithography, in *Lithography*, ed. by M. Wang (In Tech, Croatia, 2010), pp. 458–494
18. J. Elias, C. Lévy-Clément, M. Bechelany, J. Michler, G.Y. Wang, Z. Wang, L. Philippe, Hollow urchin-like ZnO thin films by electrochemical deposition. Adv. Mater. **22**, 1607–1612 (2010)
19. M. Espinoza-Castañeda, A. de la Escosura-Muñiz, A. Chamorro, C. de Torres, A. Merkoçi, Nanochannel array device operating through Prussian blue nanoparticles for sensitive label-free immunodetection of a cancer biomarker. Biosens. Bioelectron. **67**, 107–114 (2015)
20. T.R.B. Foong, A. Sellinger, X. Hu, Origin of the bottlenecks in preparing anodized aluminum oxide (AAO) templates on ITO glass. ACS Nano **2**, 2250–2256 (2008)
21. V.M. Guérin, J. Elias, T.T. Nguyen, L. Philippe, T. Pauporté, Ordered networks of ZnO-nanowire hierarchical urchin-like structures for improved dye-sensitized solar cells. Phys. Chem. Chem. Phys. **14**, 12948–12955 (2012)
22. A. Hiraki, H. Ueoka, A. Bessho, Y. Segawa, N. Takigawa, K. Kiura, K. Eguchi, T. Yoneda, M. Tanimoto, M. Harada, Parathyroid hormone-related protein measured at the time of first visit is an indicator of bone metastases and survival in lung carcinoma patients with hypercalcemia. Cancer **95**, 1706–1713 (2002)
23. C. Horn, C. Steinem, Photocurrents generated by bacteriorhodopsin adsorbed on nano-black lipid membranes. Biophys. J. **89**(2), 1046–1054 (2005)
24. X. Hou, W. Guo, L. Jiang, Biomimetic smart nanopores and nanochannels. Chem. Soc. Rev. **40**, 2385–2401 (2011)
25. S. Howorka, S. Cheley, H. Bayley, Sequence-specific detection of individual DNA strands using engineered nanopores. Nat. Biotechnol. **19**, 636–639 (2001)
26. Z. Jin, F. Meng, J. Liu, M. Li, L. Kong, J. Liu, A novel porous anodic alumina based capacitive sensor towards trace detection of PCBs. Sens. Actuators, B **157**(2), 641–646 (2011)
27. L. Juhász, J. Mizsei, Humidity sensor structures with thin film porous alumina for on-chip integration. Thin Solid Films **517**, 6198–6201 (2009)
28. K. Kant, J. Yu, C. Priest, J.G. Shapter, D. Losic, Impedance nanopore biosensor: influence of pore dimensions on biosensing performance. Analyst **139**, 1134–1140 (2014)
29. V. Khatko, G. Gorokh, A. Mozalev, D. Solovei, E. Llobet, X. Vilanova, X. Correig, Tungsten trioxide sensing layers on highly ordered nanoporous alumina template. Sens. Actuators, B **118**, 255–265 (2006)
30. A.A. Karyakin, Prussian Blue and its analogues: electrochemistry and analytical applications. Electroanalysis **13**, 813–819 (2001)
31. A.K. Kasi, J.K. Kasi, N. Afzulpurkar et al., Bending and branching of anodic aluminum oxide nanochannels and their applications. J. Vac. Sci. Technol., B **30**(031805), 1–6 (2012)
32. J. Kasianowicz, E. Brandin, D. Branton, D. Deamer, Characterization of individual polynucleotide molecules using a membrane channel. Proc. Natl. Acad. Sci. U.S.A. **93**, 13770–13773 (1996)
33. J.J. Kasianowicz, J.W.F. Robertson, E.R. Chan, J.E. Reiner, V.M. Stanford, Nanoscopic porous sensors. Annu. Rev. Anal. Chem. **1**, 737–766 (2008)
34. G. Koh, S. Agarwal, P.-S. Cheow, C.-S. Toh, Development of a membrane-based electrochemical immunosensor. Electrochim. Acta **53**, 803–810 (2007)
35. T. Kumeria, A. Santos, D. Losic, Ultrasensitive nanoporous interferometric sensor for label-free detection of gold (III) ions. ACS Appl. Mater. Interfaces. **5**, 11783–11790 (2013)
36. H. Kurz, Status and prospects of UV-nanoimprint technology. Microelectron. Eng. **83**, 827–830 (2006)
37. C.A. Lundgren, R.W. Murray, Observations on the composition of Prussian Blue films and their electrochemistry. Inorg. Chem. **27**, 933–939 (1988)

38. H. Masuda, H. Tanaka, N. Baba, Preparation of porous material by replacing microstructure of anodic alumina film with metal. Chem. Lett. **4**, 621–622 (1990)
39. D.J. Munroe, T.J.R. Harris, Third-generation sequencing fireworks at marco island. Nat. Biotechnol. **28**, 426–428 (2010)
40. G. Oukhaled, J. Mathé, A.L. Biance, L. Bacri, J.M. Betton, D. Lairez, J. Pelta, L. Auvray, Unfolding of proteins and long transient conformations detected by single nanopore recording. Phys. Rev. Lett. **98**, 158101–158104 (2007)
41. R.F. Purnell, J.J. Schmidt, Discrimination of single base substitutions in a DNA strand immobilized in a biological nanopore. ACS Nano **3**, 2533–2538 (2009)
42. L. Qian, R. Mookherjee, Convective assembly of linear gold nanoparticle arrays at the micron scale for surface enhanced Raman scattering. Nano Res. **4**, 1117–1128 (2011)
43. D. Rotem, L. Jayasinghe, M. Salichou, H. Bayley, Protein detection by nanopores equipped with aptamers. J. Am. Chem. Soc. **134**, 2781–2787 (2012)
44. A. Santos, G. Macías, J. Ferré-Borrull, J. Pallarès, L.F. Marsal, Photoluminescent enzymatic sensor based on nanoporous anodic alumina. ACS Appl. Mater. Interfaces **4**, 3584–3588 (2012)
45. A. Santos, P. Formentín, J. Ferré-Borrull, J. Pallarès, L.F. Marsal, Nanoporous anodic alumina obtained without protective oxide layer by hard anodization. Mater. Lett. **67**(1), 296–299 (2012)
46. A. Santos, T. Kumeria, D. Losic, Optically optimized photoluminescent and interferometric biosensors based on nanoporous anodic alumina: a comparison. Anal. Chem. **85**, 7904–7911 (2013)
47. A. Santos, T. Kumeria, D. Losic, Nanoporous anodic aluminum oxide for chemical sensing and biosensors. Trends Anal. Chem. **44**, 25–38 (2013)
48. A.L. Sisson, M.R. Shah, S. Bhosale, S. Matile, Synthetic ion channels and pores (2004–2005). Chem. Soc. Rev. **35**, 1269–1286 (2006)
49. Z.S. Siwy, S. Howorka, Engineered voltage-responsive nanopores. Chem. Soc. Rev. **39**, 1115–1132 (2010)
50. L.J. Suva, G.A. Winslow, R.E. Wettenhall, R.G. Hammonds, J.M. Moseley, H. Diefenbach-Jagger, C.P. Rodda, B.E. Kemp, H. Rodriguez, E.Y. Chen, A parathyroid hormone-related protein implicated in malignant hypercalcemia: cloning and expression. Science **237**, 893–896 (1987)
51. P. Takmakov, I. Vlassiouk, S. Smirnov, Sensing DNA hybridization via ionic conductance through a nanoporous electrode. Analyst **131**, 1248–1253 (2006)
52. F. Tan, P.H.M. Leung, Z. Liu, Y. Zhang, L. Xiao, W. Ye, X. Zhang, L. Yi, M. Yang, A PDMS microfluidic impedance immunosensor for E. coli O157:H7 and Staphylococcus aureus detection via antibody-immobilized nanoporous membrane. Sens. Actuators, B **159**, 328–335 (2011)
53. T. Uemura, S. Kitagawa, Prussian blue nanoparticles protected by poly(vinylpyrrolidone. J. Am. Chem. Soc. **125**, 7814–7815 (2003)
54. B.M. Venkatesan, R. Bashir, Nanopore sensors for nucleic acid analysis. Nat. Nanotechnol. **6**, 615–624 (2011)
55. W.A. Vercoutere, S. Winters-Hilt, V.S. DeGuzman, D. Deamer, S.E. Ridino, J.T. Rodgers, H. E. Olsen, A. Marziali, M. Akeson, Discrimination among individual WatsonCrick base pairs at the termini of single DNA hairpin molecules. Nucleic Acids Res. **31**, 1311–1318 (2003)
56. I. Vlassiouk, P. Takmakov, S. Smirnov, Sensing DNA hybridization via ionic conductance through a nanoporous electrode. Langmuir **21**, 4776–4778 (2005)
57. A. Walcarius, E. Sibottier, M. Etienne, J. Ghanbaja, Electrochemically assisted self-assembly of mesoporous silica thin films. Nat. Mater. **6**, 602–608 (2007)
58. A. Walcarius, A. Kuhn, Ordered porous thin films in electrochemical analysis. Trends Anal. Chem. **27**, 593–603 (2008)
59. Y. Wan, D. Zhao, On the controllable soft-templating approach to mesoporous silicates. Chem. Rev. **107**, 2821–6280 (2007)

60. W.W. Ye, J.Y. Shi, C.Y. Chan, Y. Zhang, M. Yang, A nanoporous membrane based impedance sensing platform for DNA sensing with gold nanoparticle amplification. Sens. Actuators, B **193**, 877–882 (2014)
61. L. Wen, Z. Sun, C. Han, B. Imene, D. Tian, H. Li, L. Jiang, Fabrication of layer-by-layer assembled biomimetic nanochannels for highly sensitive acetylcholine sensing. Chem. Eur. J. **19**, 7686–7690 (2013)
62. Z. Yang, S. Si, C. Zhang, Study on the activity and stability of urease immobilized onto nanoporous alumina membranes. Micropor. Mesopor. Mater. **111**, 359–366 (2008)
63. K. Zhu, D. Wang, J. Liu, Self-assembled materials for catalysis. Nano Res. **2**, 1–29 (2009)

Chapter 10
Nanoporous Alumina Membranes for Chromatography and Molecular Transporting

Tushar Kumeria and Abel Santos

Abstract Nanoporous membrane based molecular separation and transporting have gained huge attention in recent years due to an immediate need for simple, cost-competitive, and miniature ultrafiltration, chromatography and desalination devices. Nanoporous anodic alumina membranes (NAAMs) are one of the most popular materials of choice for the aforementioned applications due to its outstanding and unique set of chemical, physical, and structural properties. This chapter provides details of fabrication of NAAMs and surface modification techniques employed for enabling NAAMs for chromatography and molecular transporting applications. Next, fundamentals of chromatography and recent studies on nanopore-chromatographic systems based on nanoporous anodic alumina are summarized. Also, fundamentals of molecular transporting across NAAMs and recent examples are compiled in this chapter. The objective of this chapter is to provide a simple yet detailed overview of fabrication of NAAMs based chromatographic and molecular transporting devices with an emphasis on their performance.

10.1 Introduction

Chromatographic separation and selective molecular transport are the most widely used analytical techniques for high-resolution separation of component mixtures and analysis of proteins and biological analytes. Chromatography is a technique that not only separates the mixture components but also yields qualitative as well as quantitative information about components in complex mixtures. Chromatography process is known since early 1800s, however, then, it was only employed for artistic purposes like preparing color gradient/patterns on paper. It was first introduced for scientific

T. Kumeria (✉) · A. Santos
School of Chemical Engineering, The University of Adelaide,
North Engineering Building, 5005 Adelaide, Australia
e-mail: tushar.kumeria@adelaide.edu.au

D. Losic and A. Santos (eds.), *Nanoporous Alumina*,
Springer Series in Materials Science 219, DOI 10.1007/978-3-319-20334-8_10

purposes by a Russian scientist, Mikhail Tsvet (Michael Tswett) in 1900, for separation of plant pigments such as chlorophyll, carotenes, and xanthophylls [1, 2]. Due to the presence of color in the separated components, his study became the basis of the name chromatography, where chroma means "color" and graphy means "to write". The process starts with dissolving the component mixture into a compatible solvent called "mobile phase", which is then run through the a column/channel packed with a solid material (i.e. "stationary phase"). The speed of transport of various components of the mixture varies depending upon the difference in their interaction with stationary phase enabling their separation. As for this, the distance travelled by a mixture component is inversely proportional to its extent of interaction with stationary phase (i.e. the weaker the interaction of the component with the stationary phase the faster the component travels and vice versa) [3, 4]. During the last century, capabilities and reach of chromatography have improved significantly due to continued research and the discovery of various new types of stationary phases and their coupling with advanced detection methods. This led to division of chromatography into several subcategories such as liquid chromatography, gas chromatography, paper chromatography, thin layer chromatography, size-exclusion chromatography, ion-exchange chromatography, high performance liquid chromatography (HPLC) and others. The stationary phase for most of the aforementioned chromatography techniques is based on solid or porous particles (mainly silica, alumina, or cellulose particles) [5–7]. Furthermore, specific properties can also be imparted to these particles for specialized chromatographic techniques such as ion exchange or affinity chromatography. These techniques are primarily utilized in academics for teaching purposes, biotechnology for separation, extraction, and purification of proteins and metabolites, pharmaceutics for drug discovery, purification, and quality testing, and polymer industry for qualitative and quantitative analysis of synthetic and natural polymers and many other fields [4, 8–13]. Although these techniques have been widely adopted by industries for various applications, there are some inherent limitations to their capabilities. For example, the combined effects of bed consolidation and column binding result in very high pressure drop across a packed bed, which increases during a process. Furthermore the conventional chromatography relies on intra-particle diffusion of component mixture to their corresponding binding or pore diameter distribution site in the column that leads to increased process time since diffusion of large-molecules is slow, especially when it is hindered (i.e. due to randomly packed particles in the column) (Fig. 10.1). Due to longer process time, the volumes of elution liquid required for recovery of the separated components, also increases making the process expensive [14–17]. Also, formation of cracks in the packed bed is another major disadvantage of conventional chromatography techniques, which results in flow-passages for the components of mixtures, lowering the efficiency of the separation process. Furthermore, polydispersity of packed particles also gives rise to problems like radial and axial dispersion and the fact that the transport phenomenon is complicated, make scale-up of packed bed chromatographic processes difficult [14].

Alternatives to conventional chromatography, such as membrane chromatography have been developed, demonstrated, and marketed in last few years. Nano and

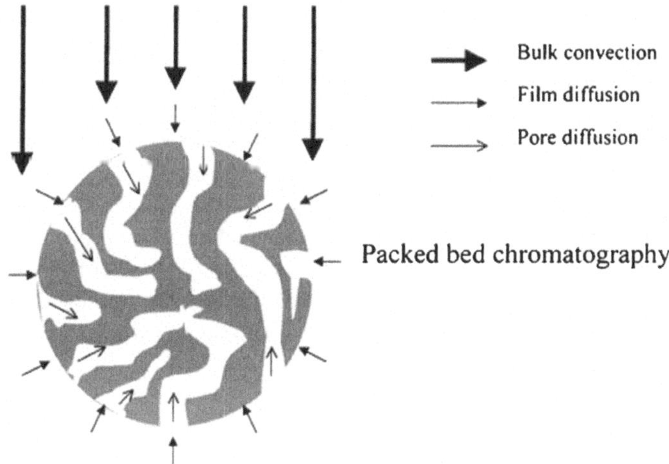

Fig. 10.1 A schematic illustration showing types of diffusion flows involved in packed bed column chromatography (adapted from [14])

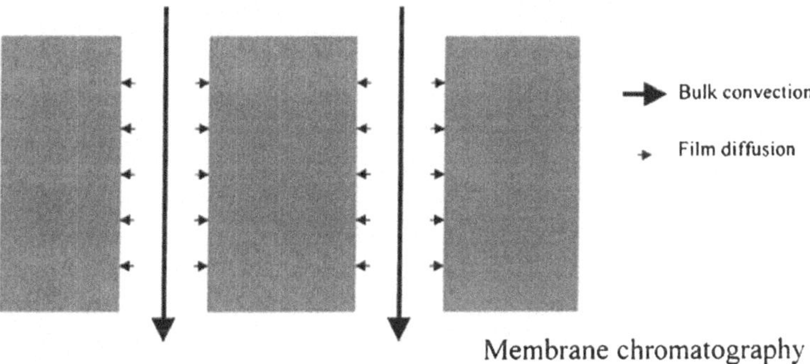

Fig. 10.2 A schematic illustration of types of diffusions involved in membrane chromatography (adapted from [14])

micro porous membranes based chromatography setups display several potential advantages. For example, in membrane based chromatography the feed solution flows through the membrane enabling a very short, wide bed type configuration hence resulting in high velocities and very short residence/separation times with modest trans-membrane pressure drops (Fig. 10.2). Furthermore, in case of membrane based affinity chromatography, the elimination of diffusional resistance enables much faster binding kinetics, thereby adsorptive separation of proteins, in one-tenth the typical time for packed columns [14, 15, 18, 19]. Notice that, most of these, membrane based chromatography systems are based on polymeric membranes [15, 20]. However, large pore diameter distribution, pore branching, tortuosity, poor mechanical strength, and thermal and chemical resistance limit their application for certain chromatographic techniques (e.g. HPLC, gas chromatography, etc.) [21].

Inorganic nanoporous membranes such as nanoporous anodic alumina (NAA) or lithographically fabricated porous silicon membranes have gained enormous attention in recent years for not only overcoming the limitations of polymer membranes but also providing additional advantages such as greater control over pore diameter and geometry and also surface chemistry [22]. Nanoporous anodic alumina membranes (NAAMs), especially are of greater interest as these can be produced by a simple and industrialized electrochemical anodization technique. The nanopores of NAAMs can be engineered through different electrochemical approaches, making it possible to design the pore shape and geometry [23–26]. Furthermore, the surface chemistry of NAAMs can be selectively tuned for specific affinity or adsorption chromatography purposes through a number of simple and well-established chemistry protocols, detailed in [23, 25].

In addition to these unique physical and chemical properties, NAAMs can act as template for fabrication of interesting nanotubular structures and composite membranes such as carbon nanotube membranes or zeolite nanotube membranes [27–29]. NAAMs based composite membranes further enhance its application frontiers to even more sophisticated chromatographic systems and a completely new field of molecular transporting. Note that, molecular transporting through membranes is a key process for several critical industrial applications as ultrafiltration, dialysis, desalination, industrial catalysis and chromatography [30–32]. In contrast, the broad pore size distributions and uncontrolled structural features of polymer membranes such as thickness lead to poor size cut-off properties, filtrate loss, and low transport flux rate. NAAMs have recently been applied for molecular transport applications to overcome the aforementioned issues [33].

This chapter briefly describes the process of fabrication of NAA membranes and most common surface modification techniques utilized for preparation of NAAMs for chromatography and molecular transporting applications. First, electrochemical anodization process for fabrication of NAA and the pore opening techniques available for converting NAA substrates to NAAMs are introduced. Second, detailed descriptions of processes and techniques used for surface functionalization of NAAMs are provided. Lastly, we will provide detailed descriptions of principles, performances and applications of NAAMs and composite NAAMs in chromatography and molecular transporting. In the conclusion, we will comment on present challenges and future outlook for application of NAAMs in chromatography and molecular transporting.

10.2 Fabrication of NAAMs

Nanoporous anodic alumina is fabricated by anodization of high purity aluminum (Al) foils under specific electrochemical conditions, which yield nanopores grown perpendicular to the Al foil surface. Electrochemical anodization has been used since early 1900s for metal surface finishing, machinery and automobile engineering, corrosion protection and others [24]. However, nanoporous structure was only discovered in the 1950s after electronic microscope studies revealed that anodization generates honeycomb like nanopores on aluminium surface [34]. It was revealed that NAA is composed of hexagonally arranged nanopores. A flood of studies characterizing physical and chemical properties and mechanism of nanoporous structure formation in NAA arrived after the initial study reporting porous structure in NAA. The most commonly adopted mechanism of nanopores formation in NAA assumes that pore nucleation starts at the defects and pits located on the surface of that oxide layer and uneven distribution of electric field which results localized heating due to Joule effect. This local increase in temperature enhances the ionic conductivity and the oxide layer is preferentially dissolved at these sites forming nanopores in the oxide film. After certain time, the pore growth achieves a steady state under which the rate of oxide growth at Al-alumina junction and rate of oxide dissolution at alumina-electrolyte junction become equal. A major milestone in the history of NAA fabrication was in 1995 when Masuda and Fukuda introduced the two-step anodization process to obtain self-organized NAA pores [35]. Then soon after that, the self-organisation conditions for the most commonly used acid electrolytes were reported (i.e. sulphuric (H_2SO_4), oxalic ($H_2C_2O_4$) and phosphoric acids (H_3PO_4)) [36–38]. In this two-step anodization process, the first anodized porous oxide layer is selectively removed by dissolving in a mixture of chromic acid (H_2CrO_4) and phosphoric acid (H_3PO_4) at 70 °C for 3 h to obtain a hexagonally arranged hemispherical pattern on the surface of the Al substrate. These self-organised pre-patterned hemispherical sites act as nucleation site for growth and propagation of cylindrical nanopores from top-to-bottom during the second anodization step. A schematic illustration of the anodization setup and process is provided in Fig. 10.3 [39].

NAA is fabricated in aqueous electrolyte solution, typically, sulphuric acid oxalic acid and phosphoric acid at voltages of 25, 40, and 195 V, respectively to obtain self organized nanopores [40]. The geometric features of the resulting NAA structures, can be defined by the pore diameter (d_p), pore length (L_p) and interpore distance (d_{int}), which can be precisely controlled by the anodization conditions within a range of 10–400 nm for d_p, from a few nanometres to hundreds of micrometres for L_p, and 50–600 nm for d_{int} [41, 42]. Figure 10.4 clearly defines these parameters and presents typical SEM images of NAA. Table 10.1 summarises the structural features of NAA produced under the most commonly used anodization conditions.

It should be noticed, that as produced NAA consists of a porous type oxide on top and a barrier type oxide layer at the pore bottom, which results in closed pores at the bottom. To obtain a NAAM it is necessary to selectively remove underlying Al left after anodization and open pores by dissolving the barrier type oxide layer. NAAMs typically possess the structural features of the parent NAA substrate such as hexagonally organized pores with narrow pore diameter distribution over the area of the membrane. These self-organized pores run vertically from top to bottom of the substrate without any branching or tortuosity. Figure 10.5 shows a schematic of NAAMs prepared from NAA substrate after chemically removing barrier layer.

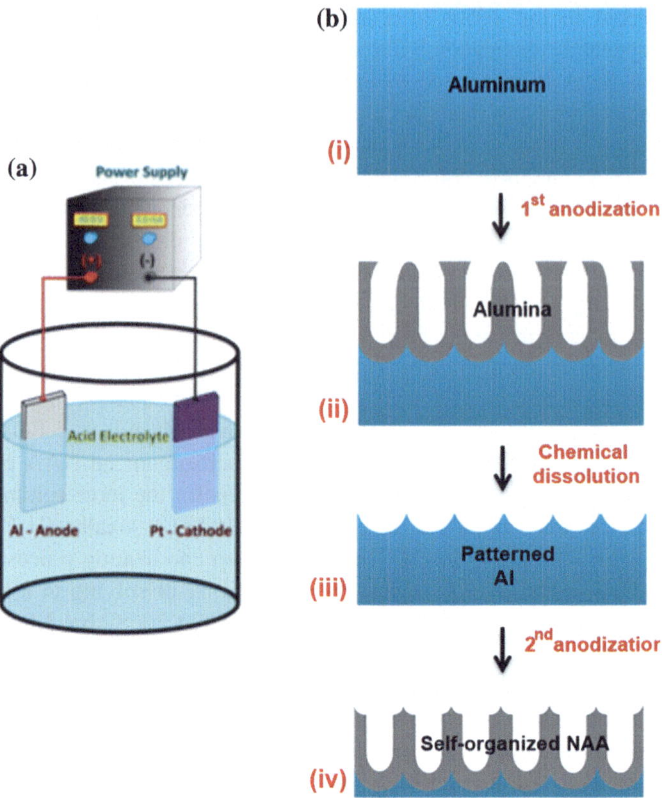

Fig. 10.3 A schematic illustration of **a** anodization setup used to prepare NAA substrates and **b** process steps involved in fabrication of self-organized NAA by two-step anodization process (adapted from [24] and [39], respectively)

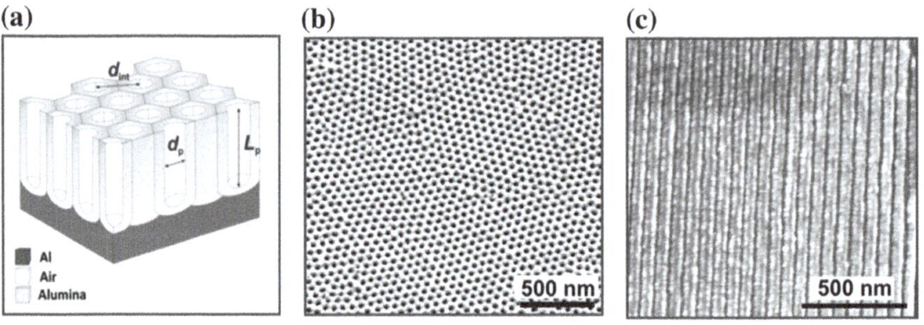

Fig. 10.4 **a** A schematic representing the structural features of NAA grown on Al substrate. **b** Top view of SEM image of NAA. **c** Cross-sectional SEM image of NAA

Table 10.1 Summary of the most widespread fabrication conditions used to produce NAA by anodization of high purity aluminum substrates along with the geometric features of the resulting nanostructures

Acid electrolyte	*V* (V)	*T* (°C)	d_p (nm)	d_{int} (nm)	Growth rate (μm h^{-1})	References
$H_2C_2O_4$ 0.3 M	140	0–1	50	280	50	[40]
$H_2C_2O_4$ 0.3 M	40	5–8	30	100	3.5	[35]
H_2SO_4 0.3 M	25	5–8	25	63	7.5	[36]
H_3PO_4 0.1 M	195	0–1	160	500	2	[37]

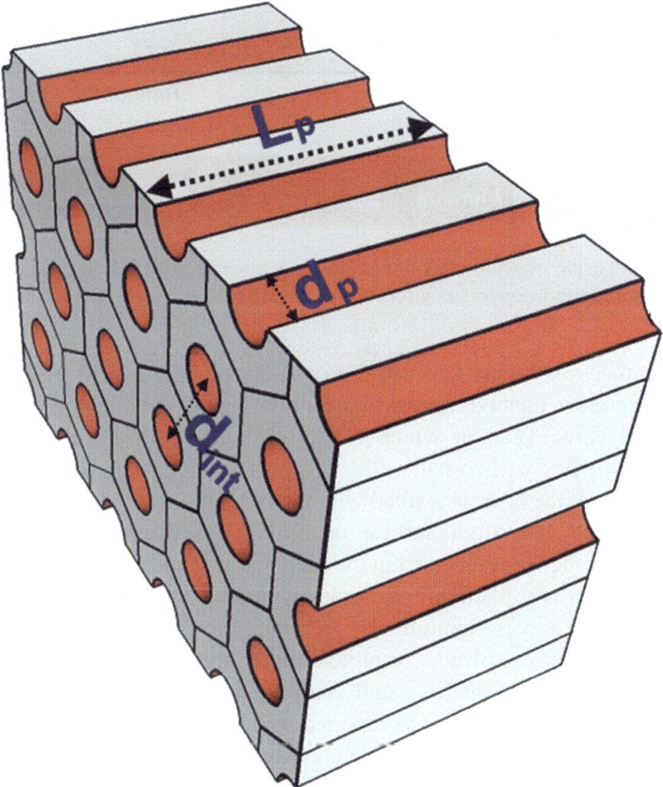

Fig. 10.5 A schematic illustration of NAAM obtained after removing the underlying Al and barrier layer from NAA substrate

To fabricate NAAMs, initially the underlying Al layer is generally selectively etched using a mixture of HCl (38 %) and $CuCl_2$ or $HgCl_2$ to obtain free standing NAA substrate with pores closed at the bottom with the barrier layer. The most common method for removing barrier type oxide layer is placing the bottom surface (i.e. barrier layer) of NAA in contact with mild acid (mostly 5–10 % H_3PO_4) [43]. However, initial reports on pore opening were very inconsistent as the barrier layer

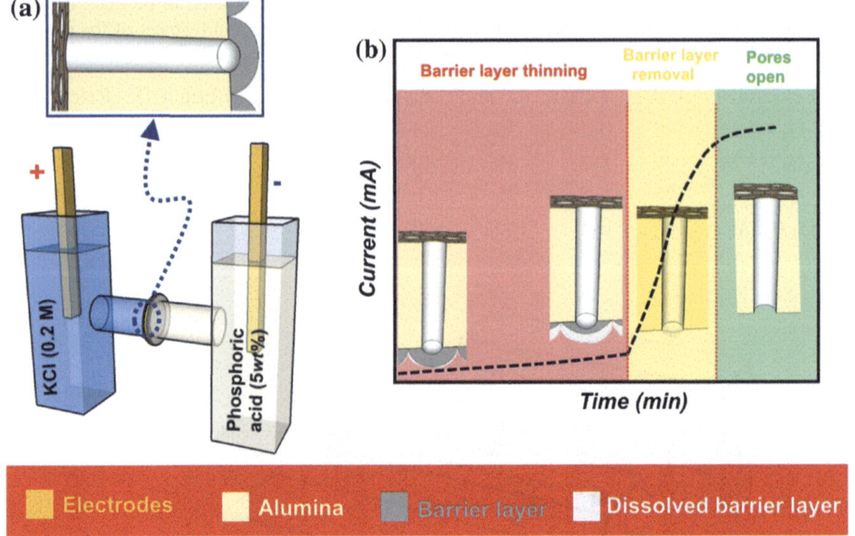

Fig. 10.6 **a** Schematic of setup used to monitor pore opening in-situ. **b** Typical pore opening curve showing increase in current as a function of time

dissolution process is highly dependent on conditions such as acid temperature and volume. Therefore, a method for real-time monitoring of pore opening was described by Losic and co-workers, in which NAA substrate was packed in a H-shaped permeation cell made of two half cells as shown in Fig. 10.6a [44]. The half-cell with barrier type oxide layer was filled with the etching solution (5 wt% H_3PO_4) and the other half-cell (i.e. front surface of NAA) was filled with KCl (0.2 M). A platinum electrode was immersed in the each half-cell and a small voltage 1.5–2 V is applied between two platinum electrodes with both the solutions under stirring. Pore opening process was monitored by following the changes in the current signal as the barrier layer dissolved. A typical pore opening curve is presented in Fig. 10.6b. A more comprehensive study on pore opening and in situ monitoring of the process using similar setup was reported by Lee and co-workers [45].

Note that, the commercially available anodized alumina membranes are prepared in a different fashion. In case of commercial NAAMs, the etching is done under hard anodization regime (i.e. high current field and low temperatures) to attain very high growth rates (i.e. in access of 50 μm h^{-1}) [46]. The pore opening and barrier layer thinning is achieved by sequentially ramping the voltage down to 0 V which results in thinning of barrier layer and eventually it completely disappears [41, 47]. This method is less time and labor intensive in comparison to the previous method. Nevertheless, the pores at the bottom of the membranes produced by this method are disorganized and branched as voltage down ramp disrupts the self-organizing voltage regime [48]. Typical SEM images of top and cross sectional view of a commercial NAAM is provided in Fig. 10.7.

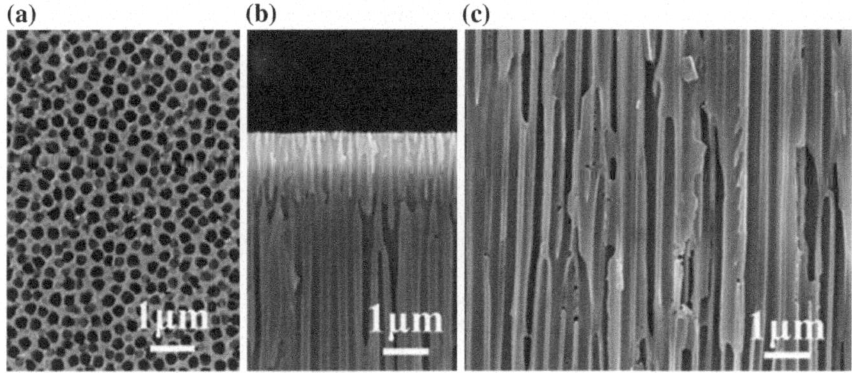

Fig. 10.7 Typical SEM images of a commercial NAA membrane. **a** Top view. **b** Cross-section view. **c** Magnified cross-section view (adapted from [48])

10.3 Surface Functionalization of NAAMs

In recent years, a wide range of techniques have been reported for surface modification of NAAMs to impart specific functionalities to its surface. The major techniques used for surface modification of NAAMs for chromatography and molecular transporting are self-assembly processes of silanes, atomic layer deposition (ALD), electrochemical and electroless deposition, chemical vapor deposition (CVD), and sol–gel processing. Comprehensive and detailed reviews on this topic are presented in reference [23, 25]. This part focuses on providing briefs about the above mentioned techniques and their recently reported examples.

The sol–gel process was developed in the 1960s due to an urgent need of new material synthesis method in the nuclear industry. Sol–gel assembly process can be defined as formation of a gel network through poly-condensation reactions of a molecular precursor in a liquid. The idea behind the sol–gel process is to dissolve a solid compound in a liquid and bring it back as a solid with a controlled structure. Sol–gel process even enables mixing of multiple sol-components in a specific stoichiometric ratio for preparation of multi component compounds. Sol–gel process has been employed with NAAMs to mainly replicate and prepare novel nanostructures as nanotubes and nanorods or reducing the pore diameter and imparting specific surface chemistries to NAAMs (Fig. 10.8) [49]. This chemistry has also been used to prepare composite sol-gel–NAAMs with a number of materials, in which, the sol–gel prepared nanostructures are embedded inside the pores. This technique basically involves hydrolysis of the sol–precursor on the NAAMs substrate by immersion or dipping or spin coating and subsequent evaporation of to form a glassy gel inside the pores [50].

The composite gel–NAAMs system is then thermally treated to remove the remaining solvent. Composite sol–gel–NAAMs with high purity homogeneous or multi-component nanostructure display unique properties such as high thermal

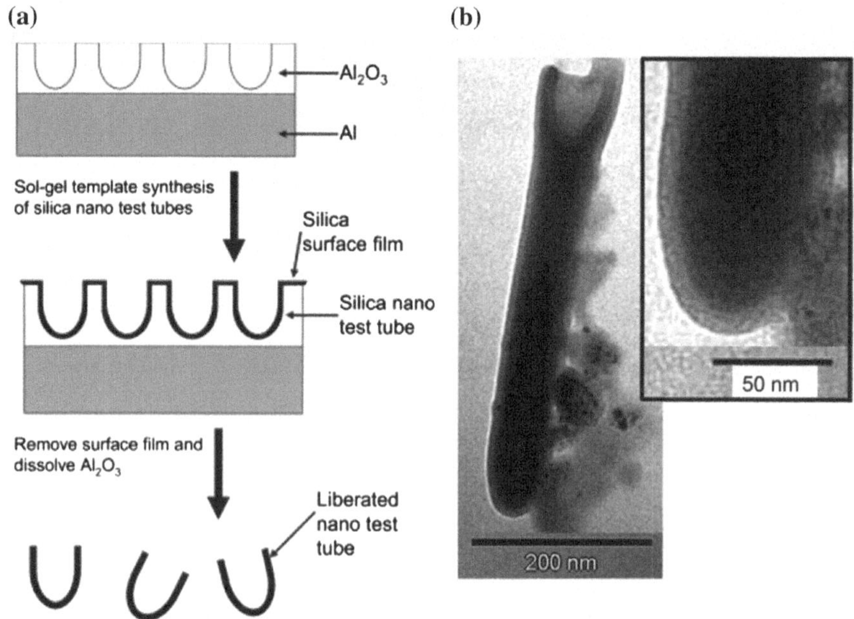

Fig. 10.8 **a** Schematic of sol–gel process for templating NAA structure fabricate silica nanotubes. **b** TEM images of the corresponding silica nanotubes (adapted from [49])

stability and surface reactivity (using precursors with additional functional groups). Notice that, such hierarchical and functional structures (i.e. containing smaller sol–gel pores within large NAAMs pores) can be of very high importance for size exclusion and affinity chromatography applications, respectively. NAAMs have been used as template to prepare sol–gel nanostructures of materials including silica, carbon, metal oxides (e.g. TiO_2, TiO_2–silica composite, SnO_2, NiO, etc.), and polymer-derived SiOC. Sol–gel–NAAMs have also been demonstrated for their ability to selectively transport molecules of choice across the membrane [24, 25].

Self-assembled monolayers (SAMs) are spontaneous assemblies of molecules formed by attachment and arrangement of molecules from the liquid solution onto solid surfaces. SAMs of organosilanes have been reported to readily form on hydroxylated surface of NAA. Organosilanization on NAA is generally employed to control its wettability, adsorption properties, and impart required terminal functional groups to its surface. For example, Ku et al. attached alkyl-trichlorosilanes with carbon chain lengths ranging between 1 to 8 to render the surface of NAA completely hydrophobic, such that its pores remain filled with air even on fully immersing into an aqueous medium [51]. Silanes with amine, epoxy, and other active functional groups are used to increase the wettability of NAA. The primary reason for imparting active functional groups to NAA surface is to covalently attach (bio)molecules, polymers, nanoparticles, DNA, cells, quantum dots, and lipid bilayers onto the surface of NAA [24]. Amino-propyltryethoxy silane (APTES) is the most commonly used silane for modification of NAA as amine terminals on APTES can be easily used to attach required molecules by simple coupling strategies.

An innovative approach was developed by our group which enabled multifunctional-layering of the surface chemistry of NAA (Fig. 10.9). This process

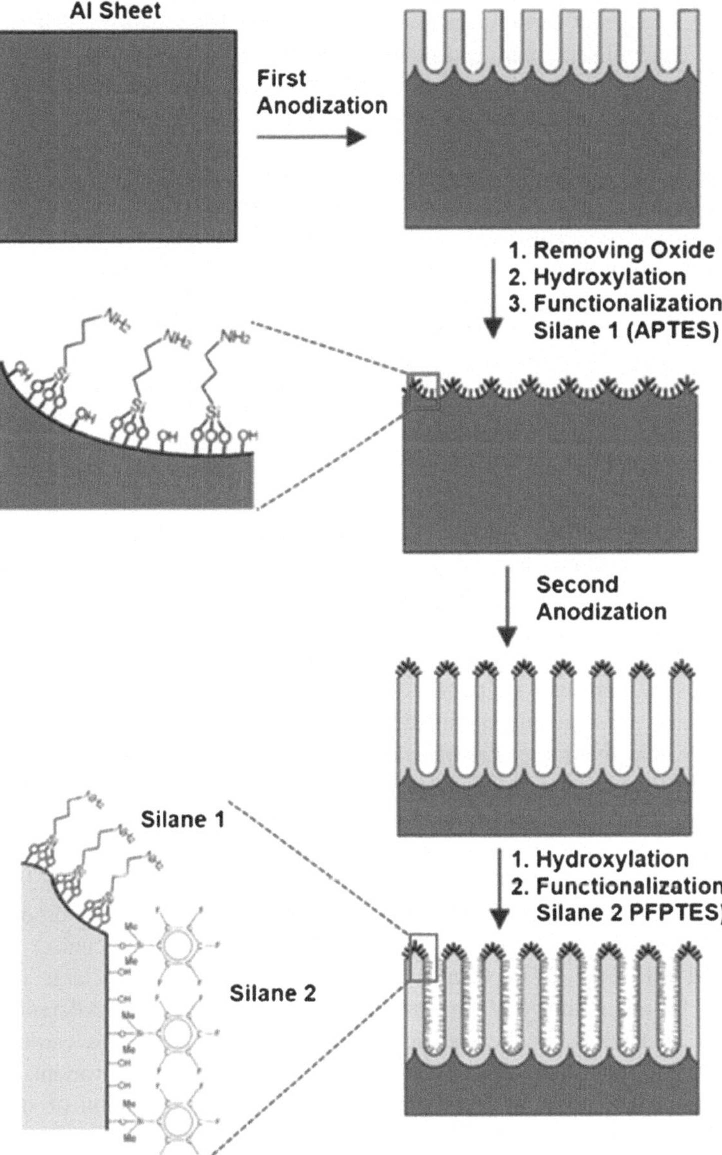

Fig. 10.9 Schematic illustrating of silanization process for functionalizing the surface of NAA with multiple silanes (adapted from [52])

involves a series of anodization and silanization cycles repeated in a sequential fashion. Therefore, this approach not only provides the control over chemistry of each layer but also the thickness of each layer. With this approach, successful functionalization of NAA surface with up to three different chemistries in layers were demonstrated (e.g. (pentafluorophenyl-dimethylpropylchloro-silane (PFPTES), APTES and N-triethoxysilylpropyl-(O-polyethyleneoxide)urethane (PEG-silane)), providing a range of surface functionalities and wettability which could be highly required for chromatography and molecular transporting applications [52, 53]. The application of this method by selective transport of hydrophobic and hydrophilic dye molecules through membranes with hydrophobic and hydrophilic layers was also demonstrated.

Atomic layer deposition (ALD) is another highly versatile and precise technique used for deposition of a broad range of oxides, nitrides, sulfides and metals coatings. ALD is a controlled layer-by-layer deposition process which deposits one atomic layer thick films at a time. For deposition, alternate pulses of precursors and reactants are purged with inert gas pulses in between and the thickness of the film is controlled by the number of cycles. ALD deposition of oxides, nitrides, and sulfides mostly use two chemicals precursors, which react on the surface of a substrate in a sequential and saturated self-controlled manner, one at a time [54]. As this technique deposits atomically thick films in one cycle, it is able to form uniform layers even in the confined spaces such as nanopores of NAAMs. The coatings prepared by ALD display high mechanical, chemical and thermal stability, and chemical functionalization ability. Thus far, ALD modification of NAAMs has been demonstrated with various materials as silica, titania and alumina for improving its catalytic, optical, molecular transport properties as well as generating 1D nanostructures [25, 55–58]. ALD is primarily employed for reducing the pore diameter of NAAMs and providing it a different surface functionality. For example, Velleman et al. deposited silica (SiO_2) inside NAA through ALD to reduce the pore diameter and subsequently modified its surface with a specific organosilane for chemically selective molecular transporting [33].

Chemical vapour deposition is a simple and adaptable technique for deposition of films on a variety of substrates. CVD involves vaporizing of the chemical precursor and subsequent deposition onto a substrate. The chemical reaction (mostly pyrolysis) that results into deposition of precursor takes place either on the substrate surface or near its surface. This process also produces chemical by-products (i.e. either undeposited precursor or degradation products of precursor) that are exhausted out of the CVD chamber along with the carrier gas. A large variety of materials for a wide range of applications can be deposited using different variants of CVD. Depending on the chemical precursor, CVD process is carried out at temperatures in the range from 200 to 1500 °C. Note that, some variants of CVD employ light or plasma to decompose the precursor for formation of stable and conformal films on the substrate. CVD process, in NAAMs technology, is mainly employed to fabricate CNTs–NAAMs composite membranes [59–61]. In this process, NAAMs act as a template for growth of vertically aligned arrays of CNTs with controlled dimensions. CNTs are grown inside NAAMs pores by high

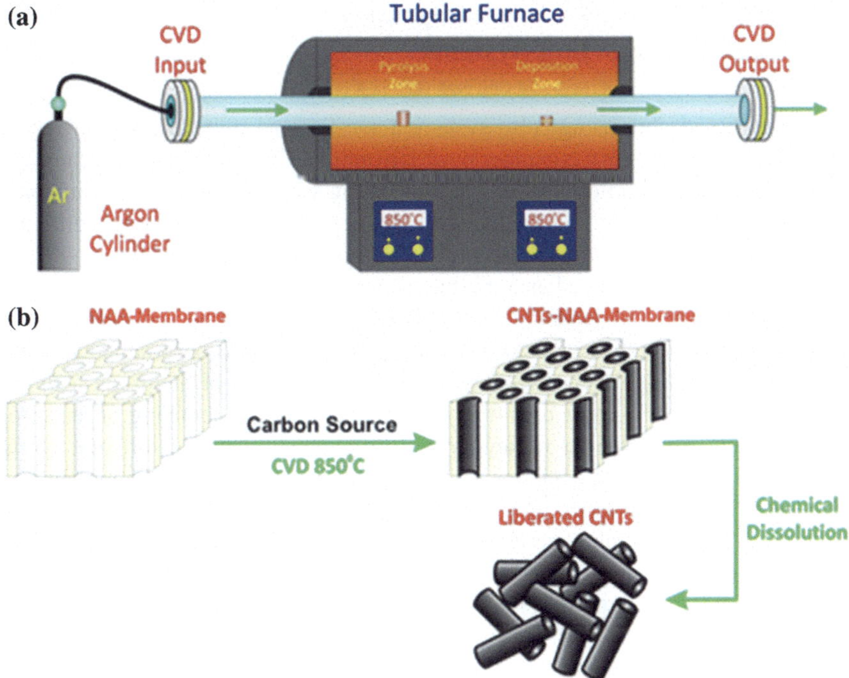

Fig. 10.10 **a** Scheme showing the chemical vapor deposition setup used to grow CNTs inside NAA templates. **b** Schematic illustration of CNTs fabrication by CVD process (adapted from [29])

temperature (>600 °C) pyrolysis of carbon precursor [62]. A large selection of carbon precursors as liquid (i.e. turpentine oil, toluene, etc.), gas (mainly acetylene) or solid (i.e. ferrocene, plastic bags, etc.) are available and have been reported in the literature for preparation of composite CNTs–NAAMs. A schematic illustration of fabrication of composite CNTs–NAAMs and free-standing CNTs from plastic waste, reported by our group, is provided in Fig. 10.10 [29]. These composite CNTs–NAAMs were employed for selective molecular transporting.

Electrochemical and electroless deposition are simple and inexpensive techniques for deposition of metal layers onto or inside the NAAMs pores. Contrary to CVD and ALD techniques, electrochemical and electroless deposition do not require expensive and specialized equipment or conditions [23, 25, 63]. These techniques are mainly employed for fabrication of a range of metal-based nanostructures (i.e. nanowires, nanorods, nanotubes and nanoparticles) using NAAMs as templates [25]. However, some studies have reported on fabrication of composite metal-NAAMs for studying selective molecular transporting through these structures.

10.4 NAAMs Based Chromatography and Transporting Systems

Membrane based chromatography systems have become quite popular in recent days due to their inherent advantages over conventional chromatography techniques as mentioned in the section above. The pioneering work in this field was initiated by Martin and co-workers in the 1990s, where they successfully separated different enantiomeric forms of a drug based on membrane affinity chromatography [64]. Most drugs that are produced as a mixture of several racemic enantiomers and one of the enantiomers is generally more efficacious than others in the mixture. Therefore, pharmaceutical industries are increasingly focused on producing, separating, and marketing the most enantiomerically pure forms of the drugs. Affinity chromatography is the most commonly used technique for obtaining an enantiomerically pure drug due to its ability is to effectively separate different chiral isomers of the drugs. In that regard, Martin et al. prepared bio-nanotube membranes to separate two enantiomers of a chiral drug. These membranes were based on nanoporous anodic alumina membranes with hexagonally organized straight pores (approximately 20 nm). The selective antibodies to bind the drug 4-[3-(4-fluorophenyl)-2-hydroxy-1- [1, 2, 4] triazol-1-yl-propyl]-benzonitrile, an inhibitor of aromatase enzyme activity, were chosen for this work. This drug molecule has two chiral centres and therefore four stereoisomers: RR, SS, SR, and RS. Silica nanotubes were synthesized within the pores of NAAMs using simple sol–gel chemistry and a specific antibody (anti-RS antibody) that only binds with one of the enantiomers of the drug was covalently immobilized onto the inner walls of the silica nanotubes. During a typical membrane chromatography process the enantiomer that specifically binds to the antibody was transported faster across the membrane, relative to the enantiomer with lower affinity for the antibody (Fig. 10.11a). The ratio of the RS flux to the SR flux through anti-RS modified

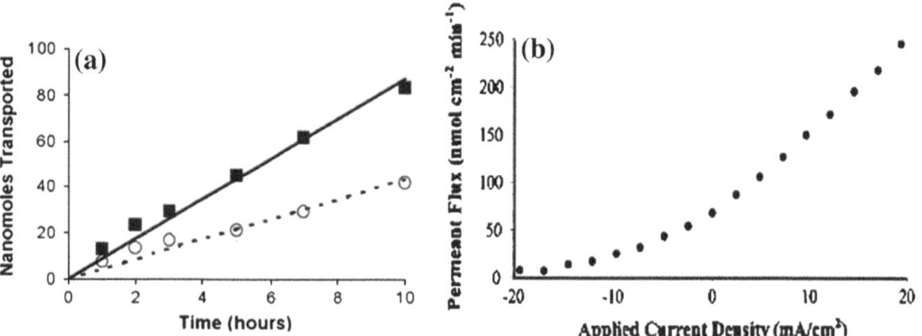

Fig. 10.11 **a** Nanomoles of RS enantiomer (*solid squares*, *solid line*) and SR enantiomer (*open circles*, *dashed line*) transported from the feed solution into the permeate solution versus time through alumina membrane modified with the anti-RS Fab fragment. **b** Phenol flux versus applied current density (adapted from [64] and [65], respectively)

membrane provides the transport selectivity coefficient (α). Their results show an average transport selectivity coefficient value around 2.0 ± 0.2 for RS enantiomer of the drug across the membranes with a pore diameter of 35 nm, which was twice as fast as SR enantiomer. They also observed and increase in α with decrease in membrane pore diameter.

They also demonstrated ability to control the electro osmotic flow of charged species through carbon nanotube (CNT) and NAAMs composite membranes [65]. For this, CNT–NAAMs composite membranes were prepared by chemical vapour deposition of carbon using ethylene gas as a precursor under inert argon environment within the pores of NAAMs. Electric potential was applied across the CNT–NAAMs to drive electro-osmotic flow (EOF) to separate two electrolyte solutions and maintain stable ionic current through the composite membrane. The anionic surface charge of CNT–NAAMs resulted in EOF in the direction of cation migration across the membrane. Phenol was used as the molecular probe for measuring the EOF across the CNT–NAAMs. Attachment of carboxylate on CNT–NAAMs surface further enhanced anionic charge on the surface resulting in expected increase in EOF. Whereas, modification of CNT–NAAMs with cationic ammonium species shows EOF in the opposite direction of the as-produced or carboxylated membranes. They also assessed the effect of applied current density by measuring the slope of the moles of phenol transported against time at different applied current density. They observed the flux of phenol decrease on applying negative potential across the CNT–NAAMs, which means EOF was in opposite direction. In contrast the phenol flux across the CNT–NAAMs increased on applying positive current density indicating EOF direction, same as diffusive transport of phenol (Fig. 10.11b).

Sano et al. demonstrated a chip based size exclusion chromatography device using NAA packed horizontally in a micro-chip. They separated DNA molecules on this chip using nanoporous anodic alumina as a separation matrix at the bottom of the microfluid channel [66]. The architecture of this NAA based chromatography chips is defined in Fig. 10.12a. This system takes advantage of the fact that smaller molecules are more frequently trapped by nanopores than larger ones and thus the larger molecules are eluted faster than the smaller molecules. This principle was previously proven on other microchip based chromatography systems using nanopillars of different gaps as the separating matrix [67, 68]. The size exclusion capabilities of this chromatography chip were demonstrated by separating two kinds of DNA samples ~3.2 and 0.3 kilobases. Two very clear bands were observed in the chromatogram presented in Fig. 10.12b, where the first band corresponds to 3.2 kilobase pair DNA and second is for 0.3 kilobase pair DNA. The authors also suggest integration of the chip with other bio-microelectro-mechanical systems for building micro total analysis system (μ-TAS) chip.

New approaches for preparation of NAAMs based size exclusion and high performance liquid chromatography device were demonstrated by Teramae and co-workers [69, 70]. They prepared a hybrid nanoporous membrane composed of silica nanocomposite deposited inside the pores of NAAMs [27]. The presence of this hierarchical pore diameter (i.e. larger pore diameter of NAA filled with silica

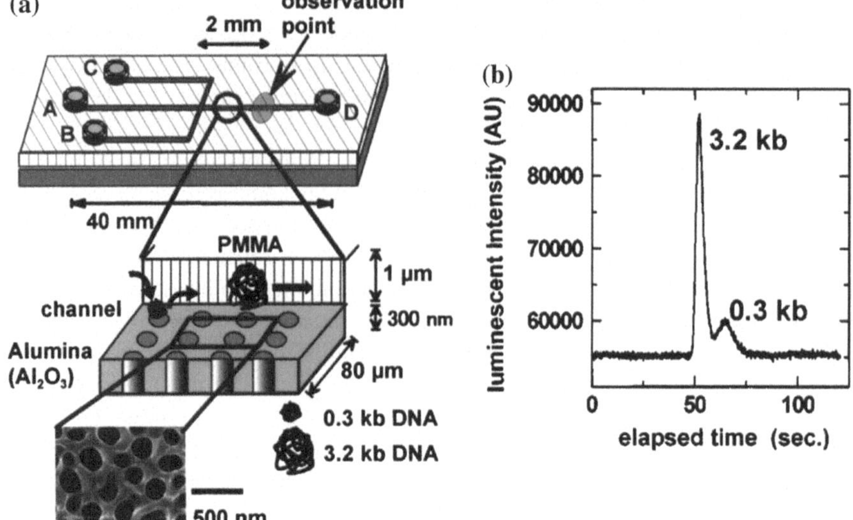

Fig. 10.12 **a** Schematic illustration of the NAA based chromatography chip. **b** Chromatogram of separation of nucleotides based on luminescent intensity measured on PMT (adapted from [66])

nanocomposites with pores of approximately 3.4 nm) makes the size-selectivity possible in this system. The nanocomposite was formed inside NAAMs pores using a surfactant template directed assembly, resulting in the nano-channels predominantly oriented along the wall of the columnar NAA pores. Molecular transporting and size exclusion capabilities of the hybrid membrane were demonstrated by measuring the transport of molecules of various sizes (such as rhodamine B (molecular size, ~1.0 nm), vitamin B12 (~2.4 nm), myoglobin (~4.0 nm), and bovine serum albumin (~7.2 nm)). They observed that the small molecules like rhodamine B and vitamin B12 were transported much faster that the larger molecules across the membrane. The proposed membrane system has a potential to be used also in chip based catalysis and other technologies.

Another study by same group fabricated mesoporous silica inside NAAMs pores and employed this hybrid membrane for membrane transport based catalytic chromatography system [71]. In this, the pores of hybrid NAAMs were modified with glucose oxidase (GOD) to catalytically process glucose (Fig. 10.13). Pluronic F127 was used as the structure guiding agent to prepare hybrid NAAMs with average 12 nm pore diameter. GOD was covalently immobilized within the pores of hybrid NAAMs using amine coupling chemistry and was utilized for oxidation of glucose to gluconic acid and hydrogen peroxide. The hybrid NAAMs glucose to gluconic acid and hydrogen peroxide conversion efficiency was 8–9 times higher than just NAAMs modified with GOD. This type of enzyme catalytic hybrid NAAMs could potentially be employed for development of applications as enzymatic sensors, biocatalysts, affinity chromatography, and bio-fuel cells. The use of

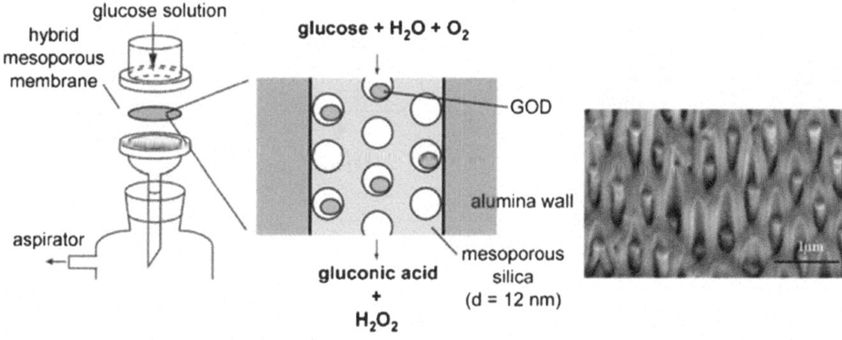

Fig. 10.13 Schematic illustrating the design of GOD immobilized hybrid NAAMs for glucose conversion and SEM image of the hybrid NAAMs columnar mesoporous silica inside its pores (adapted from [71])

NAAMs based column as normal phase for high performance liquid chromatography was also demonstrated by the same group. The performance of this NAAMs based HPLC system was evaluated using phenol and toluene with mobile phases having different solvent compositions.

Application of NAAMs as stationary phase for microchip chromatography systems was displayed by Teramae's group in 2008 [72]. NAAMs were used as a nanometre-order diameter column for chromatography chip and acetonitrile–water mixture as mobile phase. Different retention times were observed for model mixture molecules (9-anthracenemethanol (AM), 9-anthracenecarboxylic acid (AC), and dansyl–glycine (DG)), confirming the ability of this system. A schematic of the NAAMs based micro chromatography setup is provided in Fig. 10.14 along with a picture of the completely packed chip. Another microchip based chromatography system was developed by the same group to chromatographically separate adenine and two nucleosides (adenosine-50- monophosphate (AMP) and adenosine-50-triphosphate (ATP)), important coenzymes that serve as coupling agents between different enzymatic reactions in all organisms [73]. The separation of the adenine and the nucleosides is based on the interaction of hydroxyl groups on NAAMs surface with phosphate groups in the molecules. Notice that, higher number of phosphate groups in the molecule means higher interaction with NAAMs surface and longer elution time. The order of interaction between NAAMs and molecules used in this report was ATP > AMP > adenine, which is in good agreement with the number of phosphates in the molecules, indicating that nucleotides possessing more phosphates had stronger interaction with NAAMs surface. Therefore, ATP was retained more strongly than AMP, whereas adenine had the smallest retention. The order of elution time was adenine then AMP and lastly ATP.

Polymer modified NAAMs were also demonstrated for their ability towards affinity chromatography. This was achieved by functionalizing NAAMs pores with poly (2-hydroxyethyl methacrylate) (PHEMA) brushes containing nitrilotriacetate-Ni^{2+} (NTA-Ni^{2+}) terminals (Fig. 10.15) [74]. This strategy was adopted to increase the

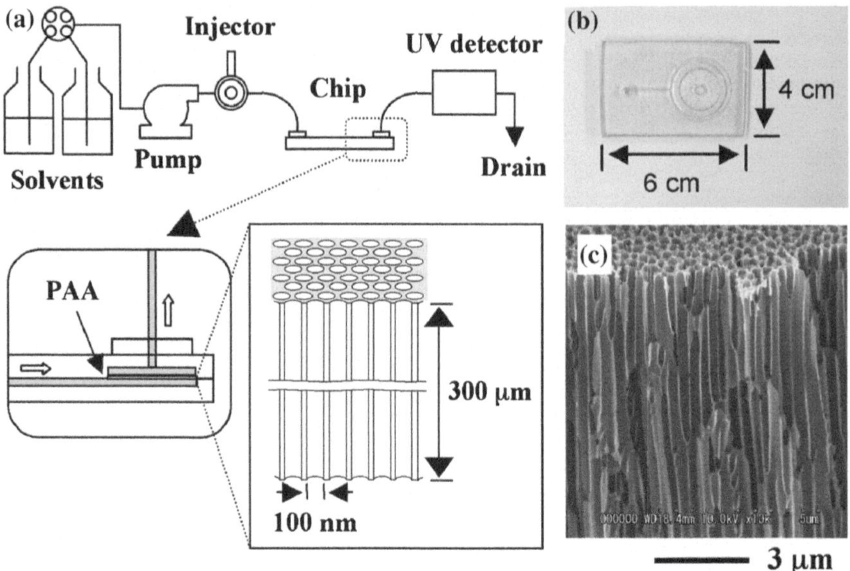

Fig. 10.14 **a** Schematic illustration of chromatographic setup. **b** Digital photograph of the chromatography chip. **c** Cross-sectional SEM image of NAA membrane used in the chip (adapted from [72])

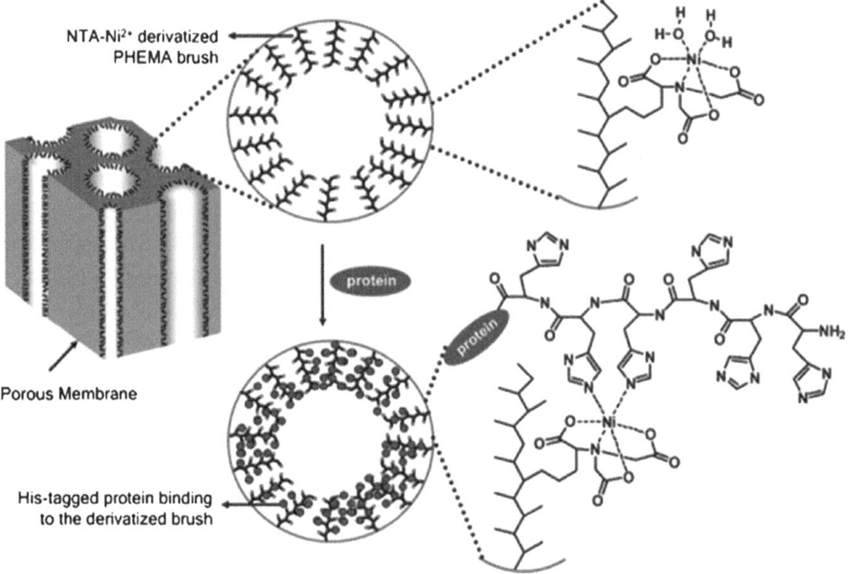

Fig. 10.15 NTA-Ni^{2+} derivatized PHEMA brush grafted inside NAAMs pores which in turn bind his-tagged protein to a NTA-Ni^{2+} moiety (adapted from [73])

protein binding capacity of NAAMs. This PHEMA modified NAAMs were able purify polyhistidine-tagged ubiquitin (HisU) in less than 30 min with a binding capacity of 120 mg of HisU/cm^3 of NAAM. Gel electrophoresis reveals high purity of the eluted protein (i.e. 99 %), even if the initial feed solution carries 10 % bovine serum or 20-fold excess of BSA.

A new hybrid membrane based on hexagonal silica nanotubes (SNT) grown inside NAAMs pores was fabricated recently by means of a direct templating method of microemulsion phases with cationic dilauryl-dimethylammonium bromide (DDAB) and cetyltrimethylammonium bromide (CTAB) surfactants [75]. The direct soft template approach was adopted for preparing densely packed silica nanotubes with a predictable order inside the pores of NAAMs (Fig. 10.16) [76]. However, removal of the surfactant results in shrinkage of the SNT structure, hence, gap between SNT and walls of NAAMs that reduces the chromatographic efficiency of the system. This issue was avoided by grafting hydrophobic trimethylsilyl (TMS) groups onto the inner pores surface of the SNT. The performance of this size exclusion chromatography system was measured by separating nanoparticles (NPs: Au, Ag, and Pt) and nanoclusters (NCs: CdS and ZnS) with different structural morphologies (i.e. spherical/pyramidal) and size distribution from 1 to 50 nm. After processing the sample through this hybrid NAAMs chromatography system, the obtained NPs or NCs have a small size distribution less than 4 nm. SNT–NAAMs hybrid system displayed high separation efficiency even after multiple reuse cycles.

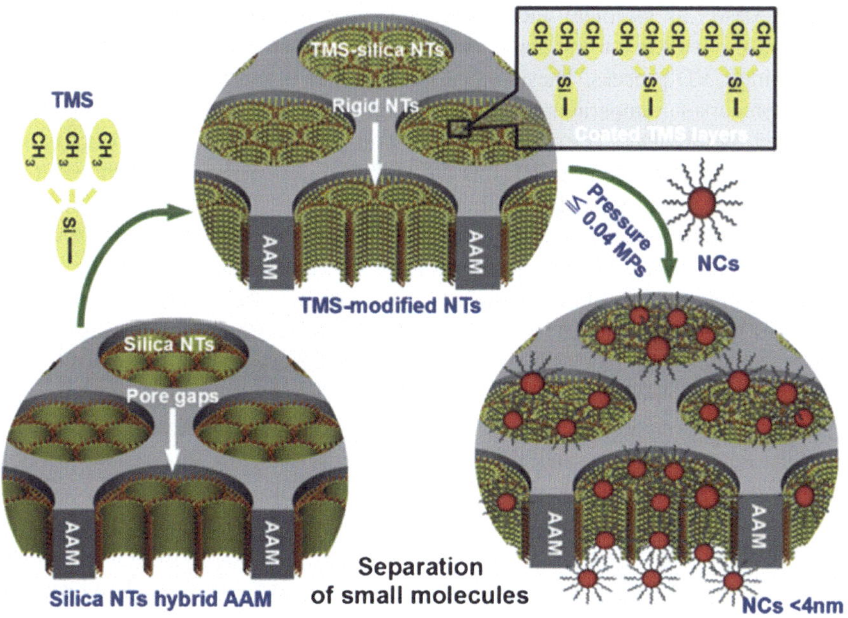

Fig. 10.16 Schematic design of hybrid NAAMs for ultrafine separation of NM NPs and SC NCs based on TMS modified silica NTs grown inside NAAMs (adapted from [75])

Interestingly, these hybrid SNT–NAAMs were also found to be suitable for separation of biomolecules such as cytochrome c (CytC). A number of studies, utilizing the same process and materials were reported by the same group for ultra-filtration of nanoparticles and molecules of biological relevance (i.e. proteins) [77–81].

Molecular transporting through nanoporous membranes has huge impact for applications such as membrane ultrafiltration, desalination, drug delivery, and bioseparation. A number of studies have been reported in this context, where molecular transporting properties of NAAMs are modified and optimized by modifying structure and surface chemistry of NAAMS. As for this, Velleman et al. studied the effect of pore diameter and surface modification of NAAMs using atomic layer deposition (ALD) [33]. The study was aimed at fine-tuning pore diameters and chemical selectivity of NAAMs. In this report, commercially available NAAMs with 20, 100, and 200 nm pore diameter were coated with silica using ALD. Silica (SiO_2) layer was deposited using tris(tert-butoxy)silanol and trimethylaluminium as the precursor couple. The silica deposition cycles were repeated ranging from 3 to 20 times to effectively adjust the pore diameter of NAAMs to tune the transporting properties. Furthermore, the surface of silica modified NAAMs was chemically functionalized with perfluorodecyldimethylchlorosilane (PFDS) to impart selectivity transport properties to NAAMs. The PFDS-modified NAAMs are hydrophobic and display enhanced sensitivity to the transport of hydrophobic dye (tris(2,20-bipyridyl)dichlororuthenium(II) hexahydrate (Rubpy), over hydrophilic dye (rose bengal (RB)). Similarly, Altalhi et al., reported on fabrication of composite membranes by growing carbon nanotubes using NAAMs as template and reported on their transport studies [29]. In this report, multi walled carbon nanotubes (MWCNT) were grown inside NAAMs pores using CVD process. They studied the influence of fabrication parameters (i.e. carbon precursor, temperature, and deposition time) CNT growth and quality of fabricated membranes. The molecular transport properties of prepared composite membranes of pore diameter 30–150 nm and thickness 5–100 μm were explored using Rose Bengal (RB) dye as model of molecule. The data shows that the hydrophilic RB dye is allowed to pass freely though composite CNT–NAAMs but with lower flux than NAAMs itself. Recently, we fabricated composite CNT–NAAMs using waste plastic bags as precursor and their molecular transport properties were investigated using Rose Bengal (RosB) and (tris(2,20-bipyridyl) dichlororuthenium(II) hexahydrate (Rubpy) dyes as model molecules.

A smart, electrically actuable molecular transporting system was recently reported by Jeon et al. for triggred drug delivery applications targeting hormone-related diseases and metabolic syndrome [82]. In this report, the top surface of NAAMs was functionalized with polypyrole doped with dodecylbenzenesulfonate anion (PPy/DBS) deposited by electropolymerization (Fig. 10.17). The system utilizes the large bulk volume change of PPy/DBS depending on its electrochemical state to manipulate the pore diameter. The pore diameter change on electrical actuation was actually demonstrated using in situ atomic force microscopy (AFM). Finally, they demonstrated actuable and pulsatile (or on-demand)

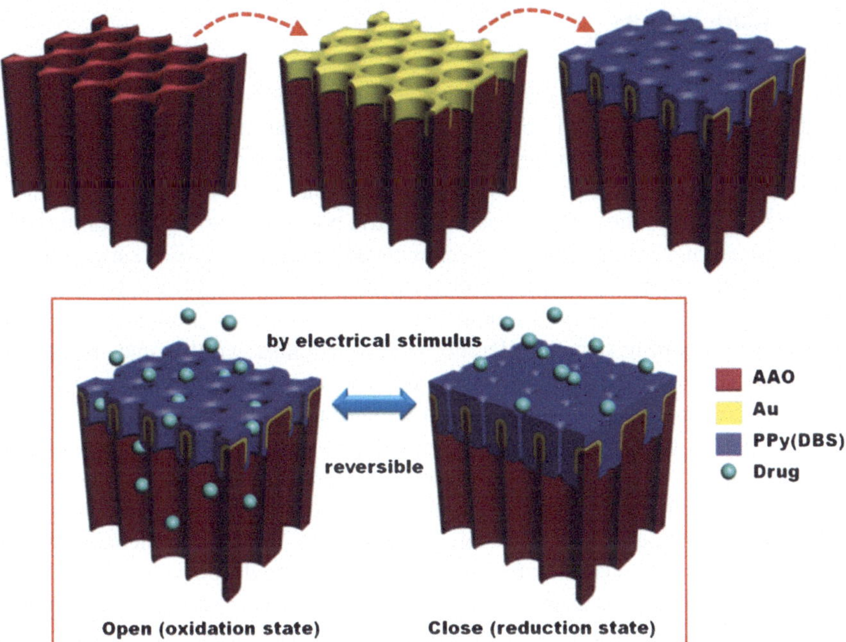

Fig. 10.17 Schematic to fabricate electrically responsive nanoporous membrane. **a** Fabrication of NAAMs starting with deposition of thin gold followed by electropolymerization of polypyrrole. *Bottom* displays the reversible change in pore diameter (and the drug release rate) according to oxidation and reduction states of polypyrole layer (adapted from [82])

drug transport trough NAAMs by using fluorescently labeled protein as a model molecule.

Another study, on ALD silica coated NAAMs was recently reported where the effect of pore diameter reduction and changes in the electric fixed charge on the membranes surface on ionic transport was studied [83]. The influence of silica modification of NAAMs on the ionic transport of NaCl through the nanoporous membranes was performed by measuring membrane potentials and electrochemical impedance spectroscopy. The results suggest a direct correlation between the effective fixed charge on NAAMs and the NaCl diffusion coefficient can be established. The silica coating on NAAMs surface reduces the effective positive fixed charge by 75 % leading to an increased counterion transport (cation transport number and diffusion coefficient) even when the pores are constrained. Another study by the same group reports on diffusive transport of neutral and charged solutes across NAAMs, where water was used as neutral solute and diffusion of charged solute was carried out using NaCl solutions labelled with a ^{22}Na radiotracer [84]. They demonstrate that the transport of ions through NAAMs is mainly controlled by both ion and pore wall friction and electrical interaction effects.

10.5 Conclusions

This chapter demonstrates the ability of NAA to transform into nanoporous membrane after removal of the underlying oxide barrier layer. NAAMs have become a highly popular material of choice for designing and templating nanostructures and composite membranes with outstanding structural and chemical properties. Owing to the properties as nanometric pore dimensions with controllable pore diameter and length, straight pores with no tortuosity, ease of surface modification, ability to form composites with other materials, NAAMs have been used in the development of a variety of unique and high performance chromatography and molecular transporting systems. Herein, we have summarized the fundamentals of conventional chromatography, the issues related to conventional chromatography, advantages of nanoporous membranes based chromatography systems, and recent developments in this field. In addition, we have also provided brief discussion and development on NAAMs based molecular transporting systems. Currently, NAAMs based chromatography and molecular transporting devices are only confined to laboratory scale, which is believed to move to industrial scale applications in the near future. Furthermore, the future aspect of these developments is believed to result in the implementation of NAAMs into lab-on-a-chip style chromatography devices with capabilities to separate, detect and quantify embedded in a single device.

References

1. J.C. Touchstone, History of chromatography. J. Liq. Chromatogr. Relat. Technol. **16**, 1647 (1993)
2. K.R. Williams, Colored bands: history of chromatography. J. Chem. Educ. **2002**(79), 922 (2002)
3. H. Schmidt-Traub, M. Schulte, A. Seidel-Morgenstern, *Preparative Chromatography* (John Wiley & Sons, New York, 2012)
4. E. Heftmann, *Chromatography: Fundamentals and Applications of Chromatography and Related Differential Migration Methods-Part B: Applications* (Elsevier, Amsterdam, 2004)
5. H. Zou, X. Huang, M. Ye, Q. Luo, Monolithic stationary phases for liquid chromatography and capillary electrochromatography. J. Chromatogr. A **954**, 5 (2002)
6. H.A. Claessens, M.A. van Straten, Review on the chemical and thermal stability of stationary phases for reversed-phase liquid chromatography. J. Chromatogr. A **1060**, 23 (2004)
7. C.J. Dunlap, P.W. Carr, C.V. McNeff, D. Stoll, Zirconia stationary phases for extreme separations. Anal. Chem. **73**, 598 (2001)
8. E. Jellum, H. Dollekamp, A. Brunsvig, R. Gislefoss, Diagnostic applications of chromatography and capillary electrophoresis. J. Chromatogr. B Biomed. Sci. Appl. **689**, 155 (1997)
9. J.C. Janson, *Protein Purification: Principles, High Resolution Methods, and Applications* (John Wiley & Sons, New York, 2012)
10. P. Brown, *High Pressure Liquid Chromatography: Biochemical and Biomedical Applications* (Elsevier, Amsterdam, 2012)

11. A.K. Mallia, *Preparative and Analytical Applications of cdi-hbdiated Affinity Chromatography, Affinity Chromatography and Biological Recognition*, vol. 191 (Elsevier, Amsterdam, 2012)
12. F. Xu, L. Zou, Y. Liu, Z. Zhang, C.N. Ong, Enhancement of the capabilities of liquid chromatography–mass spectrometry with derivatization: general principles and applications. Mass Spectrom. Rev. **30**, 1143 (2011)
13. K. Macek, Z. Deyl, J. Janák, *Liquid Column Chromatography: A Survey of Modern Techniques and Applications* (Elsevier, Amsterdam, 2011)
14. R. Ghosh, Protein separation using membrane chromatography: opportunities and challenges. J. Chromatogr. A **952**, 13 (2002)
15. X. Zeng, E. Ruckenstein, Membrane chromatography: preparation and applications to protein separation. Biotechnol. Prog. **15**, 1003 (1999)
16. H. Zou, Q. Luo, D. Zhou, Affinity membrane chromatography for the analysis and purification of proteins. J. Biochem. Biophys. Methods **49**, 199 (2001)
17. A. Tejeda, J. Ortega, I. Magaña, R. Guzmán, Optimal design of affinity membrane chromatographic columns. J. Chromatogr. A **830**, 293 (1999)
18. C. Charcosset, Review: purification of proteins by membrane chromatography. J. Chem. Technol. Biotechnol. **71**, 95 (1998)
19. C. Charcosset, Membrane processes in biotechnology: an overview. Biotechnol. Adv. **24**, 482 (2006)
20. W.J. Cheong, S.H. Yang, F. Ali, Molecular imprinted polymers for separation science: a review of reviews. J. Sep. Sci. **36**, 609 (2013)
21. P.S. Goh, A.F. Ismail, S.M. Sanip, B.C. Ng, M. Aziz, Recent advances of inorganic fillers in mixed matrix membrane for gas separation. Sep. Purif. Technol. **81**, 243 (2011)
22. P. Stroeve, N. Ileri, Biotechnical and other applications of nanoporous membranes. Trends Biotechnol. **29**, 259 (2011)
23. T. Kumeria, A. Santos, D. Losic, Nanoporous anodic alumina platforms: engineered surface chemistry and structure for optical sensing applications. Sensors **14**, 11878 (2014)
24. A. Santos, T. Kumeria, D. Losic, Nanoporous anodic alumina: a versatile platform for optical biosensors. Materials **7**, 4297 (2014)
25. A.M. Md Jani, D. Losic, N.H. Voelcker, Nanoporous anodic aluminium oxide: advances in surface engineering and emerging applications. Prog. Mater Sci. **58**, 636 (2013)
26. A. Santos, T. Kumeria, D. Losic, Nanoporous anodic aluminum oxide for chemical sensing and biosensors. TrAC Trends Anal. Chem. **44**, 25 (2013)
27. A. Yamaguchi, F. Uejo, T. Yoda, T. Uchida, Y. Tanamura, T. Yamashita, N. Teramae, Self-assembly of a silica–surfactant nanocomposite in a porous alumina membrane. Nat. Mater. **3**, 337 (2004)
28. Q. Lu, F. Gao, S. Komarneni, T.E. Mallouk, Ordered SBA-15 nanorod arrays inside a porous alumina membrane. J. Am. Chem. Soc. **126**, 8650 (2004)
29. T. Altalhi, T. Kumeria, A. Santos, D. Losic, Synthesis of well-organised carbon nanotube membranes from non-degradable plastic bags with tuneable molecular transport: towards nanotechnological recycling. Carbon **63**, 423 (2013)
30. C.R. Martin, M. Nishizawa, K. Jirage, M. Kang, Investigations of the transport properties of gold nanotubule membranes. J. Phys. Chem. B **2001**, 105 (1925)
31. M. Majumder, N. Chopra, R. Andrews, B.J. Hinds, Nanoscale hydrodynamics: enhanced flow in carbon nanotubes. Nature **438**, 44 (2005)
32. M. Majumder, N. Chopra, B.J. Hinds, Mass transport through carbon nanotube membranes in three different regimes: ionic diffusion and gas and liquid flow. ACS Nano **5**, 3867 (2011)
33. L. Velleman, G. Triani, P.J. Evans, J.G. Shapter, D. Losic, Structural and chemical modification of porous alumina membranes. Microporous Mesoporous Mater. **126**, 87 (2009)
34. F. Keller, M. Hunter, D. Robinson, Structural features of oxide coatings on aluminum. J. Electrochem. Soc. **100**, 411 (1953)
35. H. Masuda, K. Fukuda, Ordered metal nanohole arrays made by a two-step replication of honeycomb structures of anodic alumina. Science **268**, 1466 (1995)

36. H. Masuda, F. Hasegwa, S. Ono, Self-ordering of cell arrangement of anodic porous alumina formed in sulfuric acid solution. J. Electrochem. Soc. **144**, L127 (1997)
37. H. Masuda, K. Yada, A. Osaka, Self-ordering of cell configuration of anodic porous alumina with large-size pores in phosphoric acid solution. Jpn. J. Appl. Phys. **37**, L1340 (1998)
38. K. Nielsch, J. Choi, K. Schwirn, R.B. Wehrspohn, U. Gösele, Self-ordering regimes of porous alumina: the 10 porosity rule. Nano Lett. **2**, 677 (2002)
39. D. Liu, C. Zhang, G. Wang, Z. Shao, X. Zhu, N. Wang, H. Cheng, Nanoscale electrochemical metallization memories based on amorphous (La, Sr) MnO_3 using ultrathin porous alumina masks. J. Phys. D Appl. Phys. **47**, 085108 (2014)
40. W. Lee, R. Ji, U. Gösele, K. Nielsch, Fast fabrication of long-range ordered porous alumina membranes by hard anodization. Nat. Mater. **5**, 741 (2006)
41. G.D. Sulka, Highly ordered anodic porous alumina formation by self-organized anodizing, in *Nanostructured Materials in Electrochemistry*, vol. 1, ed. by A. Eftekahari (Wiley-VCH Verlag GmbH and Co. KGaA, Weinhem Germany, 2008), pp. 1–116
42. W. Lee, S.J. Park, Porous anodic aluminum oxide: anodization and templated synthesis of functional nanostructures. Chem. Rev. **114**, 7487 (2014)
43. Y. Zhao, M. Chen, Y. Zhang, T. Xu, W. Liu, A facile approach to formation of through-hole porous anodic aluminum oxide film. Mater. Lett. **59**, 40 (2005)
44. M. Lillo, D. Losic, Pore opening detection for controlled dissolution of barrier oxide layer and fabrication of nanoporous alumina with through-hole morphology. J. Membr. Sci. **327**, 11 (2009)
45. H. Han, S.J. Park, J.S. Jang, H. Ryu, K.J. Kim, S. Baik, W. Lee, in situ determination of the pore opening point during wet-chemical etching of the barrier layer of porous anodic aluminum oxide: nonuniform impurity distribution in anodic oxide. ACS Appl. Mater. Interfaces **5**, 3441 (2013)
46. K. Schwirn, W. Lee, R. Hillebrand, M. Steinhart, K. Nielsch, U. Gösele, Self-ordered anodic aluminum oxide formed by H2SO4 hard anodization. ACS Nano **2**, 302 (2008)
47. K.P. Lee, D. Mattia, Monolithic nanoporous alumina membranes for ultrafiltration applications: Characterization, selectivity–permeability analysis and fouling studies. J. Membr. Sci. **435**, 52 (2013)
48. M. He, J. Yao, Z.-X. Low, D. Yu, Y. Feng, H. Wang, A fast in situ seeding route to the growth of a zeolitic imidazolate framework-8/AAO composite membrane at room temperature. RSC Adv. **4**, 7634 (2014)
49. R. Gasparac, P. Kohli, M.O. Mota, L. Trofin, C.R. Martin, Template synthesis of nano test tubes. Nano Lett. **4**, 513 (2004)
50. L.L. Hench, J.K. West, The sol-gel process. Chem. Rev. **90**, 33 (1990)
51. A.Y. Ku, J.A. Ruud, T.A. Early, R.R. Corderman, Evidence of ion transport through surface conduction in alkylsilane-functionalized nanoporous ceramic membranes. Langmuir **22**, 8277 (2006)
52. A.M.M. Jani, E.J. Anglin, S.J. McInnes, D. Losic, J.G. Shapter, N.H. Voelcker, Nanoporous anodic aluminium oxide membranes with layered surface chemistry. Chem. Commun. **21**, 3062 (2009)
53. A.M.M. Jani, I.M. Kempson, D. Losic, N.H. Voelcker, Dressing in layers: layering surface functionalities in nanoporous aluminum oxide membranes. Angew. Chem. **122**, 8105 (2010)
54. B. Berland, I. Gartland, A. Ott, S. George, In situ monitoring of atomic layer controlled pore reduction in alumina tubular membranes using sequential surface reactions. Chem. Mater. **10**, 3941 (1998)
55. A. Ott, K. McCarley, J. Klaus, J. Way, S. George, Atomic layer controlled deposition of Al_2O_3 films using binary reaction sequence chemistry. Appl. Surf. Sci. **107**, 128 (1996)
56. G. Xiong, J.W. Elam, H. Feng, C.Y. Han, H.H. Wang, L.E. Iton, L.A. Curtiss, M.J. Pellin, M. Kung, H. Kung, Effect of atomic layer deposition coatings on the surface structure of anodic aluminum oxide membranes. J. Phys. Chem. B **109**, 14059 (2005)
57. B.H. Lee, J.K. Hwang, J.W. Nam, S.U. Lee, J.T. Kim, S.M. Koo, A. Baunemann, R.A. Fischer, M.M. Sung, Low-temperature atomic layer deposition of copper metal thin films:

self-limiting surface reaction of copper dimethylamino-2-propoxide with diethylzinc. Angew. Chem. Int. Ed. **48**, 4536 (2009)
58. J. Bachmann, J. Jing, M. Knez, S. Barth, H. Shen, S. Mathur, U. Gösele, K. Nielsch, Ordered iron oxide nanotube arrays of controlled geometry and tunable magnetism by atomic layer deposition. J. Am. Chem. Soc. **129**, 9554 (2007)
59. L. Miranda, R. Short, F. van Amerom, R. Bell, R. Byrne, Direct coupling of a carbon nanotube membrane to a mass spectrometer: contrasting nanotube and capillary tube introduction systems. J. Membr. Sci. **344**, 26 (2009)
60. A. Popp, J. Engstler, J.J. Schneider, Porous carbon nanotube-reinforced metals and ceramics via a double templating approach. Carbon **47**, 3208 (2009)
61. S. Park, Y.S. Kim, W.B. Kim, S. Jon, Carbon nanosyringe array as a platform for intracellular delivery. Nano Lett. **9**, 1325 (2009)
62. G. Che, B. Lakshmi, C. Martin, E. Fisher, R.S. Ruoff, Chemical vapor deposition based synthesis of carbon nanotubes and nanofibers using a template method. Chem. Mater. **10**, 260 (1998)
63. S.J. Hurst, E.K. Payne, L. Qin, C.A. Mirkin, Multisegmented one-dimensional nanorods prepared by hard-template synthetic methods. Angew. Chem. Int. Ed. **45**, 2672 (2006)
64. S.B. Lee, D.T. Mitchell, L. Trofin, T.K. Nevanen, H. Söderlund, C.R. Martin, Antibody-based bio-nanotube membranes for enantiomeric drug separations. Science **296**, 2198 (2002)
65. S.A. Miller, V.Y. Young, C.R. Martin, Electroosmotic flow in template-prepared carbon nanotube membranes. J. Am. Chem. Soc. **123**, 12335 (2001)
66. T. Sano, N. Iguchi, K. Iida, T. Sakamoto, M. Baba, H. Kawaura, Size-exclusion chromatography using self-organized nanopores in anodic porous alumina. Appl. Phys. Lett. **83**, 4438 (2003)
67. M. Baba, T. Sano, N. Iguchi, K. Iida, T. Sakamoto, H. Kawaura, DNA size separation using artificially nanostructured matrix. Appl. Phys. Lett. **83**, 1468 (2003)
68. L.R. Huang, E.C. Cox, R.H. Austin, J.C. Sturm, Continuous particle separation through deterministic lateral displacement. Science **304**, 987 (2004)
69. T. Yamashita, S. Kodama, M. Ohto, E. Nakayama, N. Takayanagi, T. Kemmei, A. Yamaguchi, N. Teramae, Y. Saito, Use of porous anodic alumina membranes as a nanometre-diameter column for high performance liquid chromatography. Chem. Commun. **11**, 1160 (2007)
70. T. Yamashita, S. Kodama, M. Ohto, E. Nakayama, S. Hasegawa, N. Takayanagi, T. Kemmei, A. Yamaguchi, N. Teramae, Y. Saito, Permeation flux of organic molecules through silica-surfactant nanochannels in a porous alumina membrane. Anal. Sci. **22**, 1495 (2006)
71. W. Fu, A. Yamaguchi, H. Kaneda, N. Teramae, Enzyme catalytic membrane based on a hybrid mesoporous membrane. Chem. Commun. **7**, 853 (2008)
72. T. Yamashita, S. Kodama, M. Ohto, E. Nakayama, T. Kemmei, T. Muramoto, A. Yamaguchi, N. Teramae, N. Takayanagi, Utilization of nanometre-order diameter columns inside porous anodic alumina for chromatography chip system. Chem. Lett. **37**, 18 (2008)
73. T. Yamashita, S. Kodama, T. Kemmei, M. Ohto, E. Nakayama, T. Muramoto, A. Yamaguchi, N. Teramae, N. Takayanagi, Separation of adenine, adenosine-5′-monophosphate and adenosine-5′-triphosphate by fluidic chip with nanometre-order diameter columns inside porous anodic alumina using an aqueous mobile phase. Lab-on-a-Chip **9**, 1337 (2009)
74. P. Jain, L. Sun, J. Dai, G.L. Baker, M.L. Bruening, High-capacity purification of his-tagged proteins by affinity membranes containing functionalized polymer brushes. Biomacromolecules **8**, 3102 (2007)
75. M.M. Mekawy, A. Yamaguchi, S.A. El-Safty, T. Itoh, N. Teramae, Mesoporous silica hybrid membranes for precise size-exclusive separation of silver nanoparticles. J. Colloid Interface Sci. **355**, 348 (2011)
76. S.A. El-Safty, A. Shahat, M. Mekawy, H. Nguyen, W. Warkocki, M. Ohnuma, Mesoporous silica nanotubes hybrid membranes for functional nanofiltration. Nanotechnology **21**, 375603 (2010)

77. S.A. El-Safty, M.M. Mekawy, A. Yamaguchi, A. Shahat, K. Ogawa, N. Teramae, Organic–inorganic mesoporous silica nanostrands for ultrafine filtration of spherical nanoparticles. Chem. Commun. **46**, 3917 (2010)
78. S.A. El-Safty, Designs for size-exclusion separation of macromolecules by densely-engineered mesofilters. TrAC Trends Anal. Chem. **30**, 447 (2011)
79. S.A. El-Safty, A. Shahat, W. Warkocki, M. Ohnuma, Building-block-based mosaic cage silica nanotubes for molecular transport and separation. Small **7**, 62 (2011)
80. S.A. El-Safty, A. Shahat, M.R. Awual, M. Mekawy, Large three-dimensional mesocage pores tailoring silica nanotubes as membrane filters: nanofiltration and permeation flux of proteins. J. Mater. Chem. **21**, 5593 (2011)
81. S.A. El-Safty, M.A. Shenashen, Size-selective separations of biological macromolecules on mesocylinder silica arrays. Anal. Chim. Acta **694**, 151 (2011)
82. G. Jeon, S.Y. Yang, J. Byun, J.K. Kim, Electrically actuatable smart nanoporous membrane for pulsatile drug release. Nano Lett. **11**, 1284 (2011)
83. V. Romero, V. Vega, J. García, R. Zierold, K. Nielsch, V.M. Prida, B. Hernando, J. Benavente, Changes in morphology and ionic transport induced by ALD SiO2 coating of nanoporous alumina membranes. ACS Appl. Mater. Interfaces **5**, 3556 (2013)
84. V. Romero, M.I. Vázquez, S. Cañete, V. Vega, J. García, V.M. Prida, B. Hernando, J. Benavente, Frictional and electrical effects involved in the diffusive transport through a nanoporous alumina membrane. J. Phys. Chem. C **117**, 25513 (2013)

Chapter 11
Nanoporous Anodic Alumina for Drug Delivery and Biomedical Applications

Moom Sinn Aw, Manpreet Bariana and Dusan Losic

Abstract Nanoporous anodic alumina (NAA) engineered by simple and scalable electrochemical anodization process has been extensively explored for drug delivery (DD) and biomedical applications. This review highlights the concepts of drug-releasing implants based on NAA, with a focus on recent progress in their development and future perspectives towards advanced medical therapies, including orthopaedic and dental implants, heart/coronary/vasculature stents, immunoisolation, skin healing, tissue engineering and cell culture. The specific topics discuss the advantages of NAA carrier for drug loading and release performances, due to its unique porous structure, loading capacity, chemical inertness and biocompatibility. Finally, a conclusive overview and future perspectives are provided with suggestions about future studies that are required to have this unique biomaterial and device approved for real-life biomedical applications.

11.1 Introduction

Drug delivery (DD) has achieved considerable progress in the recent decade with the introduction of new biomaterials developed from advanced nanoengineered materials and nanofabrication technologies. The ideal drug delivery system (DDS) is one that has the capacity to provide an adequate amount of therapeutic agent, safely and efficiently to specific sites for action with optimized dosage, required time and the drug release profile, to achieve the desirable therapeutic effects without being destructive to surrounding cells and tissues in the human body [1]. The critical

M.S. Aw · M. Bariana · D. Losic (✉)
School of Chemical Engineering, The University of Adelaide,
North Engineering Building, SA5005 Adelaide, Australia
e-mail: dusan.losic@adelaide.edu.au

D. Losic and A. Santos (eds.), *Nanoporous Alumina*,
Springer Series in Materials Science 219, DOI 10.1007/978-3-319-20334-8_11

challenge lies in searching for the most effective, inexpensive and less time-consuming means to deliver toxic drugs and other biomolecules to targeted sites via the shortest possible pathway with minimal damage and least complications. Unfortunately, conventional drug delivery (DD) systems fail to fulfil these criteria as they typically render rapid release of drugs into the blood circulation over a short span of time based on first-order rate kinetics [2]. Although frequent dosing can eliminate toxic peaks and sub-therapeutic troughs, it causes poor patient compliance. Conventional DD by tablets or intravenous injections using toxic drugs can cause negative side effects such as damaging healthy cells/organs, whilst less intrusive measures are not sufficiently effective in toxic manifestation when treating diseases that can spread to other areas of the body, although they pose fewer risks [3].

The development of DDS for local administration of therapeutics is garnering more attention as it reasonably resides in between the two extreme cases, which is in moderation and closest to approaching the ideal therapeutic window for the prevention of over- and under-dosing. The initial concept for localized DD was demonstrated by Folkman et al. half a century ago; when he discovered that a rabbit can be dozed off by circulating its blood inside a tube containing anaesthetic gas, thereby laying the first foundation for the concept of implantable DD [4]. Therefore, the implant as a drug-releasing carrier is considered to be a key element for the development of local DD systems where different solutions emerged from using existing metal implants or implants based on polymeric and ceramic materials. To achieve more rapid, reliable, less traumatic therapies, new and surgically less-invasive implant-based technologies are emerging with the use of biomaterials. Under the circumstances where frequent dosing and systemic drug circulation are likely to cause adverse effects, nanoporous drug-releasing materials as drug-carriers are recognized as promising candidates for localized and implantable drug delivery [5].

Among several nanomaterials used to develop drug-releasing implants, inorganic nanoporous materials produced by electrochemical anodization are especially attractive due to their simple cost-competitive, well-established scalable fabrication processes and versatility in terms of controlling their pore structure, drug-releasing performance and applicability [7]. Nanoporous anodic alumina (NAA), titania nanotubes (TNTs) and porous silicon (pSi) are three most attractive nanoporous materials used intensively to develop drug-releasing implants [8–12]. Compared to polymer-based drug-releasing systems, NAA, TNTs and pSi have better mechanical, chemical and thermal stability, resistance to erosion and/or biodegradability, which are desired properties for many clinical applications (e.g. orthopaedics) [7].

Nanoporous anodic alumina (NAA) is fabricated by electrochemical anodization to produce an inorganic framework of non-erodible porous substrates with well-defined morphology and geometry, tuneable pore volume, pore size, interpore spacing and pore thickness [13]. This lithography-free approach is an attractive nanofabrication approach because it is facile, low-cost, time-effective, convenient and can be scaled-up for industrial production [14]. The electrochemical method is

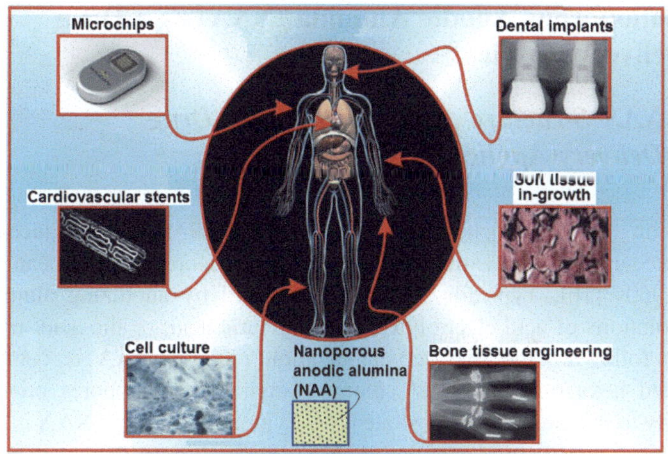

Fig. 11.1 Schematic overview summarizing the applications of nanoporous anodic alumina (*NAA*) for localized drug delivery and biomedical applications, where NAA is used to prepare drug-releasing implants, stents, microchips and materials for cell culture and tissue engineering [7]

based on self-ordering process of the formation of nanopores to enable the adjustment of nanopore dimensions (pore diameter, length and shape) with high aspect ratio, to accommodate and release a large amount of drugs [15]. Based on the inherent properties of NAA such as its bio-inertness, chemical/mechanical/thermal stability, high specific surface area, controllable pore dimensions and high porosity, they are able to provide tailorable and controlled release of pharmacological agents, without occupying a substantial amount of space or resulting in enormous inconvenience, due to its miniature size for insertion [7, 13]. These are the main reasons that NAA are considered as promising candidates for the development of drug-releasing devices for localized DD applications. In the last 10 years NAA were extensively explored for many in vitro and in vivo drug delivery applications using many drugs to address various critical medical problems [7, 8, 13].

This chapter summarizes the recent advances of the use of NAA as drug delivery implants. The main focus is to summarize the structural properties of NAA, their surface modification, the different drug-releasing concepts based on NAA, the factors influencing drug release characteristics required for developing different types of DDS for broad medical therapies and their biocompatibility. The application of NAA in orthopaedics, dentistry, cardio-vasculature, tissue engineering, cell culture, microchips and immunoisolation are specifically highlighted. Finally, this review concludes with a general overview and a prospective outlook on the future challenges and trends in this field. Figure 11.1 summarizes the most significant drug-delivery and biomedical applications of NAA.

11.2 Nanoporous Anodic Alumina (NAA) as a Drug Delivery Carrier

11.2.1 NAA Structure and Properties for Drug Delivery Applications

Research on the electrochemical fabrication of NAA to produce ordered nano-architectured NAA membranes, template and decorative layers can be traced back to 1960s [16]. Typically, NAA is fabricated by anodizing aluminium in aqueous solutions of acid electrolytes (e.g. sulphuric acid, oxalic acid, phosphoric acid, etc.) following the two-step anodization process. NAA is composed of close-packed hexagonal arrays of columnar cylindrical nanopores oriented perpendicularly to the aluminium substrate [13]. Typical structure of NAA is presented in Fig. 11.2. The structural features of NAA such as pore diameter, length and shape can be precisely controlled by the anodization parameters. The fabrication of NAA offers a very wide range of pore diameters from 5 to 300 nm, length from 0.1 to 500 μm, and with different shapes (flat, periodical, conical, funnel, etc.). This confirms the enormous versatility of this material for broad applications [12, 17]. NAA is appealing in terms of its structures for loading and releasing drugs

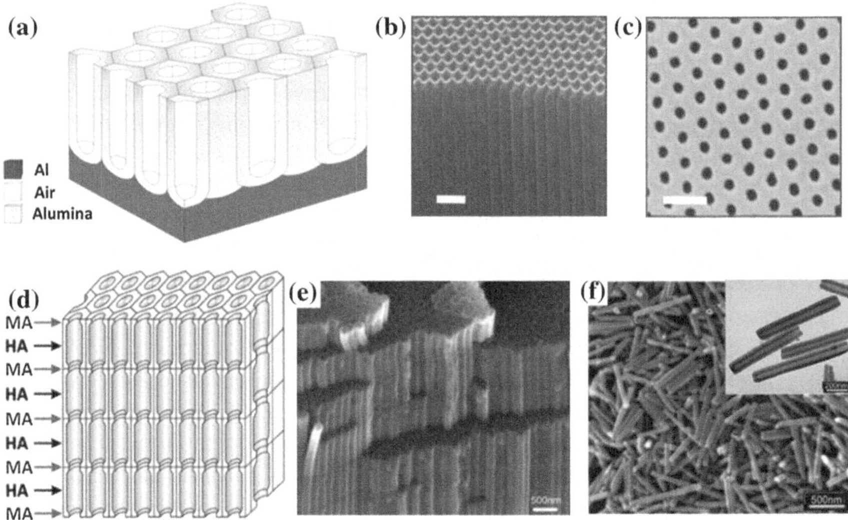

Fig. 11.2 Nanoporous anodic alumina (NAA) for developing drug-releasing implants and nanotube drug carriers. **a** Schematic illustration of the basic structure of NAA, **b** *top* and *cross-sectional* scanning electron microscopy (SEM) images of NAA (scale bars = 400 and 250 nm, respectively) (adapted from [2]), **c** schematic of cross-sectional view of NAA prepared by pulsed-anodization, **d** corresponding SEM image before the dispersion of nanotubes (*black* and *red arrows* indicate layers created by *HA* and *MA* pulses, respectively), **e** SEM images of liberated NAA nanotubes with 600 nm length [18, 19]

because of its unique ordered and high voluminous porous structure that is able to provide high payload, slow and controllable release compared to other randomly organized porous scaffold [8, 18]. NAA is also light-weighed, less bulky, robust, chemically stable, corrosion-resistive and biologically inert. These are favorable properties for implant applications [13]. The NAA nanostructures can also be fabricated on many different shapes including planar (foils, plates), three dimensional (3D), non-planar and curved surfaces, such as thin wires or needles to facilitate insertion into the body during implant surgeries. The fabrication process of NAA is scalable and inexpensive based on low cost materials and equipment, favoring their commercial mass production at industrial level. Recently, a new type of drug-delivery carrier named nanoporous anodic alumina nanotubes (NAAN) has emerged, showing new perspectives to extend the application of NAA material into targeting DD and more sophisticated localized DD without using implants [19, 20]. The NAANs with controllable lengths were prepared by a pulsed-anodization process where NAA structure was fractured into nanotubes (Fig. 11.2c, d) [19, 20].

These NAA properties are suitable for the loading of many therapeutic agents, including small molecules containing different functional groups, hydrophilic or hydrophobic, charged, uncharged, large biomolecules such as peptides, proteins, DNA, RNA, genes and also small drug-carriers such as gold or magnetic nanoparticles or polymer micelles [8, 13, 18]. Drug release from nanoporous structures such as NAA in contact with physiological solution is governed by a diffusion-controlled release process, considering that drug is completely soluble in the physiological solution [10]. This process is influenced by a number of factors, for instance, the molecular size of drugs, interfacial interaction between the drug molecules and the nanopore surface, the dissolution rate of drugs, diffusion coefficient, pore dimensions, pH of medium, polarity, or ionic charge of the pore surface. There are several options to obtain different drug release profiles from NAA including burst release, zero-ordered, combined first and zero-ordered and stimuli-responsive drug release. Zero-order type release is the most desirable release strategy as it yields the highest dosing stability and resembles the ideal situation whereby drug is released at a uniform and constant rate independent of time and concentration. Therefore NAA provides an excellent possibility to design the optimized pharmacokinetic profile required for specific therapy including burst and zero-order rate kinetics with sustained and extended drug release or their combination thereof. NAA holds the potential to generate other types of release such as on-demand, stepwise, multi-drug delivery, delayed, time-programmed, sequential to switchable release behaviors with the help of purposefully designed strategies to control drug diffusion to advance drug delivery towards achieving the aim of smart drug delivery concepts. Many studies have been reported to explore drug-releasing strategies of NAA related to different drugs and medical applications, which are briefly outlined in the following section [7, 8, 12, 18].

Structural modification of nanopores was the simplest strategy explored to extend drug-release by controlling the diffusion of drug molecules from nanopores using NAA with different pore dimensions [11]. The first use of NAA as a therapeutic device was pioneered by Gong et al., showing the influence of diffusion ofmodel molecules such as fluorescein and FITC-dextran with different molecular weights as a function of the pore size of NAA (25–55nm) [21]. The NAA membranes were converted into biofiltration capsules loaded with two types of model drugs to investigate drug release characteristics with respect to NAA pore size. Results demonstrated that the molecular size of drug is inversely proportional to its release rate. For instance, the complete release of fluorescein (the smaller sized model drug) was completed within a few hours, whereas for FITC-dextran, despite the fact that its hydrodynamic radius is less than the NAA pore on the top surface (11.6 < 55 nm), no drug release was observed. These results confirmed that drug release rate can be regulated through appropriate pore dimensions and it is dependent on the size of the drug molecules. Sustained release of drugs was demonstrated in the case for small antibiotic drug, amoxicillin ($\sim$8 nm in size) from NAA (pore size = 20 nm), showing more than 5 weeks of sustained and extended drug elution [22]. In a previous work from our group, NAA with pore diameters ranging from 65 to 160 nm were prepared to explore their impact on drug loading and drug releasing properties [11]. Indomethacin was used as the model drug for release studies from NAA, showing that drug release can be extended from 5 to 6 days, indicating that larger pore size offers more space and volume to load drugs, and the higher the drug amount stored, the more extended the release time was [11, 23]. These results are in agreement with previous work by Kang et al. using NAA stents with different pore sizes and film thickness to study the effects on drug release rates [24]. Gultepe et al. has demonstrated the comparative impact on the release of the anticancer drug, doxorubicin from NAA, in which its pores were widened to different degrees of release [6, 18]. Kipke et al. also confirmed that the release of biological molecules in NAA membranes is dependent on the NAA pore size [25]. Kwak et al. has found that the length/depth of the NAA pores had a significant influence on the release of paclitaxel that was used as the model drug and loaded inside the NAA [26]. Their results are summarized in Fig. 11.3 showing the drug release graph from NAA with different pore diameters and length.

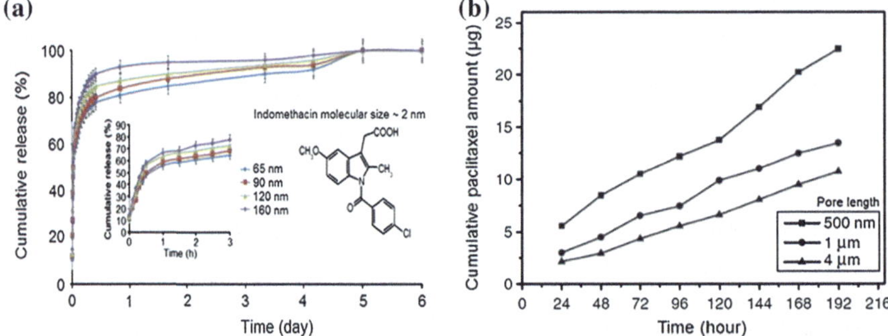

Fig. 11.3 **a** Graph showing the cumulative release (%) of indomethacin (poorly soluble drug) from NAA with different pore diameters (65–160 nm) versus release time (days) [11]; **b** cumulative release of paclitaxel release from NAA as a drug carrier with various pore lengths (0.5–4 μm) [26]

Their study showed that drug release is not completed after 400 h as the drug release occurred not only from the bottom of the NAA, but also from the side walls of the pores. It is implied that the mean diffusion distance may be smaller than the NAA pore length, and that it is not diffusion-controlled as the release time is disproportional to the square root of the NAA pore length. In conclusion, the nanopore modification strategy shows good progress, but there are still obvious limitations to significantly extend drug-release by reducing pore diameters at the same time not compromising the drug loading capacity.

Surface modifications based on the functionalization of NAA pores is the second strategy considered and widely explored to achieve sustained and controllable drug-release [11]. The main goals here are to introduce chemical modifications inside the alumina nanopores by imparting hydrophobic and hydrophilic surface properties to dynamically change the interaction with drug molecules, which will in turn alter drug loading and release kinetics. This concept is demonstrated on porous silica particles used as drug-carriers, whereby hydrophobic silica matrices loaded with poorly water soluble drug (ibuprofen) led to lower diffusion rates [7, 10]. This concept by crafting organic molecules (silanes and polymers) inside nanopores was demonstrated by several groups using hydrophilic and hydrophobic drug molecules. For example, in our work, we demonstrated that the initial burst release of indomethacin can be drastically reduced from 50 % to around 25 % by functionalizing the NAA surface with amine- or penta-fluoro-terminated silanes [11]. In addition to that, the conditions of the buffer medium such as pH and ionic strength are also factors that have to be taken into account [27]. Moreover, NAA surfaces can be decorated with both hydrophobic and hydrophilic groups, for instance, the "Janus particles" [28]. Incorporating various chemical moieties on the surface of NAA drug carrier is also highly favorable for selective functionalization, catering to a broad range of payloads and release rates. In conclusion, these results confirm the importance of interfacial, chemical and surface charge properties of nanopores for their drug loading and drug-releasing characteristics. The properties of drug molecules (size, charge, and binding affinity) should be considered to impart appropriate surface chemistry for nanoporous implants to achieve desired, optimized outcomes (drug loading and release). Hence, even though the surface modification strategy is useful for designing advanced implants with optimized performances, it is still limited to achieve ultra-long sustained drug-releasing conditions (>3 months).

To overcome this problem, our group introduced a new concept for extended drug release by using biopolymer coating on the top surface of NAA to specifically reduce the pore opening or close the pores [29, 30]. The idea of this soft-modification approach was to control drug release by controlling the pore opening on the top of NAA surface by the deposition of a thin polymer layer [11, 29–31]. The concept of pore reduction is presented in Fig. 11.4a, b where plasma polymer deposition is used to precisely reduce pore size to less than 5 nm. Plasma polymerization is an appealing approach, because it is a one-step, neat, simple (no preparation of solutions required), fast (only a couple of minutes), direct and dry technique that can be used for scalable fabrication of NAA drug-releasing implants

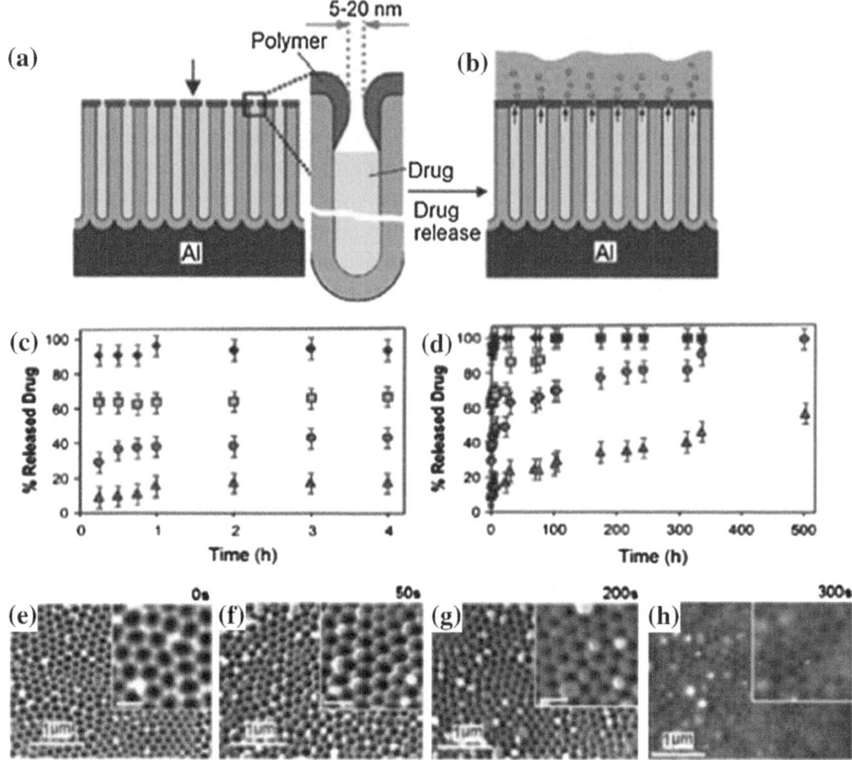

Fig. 11.4 **a** Scheme depicting the concept for controlling drug release from nanopores by the reduction of the pore opening (5–20 nm) on the *top* of NAA using plasma polymerization of biopolymer (polyallylamine); **b** suppressed diffusion of drug molecules from the reduced nanotube structures into the PBS (drug release medium); **c** controlled release of model drug vancomycin from plasma-modified NAA for 4 h and **d** 500 h. The pore opening is precisely controlled by plasma polymerization time from 50 s (*squares*), 120 s (*circles*), 200 s (*triangles*) and the control sample (no plasma deposition, *diamonds*). Scanning electron microscopy (SEM) images of NAA surface plasma-polymerized with polyallylamine using deposition times of **e** 0 s, **f** 50 s, **g** 200 s, **h** 300 s [29]

[32, 33]. The drug release profiles of model antibiotic drug, vancomycin, from NAA with and without plasma polymer deposition of polyallylamine using different deposition time (50, 120 and 200 s) are presented in Fig. 11.4c, d. These graphs clearly show that the increasing the time of deposition (film thickness) leads to slower drug release. From the uncoated NAA sample, drug was completely released within 45 min, for the deposition of 50 s, the release was extended to approximately 200 h, whereas a deposition for 120 s extends the release to almost 500 h. When polymer is deposited for 200 s, only half the amount of drug was released within 500 h [29]. Reported methods for controlled drug release by the alteration of pore dimension can extend the diffusion of small molecules for up to 2–3 weeks, and in the case of the larger ones by up to a month [29]. The release kinetics is fitted into a

zero-order equation, showing that it is possible to achieve zero-order release kinetics from NAA porous structures by this approach, which is the desired drug-releasing performance. Figure 11.4e, f present the SEM images of the top of these NAA substrates at different deposition times which confirms the reduction of pore size with the increase of polymer deposition time (that resulted in the increase in deposited film thickness). Specifically, with longer plasma deposition time, for instance, up to 500 s, drug release can be extended to more than 3 weeks. In the case of vancomycin which is a heavier molecule than that of levofloxacin, extended release can be achieved after plasma deposition of only 50 s, showing a distinct correlation between the time of deposition and drug release. The polymer dip-coating is another simpler strategy used to cover NAA nanopores with 0.5–2 μm thick polymer film (e.g. chitosan) and to replace more complicated plasma technology with the similar outcomes [11, 34]. The drug release here is controlled by the degradation rate of the polymer film covering the NAA pores with the capability to extend drug release to 3–6 months, depending on the type of polymer and polymer thickness.

The importance of responsive DD, which is one of the aspects for the smart drug delivery concept, has recently gained attention due to the need for tailorable therapy with controllable drug dosing [7, 10]. It is known that every person responds uniquely to conditions such as diabetes, cardiovascular diseases and acute pain resulting from injuries of various types. Although the concept of controlled release was introduced, releasing drugs at any arbitrary time for a desirable duration is still not a function that is available. Therefore, in order to address the problem and restrict drug level within the tolerance threshold, the concept for responsive DD using NAA was introduced. The implementation of such devices requires NAA to be susceptible to a specific physical or chemical incitement, for example, being responsive to a specific stimulus, underging protonation, a hydrolytic cleavage or a (supra)molecular conformational change to release the drug accordingly [7]. There are many possibilities for the advances in the design of nanoscale stimuli-responsive systems to control drug biodistribution in response to exogenous (variations in temperature, magnetic field, ultrasound intensity, light or electric pulses) or endogenous (changes in pH, enzyme concentration or redox gradients) stimuli. Drug delivery system using NAA membrane that is electrically responsive has been reported by Joen et al. [35]. The concept is based on polypyrrole doped with dodecylbenzenesulfonate anion (PPy/DBS) that was electro-polymerized on the upper part of a NAA membrane as shown in Fig. 11.5a, b, in which the pores on the top surface of NAA can be opened or closed, owing to the fact that the volume of the PPy/DBS changes drastically depending on the electrochemical state. The pore size of NAA was electrically actuated as proved by the atomic force microscopy (AFM) images whereby the pore size expanded and closed in Fig. 11.5c, d. This concept was implemented as a pulsatile, or on-demand release system for a model protein, i.e. FITC-labelled BSA. Owing to the fast switching time (less than 10 s) and high flux of the drugs, this membrane has considerable suitability to be applied for emergency treatment of diseases, such as angina pectoris and migraine [36]. To further develop improved drug delivery systems, viz. for

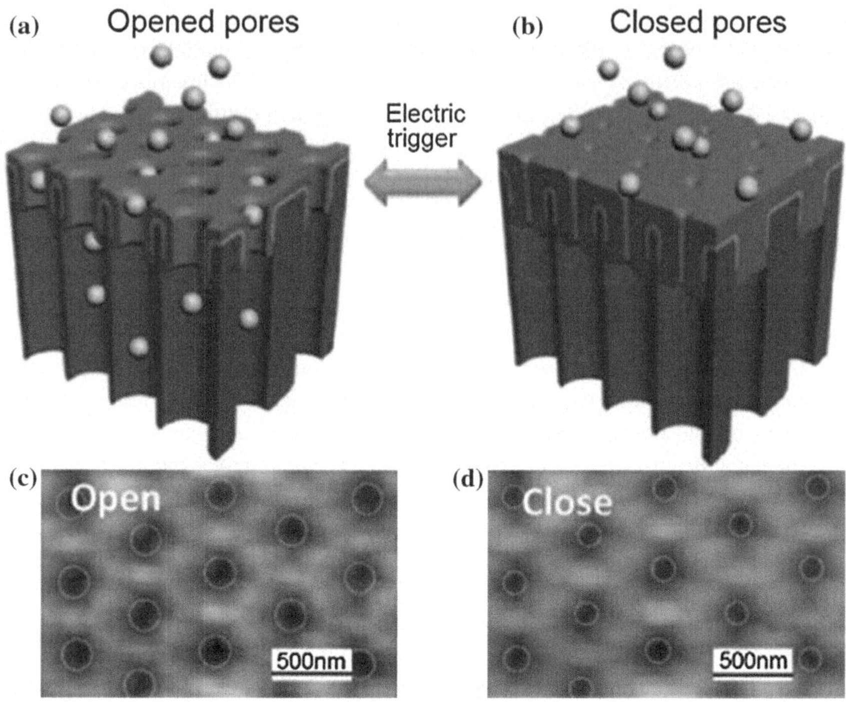

Fig. 11.5 Electrically responsive drug-releasing NAA chip. **a** Schematic diagram showing the mechanism of an electrically responsive drug-releasing chip based on NAA and PPy-DBS (reversible cycle). **b** Atomic force microscopy (AFM) images of these NAA chips before and after the application of voltage (pore mouth reduction) [35]

the aim of devising multiple or simultaneous transport of therapeutics to localized sites, it is inferred that NAA with a broad pore size distribution are more suitable than ordered mesoporous systems with single pore sizes such as MCM-silica, in terms of optimizing the delivery of more than one type of therapeutics [36, 37]. Based on various studies that have demonstrated the release of pharmacological agents in response to electric field, magnetic field, pH, temperature, or light intensity, NAA has the potential to be used as a smart DD carrier for spatial-, temporal- and dosage-controlled DD for diverse medical therapies [7]. Smart DD systems are expected to revolutionize medicine by enabling individualized therapy, to manage unpredictable condition by sensing and delivering immediate response with appropriate countermeasure, ultimately to provide patients an effective cure [7, 35–37].

Several other concepts using polymeric micelles as drug nano-carriers loaded inside NAA pores for releasing single or multiple drugs and multiple-triggered drug release methods using external sources such as applying magnetic field combining with the use of iron nanoparticles, ultrasound and radiofrequency (RF) field with the use of gold nanoparticles have been demonstrated in conjunction with studies on TNTs [11, 12, 38–41]. These studies show considerable potential of NAA to design complex drug-releasing implants with tailorable properties required for advanced localized DD.

11.3 Biocompatibility of NAA and NAA Nanotubes

Biocompatibility is the prerequisite for the application of new biomaterials and it is defined in terms of cellular response and tissue integration of implantable biomaterials. Whether or not these presented concepts for controlling drug release in NAA are feasible, their clinical applications is mainly dependent on their biocompatibility, which is the most critical question. In an orthopedic context, the inspiration for developing implants based on electrochemically anodized NAA and TNT originates from the fact that the human bone is, by nature, nanostructured and it comprises of inorganic particles within the same size range. Experiments reported in the past reveal that both nanostructured bone implants made from alumina and titania with nanophase surface topography and nano-fibers promote enhanced calcium deposition and alkaline phosphatase synthesis in osteoblast culture [42]. Aluminium and aluminium oxide are proven as biocompatible materials that are widely used in orthopedic prosthetics, dental and coronary stents [42].

The interaction between osteoblasts and NAA was reported by Desai's research team in several studies where the adhesion and proliferation of osteoblast on nanoporous anodic alumina, amorphous alumina, bare aluminium and glass were explored [43–47]. Their results proved that NAA improved the adhesion of bone cells as compared to other substrates, confirming the importance of the role of nanopore dimensions and surface chemistry on bone cell culture and growth. In recent study, using the Cercopithecus aethiops (African green monkey) kidney epithelial cell line (Vero cells), Poinern et al. demostrated the feasibility of NAA to be applied as a practical cell culture scaffold [48]. Optical microscopic observations after 24 h of cell cultivation on NAA showed good cell adhesion and coverage on NAA substrate surfaces (Fig. 11.6a) [48]. Vero cells were distributed over the entire surface and the cells were shown to generate ECM actively. The presence of filopodia distributed on the NAA surface and at the cell boundary were visibly seen from SEM (Fig. 11.6b, c). The cells were also found to be firmly attached to the NAA and ECM. It is observed that the NAA pores served as anchorage points for the cells. AFM images (Fig. 11.6d, e) disclosed not only the cells but numerous filopodia extending from the cells. To investigate if the Vero cell line can survive long-term, a cell proliferation assay was undertaken and compared to two other materials, i.e. glass (control sample) and a commercial NAA membrane (Whatman Anodisc). Results of a 72 h cell proliferation assay (Fig. 11.6f, g) showed absence of infection or toxicity on the cells. The highest number of viable cells was found on the NAA, whilst the lowest number of viable cells was found on the glass substrate (Fig. 11.6g). A direct comparison between the in-house NAA, the Whatman Anodisc and the control is presented in Fig. 11.6g, showing that in-house

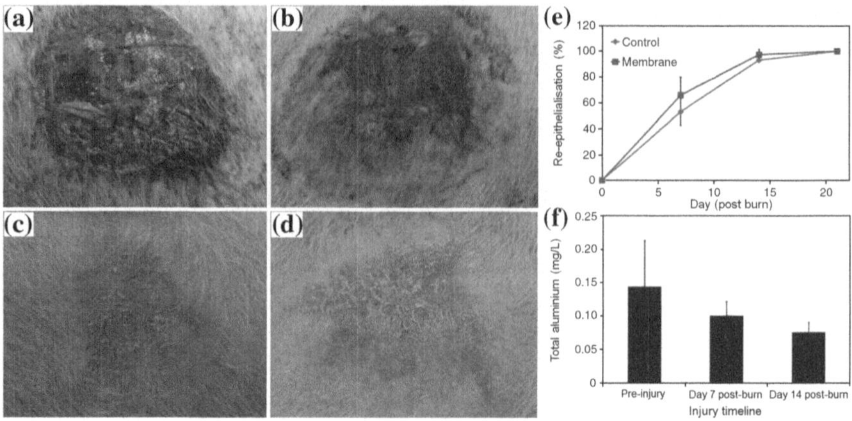

Fig. 11.6 **a** Optical microscopic image of Vero cells, **b** SEM micrograph, **c** cell attachment showing filopodia on NAA, **d** AFM images of cells and filopodia on NAA, **e** *angled view* of in-house NAA from AFM, **f** results of a 72 h cell proliferation study presented in *vertical bar charts*, and **g** comparison between the in-house NAA membrane, Whatman Anodisc membrane and the glass control in terms of viable cells over time [48]

NAA has the largest number of viable cells attached to its surface. Cell proliferation is predominantly influenced by the membrane type and surface roughness. Rougher surfaces of NAA are more favorable for cell growth. In short, the study demonstrated that the in-house NAA is suitable and was successfully used as a cell culture substrate for the Vero cell line, confirming its biocompatibility with this type of cell line.

Another study that elucidates positive cellular response was shown by Karlsson et al., where HOB cells were cultured on NAA membranes [49–51]. Two assays, Alamar Blue (Fig. 11.7a) and total DNA analysis (Fig. 11.7b) were carried out, showing osteoblast growth with increasing cellular activity for a week and increasing DNA levels for 2 weeks. After 1 and 2 weeks, cell growth of both assays decreased. Initial high levels of tritiated thymidine, [^{3}H]-TdR incorporation (per mg DNA) were observed for both controls and the NAA (Fig. 11.7c). The maximum proliferation was detected at day 3. As for the ALP expression, it peaked after 2 weeks, implying that the osteoblastic phenotype was retained on the NAA (Fig. 11.7d). After a week, a confluent cell layer was detected. It can be seen that cell filopodia attached to NAA pores from SEM and TEM micrographs in Fig. 11.7e, f, respectively. A close-up image indicates that cells in contact with the NAA membrane had entered the NAA pores (Fig. 11.7g) [49], proving that the NAA is favorable for osteoblastic cell growth, with cells rapidly proliferating, flattening and adhering firmly to the membrane surface. Nevertheless, further work remains to be carried out to confirm the absolute biocompatibility of NAA in different systems before it can be guaranteed safe for use in implantable DD.

In other studies, NAA has been proven to be non-toxic to cells and has been shown to be biocompatible in bone implants [52]. Also, numerous studies have

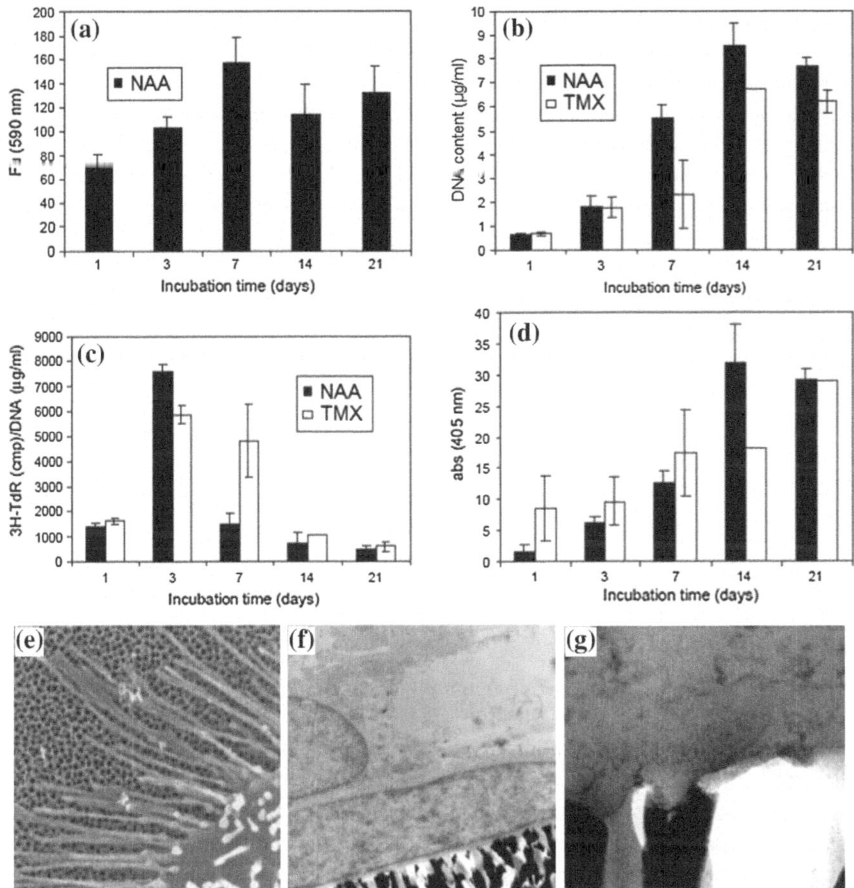

Fig. 11.7 **a** Alamar blue assay results from HOB cells cultured on NAA for up to 3 weeks, **b** DNA content for HOB cells cultured on NAA and control thermonox for up to 3 weeks, **c** [^{3}H]-TdR incorporation/DNA for HOB cells cultured on NAA and control thermonox for up to 3 weeks, **d** ALP activity of HOB cells cultured on NAA and control thermonox for up to 3 weeks. Each value represents the means of three samples, **e** SEM micrograph of HOB cells cultured on NAA for 1 day, **f** TEM micrograph of cross-section of HOB cells cultured on NAA for 3 weeks, **g** higher magnification of the image showing filipodia entering the NAA pores [49]

reported that NAA in the form of nanoporous coatings are safe for DD applications [21, 36, 53]. In some other cases of implant studies, there are still some controversies regarding the extent of the biocompatibility of NAA. A strategy to ensure non-toxicity is to impart a conformal coating of titania as a passivation layer over NAA to minimize corrosion and improve cell compatibility for implants used in dental, orthopaedic, and cardiovascular applications [43–45]. Studies involving atomic layer deposition technique, confirmed previous work carried out by A. Canabarro et al. that indicated that cells grown on TiO_2 surfaces contaminated by small amounts of NAA exhibited impaired cellular activity; for example, contamination by NAA may lead to impaired mineralization of matrices by osteoblasts bone cells [53–55].

Although alumina and NAA material are accepted as a biocompatible material, in the case of NAA nanotubes, there are potential concerns of their toxicity caused by their nano scale dimensions. Recent studies have shown that typically 1D structures that are longer (>20 μm) are more toxic than the shorter ones (<10 μm) [56]. There is a strong indication that the toxicity of nanomaterials, in particular those made from long rigid microstructures (like asbestos) in fibre or particle form is highly affected by their length, but still there are controversies [57, 58] on the key structural and chemical parameters causing toxicity on nanoscale levels. In order to explore the potential toxicity of NAA nanotubes as drug delivery carriers for cancer therapy, NAA nanotubes with 600 nm length were initally tested for their toxicity on breast cancer cells (MDA-MB231-TXSA) and immune response cells (RAW264.7 macrophages) [18, 59]. Light microscopy images showing the morphology of RAW264.7 and TXSA cell lines with NAA nanotubes at the maximum concentration (100 μg mL^{-1}) are presented in Fig. 11.8 (left panel). It was noted that NAA nanotubes tend to surround cells without affecting their growth. After 5 days of incubation, the number of cells increased significantly in both cell lines, revealing that NAA nanotubes do not affect cell viability, which confirms their biocompatibility. The interactions between AANTs and cells after incubation overnight with 100 μg mL^{-1}, NAA nanotubes were imaged by TEM in order to examine their possible internalization (Fig. 11.8c, d). TEM images proved that NAA nanotubes were readily internalized by macrophage cells and breast cancer cells. For macrophage cells, all the visible nanotubes were localized within numerous autophagic cellular vacuoles. The fusion of autophagosome with autophagic vacuoles can be found within the image area; while autophagic vacuoles containing cellular debris were also identified which demonstrated the successful

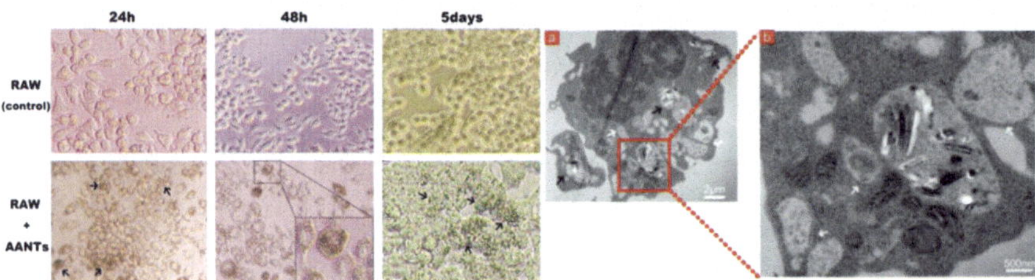

Fig. 11.8 *Left* Light microscope images of NAA nanotubes (100 μg mL^{-1}) incubated with RAW264.7 macrophages cell lines. *Black arrows* indicate that cells are closely surrounded by AANTs. Note that the cells covered by NAA nanotubes have darker color due to the optical light refraction and adsorption of anodized alumina. No cell viability inhibition was found during the 5-day experiment. *Insets* show magnified views of individual cells after 48 h. *Right* (*a, b*) TEM images of NAA nanotubes (100 μg mL^{-1}) internalized by RAW264.7 macrophage cells. The black arrows identify nanotubes inside cells; white arrows indicate the fusion of autophagic vacuoles, in which cellular debris were located inside [18]

cell uptake of AANTs (Fig. 11.8c, d). In the following study, the influence of the length of NAA nanotubes (700 nm, 2.5 μm and 5.8 μm) was studied with more details using macrophage cells and breast cancer cells as models [59]. The study confirmed that the shortest AANTs with 0.7 μm average length and tunable inner/outer surface chemistry exhibited the lowest toxicity impact, making this novel inorganic nanomaterial a promising candidate for future drug delivery.

11.4 NAA for Diabetic and Pancreatic Treatment

Pioneering work to explore NAA for in vitro drug delivery, studying their ability to transport glucose, immunoglobulin G (IgG) and insulin relevant for diabetes was reported by Desai's research team (La Flamme et al.) [45, 60, 61]. NAA biocapsules with 46–75 nm pore size were prepared by the anodization of Al tubes and loaded with insulin-secreting MIN6 insulinoma cells embedded within a collagen matrix (Fig. 11.9a). As depicted in Fig. 11.9b, results showed that glucose was effectively released but the release of IgG was considerably hindered. The diffusion coefficients were lower than the free diffusion in water for both molecules. Specifically, the decrease was much more significant for IgG. Notably, limited nutrition access to the interior of the capsule impacted the insulin release. The release of insulin from the biocapsules lasted for up to 3 h, indicating that the encapsulated cells could produce new insulin in addition to releasing the as-prepared insulin. Others also reported on micromachined silicon membranes containing NAA as biocapsules for the immunoisolation of transplanted pancreatic islet cells [62]. The pores (measuring at 10 nm wide) effectively prevented cellular and humoral immune species from transporting through the NAA. These pores

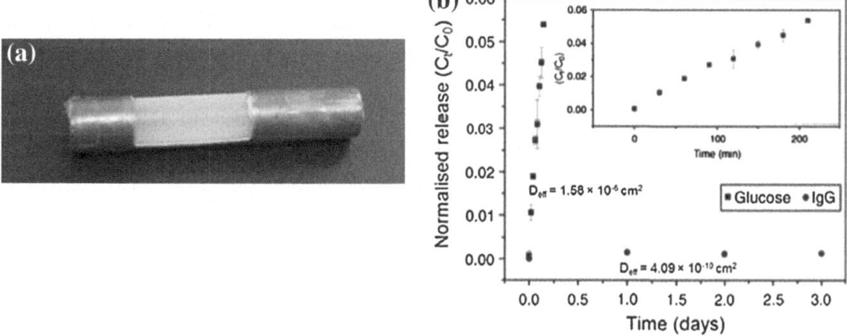

Fig. 11.9 **a** Real image of prepared NAA biocapsules and **b** normalized release of glucose (*squares*) and IgG (*circles*) with respect to time through a NAA membrane with a nominal pore size of 75 nm. C_t concentration at a point in time t, C_0 loading concentration at t = 0. *Inset* displays glucose release on a 210-min time scale [45]

permitted sufficient access of oxygen and nutrients to maintain cell viability. However, the drawback is that these devices must be surgically removed after use if implanted, similar to previously reported case study [45].

11.5 NAA Applications in Orthopaedic Prostheses and Implants

Bones play a key role in many critical functions in human physiology. Bone-related diseases and traumatic injuries not only have a direct impact on the quality of life and health of patients but also consume in total 10 % of the annual healthcare expenditure. These medical conditions are treated by complicated orthopaedic surgical procedures, including total joint replacement, spine fusion and the repair of fractures. In the biomedical field, joint bone diseases summon more and more attention, as today's estimate shows that 90 % of the population over the age of 40 suffers from a degenerative joint bone disease [62]. In 2000, the number of total hip replacement operations was about 152,000 which represents a 33 % increase from the number of operations in 1990 and also represents about half of the estimated number of operations by 2030 [63]. Typically, orthopaedic implants are based on titanium and its Al alloys in the form of screws or plates to hold the broken bone fragments together. Bone infections are the most common clinical complication associated with these implants that can lead to an implant failure and rejection. Also, it can cause substantial suffering and even the death of patients. In addition to titania nanotube (TNTs), the NAA-based implants on existing Al based metal medical implants with different shapes have been considered for biomedical applications [7, 8, 12].

To prove the orthopaedic applications of NAA, extensive studies were performed exploring the effect of NAA on osteoblast cells, which showed that human osteoblasts cultured on NAA maintain a physiological phenotype [47, 49]. Pujari-Palmer et al. investigated the effect of nanotopography on cell behavior, mainly, osteogenic differentiation on NAA, used as an inductive substrate with varied pore sizes (20, 100 and 200 nm). A murine bone marrow stromal cell line, W20-17 was used [64]. The efficacy of NAA to release an osteoconductive agent-bone morphogenic protein-2 (BMP-2) along with cell proliferation, osteogenic differentiation assays and gene expression were studied in vitro [64]. The influence of nanopore structure on cell morphology was studied by [65] Hoess et al. and summarized in Fig. 11.10, showing the different morphology of cells grown on NAA with diameters of 20 and 200 nm. It was found that the cells have more elongated morphology in NAA with 200 nm pore diameters, indicating that larger pores enhance cell spreading, whilst cells on NAA with a 20 nm pore size have a round-shaped morphology. In Fig. 11.10c, proliferation was observed to be higher on the control surfaces (wherein tissue culture polystyrene, TCPS, was used) compared to nanoporous surfaces. Figure 11.10d shows that there was no significant difference in terms of cell amount amongst the different pores, whereas in

Fig. 11.10e, the total ALP values were normalized with intracellular protein content, showing higher levels of ALP activity on 200 nm surfaces compared to 100 nm and control surfaces by 14 days of culture. A strong trend was found by day 14 with samples cultured with BMP-2, indicating higher levels of ALP activity with an increase in pore size as compared to the control. Figure 11.10f indicates an increased differentiation rate on 200 nm surfaces as compared to 100 nm surfaces ($p < 0.05$) on substrates without the addition of BMP-2. Differentiation rate seemed to be levelled for all surfaces, with the addition of BMP-2. A higher level of osteocalcin (OC) gene expression on the 200 nm surfaces compared to 20 nm surfaces ($p < 0.05$) was observed in Fig. 11.10g and the same trend can be seen with cells exposed to BMP-2 [65].

The initial in vitro interaction of osteoblasts with nano-porous alumina conducted by Karlsson et al. shows that NAA imparts no negative effect on human osteoblast cell activity, even with approximately 0.03 wt% aluminium ion leakage after 9 days of studies [50]. After 4 h of incubation, it was noticed that NAA with a grain size of less than 49 nm could yield 52 % better osteoblast adhesion than NAA with a grain size more than 67 nm. When phosphoric acid is used for the anodization of NAA, residual phosphate ions trapped within the nanopores may facilitate the mineralization of bones and the integration of implant within the tissues in orthopaedics. Moreover, NAA emerges as one of the components constituting ceramics besides calcium phosphates, silica, zirconia and titania due to their positive interactions with the human tissues [66]. It was studied alongside tricalcium phosphate (TCP) particles in casein and soybean protein scaffolds [67] used as alternative composite materials for CaP-ceramics and cements in biomedical applications due to its mechanical reinforcement.

Most of the aforementioned studies on the applications of NAA in orthopaedics and drug delivery were performed through in vitro and bone cell experiments, which are different from a real clinical and real biological environment. However, in vivo applications present many challenges such as the use of an appropriate method to monitor the distribution of drug molecules from the implant to the bone structure. To address this problem, our group pioneered a new method for ex vivo study of drug release in bones using The Zetos™ bone bioreactor. In this system, implantable NAA wires loaded with drug are inserted directly into a bovine bone core and drug release was monitored by fluorescence imaging (Fig. 11.11, left) [68]. A series of drug concentration profiles in bone were obtained from these images, which showed the capability of this approach to measure the distribution of drug molecules in bones across all directions (x, y, and z axes) from the NAA wire implant. This study also demonstrated that wire or needle-like NAA implants can be directly inserted inside bones and used as drug-releasing implants to provide the required drug dosage and release for bone therapy. More importantly, our proposed implant insertion approach and contact with the bone did not affect the adjacent bone cells viability and the robust NAA structure (Fig. 11.11, right). It is expected that this approach can be adapted in clinical conditions without surgical intervention and could provide significant advantages compared to existing drug releasing implants, which are surgically very invasive. Their use as drug-releasing

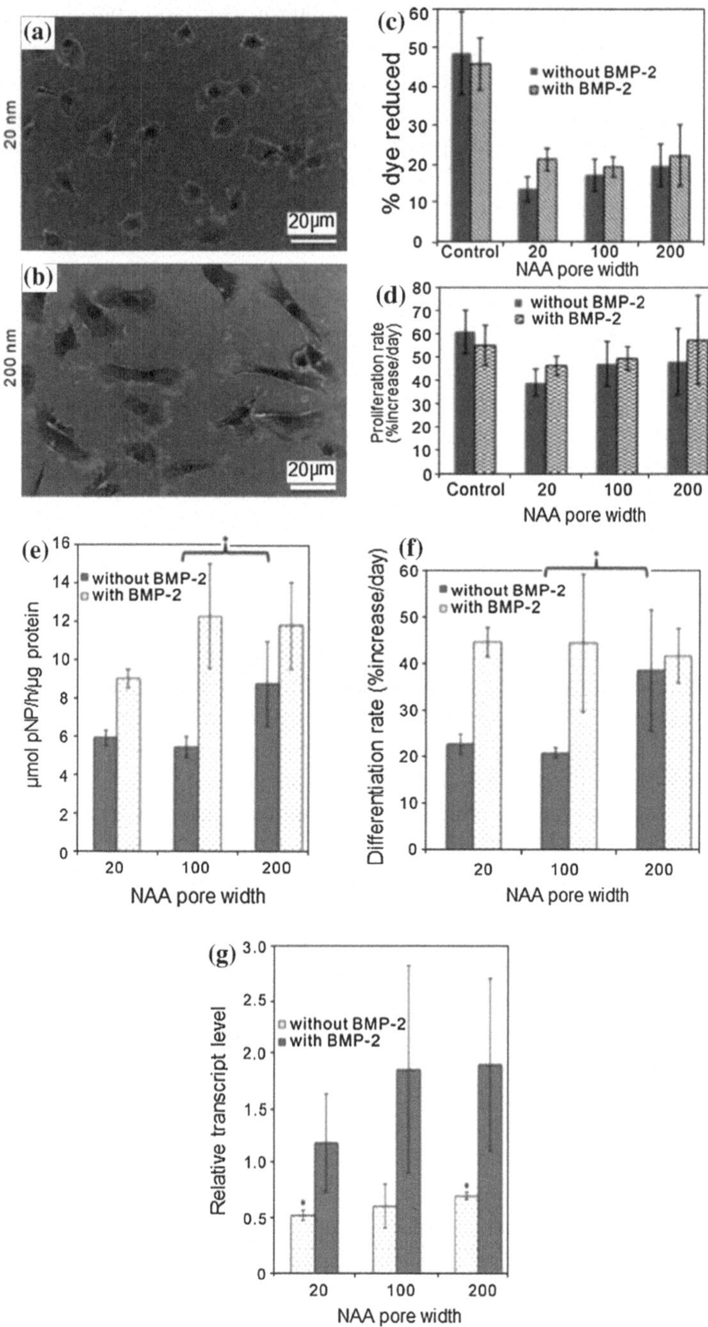
(a)
20 nm
20μm
(b)
200 nm
20μm
(c)
% dye reduced
without BMP-2
with BMP-2
Control
20
100
200
NAA pore width
(d)
Proliferation rate (%increase/day)
without BMP-2
with BMP-2
Control
20
100
200
NAA pore width
(e)
μmol pNP/h/μg protein
without BMP-2
with BMP-2
20
100
200
NAA pore width
(f)
Differentiation rate (%increase/day)
without BMP-2
with BMP-2
20
100
200
NAA pore width
(g)
Relative transcript level
without BMP-2
with BMP-2
20
100
200
NAA pore width

◄ **Fig. 11.10** Cell morphology: W20-17 cells cultured for 2 days on **a** 20 nm NAA and on **b** 200 nm NAA. **c** Cell proliferation was analyzed at day 7 using Alamar blue. **d** Cell proliferation rate was determined by analyzing the increase in cell number from 2 to 7 days. **e** Alkaline phosphatase (ALP) activity of polystyrene (control) and NAA surfaces (20-, 100-, 200-nm) with and without the addition of BMP-2 was measured after 14 days of culture. **f** Cell differentiation rate was determined by analyzing the increase of ALP enzyme activity for 12 days from day 2 to day 14. **g** OC gene expression with and without BMP-2 was analyzed at 14 days [65]

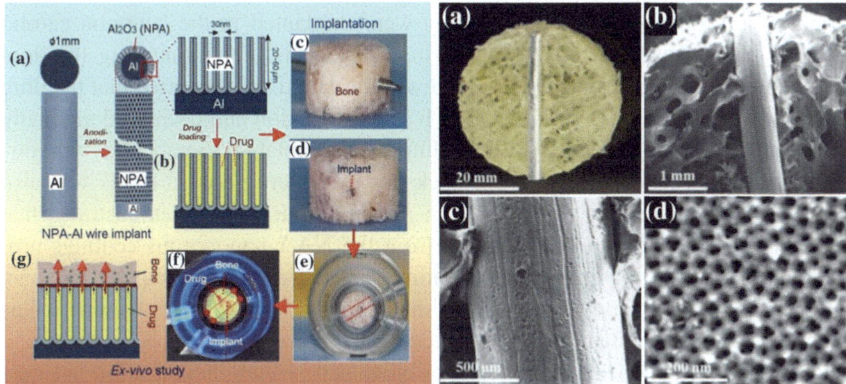

Fig. 11.11 *Left* Scheme of prepared Al wire implants with NAA layer on their surface showing their implantation into ex vivo bone model and in situ fluorescence drug release measurement using bone reactor Zetos. *Right* (**a**, **b**) SEM images showing porous bone structure with placement of the NAA-Al wire implant showing (**c**, **d**) stable and unchanged surface morphology and nanopores structures of NAA [68]

implants for clinical applications to treat a wide range of bone diseases, including bone infection, bone inflammation, fracture management, bone healing and bone cancer is very promising [7, 69, 70].

11.6 NAA Applications for Heart, Coronary and Vasculature Treatment

In interventional cardiology, coronary stent implantation has been considered superior to conventional angioplasty, but in-stent restenosis still remains a crucial challenge [70, 71]. Inflammatory infiltration and foreign body reactions in the tissue surrounding the struts, along with thrombogenesis and hyperplasia of intima are the major issues that emerge in post-implantation [72]. To overcome these problems, new stent designs and coating were explored. One proposed strategy is by imparting a layer of passive coating or to prepare an active layer on the stents to improve biocompatibility and to suppress neointima proliferation by the local delivery of antiproliferative and immunosuppressive drugs [73]. Although polymer-coated stents have been clinically approved, they can lead to delayed inflammation, which limits their use. As a result, drug-loaded inorganic NAA, amongst several other materials, have been tested to counter the inherent drawbacks of stents coated with polymers. In vivo drug delivery was studied by Wieneke et al. in the area of vasculature, especially diseases pertaining to the heart and coronary artery [71, 74]. Stainless steel coronary stents were coated with NAA and infiltrated with an immunosuppressive drug, tacrolimus, weighing 60 and 120 μg. To prepare the nanoporous coating, 316L stainless steel stents were deposited with aluminium before anodization was carried out. They were implanted in the common carotid artery of New Zealand white rabbits to treat restenosis. For monitoring the drug release, the tacrolimus level in their blood was measured. It is reported that the drug concentration in the blood peaked after 1 h of implantation and decreased gradually in the ensuing 48 h, being well within the acceptable therapeutic window. A significant reduction of neointima thickness in the drug-coated stents was also observed. Nonetheless, in vivo studies in porcine models have demonstrated that the shedding of particle debris released from the NAA nanoporous coatings produces a significant increment of neointimal hyperplasia, as compared to bare stainless steel stents. Drug-loaded stents showed significantly low inflammation as compared to the uncoated ones, independent of the drug loading. Nevertheless, a long-term investigation on the biocompatibility and re-endothelialization are necessary before the practical usage of these implants can be validated.

11.7 NAA Applications in Dentistry

Alumina is one of the most used ceramic and composite materials due to its high chemical and thermal stability, high specific area and controllable microstructure. The advantages of using NAA biofillers include higher elastic modulus and fracture toughness. For the reinforcement of dental resin composite, high elastic modulus would be useful to increase the modulus of the composite to approach the modulus of tooth dentine (12–20 GPa) [75]. With regard to the fracture toughness, alumina presents values that are three to four times higher than for silicate or glass commonly used in composite resins. In vitro studies of a novel coupling-agent-free dental restorative resin composite based on fillers made from NAA was carried out by Thorat et al. [76]. The fillers were prepared by milling NAA membranes, followed by composites prepared with standard resin at a maximum loading of 50 wt %. Silver nanoparticles were also synthesized inside the NAA pores. Results show that it has better ageing (one-year equivalent) as compared to commercial materials. The alumina particles with interconnected nanopores allow good mechanical interlocking between the fillers and matrix without the need for chemical bonding. This NAA-inclusive composite can be made bio-active with pore filling using different agents as well as with increased biocompatibility, since elution of coupling-agent degradation products is omitted. This in turn, can eliminate chemically-triggered decrease of the stiffness of the fillers. Since NAA has

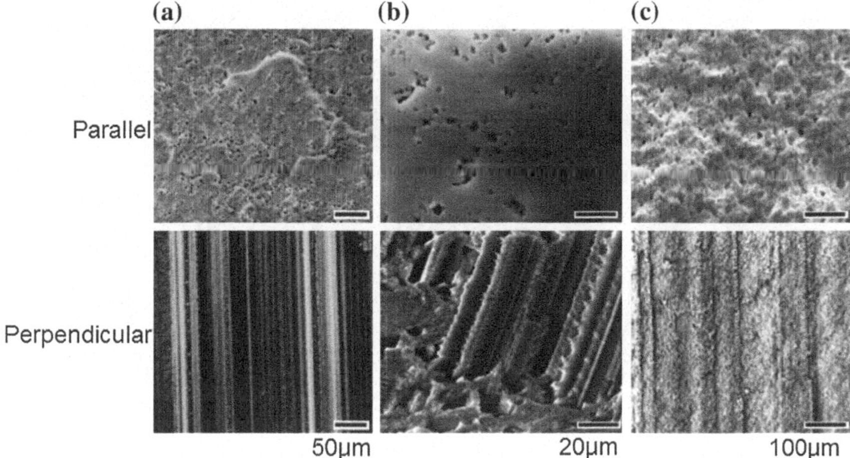

Fig. 11.12 SEM micrographs of the alumina samples with controlled specific surface (at least 10 times higher than that of the usual fillers) for the making of dental composites, with the parallel and perpendicular images (*top* to *bottom*) shown for materials (**a**) annealed at 1150 °C (**b**) and at 1300 °C (**c**) to the alumina block growth direction [77]

controllable porosity, it is suitable to serve as a drug carrier for in situ drug release besides being applied as dental fillers.

A study by Azevedo et al. reported the structural and surface treatment of experimental fillers before their use in the reinforcement of resin composites or resin-modified glass ionomer cements [77, 78]. In the study, porous alumina monoliths with high specific surface area stabilized by trimethyletoxysilane (TMES) were obtained using a novel preparation method. The treatment of the samples with TMES and their reheating at 1300 °C resulted in adsorption sites with stronger chemical bonds. The surface morphology of the materials is visualized by SEM and shown in Fig. 11.12. Results demonstrated that the extremely high porosity of the NAA monoliths enabled a rapid and homogeneous impregnation with a functionalized organometallic TMES. Nevertheless, the silanization of the mineral phase with a polymer-tuned silane and its mixture with a methacrylate or polyalkenoate matrix is required to be mechanically evaluated before it is verified to be safe for use in the reinforcement of resin composites, resin-modified glass ionomer cements, or cortical bone substitute in dentistry [77, 78].

11.8 NAA for Immunoisolation

Immunoisolation refers to the encapsulation of living metabolic cells or drug release system within a size-based semi-permeable membrane, to protect them from an immune reaction [79]. Immuno-protected cells can be used for treating hormone

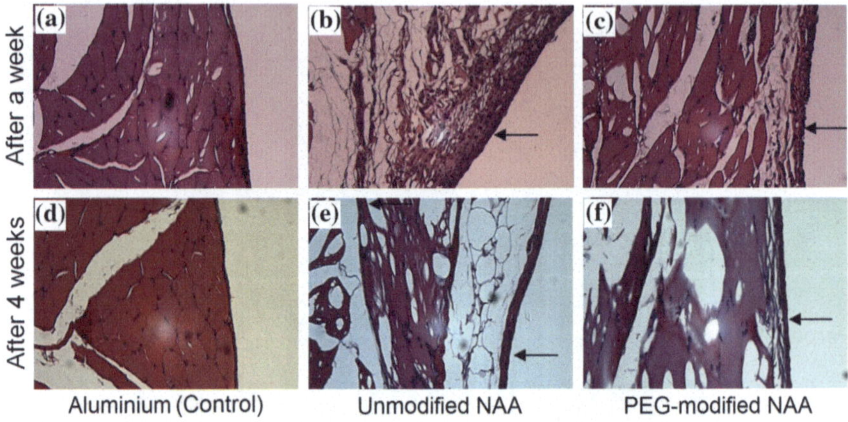

Fig. 11.13 Histological examination of tissue exposed to **a**, **d** aluminium without NAA layers, **b**, **e** unmodified NAA biocapsules and **c**, **f** PEG-modified NAA biocapsules after 1 and 4 weeks of study [84]

deficiencies arising from several diseases such as diabetes, Alzheimer's and haemophilia [80]. It was demonstrated that the NAA pores can selectively transport low-molecular-weight substances such as nutrients, electrolytes and oxygen to pass, whilst the transport of immunocytes, antibodies and other transplant-rejection effector mechanisms are impeded [81]. The immunoisolation studies on NAA applied in both in vitro and in vivo studies did not show any difference in the in vitro studies comparing aluminium and NAA substrates. However, upon comparing NAA to covalently grafted NAA with poly(ethylene glycol) (PEG) in the in vivo regime, it was found that the surface of NAA was polarized due to charged crystalline lattice defects [81]. These defects induced a net dipole moment that resulted in the adhesion of biological molecules on the NAA surface, consequently leading to pore clogging and surface fouling. However, unattached PEG groups that were freely moving in aqueous solution prevented biomolecules approaching the surface [82]. Figure 11.13 shows the histology results obtained for tissues exposed to (a, d) aluminium (control sample), (b, e) unmodified NAA and (c, f) PEG-modified NAA after 1 and 4 weeks [82]. In particular, arrows pinpointed the portion of the tissue that was exposed to the NAA biocapsule. There was moderate tissue inflammation on both the unmodified and PEG-modified NAA substrates. Slightly more inflammation was observed on the unmodified sample after 1 week of investigation. After 4 weeks, the observed decrease in the granulation layer as well as the existence of blood vessels in the tissue nearby the PEG-modified alumina infer that the inflammation was mainly due to the implantation procedure per se, not the alumina substrates [83]. In vitro testing carried out with IMR-90 lung fibroblasts on NAA biocapsules demonstrated no cytotoxicity or any significant complement activation. Extending the studies to the in vivo regime, untreated and PEG-coated NAA biocapsules were implanted into the abdominal cavity of rats. Results show that it induced a minimal transient inflammatory response (inherent to an implantation procedure) with no fibrous growth. Also, PEG coating reduced the interactions of serum albumin with the material, rendering it a perfect immunoisolation device [84]. However, one drawback in using NAA as an encapsulation device is its non-biodegradability after being exhausted and it has to be removed surgically after use.

In another study, NAA was applied for the time sequence of blood activation on platelets and complement system [76]. NAA with two pore sizes, namely, 20 and 200 nm were incubated with whole blood from 2 min to 4 h. Platelet adhesion and activation was monitored at different time points through the change in platelet number and the levels of thrombospondin-1 (TSP-1) in the fluid phase. The presence of CD41 and CD62P antigens on the material surface was evaluated by means of immunocytochemical staining, whereas complement activation was assessed by measuring C3a and sC5b-9 in plasma by means of enzyme immunoassays. Both NAA displayed similar complement activation time profiles, with levels of C3a and sC5b-9 increasing with incubation time. Platelet activation characteristics and time profiles were different between the two membranes. Platelet adhesion increased over time for the 20 nm surface, whilst the clusters of microparticles on the 200 nm surface underwent negligible change. The release of TSP-1 increased with time for both membranes, however, it was much later for the 200 nm alumina (4 h) as compared to the 20 nm membrane (60 min). To summarize, the NAA surface topography most likely affected protein transition rate, which in turn affected the material platelet activation kinetics [74].

11.9 NAA Application for Localized Chemotherapy

A major problem with the current systemic delivery of chemotherapeutics is the limited dosages of toxic anti-cancer drugs that can be achieved at the cancer site without causing serious adverse complications, such as cardio-toxicity. Surprisingly, the concept of localized drug delivery for cancer therapy has not been widely explored yet. So far, there is only one clinically proven system for the localized delivery of chemotherapeutics specifically developed for the treatment of brain tumor (glioblastoma). This implantable system is based on a biodegradable polymer impregnated with the anti-cancer drug carmustine (Gliadel, Guildford Pharmaceutics, UK) [85]. The Gliadel® wafer has a diameter of 2 cm composed of 192.3 mg of polyanhydride copolymer and 7.7 mg of anticancer drug carmustine [85]. The main disadvantages of the Gliadel® wafer are its large size which requires extensive surgical intervention preventing its use in many situations, and its single drug loading, carmustine. To overcome these limitations, our group has recently proposed two concepts of drug-releasing implants for localized treatment of cancers such as brain (glioma) and bone (osteosarcoma) using NAA wires implants and NAA nanotubes [19, 59, 68, 86].

A pioneering study on the use of NAA as nano-carrier for the drug delivery of anticancer therapeutics where Apo2L/TRAIL is used as a model drug is reported by Wang et al. [19]. The schematic is presented in Fig. 11.14 which summarizes the loading of NAA nanotubes with Apo2L/TRAIL and its mechanism of release to induce cancer cell apoptosis. NAA designed with an optimized dimension, that is, a uniform length of 600 nm and outer pore size of 100 nm were prepared to achieve high Apo2L/TRAIL loading (104 μg/mg), which is comparably higher than other nanocarriers. In vitro Apo2L/TRAIL release and delivery on cells showed that around 40 % of Apo2L/TRAIL was released after 30 min, followed by a sustained release pattern extended for 240 min (Fig. 11.14b). No burst release was evident under these conditions. Significant decrease in the viability of MDA-MB231-TXSA cancer cells due to cell apoptosis was demonstrated. Cell viability decreased significantly in a dose-dependent pattern after 165 min of incubation with Apo2L/TRAIL loaded AANTs (Fig. 11.14b). The decrease in cell viability was concomitant with a dose dependent increase in caspase-3 activity. Even at the lowest dose of 1 μg mL^{-1} AANTs, caspase-3 activity was significantly higher than in those cells treated with soluble Apo2L/TRAIL, indicating effective cell apoptosis induced by Apo2L/TRAIL delivered from AANTs. Morphological changes characteristic of cancer cells apoptosis assessed by light microscopy were clearly evident and shown in Fig. 11.14c. These findings demonstrate that NAA nanotube is a promising biomaterial carrier for drug delivery applications and localized cancer therapy.

11.10 NAA for Tissue Engineering and Skin Therapy

Tissue engineering aiming to grow new tissues or organs from cells and scaffold to produce a fully functional organ for implantation back into the donor host is a growing research area. It has shown significant progress in past years for the repair or replacement of portions of or whole tissues/organs such as bone, cartilage, blood vessels, bladder, skin, muscle etc. Among many strategies used for cell and tissue engineering, the ability to direct and guide cell behavior by topographic modulation of surface is considered as very promising. Recent studies showed that the surface topography of substrates, specifically at nano scale, on which cells are cultured can affect the cell adhesion, migration and proliferation. NAA have been extensively used as nanostructured substrates to culture different types of cells, such as osteoblast [87], neutrophil [88], hepatoma [89] and marrow stromal [50] with successful attachment, proliferation and production of extracellular matrix proteins [91]. These studies indicate the importance of interaction of nanopore structures and cell that can be used for tissue engineering and potential wound healing.

Burns are one of the most devastating injuries in medicine where delayed treatment can lead to excessive scarring that can have negative physical and psychological effects on the patient. The feasibility of NAA application in burn injury and wound healing is initially demonstrated by Parkinson et al. using NAA to

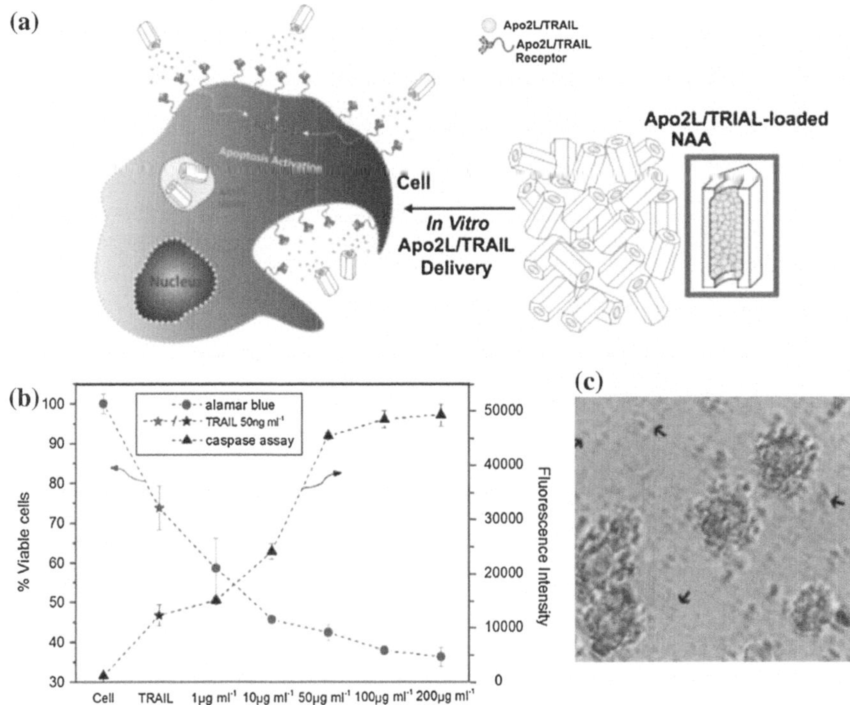

Fig. 11.14 **a** Schematic illustration showing the fabrication and in vitro drug delivery of NAA nanotubes for cancer therapy. It shows Apo2L/TRAIL loading inside nanotubes and that apoptosis mechanism kills breast cancer cells (MDA-MB231-TXSA), **b** In vitro drug release profile of NAA nanotubes loaded with Apo2L/TRAIL in PBS solution at 37 °C and Alamar blue and caspase assay after 165 min of treatment. **c** Cell apoptosis image induced by 100 μg mL^{-1} Apo2L/TRAIL released from NAA nanotubes after 120 min treatment. *Black arrows* denote AANTs in the media surrounding cells [19]

culture pig skin derived cells, namely, keratinocytes [91]. Cell morphology, proliferation and migration were found to be affected by the pore size of NAA which is summarized in Fig. 11.15. The cell attachment on NAA with smaller nanopores (41 nm) was found to have mostly spherical shaped cells. In Fig. 11.15b, numerous filopodia that are irregular (and about 60 nm) in length, together with lamellipodia, were seen extending from the cell body lying across the NAA pores. Similar observations were recorded for keratinocytes cultured on NAA with a pore diameter of 114 nm, with the filopodia seen to enter the NAA pores (Fig. 11.15c, d), confirming an unhindered interaction and an intimate contact of keratinocyte with NAA through many filopodial-pore interactions. Figure 11.15e shows that there was no significant difference in absorbance values amongst the NAA samples with different pore diameters. However, in Fig. 11.15f, cell migration was considerably faster across the 300 nm membrane as compared to NAA with smaller pore sizes. There was no difference in the rate of change for migration for the 41 and 58 nm

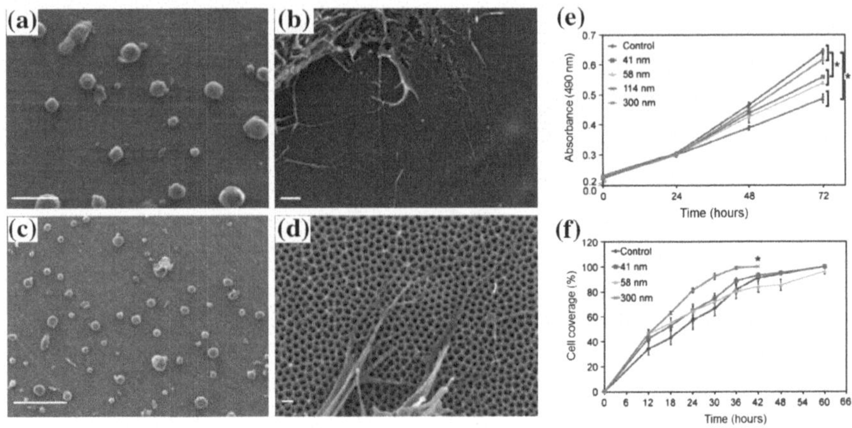

Fig. 11.15 The proliferation and migration of Keratinocyte on the NAA with nanopores (**a**, **b**) 41 nm in size (Scale bar of: 100 μm, 1 μm). **c**, **d** 114 nm in size (Scale bar of: 200 μm, 200 nm). *Graphs* of **e** absorbance versus time for NAA with different pore sizes, **f** cell coverage in percentage versus time for NAA with different pore sizes. Values are the mean of three replicates (*$p < 0.05$ as significant) [91]

NAA samples as compared to the control. The NAA with 300 nm pore diameters was found to be the optimal sample for wound repair application since it yielded increased proliferation and migration over the other samples [66]. The study also confirmed that the NAA was stable in cell culture, with no negative reactions/infections or degradation observed throughout the culture period without pore blockage caused by the cells or the cell culture medium.

To demonstrate the effect of the 300 nm NAA membrane on wound repair after burn injury, a deep dermal partial thickness (DDPT) burn model in the pig14 was explored by Parkinson et al. [91]. A DDPT burn model was selected in this study as these burns are the most common injury in pediatric hospitals and are the most difficult to manage. NAA is applied as interactive dressing for burns and changes of morphology of healing are presented in selected image in Fig. 11.16, showing the control (not treated) and NAA treated burns for 21 and 70 days. Although gross morphology of healing wounds was similar for both control and NAA treated samples, better recovery was seen in NAA as re-epithelialization occurred after a week from the wound surface and edges, with further epithelial growth after 2–3 weeks (Fig. 11.16c, d). Histology showed that wounds were re-epithelialized by day 21 and there were no signs of infection in any of the biopsies. After 8 weeks, slightly contracted and purple scars were observed for both wounds, which continued to mature to mostly flat scars by day 70 when animals were euthanized (Fig. 11.16b–e). All wounds were fully re-epithelialized (i.e. to 100 %) after 3 weeks post-burn. To evaluate NAA stability and to observe systemic absorption of Al from the NAA membranes by analyzing the total aluminium content by

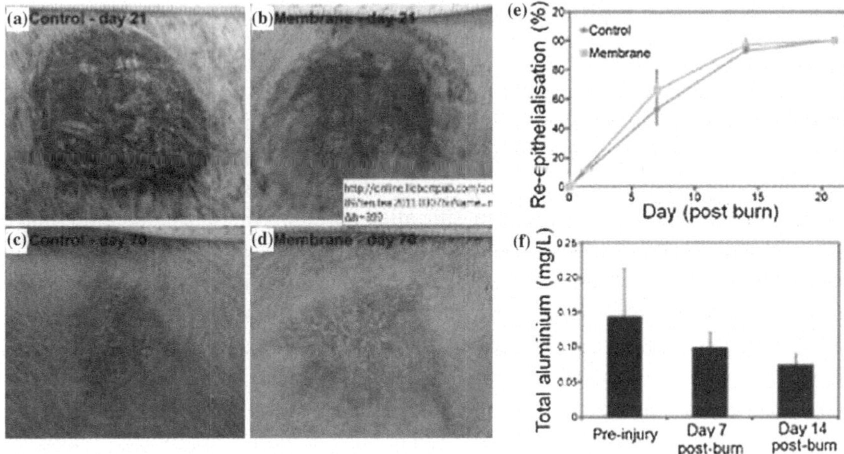

Fig. 11.16 Morphology of wound healing **a**, **c** digital photographs of a control-treated wound at time of burn injury at day 21 and day 70 post-burn respectively. **b**, **d** Matched NAA membrane-treated wound at corresponding time points [82]. *Graphs* showing the **e** average re-epithelialization (% original wound area) for control and NAA-treated wounds (data points are means of the standard error). There was no significant difference among treatments ($p > 0.05$), **f** total aluminium content (mg/L) in blood plasma collected pre-injury and $t = 7$ d and $t = 14$ d post-burn. There was no difference in aluminium concentration after contact with NAA [91]

ICP-AES, blood samples were collected during pre-injury and at days 7 and 14 post burn. Overall, there was no difference in the amount of total aluminium from pre- and post-membrane blood samples (Fig. 11.16f), indicating that NAA is non-toxic and non-immunogenic, as wounds were healed with abundant cellular proliferation and migration. Because oxygen and carbon dioxide can be maintained through the NAA membrane, the wounds treated with NAA appeared moist, indicating that wound dehydration can be prevented. An enhancement of wound healing process on pig skin with less organizing granulation tissue and more mature epidermal layers have shown that NAA together with secondary dressings are capable of creating a wound-healing environment, as no signs of bacterial infection were detected during gross examination of the wounds or from the biopsies collected during wound healing. Since many physicochemical parameters for designing NAA can be set and pre-programmed, an optimal surface shape can be facilely prepared for cellular adhesion sites for better cell attachment and proliferation [91]. In conclusion, this study has demonstrated an effect of nanotopography on keratinocyte phenotype and wound healing, and thus there is a need to consider nanoarchitecture in future wound healing therapies.

11.11 NAA for Cell Culture and Imaging

An agar plate or Petri dish is used as conventional and robust growth medium for cell cultivation. However, there are some drawbacks of this system such as poor facility for automation, slow production of results, and generation of waste (including pathogenic or genetically modified organisms), short shelf life and limited storage time of organisms in situ after culture. Alternative methods have been explored, such as porous membranes to allow for in situ staining, flexibility in nutrient environment of the cells and the culturing of intractable species [92, 93]. However, these devices cannot produce sufficient growth compartments. It is difficult to design flexible membranes at the micrometer scale and commercially, micro-engineered mechanical systems (MEMS) devices are far more costly. NAA is recognized as an excellent microbial culture support to address these limitations, due to its nonporous size (20–200 nm in diameter), remarkable porosity, inertness, thermal stability, ability to separate and retain microorganisms on its rigid surface whilst allowing nutrients to pass [88, 94, 95].

NAA was demonstrated as the versatile platform to engineer up to one million growth of discrete compartments of miniaturized microchips prepared by means of the MEMS approach to culture a high number of segregated microbial samples at an unprecedented density and a high culturing efficiency [96]. For instance, in high-throughput screening of >200,000 isolates from Rhine water based on metabolism of a fluorogenic organophosphate compound, the recovery of 22 microcolonies with the desired phenotype was procured. These isolates were envisaged to include six new species in accordance to rRNA sequence, implying significant impact of NAA as readily manufactured chips to aid full automation and multiplexing of microbial culturing, screening, counting and selection [96]. The binding kinetics can be acquired for potential drug screening operations using NAA, e.g. kinase interacting with a peptide and the inhibiting effects of kinase on the substrates [97]. To facilitate the incorporation of cell-specific detection reagents and cell-capturing agents to recruit, culture and detect selective types of bacteria or other microorganisms of similar sizes, the surface of NAA can be facilely modified or decorated with different functionalities. On the basis of the generic properties of NAA, it can be used to culture microorganisms on agar in Petri dishes as well. Bacterial proliferation shows that their growth on NAA nurtured with the appropriate nutrients is equally as effective as in liquid medium [95–98]. Moreover, microorganisms that cannot be grown on conventional growth media (because the media in which the agar is in is too difficult to prepare or unstable) can be cultured on NAA. For example, agar cannot remain gelled for the growth of the *Archaeal extremophile Sulfolobus sulfotaricus* at 80 °C and pH < 4. Contrarily, it is feasible for the microorganism to thrive in the media placed in a NAA matrix. NAA has therefore gradually become a vital part of the porous cell culture support systems for the purpose of screening new bacteria species in the soil and river [98]. Finally, it is demonstrated that NAA can be used as the substrate to grow microbes in close contact with their natural habitat and a higher cultivability can be achieved [97, 99].

Based on clinical culture of antibiotics on NAA, fluorogenic dye staining tests and image analysis have been recently proposed. They are applicable to rapidly growing pathogens, including the *Enterobacteriaceae and methicillin* resistant *Staphylococcus aureus*, in which results can be produced rapidly within hours [96]. Such benefit can be seen from the time required for susceptibility test for growing organisms such as *Mycobacterium tuberculosis* on NAA, that was reduced from >10 days to 3 days [100]. Real-time monitoring of microcolonies growth for *M. Tuberculosis* is also achievable with NAA, in which the effect of the addition or removal of drugs during the culturing process is feasible [96]. In order to culture microorganisms with discrete growth, it is vital to fabricate NAA with structured and organized divided compartments as shown in Fig. 11.17a. Since microcolonies are small, sample density can be rendered high and effectively imaged (Fig. 11.17b–h) [96]. New properties may be introduced to NAA substrate by coupling functionalized inorganic nanoparticles, polymer or organic monolayer to its surface. As shown in Fig. 11.17i, j, a modified NAA membrane was created by covalently coupling fluorophore-loaded silica nanoparticles to the surface. As the fluorophore displays pH-dependent fluorescence, the resulting surfaces can be used for the screening of strains of acid-producing bacteria, with the fluorescence intensity of the nanoparticles inversely proportional to the pH.

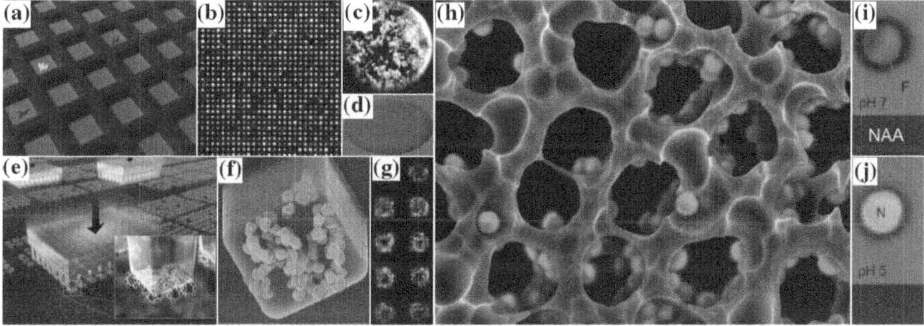

Fig. 11.17 **a** *Green* and *yellow* microcolonies of bacteria are shown on high throughput screening culture chip on NAA with a grid base, **b** microcolonies arrays in 20 × 20 μm culture wells with over a thousand microcolonies (*white*, each containing N50 bacteria) from a chip with 180,000 culture areas, **c** single well from alternative version of a culture chip showing a single 180 μm wide culture area with bacteria labelled with the fluorogenic dye Syto 9 and imaged by fluorescence microscopy, **d** SEM of culture well with the same size as panel *C*, viewed from above at a 45° angle, **e** printing method with PDMS stamp dropped from a few microns height to ensure even contact with the NAA base. Bacteria with quantum dots used to label the surface was shown, **f** PDMS pin loaded with fungal spores (*coloured green*), **g** image of 40 × 40 μm compartments printed with bacteria using a PDMS stamp, the bacteria were labelled with Syto 9 before printing and visualized by fluorescence microscopy. NAA culture chip with intrinsic nano-detection reagent, **h** SEM of NAA with up to 200 nm pores and customizable spherical nanoparticles (80 nm in size, colored *blue-green* for better visibility) loaded with pH sensitive reporter dye, **i** coupling chemistry used to attach nanoparticles to the NAA with an alkyne surface chemistry, **j** increase in fluorescence of nanoparticle by decreasing pH [96]

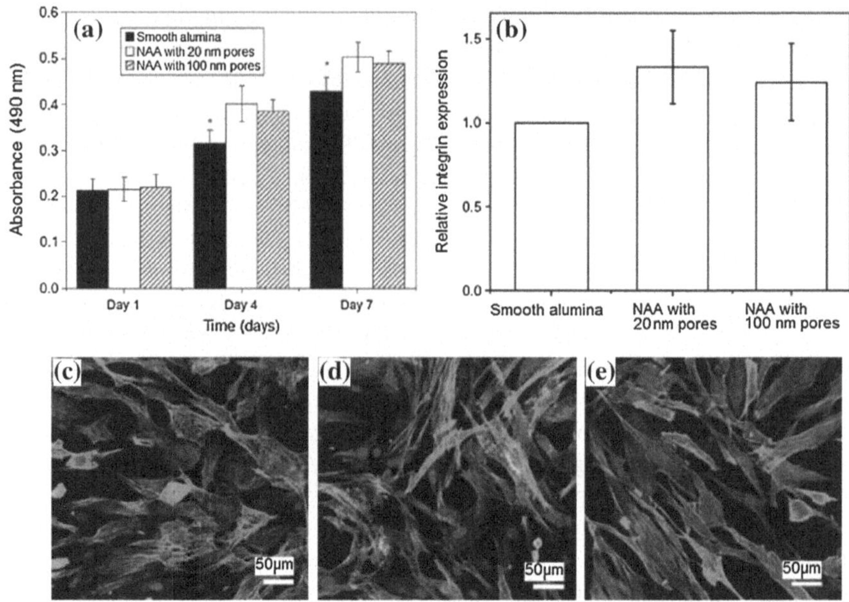

Fig. 11.18 **a** Effect of NAA substrate on cell viability. Mesenchymal stem cells (MSCs) were cultured on NAA or a smooth alumina surface for 1, 4, and 7 days and cell viability was measured using an MTT assay. $^*P < 0.05$ denotes a significant difference between the smooth alumina and nanoporous alumina. **b** Quantitative polymerase chain reaction analysis and **c–e** of integrin β1 expression in mesenchymal stem cells adhered to different NAA substrates. Immunofluorescence staining of cells on (**c**) smooth alumina (**d**) NAA with 20 nm pores (**e**) NAA with 100 nm pores, after incubation for 2 days. Cells were triple-stained with actin filaments (*green*), cell nuclei (*blue*) and integrin β1 (*red*). The expression level of integrin β1 was normalized to glyceraldehyde 3-phosphate dehydrogenase and calculated using the standard curve method [100]

The potential of NAA for orthopaedics implants is well recognized, but the cellular responses of mesenchymal stem cells (MSCs) on nanoporous structures have not been thoroughly understood yet, since literature on this topic is scarce. MSCs were cultured on smooth alumina substrates and NAA with 20 and 100 nm pore sizes to study the interaction between surface topographies of NAA and cellular behavior to evaluate the effect of their pore sizes on MSCs as measured by proliferation, morphology, expression of integrin β1 and osteogenic differentiation [100]. Figure 11.18a shows the typical absorbance of the MSCs with respect to time. MSCs cultured on NAA substrates with either 20 nm pores or 100 nm pores showed significantly higher cell viability ($P < 0.05$) than those cultured on smooth alumina after incubation for 4 days and 7 days. The expression of integrin β1 was enhanced in MSCs cultured on porous alumina (Fig. 11.18b), demonstrating that NAA was more favorable for cell growth than smooth alumina substrates. MTT assay that was used to evaluate the cell viability of MSCs shows that cell viability is inversely proportional to NAA pore size. Comparing Fig. 11.18c–e, higher levels of osteoblastic differentiation markers such as alkaline phosphatase, osteocalcin and mineralization were detected in cells cultured on NAA with 100 nm pores compared with cells cultured on smooth alumina or NAA with 20 nm pores. Extremely elongated cells and prominent cell membrane protrusions were observed in cells cultured on NAA with larger pores as seen in Fig. 11.18e. This work demonstrated that cellular behavior is affected by the variation in NAA pore size, demonstrating the potential application of NAA in tissue engineering [100].

11.12 Conclusion and Future Perspectives

This review has summarized the recent progress and key findings in the application of electrochemically engineered NAA materials for drug delivery and biomedical applications. This promising material and technology even at this early stage of development, showed many advantages and promises for translation into commercial biomedical applications. The fabrication process of NAA is based on a well-established and scalable technology using simple and low-cost electrochemical processes. As an example, NAA layers can be generated on existing medical implants (orthopaedic, stents) with no limitation in its shapes and forms, including stents, plates and wires. NAA were shown to possess a unique set of properties for drug delivery applications, including high surface area, controllable nanopore dimensions, tuneable geometries and surface chemistry, high and versatile drug loading capacity for a broad range of drug molecules with the ability to modulate drug release kinetics. NAA has shown to have many outstanding properties, including chemical and corrosion resistance, mechanical and thermal stability, rigidity and proved biocompatibility that have been explored for broad medical applications presented in this review, including orthopaedic and dental implants, heart/coronary/vasculature stents, immunoisolation, skin healing, tissue engineering and cell culture.

Nevertheless, to achieve successes of concepts presented throughout this review, more in-depth research is required to be carried out in order to make NAA based drug-releasing implants and biomedical devices feasible for real-life clinical applications. Although NAA and NAA nanotubes are reported to be bio-inert and non-toxic, more biocompatibility characterizations are required before their translation to clinical and human trial stages can be implemented. More studies are also required to confirm the efficiency and suitability of NAA for delivering other unexplored therapeutic agents including antithrombotic, antimigratory, antineoplastic, antimitotic, anticoagulants, anaesthetic agents, vascular cell growth promoters/inhibitors, vasodilation agents and their applications as biomedical devices such as a catheter, guide wires, stent/vascular grafts, vena cava filter, pacemaker and orthopaedic implant insertable into different parts of the body, e.g. oesophagus, trachea, arteries, colon, biliary tract, ureter or brain. Finally, several exciting future development are expected. This includes the integration of TNTs implants with sensors, microchips, the development of advanced triggered drug release system, the application of TNTs implants for combination therapy and the potential for localized therapy of primary solid cancers or secondary bone cancers.

Acknowledgments This research was supported by the Australian Research Council (ARC) through the grants DP120101680 and FT110100711.

References

1. R.M. Mainardes, L.P. Silva, Drug delivery systems: past, present, and future. Curr. Drug Targets **5**, 449–455 (2004)
2. J. Drews, Drug discovery: a hystorical perspective. Science **287**, 1960–1964 (2000)
3. A.S. Hoffman, The origins and evolution of "controlled" drug delivery systems. J. Control Release **132**, 153–163 (2008)
4. J. Folkman, V.H. Mark, Diffusion of anesthetics and other drugs through silicone rubber: therapeutics implications. Trans. New York Acad. Sci. **30**, 1187–1195 (1968)
5. J.B. Wolinsky, Y.L. Colson, M.W. Grinstaff, Local drug delivery strategies for cancer treatment: gels, nanoparticles, polymeric, films, rods, and wafers. J. Control Release **159**, 14–26 (2012)
6. E. Gultepe, D. Nagesha, S. Sridhar, M. Amiji, Nanoporous inorganic membranes or coatings for sustained drug delivery in implantable devices. Adv. Drug Deliv. Rev. **62**, 305–315 (2010)
7. A. Santos, M.S. Aw, M. Bariana et al., Drug-releasing implants: current progress, challenges and perspectives. J Mater Chem B **2**, 6157–6182 (2014)
8. D. Losic, S. Simovic, Self-ordered nanopore and nanotube platforms for drug delivery applications. Expert Opin. Drug Deliv. **6**, 1363–1381 (2009)
9. P. Roy, S. Berger, P. Schmuki, TiO(2) nanotubes: synthesis and applications. Angew. Chem. Int. Ed. Engl. **50**, 2904–2939 (2011)
10. S. Bauer, P. Schmuki et al., Engineering biocompatible implant surfaces, Part I: materials and surfaces. Prog. Mater. Sci. **58**, 261–326 (2013)
11. M.S. Aw, M. Kurian, D. Losic, Non-eroding drug-releasing implants with ordered nanoporous and nanotubular structures: concepts for controlling drug release. Biomater. Sci. **2**, 10–34 (2014)
12. D. Losic, M.S. Aw, A. Santos, K. Gulati, M. Bariana, Titania nanotube arrays for local drug delivery: recent advances and perspectives. Expert Opin. Drug Deliv. **12**, 103–127 (2015)
13. A.M. MdJani, D. Losic, N.H. Voelcker, Nanoporous anodic aluminium oxide: advances in surface engineering and emerging applications. Prog. Mater. Sci. **58**, 636–704 (2013)
14. H. Masuda, K. Fukuda, Ordered metal nanohole arrays made by a two-step replication of honeycomb structures of anodic alumina. Science **268**, 1466–1468 (1995)
15. W. Lee, R. Ji, U. Gosele et al., Fast fabrication of long-range ordered porous alumina membranes by hard anodization. Nat. Mater. **5**, 741–747 (2006)
16. G.C. Wood, J.P.O. Sullivan, B. Vaszko, The direct observation of barrier layers in porous anodic oxide films. J. Electrochem. Soc. **115**, 618–620 (1968)
17. A. Santos, T. Kumeria, D. Losic, Nanoporous anodic aluminium oxide (AAO) for chemical sensing and biosensors. TrAC **44**, 25–38 (2013)
18. E. Gultepe, D. Nagesha, B.D.F. Casse, Sustained drug release from non-eroding nanoporous templates. Small **6**, 213–216 (2010)
19. Y. Wang, A. Santos, G. Kaur, A. Evdokiou, D. Losic, Structurally engineered anodic alumina nanotubes as nano-carriers for delivery of anticancer therapeutics. Biomater **35**, 5517–5526 (2014)

20. Y. Wang, A. Santos, A. Evdokiou, D. Losic, Rational design of ultra-short anodic alumina nanotubes by short-time pulse anodization. Electrochim. Acta **154**, 379–386 (2015)
21. D. Gong, V. Yadavalli, M. Paulose et al., Controlled molecular release using nanoporous alumina capsules. Biomed. Microdevices **5**, 75–80 (2003)
22. K. Noh, K.S. Brammer, C. Choi et al., A new nano-platform for drug release via nanotubular aluminum oxide. J. Biomater. Nanobiotechnol. **2**, 226–233 (2011)
23. D. Losic, L. Velleman, K. Kant et al., Self-ordering electrochemistry: a simple approach for engineering nanopore and nanotube arrays for emerging applications. Aust. J. Chem. **64**, 294–301 (2011)
24. H.-J. Kang, D.J. Kim, S.-J. Park et al., Controlled drug release using nanoporous anodic aluminum oxide on stent. Thin Solid Films **515**, 5184–5187 (2007)
25. S. Kipke, G. Schmid, Nanoporous alumina membranes as diffusion controlling systems. Adv. Funct. Mater. **14**, 1184–1188 (2004)
26. D.-H. Kwak, J.-B. Yoo, D.J. Kim, Drug release behavior from nanoporous anodic aluminium oxide. J. Nanosci. Nanotech. **10**, 345–348 (2010)
27. S.K. Das, S. Kapoor, H. Yamada et al., Effects of surface acidity and pore size of mesoporous alumina on degree of loading and controlled release of ibuprofen. Mesopor. Micropor. Mater. **118**, 267–272 (2009)
28. A. Perro, S. Reculusa, S. Ravaine et al., Design and synthesis of Janus micro- and nanoparticles. J. Mater. Chem. **15**, 3745–3760 (2005)
29. S. Simovic, D. Losic, K. Vasilev, Controlled drug release from porous materials by plasma polymer deposition. Chem. Commun. **46**, 1317–1319 (2010)
30. S. Simovic, D. Losic, K. Vasilev, Controlled release from porous platforms. Pharm. Technol. **35**(8), 68–71 (2011)
31. K. Gulati, M.S. Aw, D. Losic, Controlling drug release from titania nanotube arrays using polymer nanocarriers and biopolymer coating. J. Biomater. Nanobiotechnol. **2**, 477–484 (2011)
32. D. Losic, M.A. Cole, B. Dollmann et al., Surface modification of nanoporous alumina membranes by plasma polymerization. Nanotechnology **19**, 245704 (2008). (7 pg)
33. K. Vasilev, Z. Poh, K. Kant, J. Chan, A. Michelmore, D. Losic, Tailoring the surface functionalities of titania nanotube arrays. Biomaterials **31**(3), 532–540 (2010)
34. K. Gulati, S. Ramakrishnan, M.S. Aw, G.J. Atkins, D.M. Findlay, D. Losic, Biocompatible polymer coating of titania nanotube arrays for improved drug elution and osteoblast adhesion. Acta Biomater. **8**, 449–456 (2012)
35. G. Jeon, S.Y. Yang, J. Byun et al., Electrically actuatable smart nanoporous membrane for pulsatile drug release. Nano Lett. **11**, 1284–1288 (2011)
36. S. Kapoor, R. Hegde, A.J. Bhattacharyya, Influence of surface chemistry of mesoporous alumina with wide pore distribution on controlled drug release. J. Controlled Release **140**, 34–39 (2009)
37. A.E. Abelow, K.M. Persson, E.W.H. Jager et al., Electroresponsive nanoporous membranes by coating anodized alumina with poly(3,4-ethylenedioxythiophene) and polypyrrole. Macromol. Mater. Eng. **299**, 190–197 (2014)
38. M.S. Aw, J. Addai-Mensah, D. Losic, A multi-drug delivery system with sequential release using titania nanotube arrays. Chem. Commun. **48**, 3348–3350 (2012)
39. M.S. Aw, S. Simovic, J. Addai-Mensah, D. Losic, Polymeric micelles in porous and nanotube materials as a new system for extended delivery of poorly soluble drugs. J. Mater. Chem. **21**, 7082–7089 (2011)
40. M.S. Aw, J. Addai-Mensah, D. Losic, Magnetic-responsive delivery of drug-carriers using titania nanotube arrays. J. Mater. Chem. **22**, 6561–6563 (2012)
41. M. Bariana, M.S. Aw, E. Moore, N.H. Voelcker, D. Losic, Radiofrequency-triggered release for on-demand delivery of therapeutics from titania nanotube drug-eluting implants. Nanomed **9**, 1263–1275 (2014)
42. L. Sedel, Evolution of alumina-on-alumina implants. Clin. Orthop. Relat. Res. **379**, 48–54 (2000)

43. K.C. Popat, E.E.L. Swan, V. Mukhatyar, K.I. Chatvanichkul, G.K. Mor, C.A. Grimes, T.A. Desai, Influence of nanoporous alumina membranes on long-term osteoblast response. Biomaterials **26**, 4516–4522 (2005)
44. E.E.L. Swan, K.C. Popat, C.A. Grimes, T.A. Desai, Fabrication and evaluation of nanoporous alumina membranes for osteoblast culture. J. Biomed. Mater. Res. Part A **72A**, 288–295 (2005)
45. K.E. La Flamme, K.C. Popat, L. Leoni, E. Markiewicz, T.J. La Tempa, B.B. Roman, C.A. Grimes, T.A. Desai, Biocompatibility of nanoporous alumina membranes for immunoisolation. Biomaterials **28**, 2638–2645 (2007)
46. K.C. Popat, K.I. Chatvanichkul, G.L. Barnes, T.J. Latempa, C.A. Grimes, T.A. Desai, Osteogenic differentiation of marrow stromal cells cultured on nanoporous alumina surfaces. J. Biomed. Mater. Res. Part A **80A**, 955–964 (2007)
47. E.E.L. Swan, K.C. Popat, T.A. Desai, Peptide-immobilized nanoporous alumina membranes for enhanced osteoblast adhesion. Biomaterials **26**, 1969–2197 (2005)
48. G.E.J. Poinern, X.T. Le, M. O'Dea et al, Chemical synthesis, characterisation, and biocompatibility of nanometre scale porous anodic aluminium oxide membranes for use as a cell culture substrate for the Vero cell line: a preliminary study. BioMed. Res. Int. Article ID 238762, 10 pgs (2014)
49. M. Karlsson, E. Pålsgård, P.R. Wilshaw et al., Initial in vitro interaction of osteoblasts with nano-porous alumina. Biomaterials **24**, 3039–3046 (2003)
50. M. Karlsson, A. Johansson, L. Tang, M. Boman, Nanoporous aluminum oxide affects neutrophil behaviour. Microsc. Res. Tech. **63**, 259–265 (2004)
51. M. Karlsson, L. Tang, Surface morphology and adsorbed proteins affect phagocyte responses to nano-porous alumina. J. Mater. Sci. Mater. Med. **17**, 1101–1111 (2006)
52. S. Lee, M. Park, H.-S. Park et al., A polyethylene oxide-functionalized self-organized alumina nanochannel array for an immunoprotection biofilter. Lab Chip **11**, 1049–1053 (2011)
53. A. Canabarro, M.G. Diniz, S. Paciornik et al., High concentration of residual aluminum oxide on titanium surface inhibits extracellular matrix mineralization. J. Biomed. Mater. Res. **87A**, 588–597 (2008)
54. M.A. Cameron, I.P. Gartland, J.A. Smith et al., Atomic layer deposition of SiO2 and TiO2 in alumina tubular membranes: pore reduction and effect of surface species on gas transport. Langmuir **16**, 7435–7444 (2000)
55. R.J. Narayan, N.A. Monteiro-Riviere, R.L. Brigmon et al., Atomic layer deposition of TiO2 thin films on nanoporous alumina templates: medical applications. JOM **61**, 12–16 (2009)
56. K. Kostarelos, The long and short of carbon nanotube toxicity. Nat. Biotechnol. **26**, 774–776 (2008)
57. C.A. Poland, R. Duffin, I. Kinloch et al., Carbon nanotubes introduced into the abdominal cavity of mice show asbestos-like pathogenicity in a pilot study. Nat. Nanotech. **3**, 423–428 (2008)
58. G.A. Hart, L.M. Kathman, T.W. Hesterberg, In vitro cytotoxicity of asbestos and man-made vitreous fibers: roles of fiber length, diameter and composition. Carcinogenesis **15**, 971–977 (1994)
59. Y. Wang, G. Kaur, A. Zysk, V. Liapis, S. Hay, A. Santos, D. Losic, A. Evdokiou, Systematic in vitro nanotoxicity study on anodic alumina nanotubes with engineered aspect ratio: understanding nanotoxicity by a nanomaterial model. Biomaterials **46**, 117–130 (2015)
60. K.E. La Flamme, M. Gopal, D. Gong, T. La Tempa, V.A. Fusaro, C.A. Grimes, T.A. Desai, Nanoporous alumina capsules for cellular macroencapsulation: transport and biocompatibility. Diab. Technol. Ther. **7**, 684–694 (2005)
61. T.A. Desai, W.H. Chu, G. Rasi, Microfabricated biocapsules provide short-term immunoisolation of insulinoma xenografts. Biomed. Microdevices **1**, 131–138 (1999)
62. R.J. Walczak, A. Boiarski, T. West et al., Long-term biocompatibility of NanoGATE drug delivery implant. NanoBiotechnology **1**, 35–42 (2005)

63. T.J. Webster, R.W. Siegel, R. Bizios, Osteoblast adhesion on nanophase ceramics. Biomaterials **20**, 1221–1227 (1999)
64. S. Pujari-Palmer, T. Lind, W. Xia et al., Controlling osteogenic differentiation through nanoporous alumina. J. Biomater. Nanobiotech. **5**, 98–104 (2014)
65. A. Hoess, N. Teuscher, A. Thormann et al., Cultivation of hepatoma cell line HepG2 on nanoporous aluminum oxide membranes. Acta Biomater. **3**, 43–50 (2007)
66. L.G. Parkinson, S.M. Rea, A.W. Stevenson, The effect of nano scale topography on Keratinocyte phenotype and wound healing following burn injury. Tissue Eng. Part A **18**, 703–714 (2012)
67. K. Das, S. Bose, A. Bandyopadhyay et al., Surface coatings for improvement of bone cell materials and antimicrobial activities of Ti implants. J. Biomed. Mater. Res. Part B: Appl. Biomater. **87**, 455–460 (2008)
68. S. Rahman, G. Atkins, D. Findlay, D. Losic, Nanoengineered drug releasing aluminium wire implants: a model study for localized bone therapy. J. Mater. Chem. B (2015). doi:10.1039/c5tb00150a
69. K. Gulati, M.S. Aw, D. Findlay, D. Losic, Local drug delivery to the bone by drug-releasing implants: perspectives of nano-engineered titania nanotube arrays. Ther. Deliv. **3**, 857–873 (2012)
70. I.A. Karoussos, H. Wieneke, T. Sawitowski et al., Inorganic materials as drug delivery systems in coronary artery stenting. Materialwiss. Werkstofftech. **33**, 738–746 (2002)
71. H. Wieneke, T. Sawitowski, S. Wnendt et al., Stent coating: a new approach in interventional cardiology. Herz **27**, 518–526 (2002)
72. M. Kollum, A. Farb, R. Schreiber et al., Particle debris from a nanoporous stent coating obscures potential antiproliferative effects of tacrolimus-eluting stents in a porcine model of restenosis. Catheter. Cardiovasc. Interv. **64**, 85–90 (2005)
73. G.E. Park, T.J. Webster, A review of nanotechnology for the development of better orthopedic implants. J. Biomed. Nanotechnol. **1**, 18–29 (2005)
74. H. Wieneke, O. Dirsch, T. Sawitowski et al., Synergistic effects of a novel nanoporous stent coating and tacrolimus on intima proliferation in rabbits. Catheter. Cardiovasc. Interv. **60**, 399–407 (2003)
75. L. Augker, M.V. Swain, N. Kilpatrick, Micro-mechanical characterisation of the properties of primary tooth dentine. J. Dent. **31**, 261–267 (2003)
76. S.B. Thorat, A. Diaspro, M. Salerno, In vitro investigation of coupling-agent-free dental restorative composite based on nano-porous alumina fillers. J. Dent. **42**, 279–286 (2014)
77. C. Azevedo, B. Tavernier, J.-L. Vignes et al., Design of nanoporous alumina structure and surface properties for dental composite. Key Eng. Mater. **361–363**, 809–812 (2007)
78. C. Azevedo, B. Tavernier, J.-L. Vignes et al., Structure and surface reactivity of novel nanoporous alumina fillers. J. Biomed. Mater. Res. B Appl. Biomater. **88**, 174–181 (2009)
79. S.P. Adiga, C. Jin, L.A. Curtiss, Nanoporous membranes for medical and biological applications. Wiley Interdiscip. Rev. Nanomed. Nanobiotechnol. **1**, 568–581 (2009)
80. T.A. Desai, T. West, M. Cohen, Nanoporous microsystems for islet cell replacement. Adv. Drug Deliv. Rev. **56**, 1661–1673 (2004)
81. K.C. Popat, G. Mor, C.A. Grimes et al., Surface modification of nanoporous alumina surfaces with poly(ethylene glycol). Langmuir **20**, 8035–8041 (2004)
82. T. Kaeberlein, K. Lewis, S.S. Epstein, Isolating "uncultivable" microorganisms in pure culture in a simulated natural environment. Science **296**, 1127–1129 (2002)
83. N. Ferraz, M. Karlsson Ott, J. Hong, Time sequence of blood activation by nanoporous alumina: studies on platelets and complement system. Microsc. Res. Tech. **73**, 1101–1109 (2010)
84. W.-K. Lye, K. Looi, M. Reed, Stent with nanoporous surface. US Patents 11/432,281, 30 Nov 2006 (2006)
85. F. DiMeco, H. Brem, J. Weingart, A. Olivi, Gliadel™ A new method for the treatment of malignant brain tumors. In *Drug Delivery Systems in Cancer Therapy*, ed. by D. Brown (Humana Press Inc., Totowa, NJ, 2003), pp. 215–227

86. K. Gulati, M.S. Aw, D. Losic, Nano-engineered Ti wires for local delivery of chemotherapeutics in brain. Int. J. Nanomed. **7**, 2069–2076 (2012)
87. H.M. den Besten, C.J. Ingham, J.E. van Hylckama Vlieg et al., Quantitative analysis of population herogeneity of the adaptive salt stress response and growth capability of Bacillus cereus ATCC 14579. Appl. Environ. Microbiol. **73**, 4797–4804 (2007)
88. A.L. den Hertog, D.W. Visser, C.J. Ingham et al., Simplified automated image analysis for detection and phenotyping of Mycobacterium tuberculosis on porous supports by monitoring growing microcolonies. PLoS ONE **5**, 11008 (2010)
89. C.J. Ingham, M. van den Ende, D. Pijnenburg et al., Growth and multiplexed analysis of microorganisms on a subdivided highly porous, inorganic chip manufactured from Anopore. Appl. Environ. Microbiol. **71**, 8978–8981 (2005)
90. C.J. Ingham, M. Beerthuyzen, Vlieg J. van Hylckama, Population heterogeneity of Lactobacillus plantarum WCFS1 microcolonies in response to and recovery from acid stress. Appl. Environ. Microbiol. **74**, 7750–7758 (2008)
91. L.G. Parkinson, N.L. Giles, K.F. Adcroft et al., The potential of nanoporous anodic aluminium oxide membranes to influence skin wound repair. Tissue Eng. Part A **15**, 3753–3763 (2009)
92. C.J. Ingham, A. Sprenkels, J. Bomer et al., The micro-Petri dish, a million well growth chip for the culture and high throughput screening of microorganisms. Proc. Natl. Acad. Sci. U.S. A. **13**, 18217–18222 (2007)
93. C.J. Ingham, M. van den Ende, P.C. Wever et al., Rapid antibiotic sensitivity testing and trimethoprim-mediated filamentation of clinical isolates of the Enterobacteriaceae assayed on anovel porous culture support. J. Med. Microbiol. **55**, 1511–1519 (2006)
94. P.-H. Tsou, H. Sreenivasappa, S. Hong et al., Rapid antibiotic efficiency screening with aluminium oxide nanoporous membrane filter-chip and optical detection system. Biosens. Bioelectron. **26**, 289–294 (2010)
95. C.J. Ingham, A.B. Ayad, K. Nolsen et al., Rapid drug susceptibility testing of mycobacteria by culture on a highly porous ceramic support. Int. J. Tuberc. Lung Dis. **12**, 645–650 (2008)
96. C.J. Ingham, J. ter Maat, W.M. de Vos, Where bio meets nano: the many uses for nanoporous aluminium oxide in biotechnology. Biotech. Adv. **30**, 1089–1099 (2012)
97. C.J. Ingham, A. Sprenkels, J. Bomer et al., The micro-Petri dish, a million-well growth chip for the culture and high-throughput screening of microorganisms. PNAS **104**, 18217–18222 (2007)
98. B.C. Ferrari, S.J. Binnerup, M. Gillings, Microcolony cultivation on a soil substrate membrane system selects for previously uncultured soil bacteria. Appl. Environ. Microbiol. **71**, 8714–8720 (2005)
99. I. Vivanco, D. Rohle, M. Versele et al., The phosphatase and tensin homologue regulates epidemal growth factor receptor (EGFR) inhibitor response by targeting EGFR for degradation. Proc. Natl. Acad. Sci. **107**, 6459–6464 (2010)
100. Y. Song, Y. Ju, G. Song, Morita et al., In vitro proliferation and osteogenic differentiation of mesenchymal stem cells on nanoporous alumina. Int. J. Nanomed. **8**, 2745–2756 (2013)

Index

D. Losic and A. Santos (eds.), *Nanoporous Alumina*,
Springer Series in Materials Science 219, DOI 10.1007/978-3-319-20334-8

The manufacturer's authorised representative in the EU is Springer Nature Customer Service Centre GmbH, Europaplatz 3, 69115 Heidelberg, Germany. If you have any concerns regarding our products, please contact ProductSafety@springernature.com

Printed and bound by CPI Group (UK) Ltd, Croydon, CR0 4YY
15/07/2026
02167618-0002